Birkhäuser

# Operator Theory: Advances and Applications

**Volume 233**

**Founded in 1979 by Israel Gohberg**

Ivan G. Todorov
Lyudmila Turowska
Editors

# Algebraic Methods in Functional Analysis

## The Victor Shulman Anniversary Volume

Birkhäuser

*Editors*
Ivan G. Todorov
Pure Mathematics Research Centre
Queen's University Belfast
Belfast, UK

Lyudmila Turowska
Mathematical Sciences
Chalmers University of Technology and
the University of Gothenburg
Göteborg, Sweden

ISSN 0255-0156 ISSN 2296-4878 (electronic)
ISBN 978-3-0348-0501-8 ISBN 978-3-0348-0502-5 (eBook)
DOI 10.1007/978-3-0348-0502-5
Springer Basel Heidelberg New York Dordrecht London

Mathematics Subject Classification (2010): 47A, 47B, 47D, 47L, 47N. 46A, 46B, 46C, 46H, 46L



Printed on acid-free paper

Springer Basel is part of Springer Science+Business Media (www.birkhauser-science.com)

# Contents

Operator Theory:
Advances and Applications, Vol. 233, vii–viii


# Introduction

Today, the subject of Functional Analysis is more active than ever. Celebrating the contributions of one of the leading scientists in this field, Professor Victor Shulman, has given us the opportunity to present in this volume a number of aspects of the area that are closely related to his work. His contributions to Functional Analysis are explained in greater detail in the article by E. Kissin, M. Lomonosov and Yu. Turovskii, included in the present volume.

Algebraic methods have been used with great success in Functional Analysis since its very beginnings. In the 1930s, F. Murray and J. von Neumann introduced what is known today as *von Neumann algebras* with the aim of formulating a suitable mathematical language for Quantum Physics. A decade later, I.M. Gelfand and M. Naimark developed the theory of *C*-algebras*. These two complementary and symbiotic contexts have provided the ground for many of the algebraic aspects of Operator Theory in later years. They have been the starting point for the development of hugely successful theories, such as Operator Space Theory in the 1980s and the Theory of Locally Compact Quantum Groups in the past decade. Motivation for their development has come simultaneously from "single operator theory", that is, from the study of the properties of individual operators acting on Hilbert space.

The more general setting of Banach spaces has been, both directly and indirectly, essential for the development of many new ideas in Operator Theory; indeed, it has been a motivating force in Operator Space Theory, serving at the same time as the "right" context, in terms of generality, for studying various operator theoretic questions. Large parts of Harmonic Analysis have enriched, and have benefited from, Functional Analysis; one example is the study of various classes of operators arising directly from the context of topological groups, such as Toeplitz operators, while another one is Invariant Subspace Theory for non-selfadjoint operator algebras, which has been influenced by Beurling's theorem. The successful functional analytic formulation of Non-commutative Harmonic Analysis by P. Eymard in the mid 1960s has had a decisive impact on the research in the field, and has provided the base for the aforementioned developments in the theory of locally compact quantum groups. *Banach algebras*, developed simultaneously with C*-algebras, have, on the other hand, provided a simple algebraic setting for the study of a great number of analytical problems.

The present volume reflects, to a significant extent, all of the above settings and presents the state-of-the-art of a number of subareas of Operator Theory. It is comprised of comprehensive survey articles and original research papers that concern, among others, the following domains: the theory of C*-correspondences, non-selfadjoint operator algebras and masa-bimodules, local multiplier algebras, operator ideals, Banach space geometry, the theory of linear preserves, lifting properties in C*-algebras, the theory of topological algebras and the theory of locally compact quantum groups. It also includes an article on topological radicals, which is of a larger size than the rest of the contributions due to the vast scope of its subject. The topics presented in the volume are currently undergoing intensive development, and further progress on them is expected in the future.

The papers were refereed by qualified experts in the field. The Editors would like to thank all referees for their prompt advice, which helped improve the volume.

We wish to take the opportunity to thank all participants of the conference and, in particular, the authors of the articles of these proceedings. Last, but not least, we would like to thank the Swedish Research Council and the Wenner-Gren Foundation for providing the financial support, without which the meeting would never have materialised.

8 December 2012

Ivan Todorov, Belfast
Lyudmila Turowska, Gothenburg

Operator Theory:
Advances and Applications, Vol. 233, 1–16


# Victor Shulman: The Gentle Art of Mathematics

E. Kissin, V. Lomonosov and Yu. Turovskii

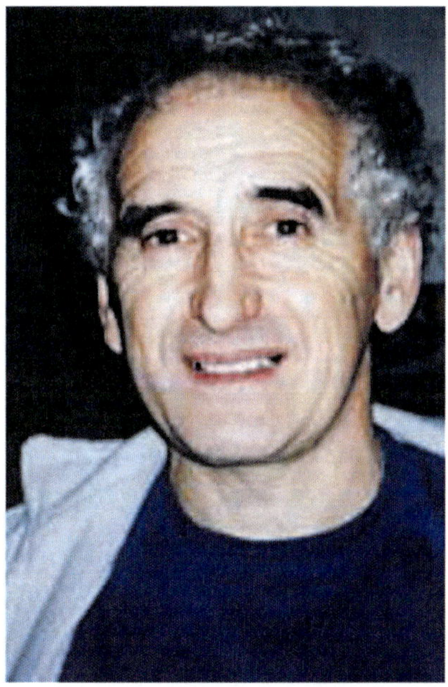

It came as a bit of a shock to all of us: Victor (or Vitya, as we affectionately call him in Russian) is 65. Hold on, this can't be true! We just played football with him and he scored a few goals. He was all over the field – full of drive and the desire to win. No, he cannot be 65! Don't tell us about his date of birth or the age of his children. It just doesn't make sense!

It all started with an email from Ivan Todorov and Lyudmila Turowska. They informed us that they were going to organize a mathematical conference in Gothenburg, Sweden in honour of the 65th anniversary of Professor Victor Shulman, and asked whether we would like to participate. Who could refuse?! Thanks to Lyudmila's and Ivan's energy, devotion and organizational skills, the conference sprang into being, the money for it materialized thanks to some magic tricks (and to the generosity of the Swedish Research Council and Wenner–Gren Foundations and to the help of the research network for Operator algebras and Dynamics supported by NordForsk), mathematicians from all over the world were invited, the topics and schedule of the talks were agreed, accommodation was found, and, most importantly, the time and place for the football match of the year was arranged.

The conference went very smoothly: the weather was nicely arranged – sunny and warm, but not too hot; the speakers delivered their talks with great gusto; the audience listened enthusiastically, asked respectful and clever questions and applauded at the end. Finally, the organizers laid on a delicious dinner for the participants and, in the intervals between courses, we listened to speeches and watched a very witty film about Victor, directed by his devoted brother Boris and Lyudmila's husband, Slava Turovski.

The conference came to an end and the participants went back to their countries, but they all took with them some particles of Victor's warmth, his gentle smile, and an indefinable sense of his inspiring presence and personality.

Let us start with some words about Victor, written by the second author of this preface, one of Victor's oldest friends and collaborators, and delivered at the conference:

> *I have known Victor Shulman for almost forty years and have admired him all that long period of time. He is a great man in all respects. He has been working successfully in many different areas of Functional Analysis and his citation index is "above the clouds". He has collaborators all over the world, some of whom he has never even met. He has an amazing feeling for the right direction in mathematics. For twenty years he and his former student Yurii Turovskii have been developing the theory of the joint spectral radius of a family of operators. Eventually two long-standing problems were solved. In particular, they proved that Volterra Lie algebras have non-trivial invariant subspaces.*
>
> *Victor is a very courageous man. In* 1968 *after the Soviet invasion of Czechoslovakia he distributed a letter of protest written by Czechoslovakian intellectuals. The consequences could have been severe. Fortunately, he was 'only' expelled from graduate school and kept under KGB close surveillance.*
>
> *Victor is a devoted husband and the father of three lovely children. Two of them – his daughters – are productive mathematicians in their own right. With the youngest of them – Tatyana – he has written two joint papers* (2009, 2011) *"On Lie submodules and tensor algebras" and "On algebras generated by inner derivations". Victor is also quite an expert on modern and classical Russian literature, and I have learned a lot from him on this subject. And what a fantastic football player he is – even better than his brother, who played in his youth for the Dinamo-Kiev junior football team! Apart from all this, Victor is a very kind man with a charming smile, always ready to help anybody who needs it. Everybody likes him and this is why all of you are here.*

Victor Semyonovich Shulman was born on 8 October 1945 in Odessa, into the family of a military officer, Semyon Moiseyevich Shulman, and his wife, Anna Yakovlevna Gutvah. Later Semyon Moiseyevich became a school teacher of mathematics and infused his sons with enthusiasm for the subject. Victor went to school in Kiev, where he not only developed his love for mathematics and Russian literature, but also learned to play football so well that he even played for the Ukrainian national youth team.

After winning a school Olympiad in physics and mathematics organized by the Moscow Institute of Physics and Technology (MFTI), Victor decided to study Theoretical Physics at MFTI. Later he started to attend the lectures of Mark Aronovich Naimark on Functional Analysis and Normed Algebras. He also attended Naimark's research seminar at the Steklov Institute of Mathematics and became his student. Even at this busy time he did not forget football and played in MFTI's football team. After graduating in 1967, he continued to study mathematics at Graduate School under Naimark's supervision. However, more important things deflected his attention from mathematics for some period of time. First was his marriage to his wife Natalia in 1965 and the birth of their daughter Ekaterina in 1966. Another distracting factor was his troubles with the KGB for distributing forbidden materials in 1968. As a result Victor was expelled from Graduate School and forced to leave Moscow in 1970. Naimark, who had a great influence on Victor and whom Victor respected very much, advised him to continue to study mathematics, commenting: "You will have to make a choice. If you want to be a mathematician, we will continue solving mathematical problems together; otherwise you will have to be prepared for various troubles for your family and forget about mathematics". Victor made his choice. He moved to Baku in Azerbaijan in 1970 and stayed there until 1975 publishing fifteen mathematical papers during those six years. In 1972 he defended his Ph.D. at the Institute of Mathematics and Mechanics in Baku.

In 1975 Victor moved to Vologda – a town situated 500 kilometres to the North-East of Moscow – and began lecturing at Vologda State Technical University. Some years later he defended his Doctor of Science (Doctor Habilitatus) dissertation on "Invariant subspaces and linear operator equations" at the People's Friendship University in Moscow. In 1995 he also started teaching at Vologda State Science and Mathematical Lyceum – a school for mathematically gifted children – where he delivers popular classes on various topics of contemporary mathematics and the methods of solving non-standard mathematical problems. Many of his pupils went on to win mathematical Olympiads at various levels. He imbues his students with an understanding and love of mathematics, passing on to them his enthusiasm and knowledge. He also befriends them and helps them to find their own way not just in mathematics but also in life; Victor enables his students to become better, decent people with integrity and high human values. And what could be better for developing friendship than a good football game? So almost every week Victor plays football with his students, and looking at the game you would never believe that he is four times older than them.

Victor's collaborative work with Edward Kissin started in 1991–1992 (although they first met in 1962 while studying at the MFTI) and kept bringing Victor to London twice a year. In 2000–2003 and 2007–2008 he was appointed a Research Fellow at London Metropolitan University. Later he became a frequent guest of the Chalmers University of Technology, Gothenburg, Sweden, where he collaborated with Lyudmila Turowska. All these travels opened challenging new horizons for Victor: improving his command of English and scoring fresh football victories in foreign lands. He learned to appreciate the writings of Dickens and Sterne more deeply, and met many different people, some of whom became good friends.

Victor is a very talented and distinguished mathematician. He is extremely active in mathematics and mathematically "sociable"; the fact that he has 28 different co-authors demonstrates his intellectual openness and ability to establish scientific and personal contacts. He combines his awesome erudition in different areas of mathematics with an amazing intuition for forming mathematical hypotheses, often proving them himself. And alongside all this he remains a very unassuming person with remarkable human qualities. His acute feeling for injustice and desire to help others draw people to him. In his cordial, sympathetic manner he is always the first to inquire about other people's health and circumstances, to find out whether they need any assistance. Being himself modest and tactful, he appreciates these qualities in others. At the same time, he is someone with whom it is great fun to converse and correspond. All his friends keep his witty, remarkable, long letters – full of great mathematics and funny stories, warmth and thoughtfulness.

Victor's interests in mathematics are unusually wide and varied. He developed work in a number of different directions, discussed below. In the early period of his mathematical life the deep ideas of Victor's teacher M.A. Naimark and of his favorite mathematician W. Arveson had a significant influence on him. Victor worked on various conjectures of Arveson and proved some of them. One of the topics that always attracted Victor was the structure of invariant subspaces of operator algebraic systems. The importance of this issue stems from the possibility of its use for the analysis of the essential properties (spectral characteristics of geometric invariants, irreducible representations) of these systems. Of particular significance is the opportunity to use these results in the study of linear operator equations and, in turn, to apply them to differential equations and pseudo-differential operators. Victor was also fascinated by the fruitful relationship between the theory of invariant subspaces and Banach algebras, spectral theory of operators, the infinite-dimensional geometry, the measure and approximation theories, harmonic analysis (in particular, the spectral synthesis), the asymptotic behaviour of operator semi-groups, structural theory of infinite-dimensional Lie algebras etc. Thanks to his erudition, he succeeded in bringing all these areas of mathematics into the orbit of his research.

In his late years M.A. Naimark worked on developing the theory of group representations and the theory of $J$-symmetric operator algebras on Pontryagin $\Pi_\kappa$-spaces, $\kappa < \infty$, with indefinite metric. Therefore it is natural that Victor's first published paper (1971) was devoted to the study of the structure of $J$-symmetric operator algebras on $\Pi_1$-spaces. In a further eleven papers Victor returned again to this subject, either by himself or with different co-authors. In 1972 he constructed six canonical models of uniformly closed $J$-symmetric operator algebras on $\Pi_1$-spaces that exhaust the variety of all such operator algebra on $\Pi_1$-spaces. Using these models, he established that if $\mathcal{A}$ has a separating vector then $\mathcal{A} = \mathcal{A}''$, and if $\mathcal{A}$ has a bicyclic vector, then there is an antilinear $J$-involution $j$ such that $\mathcal{A}' = j\mathcal{A}j$.

Victor devoted several papers to the problem of similarity of $J$-symmetric representations of *-algebras on $\Pi_\kappa$-spaces to *-representations and $J$-unitary representations of groups to unitary representations. In 1977 he proved that $J$-symmetric representations of C*-algebras on $\Pi_\kappa$-spaces are similar to *-representations. In 1978 he and A. Loginov used the brilliant theorem of J. Cuntz on locally C*-equivalent Banach *-algebras to show that any norm-closed $J$-symmetric operator algebra on a $\Pi_\kappa$-space either has an invariant subspace or is isomorphic to a C*-algebra. This result was later extended in his joint paper with E. Kissin and A. Loginov (1996), where it was established that all $J$-symmetric representations of almost Hermitian algebras on $\Pi_\kappa$-spaces have invariant subspaces.

In 1980 Victor showed that bounded $J$-unitary representations of groups on $\Pi_1$-spaces are similar to unitary representations. Some 29 years later in his joint paper with M. Ostrovskii and L. Turowska, this result was extended to bounded $J$-unitary representations of groups on $\Pi_\kappa$-spaces, for all $\kappa < \infty$. The approach of this paper was based on the study of fixed points for families of operator fractional linear transformations.

Victor told the following anecdote about his work on the fixed-point theorem. In one of his conversations with his uncle in Australia, he was asked what problems he had been working on. When the uncle heard that it was the problem of a fixed point, he was perplexed and remarked: "This is what I always suspected about mathematicians. They can study just one point which does not even move".

In 1997 Addison Wesley Longman Publishing house published a book by E. Kissin and V. Shulman "Representations on Krein spaces and derivations of C*-algebras", summarizing the material on representations of algebras and groups on Krein and $\Pi_\kappa$-spaces and their applications to derivations of C*-algebras.

In 1972–1988, together with A. Loginov, Victor studied reflexive operator algebras, hereditary and intermediate reflexivity, reductive operator algebras and the invariant subspace problem. They wrote papers on the Sarason theorem and Radjavi–Rosenthal hypothesis and on transitive operator spaces. In 1974, together with A. Loginov and M.A. Naimark, Victor published a survey "Non-selfadjoint algebras of operators on Hilbert spaces", and in 1988, together with A. Loginov, a survey "Invariant subspaces of operator algebras". These surveys were very important for mathematicians working in this area and initiated much new research, as they reviewed and summarized all known results on the subject and posed many important questions for further investigation.

In 1980 and 1982 Victor wrote two joint papers with his former student A. Fainstein on stability of Fredholm complexes of Banach spaces. Together with V. Khatskevich and Yu. Shmulian, he also wrote two papers (1984 and 1991) on hierarchy of preorders and equivalences in operator ball that arise in various operator problems. Two papers co-authored with his former student G. Mustafaev appeared in 1985 and 1989, investigating the density of vector functionals and the norms of inner derivations of operator algebras. Operator algebras with commutative symmetric lattices of invariant subspaces were explored in a joint paper with

S. Rosenoer in 1986. Operator semigroups with quasinilpotent commutators were studied by Victor in 2000, together with H. Radjavi and P. Rosenthal.

Victor's interest in theory of linear operator equations arose as early as 1973 when he wrote a joint paper with A. Loginov about the Fuglede–Putnam theorem and reflexivity. He noticed a close relationship between this theorem, operator synthesis, fine spectral structure of multiplication operators with normal coefficients and invariant subspace theory. In 1984 and 1988 Victor developed the method of approximate inverse intertwinings. In his papers of 1983 and 1990 and in his joint paper of 2004 with L. Turowska he proved a non–commutative version of the Fuglede theorem and simplified the proof of the Fuglede–Weiss theorem obtained earlier by G. Weiss. In this last paper his previous results (1983, 1988, 1996) about the fine spectral structure of multiplication operators also obtained their final form. As a corollary of a general theorem about the kernels of families of operators which he proved in 1990, Victor transferred his results about equivalence of linear operator equations with normal coefficients to the equations with quasi-nilpotent coefficients. Some applications to Varopoulos algebras and to differential operators with constant coefficients were obtained in his paper of 1990. In 1996 Victor extended the Fuglede–Weiss theorem to the Schatten ideals $\mathfrak{S}_p$, $p \in (1,2)$, of compact operators on a Hilbert space and showed that there are constants $c_p$ such that $\|[A^*,X]\|_p \leq c_p \|[A,X]\|_p$ for all normal operators $A$ and bounded operators $X$. Further extensions of this results were obtained in a series of joint papers with E. Kissin (2001–2005). Results on representations of relations of the form $i[A,B] = f(A) + f(B)$ were obtained in 1991 in his paper with Yu. Samoilenko. Semilinear relations and their representations were studied in his article of 1996, co-authored with Yu. Samoilenko and L. Turowska.

An important role in Victor's research is played by the question of triangularization – finding the maximal chains of invariant subspaces of operator algebras, Lie algebras and semigroups, and, as the first step, finding the criteria of the existence of non-trivial invariant subspaces. This question is of paramount importance for the classification of representations of these algebraic structures and for the investigation of the structure of their ideals and factors. The first infinite-dimensional results in this area are connected with the names of Hilbert, Schmidt, Weyl, von Neumann, Stone, Halmos, Pontryagin, Krein. The paper by V. Lomonosov in 1973 was a crucial step in the development of this area of research. It was followed by the work of D. Gurarie and L. Waksman, W. Wojtynski, H. Radjavi, P. Rosenthal and many other mathematicians.

To develop further the theory of invariant subspaces Victor made use of the notion of the joint spectral radius, introduced by G.-C. Rota and W.G. Strang in 1960. Using the spectral radius technique, he established in 1984 the existence of a hyperinvariant subspace for an operator algebra that has compact operators in its Jacobson radical. Later in 1999–2002 in a series of papers he and his former student Yu.V. Turovskii studied fundamental analytical properties of the joint spectral radius and applied this concept to the study of representations of infinite-dimensional Lie algebras and operator semigroups. They obtained various criteria for the existence of invariant subspaces and the triangularization of these systems. One of the first joint results they obtained was an unexpected one (that later had far-reaching generalizations) on the triangularization of infinite-dimensional solvable Lie algebras of compact operators. They also established, in particular, the existence of invariant subspaces for any Lie algebra of Volterra operators and its triangularization. Recently, together with M. Kennedy and Yu. Turovskii, Victor proved the same result for Jordan algebras of Volterra operators.

Another interesting research direction that Victor is vigorously pursuing is to determine the conditions of continuity of the map $\mathcal{A} \mapsto \operatorname{Lat} \mathcal{A}$ (the main map in the theory of invariant subspaces) that associates with each operator algebra $\mathcal{A}$ its lattice of invariant subspaces. The interest to the continuity of Lat is due to the fact that it makes it possible to apply the "approximation" method to the description of the lattices of invariant subspaces. Criteria of continuity are also useful in the study of reflexive lattices of subspaces, i.e., the lattices of the form $\operatorname{Lat} M$, for some set of operators $M$, and in determining suitable conditions for the reflexivity of lattices of subspaces. This shows the topological underpinnings of some fundamental results in the theory of operator algebras such as the Arveson theorem on reflexivity commutative lattices. The question of the continuity of Lat leads to the study of the geometry of subspace lattices and of the spatial tensor product of such lattices. In particular, it leads to the study of the Hopenwasser–Kraus problem of finding conditions on operator algebras $\mathcal{A}$ and $\mathcal{B}$, in which the lattice of invariant subspaces of their tensor product is the same as the tensor product of their lattices.

The first time the study of the continuity of Lat implicitly emerged in the work of von Neumann. Later it was developed in the writings of Apostol, Foias, Voiculescu, Halmos, Conway, Hadwin, Davidson and others. They mainly focused their research on algebras with one generator. Victor in 1989 and 1994 and in his joint papers with I. Todorov (2004, 2005) considered a much more general approach to the continuity of the map that required different and new techniques. This approach was based on the study of geometric properties of lattices of projections. They proved that

$$\operatorname{Lat}(\lim(\{\mathcal{A}_\lambda\})) = \lim \operatorname{Lat}(\{\mathcal{A}_\lambda\}),$$

for all nets of weakly * closed algebras $\{\mathcal{A}_\lambda\}$ such that all $\mathcal{A}_\lambda$ are either W*-algebras, or contain masa, or are subalgebras of a W*-algebra with properly infinite commutant. Using this result, they obtained a general approximate criterion of reflexivity of subspace lattices that allowed them to solve the reflexivity problem for a wide class of lattices. As a corollary, they obtained a positive solution of the Hopenwasser–Kraus problem for a pair of von Neumann algebras one of which is injective.

One of the focal points of Victor's work is the study of the structure of operator bimodules over maximal selfadjoint commutative operator algebras (masa) and of the corresponding bilattices of projections. The problems he has investigated can be divided into two classes: those that use purely operator technique in the study of the theory of bilattices, and those that occur when a coordinate approach is used. The investigation of the second class of problems involves the theories of measure, of capacity, of spectral analysis–synthesis and other classical areas of analysis. The coordinate approach first arose in the works of Arveson, who studied a special class of bimodules – CSL-algebras – and posed many problems, some of which have been solved by Victor. The need to study the general situation was revealed in the last decade, in response to the needs of the theory of linear operator equations that arise in the representation theory of quantum groups, integral equations and linear partial differential equations. To address this issue, Victor created and developed new means and methods: the theory of operator synthesis, the theory of approximate inverse intertwinings and the theory of pseudo topological spaces associated with the direct products of two measure spaces. These new approaches allowed him to construct subtle examples that show the difference between masa-bimodules and CSL-algebras in geometrical properties. They also allowed him to characterize the support of masa-bimodules that contain operators from Schatten ideals and to identify criteria of density of masa-bimodules. In particular, Victor has shown that transitive masa-bimodules are ultra weakly dense in the algebra of all operators – the solution of a K. Davidson's problem. The relationship between spectral and operator syntheses is itself a fruitful area of research in which Victor was able to establish a number of results that are useful for both operator theory and harmonic analysis. He started his work in this area in 1973 with A. Loginov and continued to develop it through the years involving numerous collaborators: Yu. Samoilenko and L. Turowska (1996), R. Haydon (1996), J. Erdos and A. Katavolos (1998), E. Kissin (2006) and L. Turowska (2004).

Victor's interest in the structure of differential subalgebras and unbounded derivations of Banach and C*-algebras goes back to 1993 when he published a joint paper with E. Kissin on this subject. In a subsequent series of papers (1994, 1998, 1999, 2001, 2002) they investigated various properties of differential Banach *-algebras of compact operators, their dual spaces and isomorphisms, and approximation property of differential Schatten *-algebras. Some of these results were published in their book "Representations on Krein spaces and derivations of C*–algebras" (Addison Wesley Longman, 1997). Victor's paper with E. Kissin and V. Lomonosov (2003) was also devoted to the problem of implementation of derivations and invariant subspaces.

In 2001 Victor published a paper with Yu. Turovskii about topological radicals in Banach algebras. It was the beginning of a series of papers (2005, 2011, and at least one more in preparation), where they studied radical normed algebras, their tensor products, the joint quasi-nilpotence and applications to the spectral theory of multiplication operators. They also obtained the generalized Berger–Wang formulae for calculation of the joint spectral radius.

Non-associative operator algebras also constitute a substantial part of Victor's interests. In 2005 Victor and Yu. Turovskii proved that a Lie algebra of compact operators with a non-zero Volterra Lie ideal has a non-trivial invariant subspace and applied the related results to the structure theory of Lie algebras of compact operators and normed Lie algebras with compact adjoint action. In 2009 he published two papers with M. Brešar and E. Kissin about the structure of Jordan and Lie ideals of Banach and C*-algebras. In the same year he wrote a paper with his daughter Tatyana "On Lie submodules and tensor algebras". With M. Kennedy and Yu. Turovskii in 2009 he investigated the structure and invariant subspace problem for subgraded Lie algebras of compact operators and, together with M. Brešar and Yu. Turovskii in 2010, the tractability and the ideal problem in non-associative operator algebras. In two papers with E. Kissin and Yu. Turovskii in 2009 Victor studied Banach Lie algebras with Lie subalgebras of finite codimension and proved that these Lie algebras always have closed Lie ideals. Using this as a springboard, they developed a theory of topological radicals in Banach Lie algebras and described Banach Lie algebras with sufficiently many Lie subalgebras of finite codimension.

Let $B(H)$ be the algebra of all bounded operators on a Hilbert space $H$ and let $\mathfrak{S}$ be either a Schatten ideal of $B(H)$, or $B(H)$ itself. A continuous function $f$ on $\mathbb{R}$ is an $\mathfrak{S}$-Lipschitz function, if there is $c_{\mathfrak{S}} > 0$ such that

$$\|f(A) - f(B)\|_{\mathfrak{S}} \le c_{\mathfrak{S}} \|A - B\|_{\mathfrak{S}}$$

for all selfadjoint operators $A, B$ in $\mathfrak{S}$. It is Gateaux or Fréchet $\mathfrak{S}$-differentiable, if the map $A \to f(A)$ is Gateaux or Fréchet differentiable in $\|\cdot\|_{\mathfrak{S}}$. Much work has been done on the theory of these classes of functions by Birman and Solomyak, Davies, Farforovskaya, Johnson and Williams, Peller, Aleksandrov, Arazy, Barton, Friedman, Pedersen and others. In their first paper on this subject, published in 1996, Victor and E. Kissin proved that the classes of Gateaux and Fréchet $B(H)$-differentiable functions coincide and that the functions from these classes (operator differentiable functions) act on the domains of closed derivations of C*-algebras. This work was followed by several other joint papers (2001–2007) in which the authors studied the hierarchy of the mentioned functional classes, considered the smoothness properties of the action of analytic functions on non-selfadjoint operators, introduced C*-analogues of Schur multipliers and studied many other questions on operator smoothness. In particular, it was shown that the space of operator differentiable functions coincides with the closure of the subspace of polynomials in the space of $B(H)$-Lipschitz functions and that the latter space is non-separable. For the proof of the second statement, they constructed a large variety of $B(H)$-Lipschitz functions which were not continuously differentiable (the existence of such functions was questioned by Williams). Recently, together with D. Potapov and F. Sukochev they proved for $\mathfrak{S} \neq B(H)$ that a function of many variables is an $\mathfrak{S}$-Lipschitz function if and only if it is a Lipschitz function in the usual sense. They also proved that any Lipschitz function $f$ acts on the domains of all closed *-derivations on $\mathfrak{S}$ and that $f$ is Gateaux (respectively, Fréchet) $\mathfrak{S}$-differentiable if and only if $f$ is differentiable and its derivative is bounded on all compacts (respectively, is continuous).

Well, Victor has achieved a great deal in mathematics and in life in general. Due to his talent, enthusiasm, energy and appetite for discovering mathematical mysteries and laws, he has published a book and at least 119 papers (we suspect rather more, in fact). He has brought up a wonderful family. He has helped many people to understand and enjoy mathematics. He has acquired a multitude of friends – mathematical and non-mathematical – all over the world. He has shared his warmth, his smile and good humour with many people around him, and has always stretched out a helping hand to anyone in need. We all – friends, students and colleagues – wish him "till 120" in good health and clear mind, to enjoy and appreciate life, to climb many more mathematical mountains successfully, and to continue to enrich our lives with his friendship and support.

## List of publications by V.S. Shulman

### Book

1. (with E. Kissin) Representations on Krein Spaces and Derivations of C*-algebras, Addison Wesley Longman (600 pp.), 1997.

### Papers

Papers marked with an asterisk * were written in Russian.

1. *Operator algebras in a space with indefinite metric of* $\Pi_1$*-type*, DAN SSSR 201 (1971) 44–47*.
2. *Relations between the classes of Lebesgue and Denjoi integrable functions*, Math. Notes 10 (1971) 219–28*.
3. (with V. Liberson) *Operator-irreducible symmetric operator algebras on the Pontryagin spaces*, Izv. AN SSSR 35 (1971) 1159–70*.
4. *An example of a Denjoi integrable function which is not integrable in Khintchin's sense*, Math. Notes 10 (1971) 295–300*.
5. *Symmetric Banach algebras of operators in* $\Pi_1$*-space*, Matem. Sbornik 89 (1971) 264–79*.
6. (with A. Loginov) *On Saroson Theorem and Radjavi–Rosenthal hypothesis*, DAN SSSR 205 (1972) 284–85*.
7. *On reflexive operator algebras*, Matem. Sbornik 87 (1972), 179–87*.
8. (with A. Loginov) *Fuglede–Putnam Theorem and reflexivity*, DAN SSSR 210 (1973) 543–44*.
9. (with A. Loginov) *Heredity and intermediate reflexivity of* $W^*$*-algebras*, DAN SSSR 212 (1973) 810–12*.
10. (with V. Liberson) *Non-degenerate operator algebras in indefinite metric spaces*, Izv. AN SSSR 37 (1973) 533–38*.
11. (with M.A. Naimark and A. Loginov) *Non-selfadjoint algebras of operators on Hilbert spaces*, "Math. Analysis", AN SSSR, VINITI, Moscow 12 (1974) 413–65*.

12. *Multiplication operators on $C^*$-algebras and the reflexivity problem for algebras containing masa*, Functional Analysis and its Applications 8 (1974) 92–93*.
13. *Operator algebras with strictly cyclic vectors*, Math. Notes 17 (1974) 253–57*.
14. (with A. Loginov) *Reductive operator algebras and the invariant subspace problem*, DAN SSSR 216 (1974) 36–38*.
15. *Linear operator equations with generalized scalar coefficients*, DAN SSSR 225 (1975) 56–58*.
16. (with A. Loginov) *Heredity and intermediate reflexivity*, Izv. AN SSSR, 39 (1975) 1260–73*.
17. (with A. Loginov) *Reductive operators and operator algebras*, Izv. AN SSSR 40 (1976) 845–54*.
18. *Representations of $C^*$-algebras on the spaces with indefinite metric*, Math. Notes 22(1977) 583–92*.
19. (with A. Loginov) *Irreducible $J$-symmetric operator algebras on the spaces with indefinite metric*, DAN SSSR 240 (1978) 21–23*.
20. *A fixed-point theorem*, Functional Analysis and its Applications 13 (1979) 88–89*.
21. *On fixed-points of fractional linear transformations*, Functional Analysis and its Applications 14 (1980) 93–94*.
22. (with A. Fainstein) *Fredholm complexes of Banach spaces*, Functional Analysis and its Applications 14 (1980) 87–88*.
23. (with A. Fainstein) *Stability of index for a short Fredholm complex with respect to perturbations which have small measure of non-compactness*, Spectral Theory of Operators and Applications, Baku, "Elm" 4 (1982) 189–98*.
24. *On transitive operator spaces*, Functional Analysis and its Applications 16 (1982) 91–92*.
25. (with B.P. Osilenker) *On lattices of invariant subspaces of some operators*, Functional Analysis and its Applications 17:1 (1983) 81–82*.
26. *Modules over operator algebras*, Functional Analysis and its Applications 17 (1983) 94–95*.
27. *Linear operator equations with normal coefficients*, DAN SSSR 270 (1983), 1070–73*.
28. *On vector functionals and approximation in operator spaces*, Spectral Theory of Operators and Applications, Baku, "Elm" 5 (1984) 192–225*.
29. *On invariant subspaces for Volterra operators*, Functional Analysis and its Applications 18 (1984) 85–86*.
30. *Multiplication operators and traces of commutators*, Research on linear operators and the function theory, XIII, Notes of LOMI seminars 135 (1984) 182–94*.

31. (with A. Loginov) *On invariant subspace lattices of some operators*, Research on theory of functions of one or several real variables, Yaroslavl University 16 (1984) 105–13*.

32. (with V. Khatskevich and Yu. Shmulian) *On equivalent contractions*, DAN SSSR 278 (1984) 47–49*.

33. (with G. Mustafaev) *On density of vector functionals*, DAN SSSR 280 (1985) 804–06*.

34. *On mutual disposition of some subspaces in $C^*$-algebras*, Spectral Theory of Operators and Applications, Baku, "Elm" 6 (1986) 196–216*.

35. (with S. Rosenoer) *Operator algebras with commutative symmetric lattices of invariant subspaces*, Izv. of AN Azerb. SSR 7 (1986) 21–26*.

36. (with Yu. Turovskii) *Conditions of massiveness of the image for a derivation of a Banach algebra, and related differential operators*, Math. Notes 42 (1987) 305–14*.

37. (with A. Loginov) *Invariant subspaces of operator algebras*, "Mathematical Analysis", Itogi Nauki (VINITI) 26 (1988) 65–145*.

38. *Intertwinings and linear operator equations*, DAN SSSR 301 (1988) 57–61*.

39. *Lattices of projections in Hilbert space*, Functional Analysis and its Applications 23 (1989) 86–87*.

40. (with G. Mustafaev) *Estimates of the norms for inner derivations of some operator algebras*, Math. Notes 45 (1989) 105–10*.

41. *Multiplication operators and spectral synthesis*, DAN SSSR 313 (1990), 1047–51*.

42. *Spectral synthesis and the Fuglede–Putnam–Rosenblum theorem*, Function Theory and Functional Analysis 54 (1990) 25–36*.

43. "*K.R. Davidson. Nest Algebras. Longman Sci. and Techn. Pitman Research Notes Math.* 1988, 411 pp.", Algebra and Analysis 2:3 (1990) 236–55*.

44. (with Yu. Bespalov and Yu. Samoilenko) *Families of operators related by semilinear relations*, Applications of Methods of Functional Analysis in Mathematical Physics, 1 (1991) 28–51*.

45. (with Yu. Samoilenko) *Representations of relations of the form* $i[A, B] = f(A) + g(B)$, Ukrainian Math. Journal 43 (1991) 110–14*.

46. (with V. Khatskevich and Yu. Shmulian) *Preorders and equivalence in operator ball*, Siberian Math. J. 32 (1991) 172–83*.

47. (with O.Yu. Borenshtein) *One theorem on minimax*, Math. Notes 50:1 (1991) 139–41*.

48. (with E. Kissin) *Dense Q-subalgebras of Banach and $C^*$-algebras and unbounded derivations of Banach and $C^*$-algebras*, Proc. Edinburgh Math. Soc. 36 (1993) 261–76.

49. (with A. Loginov) *Vector-valued duality for modules over Banach algebras*, Izv. RAN 57:4 (1993) 3–35*.

50. (with E. Kissin) *Differential Properties of Some Dense Subalgebras of Banach and $C^*$-algebras*, Proc. Edinburgh Math. Soc. 37 (1994) 399–422.
51. *Invariant subspaces and spectral mapping theorems*, Functional analysis and operator theory, Banach Centre Publications 30 (1994) 313–25.
52. *Operator preserving ideals in $C^*$-algebras*, Studia Math. 109:1 (1994) 67–72.
53. (with V. Khatskevich) *Operator fractional-linear transformations: convexity and compactness of image*, Studia Math. 116:2 (1995) 189–95.
54. (with E. Kissin and A. Loginov) *Derivations of $C^*$-algebras and almost Hermitian representations on $\Pi_\kappa$-spaces*, Pacific J. Math. 174 (1996) 411–30.
55. (with V. Senderov) *Operators that are not $J$-normal*, Proceedings of 26th Voronezh Math. School (1996) 85.
56. (with Yu. Samoilenko and L. Turowska) *Semilinear relations and their representations*, Methods of Functional Analysis and Topology 2:1 (1996) 55–111.
57. (with E. Kissin) *Operator differentiable functions and derivations of operator algebras*, Functional Analysis and its Applications 30:4 (1996) 75–77*.
58. *Some remarks on the Fuglede–Weiss Theorem*, Bull. London Math. Soc. 28 (1996) 385–92.
59. (with R. Haydon) *On a measure-theoretic problem of Arveson*, Proc. American Math. Soc. 124:2 (1996) 497–503.
60. *Quasivectors and Tomita–Takesaki theory for operator algebras on $\Pi_\kappa$-spaces*, Reviews of Math. Physics 31:6 (1997) 749–83.
61. *Factorization of completely positive cocycles and GNS-construction for representations on Pontryagin spaces*, Functional Analysis and its Applications 31:3 (1997) 91–94*.
62. (with E. Kissin) *Differential Banach *-algebras of compact operators associated with symmetric operators*, J. Functional Analysis 156 (1998) 1–29.
63. (with E. Kissin) *Dual spaces and isomorphisms of some differential Banach *-algebras of operators*, Pacific J. Math. 100:2 (1999) 329–60.
64. (with E. Kissin) *On the range inclusion of normal derivations: variations on a theme by Johnson, Williams and Fong*, Proc. London Math. Soc. 83 (2001) 176–98.
65. (with J. Erdos and A. Katavolos) *Rank one subspaces of Bimodules over Maximal Abelian Selfadjoint Algebras*, J. Functional Analysis, 157 (1998), 554–87.
66. (with Yu. Turovskii) *Solvable Lie algebras of compact operators have invariant subspaces*, Spectral and Evolutionary Problems, Simferopol 9 (1999) 38–44.
67. (with H. Radjavi and P. Rosenthal) *Operator semigroups and quasinilpotent commutators*, Proc. American Math. Soc. 128:8 (2000) 2413–20.
68. (with Yu. Turovskii) *Joint spectral radius and invariant subspaces*, Functional Analysis and its Applications, 34:2 (2000) 91–94*.
69. (with Yu. Turovskii) *Joint spectral radius, operator semigroups and a problem of W. Wojtynski*, J. Functional Analysis, 176 (2000) 383–441.

70. (with E. Kissin) *Differential Schatten*-algebras. Approximation property and approximate identities*, J. Operator Theory, 45 (2001) 303–34.
71. (with E. Kissin) *Some Classes of Positive Functionals on Banach *-algebras*, Proc. Conf. on Modelling and Optimization, Vologda (2001) 221–23.
72. (with L. Turowska) *On operator synthesis*, Proc. Conf. on Modelling and Optimization, Vologda (2001) 224–29.
73. *On vector functionals and minimax inequalities*, Spectral and Evolutionary Problems, Simferopol 11 (2001) 56–61.
74. *On representations of limit relations*, Methods of Functional Analysis and Topology 6:3 (2001) 73–76.
75. (with Yu. Turovskii) *Radicals in Banach algebras and some problems on radical Banach algebras*, Functional Analysis and its Applications 35:4 (2001) 95–98*.
76. (with E. Kissin) *On a Problem of J.P. Williams*, Proc. American Math. Soc. 130:12 (2002) 3605–08.
77. (with E. Kissin) *Range-inclusive maps from subspaces of $C^*$–algebras*, Quart. Journal of Mathematics Oxford, 53 (2002), no. 4, 455–65.
78. (with Yu. Turovskii) *Invariant subspaces of operator Lie algebras, and the theory of K-algebras*, Functional Analysis and its Applications 36 (2002) 88–91*; (Trans. in Funct. Anal. Appl. 36 (2002) 328–30).
79. (with Yu. Turovskii) *Formulae for joint spectral radii of sets of operators*, Studia Math. 149 (2002) 23–37.
80. (with Yu. Turovskii) *On the Kleinecke–Shirokov theorem for families of derivations*, Studia Math. 149:2 (2002) 185–90.
81. (with E. Kissin and V. Lomonosov) *Implementation of derivations and invariant subspaces*, Israel J. Math. 134 (2003) 1–28.
82. (with V. Khatskevich) *On linear operator pencils and inclusions of images of balls*, Trends in Banach spaces and operator theory (Memphis, TN, 2001), Contemp. Math., Amer. Math. Soc., Providence, RI 321 (2003) 115–24.
83. (with E. Kissin) *Classes of operator-smooth functions* II: *Operator-differentiable functions*, Integral Equations and Operator theory 49:2 (2004) 165–210.
84. (with L. Turowska). *Operator synthesis* I: *Synthetic sets, bilattices and tensor algebras,* J. Functional Analysis 209 (2004) 293–331.
85. (with I. Todorov) *On subspace lattices.* II. *Continuity of* Lat, J. Operator Theory, 52:2 (2004) 371–84.
86. (with M. Ostrovskii and V. Khatskevich) *On operator matrices generating linear fractional maps of operator balls*, Complex Analysis and Dynamical Systems, Contemp. Math., Amer. Math. Soc., Providence, RI 364 (2004) 93–102.
87. (with I. Todorov) *On subspace lattices.* I. *Closedness type properties and tensor products*, Integral Equations Operator Theory 52:4 (2005) 561–79.

88. (with E. Kissin) *Classes of operator-smooth functions. I. Operator-Lipschitz functions*, Proc. Edinburgh Math. Soc. 48 (2005) 151–73.
89. (with E. Kissin) *Classes of operator-smooth functions. III. Stable functions and Fuglede ideals*, Proc. Edinburgh Math. Soc. 48 (2005) 175–97.
90. (with Yu. Turovskii) *Topological radicals*, I. *Basic properties, tensor products and joint quasinilpotence*, Topological Algebras, their Applications, and Related Topics, Banach Center Publ., Polish Acad. Sci., Warsaw, 67 (2005) 293–333.
91. (with Yu. Turovskii) *Invariant subspaces of operator Lie algebras and Lie algebras with compact adjoint action*, J. Functional Analysis 223:2 (2005) 425–508.
92. (with M. Ostrovskii and V. Khatskevich) *Extremal problems for operators in Banach spaces arising in the study of linear operator pencils*, Integral Equations Operator Theory 51:1 (2005) 109–19.
93. *Various aspects of Fuglede's theorem*, Spectral and evolutional problems, Simferopol 16 (2005) 192–203.
94. (with E. Kissin) *Operator multipliers*, Pacific J. Math. 227:1 (2006) 109–41.
95. (with E. Kissin) *Lipschitz functions on Hermitian Banach *-algebras*, Quarterly J. Math. (Oxford), 57:2 (2006) 215–39.
96. (with E. Kissin and L. Turowska) *Extension of operator Lipschitz and commutator bounded functions*, Oper. Theory Adv. Appl., Birkhäuser, Basel 171 (2006) 225–44.
97. (with V. Khatskevich and M. Ostrovskii) *Quadratic operator inequalities and fractional linear relations*, Functional Analysis and its Applications 41 (2007) 83–87*.
98. (with E. Kissin) *On fully operator Lipschitz functions*, J. Functional Analysis 253:2 (2007) 711–28.
99. (with M. Ostrovskii and V. Khatskevich) *Quadratic inequalities for Hilbert space operators*, Integral Equations and Operator Theory 59:1 (2007) 19–34.
100. (with E. Kissin) *Functions acting on symmetrically normed ideals and on the domains of derivations on these ideals*, J. Operator Theory 58:1 (2007) 63–82.
101. (with M. Brešar and E. Kissin) *When Jordan submodules are bimodules*, Quarterly J. Mathem. (Oxford) 59:4 (2008) 409–40.
102. (with M. Brešar and E. Kissin) *Lie ideals: from pure algebras to $C^*$-algebras*, J. Reine angew Math. 623 (2008), 73–121.
103. (with M. Kennedy and Yu. Turovskii) *Invariant subspaces of subgraded Lie algebras of compact operators*, Integral Equations and Operator Theory, 63 (2009), no. 1, 47–93.
104. (with E. Kissin and Yu. Turovskii) *Banach Lie algebras with Lie subalgebras of finite codimension; their invariant subspaces and ideals*, J. Functional Analysis 256:2 (2009) 323–51.

105. (with M. Ostrovskii and L. Turowska) *Unitarizable representations and fixed points of groups of biholomorphic transformations of operator balls*, J. Functional Analysis 257:8 (2009) 2476–96.
106. (with T. Shulman) *On Lie submodules and tensor algebras*, Functional Analysis and its Applications 43 (2009) 91–96*.
107. (with M. Ostrovskii) *Weak operator topologies, operator ranges and operator equations via Kolmogorov's widths*, Integral Equations and Operator Theory 65 (2009) 551–72.
108. (with E. Kissin and Yu. Turovskii) *Banach Lie algebras with Lie subalgebras of finite codimension have Lie ideals*, J. London Math. Soc. (2) 80 (2009), 603–26.
109. (with A. Bouhrim and M. Burgos) *Linear maps preserving the minimum and reduced minimum modulus*, J. Functional Analysis 258 (2010), 50–66.
110. (with M. Brešar and Yu. Turovskii), *On tractability and Ideal Problem in non–associative operator algebras*, Integral Equations and Operator Theory 67 (2010) 279–300.
111. (with I. Todorov and L. Turowska) *Closable multipliers*, Integral Equations and Operator Theory 69 (2011) 29–62.
112. (with Yu. Turovskii) *Topological radicals*, II. *Applications to spectral theory of multiplication operators*, in book "Elementary operators and their applications", Operator Theory Advances and Applications 212 (2011) 45–114.
113. (with L. Turowska) *An elementary approach to elementary operators*, in book "Elementary operators and their applications", Operator Theory Advances and Applications 212 (2011), 115–32.
114. (with M. Ostrovskii and L. Turowska) *Fixed points of holomorphic transformations of operator balls*, Quarterly J. Math. (Oxford) 62 (2011), 173–87.
115. (with T. Shulman) *On algebras generated by inner derivations*, J. Operator Theory 65 (2011), 101–25.
116. (with E. Kissin, D. Potapov and F. Sukochev), *Lipschitz functions, Schatten ideals and unbounded derivations*, Functional Analysis and its Applications 45:2 (2011) 93–96*.
117. (with E. Kissin, D. Potapov and F. Sukochev), *Operator smoothness in Schatten norms for functions of several variables: Lipschitz conditions, differentiability and unbounded derivations*, Proc. London Math. Soc. 105 (2012), 661–702.
118. (with Yu. Turovskii) *Topological radicals and joint spectral radius*, Functional Analysis and its Applications, 46:4 (2012), 287–304.*
119. (with E. Kissin and Yu. Turovskii) *Topological radicals and Frattini theory for Banach Lie algebras*, Integral Equations and Operator Theory, 74:1 (2012), 51–121.

Operator Theory:
Advances and Applications, Vol. 233, 17–32


# Operators Splitting the Arveson Spectrum

J. Alaminos, J. Extremera and A.R. Villena

**Abstract.** Let $\sigma_1$ and $\sigma_2$ be commuting representations of a locally compact abelian group $G$ on a complex Banach space $X$ and let $\tau$ be a representation of $G$ on another complex Banach space $Y$. This paper is devoted to study the continuous linear operators $A\colon X \to Y$ with the property that $\mathrm{sp}(\tau, Ax) \subset \mathrm{sp}(\sigma_1, x) \cup \mathrm{sp}(\sigma_2, x)$ $(x \in X)$. We are also concerned with those operators that satisfy the preceding property only approximately.

**Mathematics Subject Classification (2010).** Primary: 47A11, Secondary: 47D03.

**Keywords.** Arveson spectrum, local spectrum, doubly power bounded operator, operator with polynomial growth.

## 1. Introduction and preliminaries

### 1.1. Introduction

It is well known that local spectral inclusions can be used for analysing the link between two given operators $S$ and $T$, defined on Banach spaces $X$ and $Y$, respectively (see [8, Chapter 3]). Through a series of papers [1–4] we have pointed out that the analysis of operators preserving zero products may be reduced to the analysis of bilinear maps on the weighted Fourier algebra on the circle group which have the property of taking functions with disjoint support into zero. Further, we realised in [5] that the above-mentioned reduction still works for analysing the behaviour of operators $A\colon X \to Y$ between Banach spaces, which have the property of shrinking the Arveson spectrum with respect to appropriate actions of a locally compact abelian group $G$ on $X$ and $Y$. This means that there exist representations $\sigma$ and $\tau$ of $G$ on $X$ and $Y$, respectively, such that $\mathrm{sp}\big(\tau, Ax\big) \subset \mathrm{sp}\big(\sigma, x\big)$ for each $x \in X$. Moreover we used this analysis for exploring the intertwining condition $AS = TA$ of the operators $S \in \mathcal{B}(X)$ and $T \in \mathcal{B}(Y)$ by $A \in \mathcal{B}(X, Y)$ through the spectral inclusions $\mathrm{sp}(T, Ax) \subset \mathrm{sp}(S, x)$ $(x \in X)$. Here and subsequently, $\mathrm{sp}(R, z)$

The authors were supported by MICINN Grant MTM2012-31755 and Junta de Andalucía Grants FQM-185 and FQM-4911.

stands for the local spectrum of the operator $R$ at the point $z$ (see Example 1.1 for the definition).

It soon became evident that the methods involved in [5] could be applied to a variety of problems concerning operators which are well behaved with the Arveson spectrum. This paper is devoted to study the case when the operator $A$ *splits the Arveson spectrum* in the sense that

$$\operatorname{sp}(\tau, Ax) \subset \operatorname{sp}(\sigma_1, x) \cup \operatorname{sp}(\sigma_2, x) \quad (x \in X), \tag{1.1}$$

where $\sigma_1$ and $\sigma_2$ are commuting representations of $G$ on $X$ and $\tau$ is a representation of $G$ on $Y$. In Section 2 we show that if $A$ satisfies (1.1) and the representations have polynomial growth of degree $\alpha$, then

$$\sum_{n_1,n_2=0}^{N} (-1)^{n_1+n_2} \binom{N}{n_1}\binom{N}{n_2} \tau(t)^{2N-(n_1+n_2)} A\sigma_1(t)^{n_1}\sigma_2(t)^{n_2} = 0 \quad (t \in G) \tag{1.2}$$

for each $N > 2\alpha$. We apply the preceding result to the problem of finding the intertwining link of some operators $S_1, S_2 \in \mathcal{B}(X)$ and $T \in \mathcal{B}(Y)$ in the case when

$$\operatorname{sp}(T, Ax) \subset \operatorname{sp}(S_1, x) \cup \operatorname{sp}(S_2, x) \quad (x \in X) \tag{1.3}$$

for some operator $A \in \mathcal{B}(X, Y)$. As a matter of fact, we show that if the operators $S_1, S_2$, and $T$ are invertible and have polynomial growth of degree $\alpha$, then

$$\sum_{n_1,n_2=0}^{N} (-1)^{n_1+n_2} \binom{N}{n_1}\binom{N}{n_2} T^{2N-(n_1+n_2)} AS_1^{n_1}S_2^{n_2} = 0 \tag{1.4}$$

for each $N > 2\alpha$.

In Section 3 we provide a quantitative estimate of the above-mentioned result that (1.1) implies (1.2) by considering the perturbed spectral inclusion

$$\operatorname{sp}(\tau, Ax) \subset \operatorname{sp}(\sigma_1, x)U(K, \varepsilon_1) \cup \operatorname{sp}(\sigma_2, x)U(K, \varepsilon_2) \quad (x \in X), \tag{1.5}$$

for some $K \subset G$ compact neighbourhood of the identity and some $\varepsilon_1, \varepsilon_2 > 0$, where $U(K, \varepsilon_i) = \{\gamma \in \widehat{G} \colon \ \gamma(K) \subset \{e^{i\theta} \colon \ |\theta| < \varepsilon_i\}\}$ $(i = 1, 2)$. We show that if (1.5) holds and the representations have polynomial growth of degree $\alpha \in \mathbb{Z}$, then

$$\begin{aligned} &\left\| \sum_{n_1,n_2=0}^{N} (-1)^{n_1+n_2} \binom{N}{n_1}\binom{N}{n_2} \tau(t)^{2N-(n_1+n_2)} A\sigma_1(t)^{n_1}\sigma_2(t)^{n_2} \right\| \qquad (1.6)\\ &\leq \Big(2\tan\left(\tfrac{N}{2}\varepsilon_1\right) + 2\tan\left(\tfrac{N}{2}\varepsilon_2\right) + 4\tan\left(\tfrac{N}{2}\varepsilon_1\right)\tan\left(\tfrac{N}{2}\varepsilon_2\right)\Big)\kappa(t)\|A\|C(N-1) \end{aligned}$$

for each $t \in K$, where $N = 3\alpha + 1$,

$$\kappa(t) = \sup_{k\in\mathbb{Z}} \frac{\|\sigma_1(kt)\|}{(1+|k|)^\alpha} \sup_{k\in\mathbb{Z}} \frac{\|\sigma_2(kt)\|}{(1+|k|)^\alpha} \sup_{k\in\mathbb{Z}} \frac{\|\tau(kt)\|}{(1+|k|)^\alpha} \quad (t \in G)$$

and

$$C(N) = 3^{N-1}(2N+1)^{N-1} \sum_{n_1=1}^{N}\sum_{n_2=0}^{N} \binom{N}{n_1}\binom{N}{n_2} {n_1}^{N-1}(n_2+1)^{N-1}.$$

We then apply this result for giving a quantitative estimate of the property that (1.3) implies (1.4).

Finally, we emphasize that the method we are using throughout this paper still goes when we consider (1.1) or (1.5) for finitely many representations $\sigma_1, \ldots, \sigma_n$ instead of just two of them. Nevertheless, we leave it to the reader to obtain the analogous to (1.2) and (1.6) in such a case.

### 1.2. Preliminaries

All Banach spaces and Banach algebras which we consider throughout this paper are assumed to be complex.

Let $X$ be a non-zero complex Banach space. Let $X^*$ denote the topological dual space of $X$. We write $\langle \cdot, \cdot \rangle$ for the dual pairing of $X$ and $X^*$. If $X_*$ is any linear subspace of $X^*$, then $\sigma(X, X_*)$ stands for the coarsest topology on $X$ for which each of the functionals in $X_*$ are continuous. For a Banach space $Y$, let $\mathcal{B}(X, Y)$ denote the space of all continuous linear operators from $X$ into $Y$. As usual, we write $\mathcal{B}(X)$ instead of $\mathcal{B}(X, X)$.

We are given, along with a complex Banach space $X$, a linear subspace $X_*$ of the dual $X^*$ of $X$. Assume further

$$\|x\| = \sup\{|\langle x, \varphi \rangle| \colon \ \varphi \in X_*, \ \|\varphi\| \leq 1\} \quad (x \in X), \tag{1.7}$$

and

$$\text{the } \sigma(X, X_*)\text{-closed convex hull of every } \sigma(X, X_*)\text{-compact set in } X \text{ is } \sigma(X, X_*)\text{-compact.} \tag{1.8}$$

For example, if we take $X_* = X^*$, then (1.7) holds by the Hahn–Banach theorem and (1.8) follows from the Krein–Smulian theorem on weak compactness. For a second example, if $X$ is the dual of a Banach space $X_*$, then (1.7) holds by definition and (1.8) holds by the Banach–Alaoglu theorem. In [6] there are some other examples.

Throughout this paper, $G$ stands for a locally compact abelian group with dual group $\widehat{G}$. By a *representation* of $G$ on $X$ we mean a group homomorphism $\tau$ from $G$ into the group of all invertible $\sigma(X, X_*)$-continuous linear operators on $X$ such that

$$\text{the map } t \mapsto \langle \tau(t)x, \varphi \rangle \text{ is continuous for all } x \in X \text{ and } \varphi \in X_*. \tag{1.9}$$

We will restrict our attention to those representations with *polynomial growth.* This means that

$$\|\tau(kt)\| = O(|k|^\alpha) \text{ as } |k| \to \infty \ (t \in G) \text{ for some } \alpha \geq 0. \tag{1.10}$$

Let us also recall that $\tau$ is said to be *bounded* if $\sup_{t \in G} \|\tau(t)\| < \infty$.

Let $\tau$ be a representation of $G$ on $X$ satisfying all the preceding requirements (1.7)–(1.10). Then the map $\omega \colon G \to \mathbb{R}$ defined by $\omega(t) = \|\tau(t)\|$ $(t \in G)$ gives a *non-quasianalytic weight* on $G$. The corresponding Beurling algebra consists of the Banach space $L^1(G, \omega)$ of those (equivalence classes of) Borel functions on $G$ for which $\|f\| := \int_G |f(t)|\omega(t)dt < \infty$ endowed with convolution as multiplication.

This Banach algebra can be thought of as a closed ideal of the Banach algebra $M(G,\omega)$ of all locally finite regular Borel measures $\mu$ on $G$ for which the weighted measure $\omega\mu$ is of finite variation. Moreover, the representation $\tau$ gives rise to a norm-decreasing algebra homomorphism $\widetilde{\tau}\colon M(G,\omega)\to\mathcal{B}(X)$ which is defined by

$$\langle\widetilde{\tau}(\mu)x,\varphi\rangle=\int_G\langle\tau(t)x,\varphi\rangle\,d\mu(t)\quad(\mu\in M(G,\omega),\ x\in X,\ \varphi\in X_*)$$

(see [8, pp. 446–448]). For every $x\in X$, the *Arveson spectrum* of $\tau$ at $x$ is defined as

$$\operatorname{sp}(\tau,x)=\bigl\{\gamma\in\widehat{G}\colon\ \widehat{f}(\gamma)=0\text{ for each }f\in L^1(G,\omega)\text{ such that }\ \widetilde{\tau}(f)x=0\bigr\}.$$

The basics of the theory of the Arveson spectrum may be found in [6], [7, Section 3.2.3], and [8, Section 4.12].

*Example* 1.1. Let $X$ be a complex Banach space. Given $T\in\mathcal{B}(X)$ and $x\in X$, the *local resolvent* of $T$ at $x\in X$ is the union of all open subsets $U$ of $\mathbb{C}$ for which there is an analytic function $f\colon U\to X$ satisfying $(T-z)f(z)=x$ for each $z\in U$. The *local spectrum* of $T$ at $x$, from now denoted by $\operatorname{sp}(T,x)$, is defined as the complement of the local resolvent of $T$ at $x$. We refer the reader to [8] for a full account of the local spectral theory.

We now assume that $T\in\mathcal{B}(X)$ is an invertible operator with polynomial growth, in the sense that $\|T^k\|=O(|k|^\alpha)$ as $|k|\to\infty$ for some $\alpha\geq 0$. Then the map $\tau_T\colon\mathbb{Z}\to\mathcal{B}(X)$ defined by $\tau_T(k)=T^k$ $(k\in\mathbb{Z})$ is a representation of $\mathbb{Z}$ on $X$ and [8, Example 4.12.7] shows that $\operatorname{sp}(\tau_T,x)=\operatorname{sp}(T,x)$ for each $x\in X$.

## 2. Exact splitting

In this section we are concerned with the question of showing that (1.1) implies (1.2). Our method consists in giving rise in a natural way to a trilinear map $\varphi$ on the weighted Fourier algebra $A_\alpha(\mathbb{T})$ on the circle group $\mathbb{T}$ which encodes the personality of the operator $A$. Specifically, the map $\varphi$ has the property that $\varphi(f,g,h)=0$ whenever $f,g,h\in A_\alpha(\mathbb{T})$ are such that $\operatorname{supp}(f)\cap\operatorname{supp}(g)=\emptyset$ and $\operatorname{supp}(f)\cap\operatorname{supp}(h)=\emptyset$.

For $n\in\mathbb{N}$ and $\alpha\geq 0$, let $A_\alpha(\mathbb{T}^n)$ denote the *weighted Fourier algebra* consisting of all functions $f\in C(\mathbb{T}^n)$ such that

$$\|f\|_{A_\alpha(\mathbb{T}^n)}:=\sum_{k\in\mathbb{Z}^n}\left|\widehat{f}(k)\right|\left(1+|k|\right)^\alpha\ <\ \infty,$$

where for $k\in\mathbb{Z}^n$ we write $|k|=|k_1|+\cdots+|k_n|$. We abbreviate $A_0(\mathbb{T}^n)$ to $A(\mathbb{T}^n)$. From now on, $\mathbf{z}$ stands for the function in $A(\mathbb{T})$ defined by $\mathbf{z}(z)=z$ $(z\in\mathbb{T})$. For $n\in\mathbb{N}$ and $j\in\{1,\dots,n\}$ we will denote by $\mathbf{z_j}$ the function in $A(\mathbb{T}^n)$ given by $\mathbf{z_j}(z)=z_j$ $(z\in\mathbb{T}^n)$.

Given $f\in C(\mathbb{T}^m)$ and $g\in C(\mathbb{T}^n)$ we define $f\otimes g\in C(\mathbb{T}^{m+n})$ by

$$(f\otimes g)(z)=f(z_1,\dots,z_m)g(z_{m+1},\dots,z_{m+n})\quad(z\in\mathbb{T}^{m+n}).$$

It is straightforward to check that, if $f \in A_\alpha(\mathbb{T}^m)$ and $g \in A_\alpha(\mathbb{T}^n)$, then $f \otimes g \in A_\alpha(\mathbb{T}^{m+n})$ with

$$\|f \otimes g\|_{A_\alpha(\mathbb{T}^{m+n})} \leq \|f\|_{A_\alpha(\mathbb{T}^m)} \|g\|_{A_\alpha(\mathbb{T}^n)}.$$

In order to address the problem of describing the operators shrinking the Arveson spectrum we involved in [5] the following key result.

**Theorem 2.1 ([4, Theorem 2.2]).** *Let $\alpha \geq 0$ and let $\varphi \colon A_\alpha(\mathbb{T}) \times A_\alpha(\mathbb{T}) \to X$ be a continuous bilinear map into some Banach space $X$ with the property that*

$$f, g \in A_\alpha(\mathbb{T}),\ \operatorname{supp}(f) \cap \operatorname{supp}(g) = \emptyset \quad \Rightarrow \quad \varphi(f,g) = 0. \tag{2.1}$$

*Then*

$$\sum_{n=0}^{N} (-1)^n \binom{N}{n} \varphi(\mathbf{z}^{N-n}, \mathbf{z}^n) = 0$$

*for each $N > 2\alpha$.*

We now introduce a trilinear version of Theorem 2.1 which can be immediately applied for analysing the operators satisfying (1.1).

**Theorem 2.2.** *Let $\alpha \geq 0$ and let $\varphi \colon A_\alpha(\mathbb{T}) \times A_\alpha(\mathbb{T}) \times A_\alpha(\mathbb{T}) \to X$ be a continuous trilinear map into some Banach space $X$ with the property that*

$$\begin{aligned} &f, g, h \in A_\alpha(\mathbb{T}), \\ &\quad \operatorname{supp}(f) \cap \operatorname{supp}(g) = \operatorname{supp}(f) \cap \operatorname{supp}(h) = \emptyset \quad \Rightarrow \varphi(f,g,h) = 0. \end{aligned} \tag{2.2}$$

*Then*

$$\sum_{n_1, n_2 = 0}^{N} (-1)^{n_1+n_2} \binom{N}{n_1} \binom{N}{n_2} \varphi(\mathbf{z}^{2N-(n_1+n_2)}, \mathbf{z}^{n_1}, \mathbf{z}^{n_2}) = 0 \tag{2.3}$$

*for each $N > 2\alpha$.*

*Proof.* Pick $u, v \in A_\alpha(\mathbb{T})$ such that $\operatorname{supp}(u) \cap \operatorname{supp}(v) = \emptyset$. We define a continuous bilinear map $\varphi_{u,v} \colon A_\alpha(\mathbb{T}) \times A_\alpha(\mathbb{T}) \to X$ by

$$\varphi_{u,v}(f,g) = \varphi(fu, g, v) \quad (f, g \in A_\alpha(\mathbb{T})).$$

Taking into account (2.2), we see immediately that $\varphi_{u,v}$ satisfies (2.1). Theorem 2.1 now gives

$$0 = \sum_{n=0}^{N} (-1)^n \binom{N}{n} \varphi_{u,v}(\mathbf{z}^{N-n}, \mathbf{z}^n) = \sum_{n=0}^{N} (-1)^n \binom{N}{n} \varphi(\mathbf{z}^{N-n} u, \mathbf{z}^n, v) \tag{2.4}$$

for each $N > 2\alpha$. We now consider the continuous bilinear map $\psi \colon A_\alpha(\mathbb{T}) \times A_\alpha(\mathbb{T}) \to X$ defined by

$$\psi(f,g) = \sum_{n=0}^{N} (-1)^n \binom{N}{n} \varphi(\mathbf{z}^{N-n} f, \mathbf{z}^n, g) \quad (f, g \in A_\alpha(\mathbb{T})).$$

On account of (2.4), the map $\psi$ satisfies (2.1) and Theorem 2.1 finally gives (2.3). □

**Theorem 2.3.** *Let $G$ be a locally compact abelian group. Let $X$ and $Y$ be Banach spaces and let $X_*$ and $Y_*$ be linear subspaces of the duals $X^*$ and $Y^*$, respectively, satisfying* (1.7) *and* (1.8). *Suppose that $\sigma_1$ and $\sigma_2$ are commuting representations of $G$ on $X$ and $\tau$ is a representation of $G$ on $Y$ such that*

$$\|\sigma_1(kt)\|, \|\sigma_2(kt)\|, \|\tau(kt)\| = O(|k|^\alpha) \text{ as } |k| \to \infty \ (t \in G)$$

*for some $\alpha \geq 0$. If $A \in \mathcal{B}(X,Y)$ is such that*

$$\operatorname{sp}(\tau, Ax) \subset \operatorname{sp}(\sigma_1, x) \cup \operatorname{sp}(\sigma_2, x) \quad (x \in X),$$

*then*

$$\sum_{n_1,n_2=0}^{N} (-1)^{n_1+n_2} \binom{N}{n_1}\binom{N}{n_2} \tau(t)^{2N-(n_1+n_2)} A\sigma_1(t)^{n_1}\sigma_2(t)^{n_2} = 0 \quad (t \in G)$$

*for each $N > 2\alpha$.*

*Proof.* Let $t \in G$. We consider the continuous linear map $\Phi_t\colon A_\alpha(\mathbb{T}) \to M(G,\omega)$ and the continuous trilinear map $\varphi\colon A_\alpha(\mathbb{T}) \times A_\alpha(\mathbb{T}) \times A_\alpha(\mathbb{T}) \to \mathcal{B}(X,Y)$ defined by

$$\Phi_t(f) = \sum_{k=-\infty}^{+\infty} \widehat{f}(k)\delta_t^k \quad (f \in A_\alpha(\mathbb{T}))$$

and

$$\varphi_t(f,g,h) = \widetilde{\tau}(\Phi_t(f)) \circ A \circ \widetilde{\sigma_1}(\Phi_t(g)) \circ \widetilde{\sigma_2}(\Phi_t(h)) \quad (f,g,h \in A_\alpha(\mathbb{T})),$$

respectively.

It is immediate to check that

$$\widehat{\Phi_t(f)}(\gamma) = f(\gamma(t)) \quad (f \in A_\alpha(\mathbb{T}),\ \gamma \in \widehat{G})$$

and so

$$\operatorname{supp}\big(\widehat{\Phi_t(f)}\big) \subset \big\{\gamma \in \widehat{G}\colon\ \gamma(t) \in \operatorname{supp}(f)\big\}. \tag{2.5}$$

Let $f,g,h \in A_\alpha(\mathbb{T})$ and $x \in X$. On account of [8, Lemma 4.12.6] and (2.5), we have

$$\begin{aligned}
\operatorname{sp}(\tau, \varphi_t(f,g,h)x)) &\subset \operatorname{supp}\big(\widehat{\Phi_t(f)}\big) \cap \operatorname{sp}\Big(\tau, A\big(\widetilde{\sigma_1}(\Phi_t(g))\widetilde{\sigma_2}(\Phi_t(h))x\big)\Big)\\
&\subset \operatorname{supp}\big(\widehat{\Phi_t(f)}\big) \cap \operatorname{sp}(\sigma_1, \widetilde{\sigma_1}(\Phi_t(g))\widetilde{\sigma_2}(\Phi_t(h))x)\\
&\quad \cup \operatorname{supp}\big(\widehat{\Phi_t(f)}\big) \cap \operatorname{sp}(\sigma_2, \widetilde{\sigma_1}(\Phi_t(g))\widetilde{\sigma_2}(\Phi_t(h))x)\\
&\subset \operatorname{supp}\big(\widehat{\Phi_t(f)}\big) \cap \operatorname{supp}\big(\widehat{\Phi_t(g)}\big) \cap \operatorname{sp}(\sigma_1, \widetilde{\sigma_2}(\Phi_t(h))x)\\
&\quad \cup \operatorname{supp}\big(\widehat{\Phi_t(f)}\big) \cap \operatorname{supp}\big(\widehat{\Phi_t(h)}\big) \cap \operatorname{sp}(\sigma_2, \widetilde{\sigma_1}(\Phi_t(g))x).
\end{aligned}$$

Hence,

$$\begin{aligned}
&\operatorname{sp}(\tau, \varphi_t(f,g,h)x)\\
&\quad \subset \Big\{\gamma \in \widehat{G}\colon\ \gamma(t) \in \Big(\operatorname{supp}(f) \cap \operatorname{supp}(g)\Big) \cup \Big(\operatorname{supp}(f) \cap \operatorname{supp}(h)\Big)\Big\}.
\end{aligned}$$

Accordingly, if $f, g, h \in A(\mathbb{T})$ are such that $\operatorname{supp}(f) \cap \operatorname{supp}(g) = \operatorname{supp}(f) \cap \operatorname{supp}(h) = \emptyset$, then we see that $\operatorname{sp}(\tau, \varphi_t(f, g, h)x) = \emptyset$ and [8, Proposition 4.12.4] now yields $\varphi_t(f, g, h)x = 0$ for each $x \in X$. Theorem 2.2 then gives

$$\sum_{n_1, n_2 = 0}^{N} (-1)^{n_1 + n_2} \binom{N}{n_1} \binom{N}{n_1} \varphi_t(\mathbf{z}^{2N-(n_1+n_2)}, \mathbf{z}^{n_1}, \mathbf{z}^{n_2}) = 0.$$

for each $N > 2\alpha$. On the other hand, it is easily seen that

$$\varphi_t(\mathbf{z}^i, \mathbf{z}^j, \mathbf{z}^k) = \tau(t)^i A \sigma_1(t)^j \sigma_2(t)^k \quad (t \in G,\ i, j, k \in \mathbb{Z}),$$

which completes the proof. □

Our next goal consists in applying Theorem 2.3 to derive some intertwining link of three given operators $S_1, S_2 \in \mathcal{B}(X)$ and $T \in \mathcal{B}(Y)$ from its local spectral properties.

Let $X$ and $Y$ be complex Banach spaces. Given operators $S \in \mathcal{B}(X)$ and $T \in \mathcal{B}(Y)$, the *commutator* $C(T, S)$ is defined as the mapping

$$C(T, S)\colon \mathcal{B}(X, Y) \to \mathcal{B}(X, Y), \quad C(T, S) = TA - AS, \quad (A \in \mathcal{B}(X, Y)).$$

The iterates $C(T, S)^n$ are often called *higher-order commutators* and it is easily seen that

$$C(T, S)^n(A) = \sum_{k=0}^{n} (-1)^k \binom{n}{k} T^{n-k} A S^k, \quad (A \in \mathcal{B}(X, Y)).$$

If $S$, $T$, and $A$ are pairwise commuting operators on $X$, then

$$C(T, S)^n(A) = (T - S)^n A, \quad (n \in \mathbb{N}).$$

**Corollary 2.4.** *Let $X$ and $Y$ be complex Banach spaces and let $S_1, S_2 \in \mathcal{B}(X)$ and $T \in \mathcal{B}(Y)$ be invertible operators with $S_1$ and $S_2$ commuting and*

$$\|S_1^k\|, \|S_2^k\|, \|T^k\| = O(|k|^\alpha) \text{ as } |k| \to \infty$$

*for some $\alpha \geq 0$. If $A \in \mathcal{B}(X, Y)$ is such that*

$$\operatorname{sp}(T, Ax) \subset \operatorname{sp}(S_1, x) \cup \operatorname{sp}(S_2, x) \quad \forall x \in X,$$

*then $C(T, S_2)^N C(T, S_1)^N(A) = 0$ for each $N > 2\alpha$. Consequently,*

1. *If $S_1$, $S_2$, and $T$ are doubly power bounded, then $C(T, S_2)C(T, S_1)(A) = 0$.*
2. *If $X = Y$, and if $S_1$, $S_2$, $T$, and $A$ are pairwise commuting, then $(T - S_2)^N (T - S_1)^N A = 0$.*

*Proof.* It is sufficient to apply Theorem 2.3 by taking the representations $\tau = \tau_T$, $\sigma_1 = \tau_{S_1}$, and $\sigma_2 = \tau_{S_2}$, which are given in Example 1.1. □

**Corollary 2.5.** *Let $X$ be a complex Banach space and let $S_1, S_2, T \in \mathcal{B}(X)$ be pairwise commuting invertible operators with $\|S_1^k\|, \|S_2^k\|, \|T^k\| = O(|k|^\alpha)$ as $|k| \to \infty$ for some $\alpha \geq 0$. If $\operatorname{sp}(T, x) \subset \operatorname{sp}(S_1, x) \cup \operatorname{sp}(S_2, x)$ for each $x \in X$, then $(T - S_2)^N (T - S_1)^N = 0$ for each $N > 2\alpha$.*

*Proof.* We are reduced to apply Corollary 2.4 with $X = Y$ and $A$ being the identity operator on $X$. □

**Corollary 2.6.** *Let $X$ be a complex Banach space and let $T \in \mathcal{B}(X)$ invertible and such that $\|T^k\| = O(|k|^\alpha)$ as $|k| \to \infty$ for some $\alpha \geq 0$. If $\operatorname{sp}(T) \subset \{\lambda_1, \lambda_2\}$ for some $\lambda_1, \lambda_2 \in \mathbb{C}$, then $(T - \lambda_1 I_X)^N (T - \lambda_2 I_X)^N = 0$ for each $N > 2\alpha$.*

*Proof.* Since $\operatorname{sp}(T, x) \subset \operatorname{sp}(T)$ for each $x \in X$, it suffices to apply Corollary 2.5 with $S_1 = \lambda_1 I_X$ and $S_2 = \lambda_2 I_X$. □

## 3. Approximate splitting

The purpose of this section consists in providing quantitative estimates of the results given in the preceding section in the same vein as in [5]. The key result for the approach given in [5] is the following lemma, which requires to introduce some notation. Throughout this section we shall consider sets of the form

$$E_\varepsilon = \{e^{i\theta} : \ |\theta| \leq \varepsilon\},$$

where $0 \leq \varepsilon < \pi$. Given a nonempty closed set $\mathcal{S} \subset \mathbb{T}^n$ we define

$$I_{A_\alpha(\mathbb{T}^n)}(\mathcal{S}) = \{f \in A_\alpha(\mathbb{T}^n) : \ f(\mathcal{S}) = \{0\}\}.$$

Then $I_{A_\alpha(\mathbb{T}^n)}(\mathcal{S})$ is a closed ideal of $A_\alpha(\mathbb{T}^n)$.

**Lemma 3.1** (**[5, Lemma 3.1]**). *Let $N \in \mathbb{Z}$ with $N \geq 0$ and $0 \leq \varepsilon < \frac{\pi}{N+1}$. Then*

$$\operatorname{dist}\Big((\mathbf{z} - \mathbf{1})^{N+1}, I_{A_N(\mathbb{T})}(E_\varepsilon)\Big) \leq 2\tan\left(\tfrac{N+1}{2}\varepsilon\right) C_1(N),$$

*where, here and subsequently,*

$$C_1(N) = 3^N \sum_{n=1}^{N+1} \binom{N+1}{n} n^N.$$

By using Lemma 3.1 together with some more preliminary lemmas we shall obtain a quantitative estimate of Theorem 2.2 and then of Theorem 2.3.

**Lemma 3.2.** *Let $N \in \mathbb{Z}$ with $N \geq 0$ and $0 \leq \varepsilon_1, \varepsilon_2 < \frac{\pi}{N+1}$. Then*

$$\begin{aligned} &\operatorname{dist}\Big((\mathbf{z_1} - \mathbf{1})^{N+1}(\mathbf{z_2} - \mathbf{1})^{N+1}, I_{A_N(\mathbb{T}^2)}(E_{\varepsilon_1} \times \mathbb{T} \cup \mathbb{T} \times E_{\varepsilon_2})\Big) \\ &\quad \leq \Big(2\tan\left(\tfrac{N+1}{2}\varepsilon_1\right) + 2\tan\left(\tfrac{N+1}{2}\varepsilon_2\right) + 4\tan\left(\tfrac{N+1}{2}\varepsilon_1\right)\tan\left(\tfrac{N+1}{2}\varepsilon_2\right)\Big) C_2(N). \end{aligned}$$

*Here and subsequently,*

$$C_2(N) = C_1(N) \sum_{n=0}^{N+1} \binom{N+1}{n} (n+1)^N.$$

*Proof.* Let $\alpha_1 > 2\tan\left(\frac{N+1}{2}\varepsilon_1\right)C_1(N)$ and $\alpha_2 > 2\tan\left(\frac{N+1}{2}\varepsilon_2\right)C_1(N)$. According to Lemma 3.1, there are functions $f \in I_{A_N(\mathbb{T})}(E_{\varepsilon_1})$ and $g \in I_{A_N(\mathbb{T})}(E_{\varepsilon_2})$ such that $\|(\mathbf{z}-\mathbf{1})^{N+1} - f\|_{A_N(\mathbb{T})} < \alpha_1$ and $\|(\mathbf{z}-\mathbf{1})^{N+1} - g\|_{A_N(\mathbb{T})} < \alpha_2$. Since $f \otimes g \in I_{A_N(\mathbb{T}^2)}(E_{\varepsilon_1} \times \mathbb{T} \cup \mathbb{T} \times E_{\varepsilon_2})$,

$$\begin{aligned}(\mathbf{z_1}-\mathbf{1})^{N+1}(\mathbf{z_2}-\mathbf{1})^{N+1} - f\otimes g = (\mathbf{z}-\mathbf{1})^{N+1} &\otimes \left((\mathbf{z}-\mathbf{1})^{N+1} - g\right) \\ &+ \left((\mathbf{z}-\mathbf{1})^{N+1} - f\right)\otimes(\mathbf{z}-\mathbf{1})^{N+1} \\ &- \left((\mathbf{z}-\mathbf{1})^{N+1} - f\right)\otimes\left((\mathbf{z}-\mathbf{1})^{N+1} - g\right),\end{aligned}$$

and

$$\left\|(\mathbf{z}-\mathbf{1})^{N+1}\right\|_{A_N(\mathbb{T})} \leq \sum_{n=0}^{N+1}\binom{N+1}{n}(n+1)^N,$$

it follows that

$$\begin{aligned}&\operatorname{dist}\Big((\mathbf{z_1}-\mathbf{1})^{N+1}(\mathbf{z_2}-\mathbf{1})^{N+1}, I_{A_N(\mathbb{T}^2)}(E_{\varepsilon_1}\times\mathbb{T}\cup\mathbb{T}\times E_{\varepsilon_2})\Big) \\ &\leq \left\|(\mathbf{z_1}-\mathbf{1})^{N+1}(\mathbf{z_2}-\mathbf{1})^{N+1} - f\otimes g\right\|_{A_N(\mathbb{T}^2)} \\ &\leq \left\|(\mathbf{z}-\mathbf{1})^{N+1}\right\|_{A_N(\mathbb{T})}\left\|(\mathbf{z}-\mathbf{1})^{N+1} - g\right\|_{A_N(\mathbb{T})} \\ &\quad + \left\|(\mathbf{z}-\mathbf{1})^{N+1} - f\right\|_{A_N(\mathbb{T})}\left\|(\mathbf{z}-\mathbf{1})^{N+1}\right\|_{A_N(\mathbb{T})} \\ &\quad + \left\|(\mathbf{z}-\mathbf{1})^{N+1} - f\right\|_{A_N(\mathbb{T})}\left\|(\mathbf{z}-\mathbf{1})^{N+1} - g\right\|_{A_N(\mathbb{T})} \\ &\leq \left(\sum_{n=0}^{N+1}\binom{N+1}{n}(n+1)^N\right)(\alpha_1+\alpha_2+\alpha_1\alpha_2).\end{aligned}$$

By taking $\alpha_1 \to 2\tan\left(\frac{N+1}{2}\varepsilon_1\right)C_1(N)$ and $\alpha_2 \to 2\tan\left(\frac{N+1}{2}\varepsilon_2\right)C_1(N)$ in the preceding inequality we arrive at the inequality claimed in the lemma. □

**Lemma 3.3.** *Let $N \in \mathbb{Z}$ with $N \geq 0$ and $0 \leq \varepsilon_1, \varepsilon_2 < \frac{\pi}{N+1}$. Then*

$$\begin{aligned}&\operatorname{dist}\Big((\mathbf{z_1}-\mathbf{z_2})^{N+1}(\mathbf{z_1}-\mathbf{z_3})^{N+1}, I_{A_N(\mathbb{T}^3)}(E^3_{\varepsilon_1,\varepsilon_2})\Big) \leq \\ &\quad\Big(2\tan\left(\tfrac{N+1}{2}\varepsilon_1\right) + 2\tan\left(\tfrac{N+1}{2}\varepsilon_2\right) + 4\tan\left(\tfrac{N+1}{2}\varepsilon_1\right)\tan\left(\tfrac{N+1}{2}\varepsilon_2\right)\Big)C_3(N),\end{aligned}$$

*where*

$$E^3_{\varepsilon_1,\varepsilon_2} = \left\{(z_1,z_2,z_3)\in\mathbb{T}^3\colon\ z_1z_2^{-1}\in E_{\varepsilon_1}\right\} \cup \left\{(z_1,z_2,z_3)\in\mathbb{T}^3\colon\ z_1z_3^{-1}\in E_{\varepsilon_1}\right\}.$$

*Here and subsequently,*

$$C_3(N) = C_2(N)(2N+3)^N.$$

*Proof.* We begin by defining a continuous linear operator

$$\begin{aligned}&\Psi\colon A_N(\mathbb{T}^2)\to A_N(\mathbb{T}^3),\\ &(\Psi f)(z_1,z_2,z_3) = f(z_1z_2^{-1}, z_1z_3^{-1})z_2^{N+1}z_3^{N+1}\quad (z_1,z_2,z_3\in\mathbb{T}).\end{aligned}$$

We now proceed to estimate the norm of $\Psi$. Let $f \in A_N(\mathbb{T}^2)$. Then

$$\begin{aligned}\|\Psi f\|_{A_N(\mathbb{T}^2)} &= \sum_{k_1,k_2\in\mathbb{Z}} \left|\widehat{f}(k_1,k_2)\right| \left(1+|k_1+k_2|+|-k_1+N+1|+|-k_2+N+1|\right)^N \\ &\leq \sum_{k_1,k_2\in\mathbb{Z}} \left|\widehat{f}(k_1,k_2)\right| \left(2+2|k_1|+2|k_2|+2N+1\right)^N \\ &= \sum_{k_1,k_2\in\mathbb{Z}} \left|\widehat{f}(k_1,k_2)\right| \sum_{n=0}^{N} \binom{N}{n} \left(1+|k_1|+|k_2|\right)^n 2^n (2N+1)^{N-n} \\ &\leq \|f\|_{A_N(\mathbb{T}^2)} \sum_{n=0}^{N} \binom{N}{n} 2^n (2N+1)^{N-n} = \|f\|_{A_N(\mathbb{T}^2)} (2N+3)^N.\end{aligned}$$

Consequently, $\|\Psi\| \leq (2N+3)^N$. On the other hand, we check at once that

$$\Psi\left(I_{A_N(\mathbb{T}^2)}(E_{\varepsilon_1}\times\mathbb{T}\cup\mathbb{T}\times E_{\varepsilon_2})\right) \subset I_{A_N(\mathbb{T}^3)}(E^3_{\varepsilon_1,\varepsilon_2})$$

and that

$$\Psi\left((\mathbf{z_1}-\mathbf{1})^{N+1}(\mathbf{z_2}-\mathbf{1})^{N+1}\right) = (\mathbf{z_1}-\mathbf{z_2})^{N+1}(\mathbf{z_1}-\mathbf{z_3})^{N+1}.$$

Hence

$$\begin{aligned}&\operatorname{dist}\Big((\mathbf{z_1}-\mathbf{z_2})^{N+1}(\mathbf{z_1}-\mathbf{z_3})^{N+1}, I_{A_N(\mathbb{T}^3)}(E^3_{\varepsilon_1,\varepsilon_2})\Big) \\ &\quad\leq \operatorname{dist}\Big(\Psi\left((\mathbf{z_1}-\mathbf{1})^{N+1}(\mathbf{z_2}-\mathbf{1})^{N+1}\right), \Psi\left(I_{A_N(\mathbb{T}^2)}(E_{\varepsilon_1}\times\mathbb{T}\cup\mathbb{T}\times E_{\varepsilon_2})\right)\Big) \\ &\quad\leq (2N+3)^N \operatorname{dist}\Big((\mathbf{z_1}-\mathbf{1})^{N+1}(\mathbf{z_2}-\mathbf{1})^{N+1}, I_{A_N(\mathbb{T}^2)}(E_{\varepsilon_1}\times\mathbb{T}\cup\mathbb{T}\times E_{\varepsilon_2})\Big)\end{aligned}$$

and finally Lemma 3.2 establishes the result. □

**Theorem 3.4.** *Let $\alpha \in \mathbb{Z}$ with $\alpha \geq 0$, and let $\varphi\colon A_\alpha(\mathbb{T})\times A_\alpha(\mathbb{T})\times A_\alpha(\mathbb{T}) \to X$ be a continuous trilinear map into some Banach space $X$ with the property that*

$$\begin{aligned}&f,g,h \in A_\alpha(\mathbb{T}), \\ &\quad \operatorname{supp}(f)\big(\operatorname{supp}(g)\big)^{-1}\cap E_{\varepsilon_1} = \operatorname{supp}(f)\big(\operatorname{supp}(h)\big)^{-1}\cap E_{\varepsilon_2} = \emptyset \\ &\quad \Rightarrow \varphi(f,g,h) = 0\end{aligned} \tag{3.1}$$

*for some $0 \leq \varepsilon_1, \varepsilon_2 < \frac{\pi}{3\alpha+1}$. Then*

$$\begin{aligned}&\left\| \sum_{n_1,n_2=0}^{N} (-1)^{n_1+n_2} \binom{N}{n_1}\binom{N}{n_2} \varphi(\mathbf{z}^{2N-(n_1+n_2)}, \mathbf{z}^{n_1}, \mathbf{z}^{n_2}) \right\| \\ &\quad \leq \Big(2\tan\left(\tfrac{N}{2}\varepsilon_1\right) + 2\tan\left(\tfrac{N}{2}\varepsilon_2\right) + 4\tan\left(\tfrac{N}{2}\varepsilon_1\right)\tan\left(\tfrac{N}{2}\varepsilon_2\right)\Big)\|\varphi\| C_3(N-1)\end{aligned} \tag{3.2}$$

*for $N = 3\alpha+1$.*

*Proof.* The map $\varphi$ gives rise to a continuous linear operator $\Phi\colon A_{3\alpha}(\mathbb{T}^3) \to X$ by defining

$$\Phi(f) = \sum_{k_1,k_2,k_3\in\mathbb{Z}} \widehat{f}(k_1,k_2,k_3)\varphi(\mathbf{z}^{k_1},\mathbf{z}^{k_2},\mathbf{z}^{k_3}) \quad (f \in A_{3\alpha}(\mathbb{T}^3)).$$

If $f \in A_{3\alpha}(\mathbb{T}^2)$, then

$$\begin{aligned}
\|\Phi(f)\| &\le \sum_{k_1,k_2,k_3\in\mathbb{Z}} \left|\widehat{f}(k_1,k_2,k_3)\right| \|\varphi(\mathbf{z}^{k_1},\mathbf{z}^{k_2},\mathbf{z}^{k_3})\| \\
&\le \sum_{k_1,k_2,k_3\in\mathbb{Z}} \left|\widehat{f}(k_1,k_2,k_3)\right| \|\varphi\|(1+|k_1|)^\alpha(1+|k_2|)^\alpha(1+|k_3|)^\alpha \\
&\le \|\varphi\| \sum_{k_1,k_2,k_3\in\mathbb{Z}} \left|\widehat{f}(k_1,k_2,k_3)\right| (1+|k_1|+|k_2|+|k_3|)^{3\alpha} \\
&= \|\varphi\|\|f\|_{A_{3\alpha}(\mathbb{T}^2)}
\end{aligned}$$

and therefore $\|\Phi\| \le \|\varphi\|$.

We now claim that

$$\Phi(f\otimes g\otimes h) = \varphi(f,g,h) \quad (f,g,h \in A_{3\alpha}(\mathbb{T})). \tag{3.3}$$

Indeed, on account of the continuity of $\varphi$, we have

$$\begin{aligned}
\Phi(f\otimes g\otimes h) &= \sum_{k_1,k_2,k_3\in\mathbb{Z}} \widehat{f}(k_1)\widehat{g}(k_2)\widehat{h}(k_3)\varphi(\mathbf{z}^{k_1},\mathbf{z}^{k_2},\mathbf{z}^{k_3}) \\
&= \varphi\left(\sum_{k_1\in\mathbb{Z}} \widehat{f}(k_1)\mathbf{z}^{k_1}, \sum_{k_2\in\mathbb{Z}} \widehat{g}(k_2)\mathbf{z}^{k_2}, \sum_{k_3\in\mathbb{Z}} \widehat{h}(k_3)\mathbf{z}^{k_3}\right) \\
&= \varphi(f,g,h)
\end{aligned}$$

for all $f,g,h \in A_{3\alpha}(\mathbb{T})$.

Our next claim is that $\Phi$ has the following property

$$\delta > 0,\ f \in I_{A_{3\alpha}(\mathbb{T}^3)}(E^3_{\varepsilon_1+\delta,\varepsilon_2+\delta}) \ \Rightarrow\ \Phi(f) = 0. \tag{3.4}$$

Let $\delta > 0$ and $f \in I_{A_{3\alpha}(\mathbb{T}^3)}(E^3_{\varepsilon_1+\delta,\varepsilon_2+\delta})$. For each $z \in \mathbb{T}$ let $U_z = \{ze^{i\theta}\colon\ |\theta| < \delta/4\}$. By compactness, there exist $z_1,\dots,z_m \in \mathbb{T}$ such that $\mathbb{T} = \cup_{p=1}^m U_{z_j}$. There are functions $\omega_1,\dots,\omega_m \in C^\infty(\mathbb{T})$ with $\omega_1+\dots+\omega_m = \mathbf{1}$ and $\operatorname{supp}(\omega_p) \subset U_{z_p}$ for $p = 1,\dots,m$. Since

$$\sum_{p_1,p_2,p_3=1}^m \omega_{p_1}\otimes\omega_{p_2}\otimes\omega_{p_3} = \mathbf{1},$$

it follows that

$$
\begin{aligned}
f &= \sum_{p_1,p_2,p_3=1}^{m} f\,(\omega_{p_1}\otimes\omega_{p_2}\otimes\omega_{p_3}) \\
&= \sum_{U_{z_{p_1}}U_{z_{p_2}}^{-1}\cap E_{\varepsilon_1}=U_{z_{p_1}}U_{z_{p_3}}^{-1}\cap E_{\varepsilon_2}=\emptyset} f\,(\omega_{p_1}\otimes\omega_{p_2}\otimes\omega_{p_3}) \\
&\quad + \sum_{U_{z_{p_1}}U_{z_{p_2}}^{-1}\cap E_{\varepsilon_1}\neq\emptyset \text{ or } U_{z_{p_1}}U_{z_{p_3}}^{-1}\cap E_{\varepsilon_2}\neq\emptyset} f\,(\omega_{p_1}\otimes\omega_{p_2}\otimes\omega_{p_3}).
\end{aligned}
$$

Assume that $U_{z_{p_1}}U_{z_{p_2}}^{-1}\cap E_{\varepsilon_1}\neq\emptyset$ and let $z_0\in U_{z_{p_1}}$, $w_0\in U_{z_{p_2}}$ with $z_0w_0^{-1}\in E_{\varepsilon_1}$. If $z\in U_{z_{p_1}}$ and $w\in U_{z_{p_2}}$, then

$$
zw^{-1}=(zz_0^{-1})(ww_0^{-1})(z_0w_0^{-1})\in E_{\delta/2}E_{\delta/2}E_{\varepsilon_1}\subset E_{\varepsilon_1+\delta}.
$$

This entails that

$$
\sum_{U_{z_{p_1}}U_{z_{p_2}}^{-1}\cap E_{\varepsilon_1}\neq\emptyset} f\,(\omega_{p_1}\otimes\omega_{p_2}\otimes\omega_{p_3})=0.
$$

In the same manner we can see that

$$
\sum_{U_{z_{p_1}}U_{z_{p_3}}^{-1}\cap E_{\varepsilon_2}\neq\emptyset} f\,(\omega_{p_1}\otimes\omega_{p_2}\otimes\omega_{p_3})=0.
$$

We thus get

$$
f=\sum_{U_{z_{p_1}}U_{z_{p_2}}^{-1}\cap E_{\varepsilon_1}=U_{z_{p_1}}U_{z_{p_3}}^{-1}\cap E_{\varepsilon_2}=\emptyset} f\,(\omega_{p_1}\otimes\omega_{p_2}\otimes\omega_{p_3}).
$$

On the other hand, we have

$$
f=\sum_{k_1,k_2,k_3\in\mathbb{Z}}\widehat{f}(k_1,k_2,k_3)\mathbf{z}^{k_1}\otimes\mathbf{z}^{k_2}\otimes\mathbf{z}^{k_3}.
$$

Hence

$$
f=\sum_{\substack{U_{z_{p_1}}U_{z_{p_2}}^{-1}\cap E_{\varepsilon_1}=\emptyset\\ U_{z_{p_1}}U_{z_{p_3}}^{-1}\cap E_{\varepsilon_2}=\emptyset}}\ \sum_{k_1,k_2,k_3\in\mathbb{Z}}\widehat{f}(k_1,k_2,k_3)(\mathbf{z}^{k_1}\omega_{p_1})\otimes(\mathbf{z}^{k_2}\omega_{p_2})\otimes(\mathbf{z}^{k_3}\omega_{p_3})
$$

and therefore

$$
\begin{aligned}
\Phi(f) &= \sum_{\substack{U_{z_{p_1}}U_{z_{p_2}}^{-1}\cap E_{\varepsilon_1}=\emptyset\\ U_{z_{p_1}}U_{z_{p_3}}^{-1}\cap E_{\varepsilon_2}=\emptyset}}\ \sum_{k_1,k_2,k_3\in\mathbb{Z}}\widehat{f}(k_1,k_2,k_3)\Phi\Big((\mathbf{z}^{k_1}\omega_{p_1})\otimes(\mathbf{z}^{k_2}\omega_{p_2})\otimes(\mathbf{z}^{k_3}\omega_{p_3})\Big) \\
&= \sum_{\substack{U_{z_{p_1}}U_{z_{p_2}}^{-1}\cap E_{\varepsilon_1}=\emptyset\\ U_{z_{p_1}}U_{z_{p_3}}^{-1}\cap E_{\varepsilon_2}=\emptyset}}\ \sum_{k_1,k_2,k_3\in\mathbb{Z}}\widehat{f}(k_1,k_2,k_3)\varphi(\mathbf{z}^{k_1}\omega_{p_1},\mathbf{z}^{k_2}\omega_{p_2},\mathbf{z}^{k_3}\omega_{p_3}).
\end{aligned}
$$

Since

$$\operatorname{supp}(\mathbf{z}^{k_1}\omega_{p_1})\big(\operatorname{supp}(\mathbf{z}^{k_2}\omega_{p_2})\big)^{-1}\cap E_{\varepsilon_1}\subset U_{z_{p_1}}U_{z_{p_2}}^{-1}\cap E_{\varepsilon_1}=\emptyset$$

and

$$\operatorname{supp}(\mathbf{z}^{k_1}\omega_{p_1})\big(\operatorname{supp}(\mathbf{z}^{k_3}\omega_{p_3})\big)^{-1}\cap E_{\varepsilon_1}\subset U_{z_{p_1}}U_{z_{p_3}}^{-1}\cap E_{\varepsilon_2}=\emptyset$$

for all the terms appearing in the preceding identity for $\Phi(f)$, it follows that $\Phi(f)=0$, as claimed.

Since $I_{A_{3\alpha}(\mathbb{T}^3)}(E^3_{\varepsilon_1+\delta,\varepsilon_2+\delta})\subset\ker\Phi$, we conclude that $\Phi$ drops to a continuous linear operator on the quotient $A_{3\alpha}(\mathbb{T}^3)/I_{A_{3\alpha}(\mathbb{T}^3)}(E^3_{\varepsilon_1+\delta,\varepsilon_2+\delta})$

$$\bar{\Phi}_\delta\colon A_{3\alpha}(\mathbb{T}^3)/I_{A_{3\alpha}(\mathbb{T}^3)}(E^3_{\varepsilon_1+\delta,\varepsilon_2+\delta})\to X,\quad \bar{\Phi}_\delta(f)=\Phi(f),\quad (f\in A_{3\alpha}(\mathbb{T}^3)),$$

with the property that $\|\bar{\Phi}_\delta\|\le\|\Phi\|\le\|\varphi\|$. This implies that

$$\|\Phi(f)\|\le\|\varphi\|\operatorname{dist}\big(f,I_{A_{3\alpha}(\mathbb{T}^3)}(E^3_{\varepsilon_1+\delta,\varepsilon_2+\delta})\big),\quad (f\in A_{3\alpha}(\mathbb{T}^3)).\tag{3.5}$$

We check at once that

$$\Phi\left((\mathbf{z_1}-\mathbf{z_2})^N(\mathbf{z_1}-\mathbf{z_2})^N\right)=\sum_{n_1,n_2=0}^{N}(-1)^{n_1+n_2}\binom{N}{n_1}\binom{N}{n_2}\varphi(\mathbf{z}^{2N-(n_1+n_2)},\mathbf{z}^{n_1},\mathbf{z}^{n_2}).$$

On account of (3.5) and Lemma 3.3, we have

$$\begin{aligned}&\left\|\sum_{n_1,n_2=0}^{N}(-1)^{n_1+n_2}\binom{N}{n_1}\binom{N}{n_2}\varphi(\mathbf{z}^{2N-(n_1+n_2)},\mathbf{z}^{n_1},\mathbf{z}^{n_2})\right\|\\&\quad\le\Big(2\tan\left(\tfrac{N}{2}(\varepsilon_1+\delta)\right)+2\tan\left(\tfrac{N}{2}(\varepsilon_2+\delta)\right)\\&\qquad+4\tan\left(\tfrac{N}{2}(\varepsilon_1+\delta)\right)\tan\left(\tfrac{N}{2}(\varepsilon_2+\delta)\right)\Big)C_3(N-1)\|\varphi\|,\end{aligned}\tag{3.6}$$

whenever $\delta>0$ is such that $\varepsilon_1+\delta,\varepsilon_2+\delta<\frac{\pi}{N}$. Finally, taking $\delta\to0$ in (3.6) we arrive at (3.2). □

In order to introduce a quantitative version of Theorem 2.3 we are required to consider the sets of the form

$$U(K,\varepsilon)=\{\gamma\in\widehat{G}\colon\ \gamma(K)\subset E_\varepsilon\},$$

where $0\le\varepsilon<\pi$ and $K$ is a compact neighbourhood of the identity in $G$. It is worth pointing out that the family consisting of all those $U(K,\varepsilon)$ is a basis of neighbourhoods of the identity in $\widehat{G}$.

**Theorem 3.5.** *Let $G$ be a locally compact abelian group. Let $X$ and $Y$ be Banach spaces and let $X_*$ and $Y_*$ be linear subspaces of the duals $X^*$ and $Y^*$, respectively, satisfying* (1.7) *and* (1.8). *Suppose that $\sigma_1$ and $\sigma_2$ are commuting representations of $G$ on $X$ and $\tau$ is a representation of $G$ on $Y$ such that*

$$\|\sigma_1(kt)\|,\|\sigma_2(kt)\|,\|\tau(kt)\|=O(|k|^\alpha)\ \text{as}\ |k|\to\infty\ (t\in G)$$

*for some $\alpha\in\mathbb{Z}$ with $\alpha\ge0$. If $A\in\mathcal{B}(X,Y)$ is such that*

$$\operatorname{sp}(\tau,Ax)\subset\operatorname{sp}(\sigma_1,x)U(K,\varepsilon_1)\cup\operatorname{sp}(\sigma_2,x)U(K,\varepsilon_2)\quad(x\in X),$$

*for some* $0 \le \varepsilon_1, \varepsilon_2 < \frac{\pi}{3\alpha+1}$ *and some* $K \subset G$ *compact neighbourhood of the identity in* $G$, *then*

$$\left\| \sum_{n_1,n_2=0}^{N} (-1)^{n_1+n_2} \binom{N}{n_1}\binom{N}{n_2} \tau(t)^{2N-(n_1+n_2)} A\sigma_1(t)^{n_1}\sigma_2(t)^{n_2} \right\|$$
$$\le \Big(2\tan\big(\tfrac{N}{2}\varepsilon_1\big) + 2\tan\big(\tfrac{N}{2}\varepsilon_2\big) + 4\tan\big(\tfrac{N}{2}\varepsilon_1\big)\tan\big(\tfrac{N}{2}\varepsilon_2\big)\Big)\kappa(t)\|A\|C_3(N-1) \quad (t \in K)$$

*for* $N = 3\alpha + 1$, *where*

$$\kappa(t) = \sup_{k\in\mathbb{Z}} \frac{\|\sigma_1(kt)\|}{(1+|k|)^\alpha} \sup_{k\in\mathbb{Z}} \frac{\|\sigma_2(kt)\|}{(1+|k|)^\alpha} \sup_{k\in\mathbb{Z}} \frac{\|\tau(kt)\|}{(1+|k|)^\alpha} \quad (t \in G).$$

*Proof.* This follows by the same method as in the proofs of Theorem 2.3. Pick $t \in K$. We define a continuous linear map $\Phi_t \colon A_\alpha(\mathbb{T}) \to M(G,\omega)$ and a continuous trilinear map $\varphi \colon A_\alpha(\mathbb{T}) \times A_\alpha(\mathbb{T}) \times A_\alpha(\mathbb{T}) \to \mathcal{B}(X,Y)$ by

$$\Phi_t(f) = \sum_{k=-\infty}^{+\infty} \widehat{f}(k)\delta_t^k \quad (f \in A_\alpha(\mathbb{T}))$$

and

$$\varphi_t(f,g,h) = \widetilde{\tau}\big(\Phi_t(f)\big) \circ A \circ \widetilde{\sigma_1}\big(\Phi_t(g)\big) \circ \widetilde{\sigma_2}\big(\Phi_t(h)\big) \quad (f,g,h \in A_\alpha(\mathbb{T})),$$

respectively. It is immediate to check that $\|\varphi_t\| \le \kappa(t)\|A\|$.

By using [8, Lemma 4.12.6] and (2.5), we can prove as in the proof of Theorem 2.3 that, if $f,g,h \in A_\alpha(\mathbb{T})$ and $x \in X$, then

$$\begin{aligned}&\operatorname{sp}(\tau, \varphi_t(f,g,h)x)) \\ &\quad \subset \Big\{\gamma \in \widehat{G} \colon \ \gamma(t) \in \big(\operatorname{supp}(f) \cap \big(\operatorname{supp}(g)E_{\varepsilon_1}\big)\big) \cup \big(\operatorname{supp}(f) \cap \big(\operatorname{supp}(h)E_{\varepsilon_2}\big)\big)\Big\}.\end{aligned}$$

Accordingly, $\varphi$ satisfies (3.1) and therefore Theorem 3.4 completes the proof. □

We can now apply Theorem 3.5 to obtain a quantitative estimate of Corollary 2.4 and then we can derive quantitative estimates of Corollaries 2.5 and 2.6.

**Corollary 3.6.** *Let* $X$ *and* $Y$ *be complex Banach spaces and let* $S_1, S_2 \in \mathcal{B}(X)$ *and* $T \in \mathcal{B}(Y)$ *be invertible operators with* $S_1$ *and* $S_2$ *commuting and*

$$\|S_1^k\|, \|S_2^k\|, \|T^k\| = O(|k|^\alpha) \ \text{as} \ |k| \to \infty$$

*for some* $\alpha \in \mathbb{Z}$ *with* $\alpha \ge 0$. *If* $A \in \mathcal{B}(X,Y)$ *is such that*

$$\operatorname{sp}(T, Ax) \subset \operatorname{sp}(S_1, x)E_{\varepsilon_1} \cup \operatorname{sp}(S_2, x)E_{\varepsilon_2} \quad \forall x \in X,$$

*for some* $0 \le \varepsilon_1, \varepsilon_2 < \frac{\pi}{3\alpha+1}$, *then*

$$\begin{aligned}\|C(T,S_2)^N C(T,S_1)^N(A)\| &\le \Big(2\tan\big(\tfrac{N}{2}\varepsilon_1\big) + 2\tan\big(\tfrac{N}{2}\varepsilon_2\big) + 4\tan\big(\tfrac{N}{2}\varepsilon_1\big)\tan\big(\tfrac{N}{2}\varepsilon_2\big)\Big) \\ &\qquad \sup_{k\in\mathbb{Z}} \tfrac{\|S_1^k\|}{(1+|k|)^\alpha} \sup_{k\in\mathbb{Z}} \tfrac{\|S_2^k\|}{(1+|k|)^\alpha} \sup_{k\in\mathbb{Z}} \tfrac{\|T^k\|}{(1+|k|)^\alpha} C_3(N-1)\|A\|\end{aligned}$$

*for* $N = 3\alpha + 1$.

*Consequently,*

1. *If $S_1, S_2, T$ are doubly power bounded, then*

$$\|C(T,S_2)C(T,S_1)(A)\| \\ \leq \big(2\tan(\varepsilon_1/2) + 2\tan(\varepsilon_2/2) + 4\tan(\varepsilon_1/2)\tan(\varepsilon_2/2)\big) \\ \sup_{k\in\mathbb{Z}} \|S_1^k\| \sup_{k\in\mathbb{Z}} \|S_2^k\| \sup_{k\in\mathbb{Z}} \|T^k\| \|A\|.$$

2. *If $X = Y$, and if $S_1, S_2, T, A$ are pairwise commuting, then*

$$\left\|(T-S_2)^N(T-S_1)^N A\right\| \\ \leq \left(2\tan\left(\tfrac{N}{2}\varepsilon_1\right) + 2\tan\left(\tfrac{N}{2}\varepsilon_2\right) + 4\tan\left(\tfrac{N}{2}\varepsilon_1\right)\tan\left(\tfrac{N}{2}\varepsilon_2\right)\right) \\ \sup_{k\in\mathbb{Z}} \tfrac{\|S_1^k\|}{(1+|k|)^\alpha} \sup_{k\in\mathbb{Z}} \tfrac{\|S_2^k\|}{(1+|k|)^\alpha} \sup_{k\in\mathbb{Z}} \tfrac{\|T^k\|}{(1+|k|)^\alpha} C_3(N-1)\|A\|.$$

*Proof.* Take the representations $\tau = \tau_T$, $\sigma_1 = \tau_{S_1}$, and $\sigma_2 = \tau_{S_2}$ introduced in Example 1.1 and take $K = \{0,1\}$. Then $U(K,\varepsilon) = E_\varepsilon$ for each $\varepsilon > 0$ and the requirements in Theorem 3.5 are satisfied. Consequently, the inequality stated in the corollary follows from Theorem 3.5 taking into account that

$$\sum_{n_1,n_2=0}^{N} (-1)^{n_1+n_2} \binom{N}{n_1}\binom{N}{n_2} \tau(t)^{2N-(n_1+n_2)} A\sigma_1(t)^{n_1}\sigma_2(t)^{n_2} \\ = C(T,S_2)^N C(T,S_1)^N(A).$$

□

**Corollary 3.7.** *Let $X$ be a complex Banach space and let $S_1, S_2, T \in \mathcal{B}(X)$ be pairwise commuting invertible operators with $\|S_1^k\|, \|S_2^k\|, \|T^k\| = O(|k|^\alpha)$ as $|k| \to \infty$ for some $\alpha \in \mathbb{Z}$ with $\alpha \geq 0$. If*

$$\operatorname{sp}(T,x) \subset \operatorname{sp}(S_1,x)E_{\varepsilon_1} \cup \operatorname{sp}(S_1,x)E_{\varepsilon_2} \quad \forall x \in X,$$

*for some $0 \leq \varepsilon_1, \varepsilon_2 < \frac{\pi}{3\alpha+1}$, then*

$$\left\|(T-S_2)^N(T-S_1)^N\right\| \\ \leq \left(2\tan\left(\tfrac{N}{2}\varepsilon_1\right) + 2\tan\left(\tfrac{N}{2}\varepsilon_2\right) + 4\tan\left(\tfrac{N}{2}\varepsilon_1\right)\tan\left(\tfrac{N}{2}\varepsilon_2\right)\right) \\ \sup_{k\in\mathbb{Z}} \tfrac{\|S_1^k\|}{(1+|k|)^\alpha} \sup_{k\in\mathbb{Z}} \tfrac{\|S_2^k\|}{(1+|k|)^\alpha} \sup_{k\in\mathbb{Z}} \tfrac{\|T^k\|}{(1+|k|)^\alpha} C_3(N-1).$$

*for $N = 3\alpha + 1$.*

*Proof.* It suffices to apply Corollary 3.6 with $X = Y$ and $A = I_X$. □

**Corollary 3.8.** *Let $X$ be a complex Banach space and let $T \in \mathcal{B}(X)$ be invertible and such that $\|T^k\| = O(|k|^\alpha)$ as $|k| \to \infty$ for some $\alpha \in \mathbb{Z}$ with $\alpha \geq 0$. If*

$\mathrm{sp}(T) \subset \lambda_1 E_{\varepsilon_1} \cup \lambda_2 E_{\varepsilon_2}$, *for some* $\lambda_1, \lambda_2 \in \mathbb{C}$ *and* $0 \leq \varepsilon_1, \varepsilon_2 < \frac{\pi}{3\alpha+1}$, *then*

$$\begin{aligned} &\left\|(T - \lambda_2 I_X)^N (T - \lambda_1 I_X)^N\right\| \\ &\quad \leq \left(2\tan\left(\tfrac{N}{2}\varepsilon_1\right) + 2\tan\left(\tfrac{N}{2}\varepsilon_2\right) + 4\tan\left(\tfrac{N}{2}\varepsilon_1\right)\tan\left(\tfrac{N}{2}\varepsilon_2\right)\right) \\ &\qquad \sup_{k\in\mathbb{Z}} \tfrac{\|T^k\|}{(1+|k|)^\alpha} C_3(N-1). \end{aligned}$$

*for* $N = 3\alpha + 1$.

*Proof.* Apply Corollary 3.7 with $S_1 = \lambda_1 I_X$ and $S_2 = \lambda_2 I_X$. □

## References


[1] J. Alaminos, M. Brešar, M. Černe, J. Extremera, and A.R. Villena, Zero product preserving maps on $C^1[0,1]$, *J. Math. Anal. Appl.* **347** (2008), 472–481.

[2] J. Alaminos, M. Brešar, J. Extremera, and A.R. Villena, Maps preserving zero products, *Studia Math.* **193** (2) (2009), 131–159.

[3] J. Alaminos, M. Brešar, J. Extremera, and A.R. Villena, Characterizing Jordan maps on $C^*$-algebras through zero products, *Proc. Edinb. Math. Soc.* **53**(3) (2010), 543–555.

[4] J. Alaminos, J. Extremera, and A.R. Villena, Zero product preserving maps on Banach algebras of Lipschitz functions, *J. Math. Anal. Appl.* **369** (2010), 94–100.

[5] J. Alaminos, J. Extremera, and A.R. Villena, Operators shrinking the Arveson spectrum, *Publ. Math. Debrecen* (to appear).

[6] W. Arveson, On groups of automorphisms of operator algebras, *J. Funct. Anal.* **15** (1974), 217–243.

[7] O. Bratteli and D.W. Robinson, *Operator algebras and quantum statistical mechanics. 1. $C^*$- and $W^*$-algebras, symmetry groups, decomposition of states.* Second edition. Texts and Monographs in Physics. Springer-Verlag, New York, 1987. xiv+505 pp.

[8] K.B. Laursen and M.M. Neumann, *An introduction to local spectral theory.* London Mathematical Society Monographs. New Series, 20. The Clarendon Press, Oxford University Press, New York, 2000. xii+591 pp.



J. Alaminos, J. Extremera and A.R. Villena
Departamento de Análisis Matemático
Facultad de Ciencias
Universidad de Granada
E-18071 Granada, Spain
e-mail: `alaminos@ugr.es`
`jlizana@ugr.es`
`avillena@ugr.es`

Operator Theory:
Advances and Applications, Vol. 233, 33–44


# Singly Generated Operator Algebras Satisfying Weakened Versions of Amenability

Yemon Choi

*Dedicated to V.S. Shulman on the occasion of his 65th birthday*

**Abstract.** We construct a singly generated subalgebra of $\mathcal{K}(\mathcal{H})$ which is non-amenable, yet is boundedly approximately contractible. The example embeds into a homogeneous von Neumann algebra. We also observe that there are singly generated, biflat subalgebras of finite Type I von Neumann algebras, which are not amenable (and hence are not isomorphic to C*-algebras). Such an example can be used to show that a certain extension property for commutative operator algebras, which is shown in [3] to follow from amenability, does not necessarily imply amenability.

**Mathematics Subject Classification (2010).** 47L75 (primary); 46J40 (secondary).

**Keywords.** Approximate amenability, biflatness, compact operators, finite von Neumann algebra, monogenic Banach algebra, Type I von Neumann algebra.

## 1. Introduction

In his pioneering 1972 monograph [14], which formally introduced the notion of amenability for Banach algebras, Johnson observed that commutative C*-algebras are amenable. The problem of characterizing the commutative, amenable operator algebras – that is, the norm-closed subalgebras of $\mathcal{B}(\mathcal{H})$, $\mathcal{H}$ a Hilbert space, which are amenable as Banach algebras – remains unsolved, even in the commutative case. There are some partial results which suggest that every commutative, amenable operator algebra is isomorphic as a Banach algebra to a commutative C*-algebra: see recent work of Marcoux [15] and the present author [3].

Like many other concepts which have been much studied, amenability has spawned several weaker versions, some studied in more depth than others. In trying to assess the relative worth, importance or interest of these notions, there are two natural criteria to consider. Firstly, if a Banach algebra satisfies one of these weaker versions of amenability, does that tell us something about its internal structure (e.g., properties of its closed ideals) or about the behaviour of modules/representations of the algebra? Secondly, is there a good supply of examples which *are non-amenable* yet satisfy the weaker version?

It is this second question, in the context of operator algebras, which motivates the present note. We consider two of these weaker versions of amenability, namely bounded approximate contractibility and biflatness, in the context of commutative operator algebras. In each case, we construct an explicit example which satisfies that property, and is singly generated and semisimple, but is non-amenable; the boundedly approximately contractible example is contained in $\mathcal{K}(\mathcal{H})$.

Our examples can be embedded inside finite von Neumann algebras of Type I (more precisely, inside countable products of matrix algebras). This shows that in the results of [3], one cannot replace amenability with biflatness, nor with bounded approximate contractibility. In passing, we take the opportunity to show that certain technical arguments from [3] have rather sensitive hypotheses. We remark that these examples therefore embed inside Type $\mathrm{II}_1$ algebras, since any finite Type I algebra with separable predual can be embedded into the hyperfinite $\mathrm{II}_1$ factor (J. Peterson, personal communication, [16]).

## 2. General preliminaries

We refer the reader to [2] for basics on Banach algebras, and to [2, §43] in particular for the basic definitions and characterizations of amenable Banach algebras.

The examples we will construct in this article are each generated by a countable set of pairwise-orthogonal idempotents. We note that this ensures they are singly generated as Banach algebras; this is a general fact, which is not restricted to the setting of subalgebras of $\mathcal{B}(\mathcal{H})$.

**Lemma 2.1.** *Let $A$ be a commutative Banach algebra generated by a countable set of pairwise-orthogonal idempotents. Then $A$ is singly generated as a Banach algebra.*

This is presumably a standard result, but for convenience, we provide a full proof, inspired by a similar argument in [1, Proposition 1.1] for certain subalgebras of $\mathcal{B}(\mathcal{H})$.

*Proof.* Enumerate the given set of idempotents as $(e_j)_{j\geq 1}$, and fix a strictly decreasing sequence of strictly positive reals $\lambda_1 > \lambda_2 > \cdots$ with the property that $\sum_{j\geq 1} \lambda_j \|e_j\| < \infty$. Set $b = \sum_{j\geq 1} \lambda_j e_j$, and let $B$ be the norm-closed subalgebra of $A$ generated by $b$. We claim that $e_n \in B$ for all $n \geq 1$, which will clearly imply $B = A$ since $A$ is generated by the set $\{e_j : j \in \mathbb{N}\}$.

The proof is by strong induction on $n$. We start by noting that for all $r \in \mathbb{N}$, we have $b^r = \sum_{j\geq 1} \lambda_j^r e_j$, the sum converging absolutely. As $0 < \lambda_j \leq \lambda_2 < \lambda_1$ for all $j \geq 2$, it follows that

$$\|e_1 - (\lambda_1^{-1}b)^r\| \leq \sum_{j\geq 2} \left(\frac{\lambda_j}{\lambda_1}\right)^r \|e_j\| \leq \frac{1}{\lambda_1}\left(\frac{\lambda_2}{\lambda_1}\right)^{r-1} \sum_{j\geq 2} \lambda_j \|e_j\| \to 0 \quad \text{as } r \to \infty.$$

Thus $e_1 \in B$, and so the claim holds for $n = 1$.

Now suppose the claim holds for all $n \in \{1, \ldots, m-1\}$ for some $m \geq 2$. Let

$$b_m = b - \left(\sum_{j=1}^{m-1} \lambda_j e_j\right) = \sum_{j\geq m} \lambda_j e_j;$$

by the inductive hypothesis, $b_m \in B$. For all $r \in \mathbb{N}$ we have $b_m^r = \sum_{j\geq m} \lambda_j^r e_j$, the sum converging absolutely. As $0 < \lambda_j \leq \lambda_{m+1} < \lambda_m$ for all $j \geq m+1$, a similar argument to the one used above shows that $e_m = \lim_r (\lambda_m^{-1} b_m)^r \in B$. So the claim holds for $n = m$, and this completes the inductive step. □

If $\mathfrak{A}$ is a Banach algebra, isomorphic to a C*-algebra, then the set of central idempotents in $\mathfrak{A}$ will be a bounded subset of $\mathfrak{A}$. (This is because the centre of a C*-algebra is itself a commutative C*-algebra, and hence all its non-zero idempotent elements have norm 1.) In particular, a commutative operator algebra which is isomorphic to a C*-algebra will have uniformly bounded idempotents. Gifford observed that the same will be true for any amenable, commutative operator algebra. More precisely, we have the following result (in whose statement the prime denotes the commutant of a subset of $\mathcal{B}(\mathcal{H})$).

**Theorem 2.2 (Gifford, [12]).** *If $\mathfrak{A}$ is an amenable operator algebra, then the set of idempotents in $\mathfrak{A}'' \cap \mathfrak{A}'$ is norm-bounded.*

Gifford actually proved something more general, namely that the conclusion of Theorem 2.2 holds whenever $\mathfrak{A}$ is an operator algebra with the so-called *total reduction property*. Since we shall not discuss the total reduction property in this article, we refer the reader to [12] for further details, and provide a more direct proof of Theorem 2.2 as follows.

*Proof.* Regard $\mathfrak{A}$ as a closed subalgebra of some $\mathcal{B}(\mathcal{H})$. Since the unitization of an amenable Banach algebra is amenable, we may assume without loss of generality that $\mathfrak{A}$ contains the identity operator $I$.

Let $(\Delta_\alpha)$ be a bounded approximate diagonal for $\mathfrak{A}$. Define $T : \mathfrak{A} \widehat{\otimes} \mathfrak{A} \to \mathcal{B}(\mathcal{B}(\mathcal{H}))$ by $T(c \otimes d)(x) = cxd$, and let $E$ be a point-to-weak* cluster point of the net $T(\Delta_\alpha) \subset \mathcal{B}(\mathcal{B}(\mathcal{H}))$. As $(\Delta_\alpha)$ is a bounded approximate diagonal for $\mathfrak{A}$, it follows from routine estimates and convergence arguments that the following properties hold:

(i) $E(x) \in \mathfrak{A}'$ for all $x \in \mathcal{B}(\mathcal{H})$;
(ii) $E(u) = u$ for all $u \in \mathfrak{A}'$;
(iii) $E(uxv) = uE(x)v$ for all $u, v \in \mathfrak{A}'$ and all $x \in \mathcal{B}(\mathcal{H})$.

Let $e \in \mathfrak{A}'' \cap \mathfrak{A}'$ be an idempotent, and let $p$ be the orthogonal projection from $\mathcal{H}$ onto the closed subspace $e\mathcal{H}$. We have $ep = p$ and $pe = e$. Therefore,

$$\begin{aligned} E(p) &= E(ep) \\ &= eE(p) \quad \text{(by (iii) and } e \in \mathfrak{A}'\text{)} \\ &= E(p)e \quad \text{(by (i) and } e \in \mathfrak{A}''\text{)} \\ &= E(pe) \quad \text{(by (iii) and } e \in \mathfrak{A}'\text{)} \\ &= E(e) \\ &= e \quad \text{(by (ii) and } e \in \mathfrak{A}'\text{).} \end{aligned}$$

In particular, $\|e\| = \|E(p)\| \leq \|E\|$. Since $E$ is independent of the choice of $e$, we are done. □

**Remark 2.3.** The proof just given requires little background from the theory of amenability, but may seem somewhat unmotivated. For a more conceptual but less self-contained approach, the reader is encouraged to consult Gifford's original article [12].

**Corollary 2.4.** *Let $\mathfrak{A}$ be a commutative operator algebra which contains an unbounded family of idempotents. Then $\mathfrak{A}$ is not amenable.*

## 3. Bounded approximate contractibility

In [9], Ghahramani and Loy began the systematic study of certain "approximate" versions of amenability. These have since been pursued in various directions by several different groups of authors; see [20] for an overview of some of the results to date.

In this article, we are only concerned with one such variant, which we now briefly describe. Given a Banach algebra $A$ and a Banach $A$-bimodule $X$ and $\xi \in X$, we denote by $\operatorname{ad}\xi$ the inner derivation $a \mapsto a \cdot \xi - \xi \cdot a$.

**Definition 3.1.** A Banach algebra $A$ is *boundedly approximately contractible* if for each Banach $A$-bimodule $X$ and each continuous derivation $D : A \to X$, there exists a net $(\xi_i) \subset X$, not necessarily bounded, such that the net $(\operatorname{ad}\xi_i)$ is norm bounded (as a subset of $\mathcal{B}(A, X)$) and converges in the strong operator topology of $\mathcal{B}(A, X)$ to $D$.

Note that since there are no non-zero, inner derivations from a Banach algebra $A$ into a symmetric $A$-bimodule, every *commutative*, boundedly approximately contractible Banach algebra is weakly amenable.

**Definition 3.2.** Let $A$ be a Banach algebra, and let $\pi : A \widehat{\otimes} A \to A$ denote the linearized multiplication map. A *multiplier-bounded approximate diagonal* for $A$ is a net $(M_i) \subset A \widehat{\otimes} A$ such that

(i) for each $a \in A$, $\lim_i \|a\pi(\Delta_i) - a\| = 0$;
(ii) for each $a \in A$, $\lim_i \|a \cdot \Delta_i - \Delta_i \cdot a\| = 0$;

(iii) there exists a constant $C > 0$ such that $\sup_i \|a \cdot \Delta_i - \Delta_i \cdot a\| \leq C\|a\|$ for all $a \in A$.

**Lemma 3.3.** *Let $A$ be a Banach algebra. Suppose there exists a net $(\Delta_i) \subseteq A \otimes A$ with the following properties:*

(a) $\pi(\Delta_i)$ *is a* central, bounded *approximate identity for $A$;*

(b) $(\Delta_i)$ *is a multiplier-bounded approximate diagonal for $A$.*

*Then $A$ is boundedly approximately contractible.*

(One can weaken condition (a), but the proof becomes technically more demanding, and the version given here will be enough for our purposes.)

*Proof.* As observed in [4, Proposition 2.2], it suffices to show that the unitization $A^\sharp$ has a multiplier-bounded approximate diagonal. Let $u_i = \pi(\Delta_i)$, and define

$$M_i = 2\Delta_i - u_i \cdot \Delta_i + (1 - u_i) \otimes (1 - u_i).$$

We claim that $(M_i)$ is a multiplier-bounded approximate diagonal for $A^\sharp$.

Firstly, let $\pi^\sharp : A^\sharp \widehat{\otimes} A^\sharp \to A^\sharp$ be the product map, which clearly satisfies $\pi^\sharp|_{A \widehat{\otimes} A} = \pi$. Then, using the identity $\pi(u \cdot x) = u\pi(x)$ for all $u \in A$ and $x \in A \widehat{\otimes} A$, a direct calculation yields $\pi^\sharp(M_i) = 1$ for all $i$. So condition (i) of Definition 3.2 is satisfied.

Secondly, let $K = \sup_i \|u_i\| < \infty$, and observe that as $(u_i)$ lies in the centre of $A$, we have for each $a \in A$

$$\begin{aligned} a \cdot M_i - M_i \cdot a &= 2(a \cdot \Delta_i - \Delta_i \cdot a) - u_i \cdot (a \cdot \Delta_i - \Delta_i \cdot a) \\ &\quad + (a - au_i) \otimes (1 - u_i) - (1 - u_i) \otimes (a - au_i). \end{aligned}$$

Hence for each $a \in A$ and $\lambda \in \mathbb{C}$,

$$\begin{aligned} \|(a + \lambda 1) \cdot M_i - M_i \cdot (a + \lambda 1)\| &= \|a \cdot M_i - M_i \cdot a\| \\ &\leq (2 + K)\|a \cdot \Delta_i - \Delta_i \cdot a\| + 2(1 + K)\|a - au_i\|. \end{aligned}$$

Thus condition (ii) of Definition 3.2 is satisfied. Finally: by assumption, there exists $C > 0$ such that $\|a \cdot \Delta_i - \Delta_i \cdot a\| \leq C\|a\|$ for all $a \in A$. Hence,

$$\begin{aligned} \sup_i \|(a + \lambda 1) \cdot M_i - M_i \cdot (a + \lambda 1)\| &\leq (2 + K)C\|a\| + 2(1 + K)^2\|a\| \\ &\leq ((2 + K)C + 2(1 + K)^2)\, \|a + \lambda 1\|. \end{aligned}$$

Thus condition (iii) of Definition 3.2 is satisfied. This completes the proof of the claim, and hence of the lemma. □

To construct explicit examples, we use the following result. It is a slightly more abstract version of known results for Banach sequence algebras (cf. [10, Theorem 4.4] or [5, Corollary 3.5]), and it extends [10, Example 4.6], although our approach is slightly different from the argument there.

**Proposition 3.4.** *Let $A$ be a Banach algebra, containing a sequence $(e_n)_{n \geq 1}$ with the following properties:*

(i) $e_m e_n = e_{\min(m,n)} = e_n e_m$ *for all* $m, n \in \mathbb{N}$*;*
(ii) *the set* $\{e_n : n \in \mathbb{N}\}$ *has dense linear span in* $A$*;*
(iii) *there is a subsequence* $n(1) < n(2) < \cdots$ *such that* $(e_{n(j)})_{j\geq 1}$ *is a bounded approximate identity for* $A$.

*Then* $A$ *is boundedly approximately contractible.*

*Proof.* This is similar to the proof of [4, Theorem 6.1]. For $n \geq 2$, define

$$\Delta_n = e_1 \otimes e_1 + \sum_{j=2}^{n} (e_j - e_{j-1}) \otimes (e_j - e_{j-1}).$$

We then have the following identities:

$$\pi(\Delta_n) = e_n \quad \text{for all } n; \tag{1}$$

$$a \cdot \Delta_n = \Delta_n \cdot a \quad \text{for all } n \text{ and all } a \in A. \tag{2}$$

The identity (1) can be shown by direct calculation, using property (i). The identity (2) is true for $a = e_m$ ($m$ arbitrary); this is another direct calculation using (i), which is most easily done by treating the cases $m \leq n$ and $m > n$ separately. Hence, by linearity and continuity (using property (ii)), this identity holds for all $a \in A$, as claimed.

It follows immediately from (1), (2) and property (iii) that:

(a) $\pi(\Delta_{n(j)})$ is a bounded approximate identity for $A$; and
(b) the sequence $(\Delta_{n(j)})$ is a multiplier-bounded approximate diagonal for $A$.

Therefore $A$ is boundedly approximately contractible, by Lemma 3.3. □

**Remark 3.5.** The proof of Proposition 3.4 shows slightly more, namely that $A$ is not only boundedly approximately contractible, but also *pseudo-contractible* in the sense of [11]. (There are pseudo-contractible Banach algebras which are not boundedly approximately contractible, and vice versa.)

Now, fix a Hilbert space $\mathcal{H}$ and a strictly ascending chain of non-zero subspaces $\mathcal{H}_1 \subset \mathcal{H}_2 \subset \mathcal{H}_3 \subset \cdots$; for each $n \in \mathbb{N}$, let $p_n$ be the orthogonal projection of $\mathcal{H}$ onto $\mathcal{H}_n$. For each $k \in \mathbb{N}$, choose a bounded operator $b_{2k} \in \mathcal{B}(\mathcal{H}_{2k+1} \ominus \mathcal{H}_{2k}, \mathcal{H}_{2k} \ominus \mathcal{H}_{2k-1})$, such that $\|b_{2k}\| \to \infty$ as $k \to \infty$, and define a sequence $(e_n)_{n\geq 1} \subset \mathcal{B}(\mathcal{H})$ by

$$e_{2k-1} := p_{2k-1} \quad \text{and} \quad e_{2k} := p_{2k} + b_{2k}(p_{2k+1} - p_{2k}) \quad \text{for } k = 1, 2, \ldots$$

Thus, in block matrix form,

$$e_{2k-1} = \begin{bmatrix} I & 0 & 0 & 0 \\ 0 & 0 & 0 & 0 \\ 0 & 0 & 0 & 0 \\ 0 & 0 & 0 & 0 \end{bmatrix} \begin{matrix} \mathcal{H}_{2k-1} \\ \mathcal{H}_{2k} \ominus \mathcal{H}_{2k-1} \\ \mathcal{H}_{2k+1} \ominus \mathcal{H}_{2k} \\ \mathcal{H} \ominus \mathcal{H}_{2k+1} \end{matrix} \quad \text{and} \quad e_{2k} = \begin{bmatrix} I & 0 & 0 & 0 \\ 0 & I & b_{2k} & 0 \\ 0 & 0 & 0 & 0 \\ 0 & 0 & 0 & 0 \end{bmatrix} \begin{matrix} \mathcal{H}_{2k-1} \\ \mathcal{H}_{2k} \ominus \mathcal{H}_{2k-1} \\ \mathcal{H}_{2k+1} \ominus \mathcal{H}_{2k} \\ \mathcal{H} \ominus \mathcal{H}_{2k+1} \end{matrix}.$$

It is easily checked that for each $n$, we have $e_n^2 = e_n$ and $e_n e_{n+1} = e_n = e_{n+1} e_n$ (consider the cases of odd and even $n$ separately). The latter property implies, by induction, that $e_m e_n = e_{\min(m,n)} = e_n e_m$ for all $m, n \in \mathbb{N}$. Since

$\|e_{2k-1}\| = 1$ for all $k$, we see that the algebra $\mathfrak{A} = \overline{\mathrm{lin}}\{e_n : n \in \mathbb{N}\}$ satisfies the conditions of Proposition 3.4. Hence it is a boundedly approximately contractible Banach algebra. It is also singly generated, by Lemma 2.1. On the other hand, since

$$\|e_{2k}\| \geq \|b_{2k}\| \to \infty \qquad \text{as } k \to \infty,$$

$\mathfrak{A}$ is non-amenable, by Corollary 2.4.

In particular, suppose that we take $\mathcal{H} = \ell^2(\mathbb{Z}_+)$, with its standard o.n. basis $(\delta_n)_{n\geq 0}$, and define $\mathcal{H}_k = \mathrm{lin}(\delta_0, \ldots, \delta_k)$, so that each $b_{2j}$ is a scalar. Then since each idempotent $e_n$ has finite rank, $\mathfrak{A} \subset \mathcal{K}(\mathcal{H})$; and $\mathfrak{A}$ is singly generated, by Lemma 2.1. Using that lemma, we thus obtain an example of a compact operator on Hilbert space which generates a non-amenable, boundedly approximately contractible algebra. (We recall that by [19], a compact operator on Hilbert space generates an amenable algebra if and only if it is similar to a normal compact operator.)

Furthermore, it is clear that $\mathfrak{A}$ is contained in the subalgebra of $\mathcal{B}(\ell^2(\mathbb{Z}_+))$ formed by the block-diagonal matrices of block size 2, i.e., it embeds into the finite, homogeneous von Neumann algebra $\ell^\infty(\mathbb{Z}_+)\otimes\mathbb{M}_2$. We finish this section by noting that one can obtain many examples with the same properties, by choosing different sequences $(\mathcal{H}_n)$ and $(b_{2k})$. It may be interesting to study how these different examples might differ in their finer structure, or share other common features beyond being boundedly approximately contractible, singly and compactly generated, etc.

## 4. Biflatness

The notion of a biflat Banach algebra is due to Helemskiĭ. It should be emphasized that it is not an ad hoc weakening of amenability, obtained by randomly omitting certain conditions and "seeing what happens". Rather, it is linked to the notion of (homologically) flat Banach modules over a given Banach algebra, which is itself a key concept in Helemskiĭ's versions of Ext and Tor in the Banach algebraic setting.

That said, we shall not concern ourselves with the deeper homological implications of biflatness, and we will not give the original homological definition. For that, the reader should consult [13]. Instead, we take the following, well-known characterization of biflatness as a working definition.

**Lemma 4.1 (see [13, Exercise VII.2.8]).** *A Banach algebra $A$ is biflat if and only if there exists a continuous, linear $A$-bimodule map $\sigma : A \to (A\widehat{\otimes} A)^{**}$ such that $\pi^{**}\sigma = \kappa_A$, the natural embeddding $A \to A^{**}$.*

It is easy to see, from this characterization, that:

(i) every amenable Banach algebra is biflat (if $M \subset (A\widehat{\otimes} A)^{**}$ is a virtual diagonal for $A$, define $\sigma(a) = a \cdot M$);
(ii) a biflat Banach algebra with a bounded approximate identity is amenable (if $(u_i)$ is a BAI for $A$, let $M$ be any w*-cluster point in $(A\widehat{\otimes} A)^{**}$ of the net $\sigma(u_i)$).

**Remark 4.2.** Biflat Banach algebras are simplicially trivial, i.e., if $A$ is biflat then the continuous Hochschild cohomology groups $\mathcal{H}^n(A, A^*)$ vanish for all $n \geq 1$. In particular, biflat algebras are weakly amenable.

Let $A$ denote the Banach algebra obtained by equipping $\ell^1$ with pointwise multiplication. $A$ is a standard example of a commutative, semisimple, biflat Banach algebra that has no bounded approximate identity, and hence is non-amenable. Moreover, by Lemma 2.1, it is singly generated as a Banach algebra.

It turns out that there is a continuous algebra homomorphism $\theta : A \to \mathcal{B}(\mathcal{H})$ whose range is closed, so that $\theta(A)$ is a singly generated, biflat operator algebra. Moreover, given such an embedding $\theta$, one can construct an embedding of $A$ as a closed subalgebra of a finite, Type I von Neumann algebra. (The basic idea is as follows. If $(e_n)$ denotes the standard unit basis of $A = \ell^1$, let $A_n = \operatorname{lin}(e_1, \ldots, e_n)$, observe that $\theta|_{A_n}$ can be viewed as a representation of $A_n$ on a finite-dimensional Hilbert space, and then take the direct product of the representations $\theta\pi_n$ where $\pi_n : A \to A_n$ is the obvious truncation homomorphism.)

The existence of a map $\theta$ with these properties can be shown by combining the following two results: $A$ is a *$Q$-algebra* in the sense of Varopoulos (proved independently by Davie [6] and Varopoulos [18], see also Example 18.3 and Theorem 18.7 in [8]); and every $Q$-algebra is isomorphic to some closed subalgebra of $\mathcal{B}(\mathcal{H})$ (this is a theorem of Cole; see [2, §50] or [8, Theorem 18.8]). This approach is somewhat indirect, and does not seem to give an explicit description of an embedding. It is therefore desirable to have an explicit construction of an embedding of $A$ as a closed subalgebra of a product of matrix algebras. This can be done with the following construction, which was shown to me by M. de la Salle [7], and is included here with his kind permission. The presentation here is paraphrased slightly from his original wording. It seems likely that similar embeddings were known previously, but I was unable to find an explicit description in the literature.

**Lemma 4.3.** *Consider the Hilbert space $\ell^2(\mathbb{N} \cup \{\alpha, \omega\})$, where $\alpha$ and $\omega$ are formal symbols. There exists a sequence of rank-one idempotents $(E_n)_{n\geq 1} \subset \mathcal{B}(\ell^2(\mathbb{N} \cup \{\alpha, \omega\}))$, with the following properties:*

(i) $\|E_n\| \leq 3$ *for all* $n$;
(ii) $E_j E_k = 0$ *whenever* $j \neq k$;
(iii) *for each* $n$, $\operatorname{ran}(E_n)$ *and* $\operatorname{ran}(E_n^*)$ *are both contained in* $\operatorname{lin}(e_\alpha, e_\omega, e_n)$;
(iv) $\|\sum_j a_j E_j\| \geq |\sum_j a_j|$ *for any* $a \in c_{00}$.

*Proof* (*de la Salle, personal communication*). Let $e_\alpha$, $e_\omega$, $e_1$, $e_2, \ldots$ denote the standard basis vectors, and for each $n \in \mathbb{N}$ put $x_n = e_\omega + e_\alpha + e_n$, $y_n = e_\omega - e_\alpha + e_n$. Define $E_n$ by taking $E_n(\xi) = y_n \langle \xi, x_n \rangle$. Clearly each $E_n$ is a rank-one operator; direct calculation shows that $E_n^2 = E_n$. Properties (i)–(iii) are also easily verified, and (iv) follows from observing that $\langle E_n e_\omega, e_\omega \rangle = 1$. ☐

**Constructing the desired embedding.** Let $\mathcal{F}$ denote the family of finite, non-empty subsets of $\mathbb{N}$. For each $F \in \mathcal{F}$ let $\mathbb{M}_{F\cup\{\alpha,\omega\}}$ be the algebra of square matrices indexed by $F \cup \{\alpha, \omega\}$, given the usual (C*-algebra) norm; then if $j \in F$ we can identify $E_j$ and $E_j^*$ with elements of $\mathbb{M}_{F\cup\{\alpha,\omega\}}$.

Let $\mathcal{M}$ be the finite, Type I von Neumann algebra $\prod_{F\in\mathcal{F}} \mathbb{M}_{F\cup\{\alpha,\omega\}}$ – we remark that this has separable predual, since $\mathcal{F}$ is countable – and define $\phi : A \to \mathcal{M}$ by

$$\phi(\delta_j)_F = \begin{cases} 0 & \text{if } j \notin F \\ E_j & \text{if } j \in F. \end{cases} \tag{3}$$

Clearly $\phi$ is bounded linear with $\|\phi\| \leq \sup_j \|E_j\| \leq 3$, and since

$$\phi(\delta_j)_F \phi(\delta_k)_F = \begin{cases} \phi(\delta_j) & \text{if } j = k, \\ 0 & \text{if } j \neq k, \end{cases}$$

it follows by continuity that $\phi$ is an algebra homomorphism. To see that $\phi$ is bounded below, we use the estimate

$$\sup_{F\in\mathcal{F}} |\sum_{j\in F} a_j| \geq \frac{1}{\pi}\|a\|_1 . \tag{4}$$

(This can be found as [17, Lemma 6.3]. An earlier version of this paper had a weaker estimate, with 4 instead of $\pi$; my thanks to the referee for providing a reference for the sharper estimate.) We then have, for each $a \in A$,

$$\begin{aligned} \|\phi(a)\| = \sup_{F\in\mathcal{F}} \|\phi(a)_F\| &= \sup_{F\in\mathcal{F}} \|\sum_{j\in\mathbb{N}} a_j \phi(\delta_j)_F\| \\ &= \sup_{F\in\mathcal{F}} \|\sum_{j\in F} a_j E_j\| \\ &\geq \sup_{F\in\mathcal{F}} |\sum_{j\in F} a_j| && \text{(by Proposition 4.3(iv))} \\ &\geq \frac{1}{\pi}\|a\|_1 && \text{(by the inequality (4)).} \end{aligned}$$

Thus $\phi$ has closed range, as required.

## 5. An extension result

As previously mentioned: it has been asked if every amenable commutative subalgebra $\mathfrak{A} \subseteq \mathcal{B}(\mathcal{H})$ is isomorphic to a commutative C*-algebra, which is equivalent to asking if the Gelfand transform $\mathsf{G}_\mathfrak{A} : \mathfrak{A} \to C_0(\Phi_\mathfrak{A})$ is an isomorphism of Banach algebras. The answer is positive if $\mathfrak{A}$ is contained in a finite von Neumann algebra $\mathcal{M}$. A key part of the proof of this in [3] is the following technical result.

**Proposition 5.1.** *Let $\mathcal{M}$ be a von Neumann algebra with a faithful, finite, normal trace $\tau$, and let $\mathfrak{A} \subseteq \mathcal{M}$ be a closed subalgebra which is commutative and amenable. Let $\Phi_\mathfrak{A}$ denote the character space of $\mathfrak{A}$, and let $\mathsf{G}_\mathfrak{A} : \mathfrak{A} \to C_0(\Phi_\mathfrak{A})$ denote the Gelfand transform. Then $\mathsf{G}_\mathfrak{A}$ is injective with dense range, and there is a bounded* *linear map* $\theta : C_0(\Phi_{\mathfrak{A}}) \to L^1(\mathcal{M},\tau)$ *which extends the inclusion* $\mathfrak{A} \to \mathcal{M}$*, in the sense that the diagram*

$$\begin{array}{ccc} \mathfrak{A} & \xrightarrow{\text{inclusion}} & \mathcal{M} \\ {\scriptstyle \mathsf{G}_{\mathfrak{A}}}\downarrow & & \downarrow{\scriptstyle \text{inclusion}} \\ C_0(\Phi_{\mathfrak{A}}) & \xrightarrow{\theta} & L^1(\mathcal{M},\tau) \end{array} \tag{5}$$

*commutes. Here,* $L^1(\mathcal{M},\tau)$ *denotes the completion of* $\mathcal{M}$ *with respect to the norm* $\|x\|_{L^1(\tau)} = \sup\{|\tau(xy)| : y \in \mathcal{M}, \|y\| \leq 1\}$.

One might wonder if the converse of this proposition holds, perhaps with the addition of some side-conditions on $\mathfrak{A}$. More precisely:

**Question.** Let $\mathcal{M}$ be as in Proposition 5.1, and let $\mathfrak{A} \subseteq \mathcal{M}$ be a closed subalgebra, such that

(i) $\mathsf{G}_{\mathfrak{A}} : \mathfrak{A} \to C_0(\Phi_{\mathfrak{A}})$ is injective with dense range;
(ii) there exists a bounded linear map $\theta : C_0(\Phi_{\mathfrak{A}}) \to L^1(\mathcal{M},\tau)$ making the diagram (5) commute.

If $\mathfrak{A}$ furthermore satisfies some weak version of amenability, must it automatically be amenable?

If we take "weak version of amenability" to be biflatness (and therefore, anything strictly weaker than biflatness), then the answer is negative. Our example will be the algebra $\mathfrak{A}$ obtained as an embedding of $\ell^1$. Let $\phi$ be the embedding constructed in (3), and let $\mathsf{G}_A : A \to c_0(\mathbb{N})$ be the Gelfand transform (which is just the usual injection from $\ell^1(\mathbb{N})$ into $c_0(\mathbb{N})$.)

**Proposition 5.2.** *Let* $\tau$ *be a faithful, normal, finite, tracial state* $\tau$ *on* $\mathcal{M}$*. Then*

$$\|\phi(a)\|_{L^1(\tau)} \leq 3\|a\|_\infty \quad \textit{for all } a \in \ell^1(\mathbb{N}). \tag{6}$$

*Hence there is a bounded linear map* $\theta : c_0(\mathbb{N}) \to L^1(\tau)$ *which makes the diagram (5) commute.*

*Proof.* By normality, any such trace $\tau$ is determined by its values on each summand $\mathbb{M}_{F\cup\{\alpha,\omega\}}$, and therefore must have the form

$$\tau(x) = \sum_{F\in\mathcal{F}} \lambda_F \operatorname{Tr}(x_F) \qquad (x \in \mathcal{M}),$$

where the family $(\lambda_F)_{F\in\mathcal{F}}$ is strictly positive and absolutely summable with

$$\sum_{F\in\mathcal{F}} \lambda_F = 1,$$

and where Tr denotes the *normalized* trace on $\mathbb{M}_{F\cup\{\alpha,\omega\}}$. Hence

$$\|y\|_{L^1(\tau)} = \sum_{F\in\mathcal{F}} \frac{\lambda_F}{|F|+2}\|y_F\|_{(1)} \qquad (y \in \mathcal{M}),$$

where $\|\cdot\|_{(1)}$ denotes the usual, *unnormalized* Schatten 1-norm on matrices.

It suffices to prove the inequality (6); the rest is routine. So, let $a \in c_{00}(\mathbb{N})$. Then

$$\begin{aligned}\|\phi(a)\|_{L^1(\tau)} = \sum_{F\in\mathcal{F}} \frac{\lambda_F}{|F|+2}\|\phi(a)_F\|_{(1)} &\leq \sup_{F\in\mathcal{F}} \frac{1}{|F|+2}\|\phi(a)_F\|_{(1)} \\ &= \sup_{F\in\mathcal{F}} \frac{1}{|F|+2}\left\|\sum\nolimits_{j\in F} a_j E_j\right\|_{(1)} \\ &\leq \|a\|_\infty \sup_{F\in\mathcal{F}} \frac{1}{|F|+2}\sum_{j\in F}\|E_j\|_{(1)}.\end{aligned}$$

Since each $E_j$ is a rank-one operator, $\|E_j\|_{(1)} = \|E_j\| \leq 3$, and the desired inequality (6) now follows. □

**Acknowledgments**

The construction in Proposition 5.2, and some of the results from [3], formed part of a lecture presented at the conference *Operator theory and applications*, held at Chalmers University, 26th–29th April 2011, in honour of V. S. Shulman. The author thanks the conference organizers for an enjoyable and stimulating meeting. He would also like to thank Mikael de la Salle and Jesse Peterson for useful discussions on the *MathOverflow* website.

Finally, the author thanks the anonymous referee for his or her suggestions and corrections: in particular, the requests for clarification of the original version of Section 2, which led to the repair and improvement of some incomplete arguments.

The work here was partially supported by NSERC Discovery Grant 402153-2011.

## References

[1] E. Azoff and H. Shehada, *Algebras generated by mutually orthogonal idempotent operators*, J. Operator Theory, 29 (1993), pp. 249–267.

[2] F.F. Bonsall and J. Duncan, *Complete normed algebras*, Ergebnisse der Mathematik und ihrer Grenzgebiete, Band 80, Springer-Verlag, New York, 1973.

[3] Y. Choi, *On commutative, operator amenable subalgebras of finite von Neumann algebras*. J. Reine. Angew. Math. 678 (2013), 201–222.

[4] Y. Choi, F. Ghahramani, and Y. Zhang, *Approximate and pseudo-amenability of various classes of Banach algebras*, J. Funct. Anal., 256 (2009), pp. 3158–3191.

[5] H.G. Dales, R.J. Loy, and Y. Zhang, *Approximate amenability for Banach sequence algebras*, Studia Math., 177 (2006), pp. 81–96.

[6] A.M. Davie, *Quotient algebras of uniform algebras*, J. London Math. Soc. (2), 7 (1973), pp. 31–40.

[7] M. de la Salle, *personal communication via MathOverflow*. See `http://mathoverflow.net/questions/49788` (version: 2010-12-18).

[8] J. Diestel, H. Jarchow, and A. Tonge, *Absolutely summing operators*, vol. 43 of Cambridge Studies in Advanced Mathematics, Cambridge University Press, Cambridge, 1995.

[9] F. Ghahramani and R.J. Loy, *Generalized notions of amenability*, J. Funct. Anal., 208 (2004), pp. 229–260.

[10] F. Ghahramani, R.J. Loy, and Y. Zhang, *Generalized notions of amenability, II.*, J. Funct. Anal., 254 (2008), pp. 1776–1810.

[11] F. Ghahramani and Y. Zhang, *Pseudo-amenable and pseudo-contractible Banach algebras*, Math. Proc. Camb. Phil. Soc., 142 (2007), pp. 111–123.

[12] J.A. Gifford, *Operator algebras with a reduction property*, J. Aust. Math. Soc., 80 (2006), pp. 297–315.

[13] A.Ya. Helemskiĭ, *The Homology of Banach and Topological Algebras*, vol. 41 of Mathematics and its Applications (Soviet Series), Kluwer Academic Publishers Group, Dordrecht, 1989.

[14] B.E. Johnson, *Cohomology in Banach algebras*, American Mathematical Society, Providence, R.I., 1972.

[15] L.W. Marcoux, *On abelian, triangularizable, total reduction algebras*, J. Lond. Math. Soc. (2), 77 (2008), pp. 164–182.

[16] J. Peterson, *personal communication via MathOverflow*. See `http://mathoverflow.net/questions/85878` (version: 2012-01-17).

[17] W. Rudin, *Real and complex analysis*, McGraw-Hill Book Co., New York, third ed., 1987.

[18] N.T. Varopoulos, *Some remarks on Q-algebras*, Ann. Inst. Fourier (Grenoble), 22 (1972), pp. 1–11.

[19] G.A. Willis, *When the algebra generated by an operator is amenable*, J. Operator Theory, 34 (1995), pp. 239–249.

[20] Y. Zhang, *Solved and unsolved problems in generalized notions of amenability for Banach algebras*, in Banach algebras 2009, vol. 91 of Banach Center Publ., Polish Acad. Sci. Inst. Math., Warsaw, 2010, pp. 441–454.

Yemon Choi
Department of Mathematics and Statistics
McLean Hall
University of Saskatchewan
106 Wiggins Road, Saskatoon, SK
Canada, S7N 5E6
e-mail: `choi@math.usask.ca`

Operator Theory:
Advances and Applications, Vol. 233, 45–73


# Operator Algebras and C*-correspondences: A Survey

Evgenios T.A. Kakariadis and Elias G. Katsoulis

**Abstract.** In this paper we survey our recent work on C*-correspondences and their associated operator algebras; in particular, on adding tails, the Shift Equivalence Problem and Hilbert bimodules.

**Mathematics Subject Classification (2010).** 46L08, 47L55.

**Keywords.** C*-correspondences, C*-envelope, shift equivalence.

## Introduction

Traditionally, the theory of C*-correspondences has been used to generalize concrete results either from the theory of Cuntz–Krieger C*-algebras or from the theory of crossed product C*-algebras. Our goal in this project has been to discover a path in the opposite direction: to use the theory of C*-correspondences in order to obtain results which are new even for the Cuntz–Krieger or the crossed product C*-algebras. For instance in Theorem 2.1 we describe a general process for adding tails to a C*-correspondence that has lead to new results in the theory of crossed product C*-algebras and Cuntz's twisted crossed products. In the same spirit, our Theorem 3.8 on the concept of shift equivalence of C*-correspondences leads to a new result regarding the strong Morita equivalence of Cuntz–Krieger C*-algebras (Theorem 3.10(4)). Therefore, even though the next few pages may seem rather abstract or encyclopedic, we believe that this abstraction will benefit even the reader who is interested only in the special classes of operator algebras mentioned above.

## 1. Preliminaries

Let $A$ be a C*-algebra. An *inner-product right $A$-module* is a linear space $X$ which is a right $A$-module together with a map

$$(\cdot,\cdot) \mapsto \langle \cdot,\cdot \rangle_X : X \times X \to A$$

such that

$$\begin{aligned} \langle \xi, \lambda y + \eta \rangle_X &= \lambda \langle x, y \rangle_X + \langle x, \eta \rangle_X && (\xi, y, \eta \in X, \lambda \in \mathbb{C}), \\ \langle \xi, \eta a \rangle_X &= \langle \xi, \eta \rangle_X a && (\xi, \eta \in X, a \in A), \\ \langle \eta, \xi \rangle_X &= \langle \xi, \eta \rangle_X^* && (\xi, \eta \in X), \\ \langle \xi, \xi \rangle_X &\geq 0; \text{ if } \langle \xi, \xi \rangle_X = 0 \text{ then } \xi = 0. \end{aligned}$$

A compatibility relation for the scalar multiplication is required, that is $\lambda(\xi a) = (\lambda \xi) a = \xi(\lambda a)$, for all $\lambda \in \mathbb{C}, \xi \in X, a \in A$. For $\xi \in X$ we write $\|\xi\|_X := \|\langle \xi, \xi \rangle_A\|_A$ and one can deduce that $\|\cdot\|_X$ is actually a norm. $X$ equipped with that norm will be called *right Hilbert $A$-module* if it is complete and will be denoted as $X_A$.

For $X, Y$ Hilbert $A$-modules we define the set $\mathcal{L}(X,Y)$ of the *adjointable maps* that consists of all maps $s : X \to Y$ for which there is a map $s^* : Y \to X$ such that

$$\langle s\xi, y \rangle_Y = \langle \xi, s^* y \rangle_Y, \qquad (x \in X, y \in Y).$$

Every element of $\mathcal{L}(X,Y)$ is automatically a bounded $A$-linear map. An element $u \in \mathcal{L}(X,Y)$ is called *unitary* if it is onto $Y$ and $\langle u\xi, u\zeta \rangle_Y = \langle \xi, \zeta \rangle_X$, for all $\xi, \zeta \in X$.

In particular, for $\xi \in X$ and $y \in Y$ we define $\Theta_{y,\xi} : X \to Y$ such that $\Theta_{y,\xi}(\zeta) = y \langle \xi, \zeta \rangle_X$, for all $\zeta \in X$. It is easy to check that $\Theta_{y,\xi} \in \mathcal{L}(X,Y)$ with $\Theta_{y,\xi}^* = \Theta_{\xi,y}$. We denote by $\mathcal{K}(X,Y)$ the closed linear space of $\mathcal{L}(X,Y)$ spanned by $\{\Theta_{y,\xi} : \xi \in X, y \in Y\}$. If $X = Y$ then $\mathcal{K}(X,X) \equiv \mathcal{K}(X)$ is a closed ideal of the C*-algebra $\mathcal{L}(X,X) \equiv \mathcal{L}(X)$.

In a dual way we call $X$ a *left Hilbert $A$-module* if it is complete with respect to the norm induced by an *inner-product left $A$-module* $[\cdot,\cdot]_X$. The term *Hilbert module* is reserved for the right Hilbert modules, whereas the left case will be clearly stated.

Given a Hilbert $A$-module $X$ over $A$, let $X^* = \{\xi^* \in \mathcal{L}(X,A) : \xi^*(\zeta) = \langle \xi, \zeta \rangle_X\}$ be the *dual left Hilbert $A$-module*, with

$$\begin{aligned} a \cdot \xi^* &= (\xi a^*)^*, && (\xi \in X, a \in A), \\ [\xi^*, \zeta^*]_{X^*} &= \langle \xi, \zeta \rangle_X, && (\xi, \zeta \in X). \end{aligned}$$

A Hilbert $A$-module $X$ is called *self-dual* when $X^*$ coincides with the set of bounded (not necessarily adjointable) $A$-linear mappings from $X$ to $A$, hence a Riesz–Fréchet Theorem is valid.

**Example 1.1.** A C*-algebra $A$ is a (trivial) Hilbert $A$-module, when it is viewed as a Banach space with $a \cdot b := ab$ and $\langle a, b \rangle_A := a^* b$ for all $a, b \in A$. It is a left Hilbert $A$-module when it is endowed with the left inner product $[a,b]_A := ab^*$, for all $a, b \in A$. Finally $\mathcal{K}(A) \simeq A$ and $\mathcal{L}(A_A) \simeq \mathcal{M}(A)$, i.e., the multiplier algebra of $A$.

### 1.1. C*-correspondences

Even though the class of C*-correspondences has been thoroughly investigated for the last 25 years, the terminology still differs from author to author. We therefore present the terminology that we will be using in this paper.

**Definition 1.2.** An *A-B-correspondence* $X$ is a right Hilbert $B$-module together with a $*$-homomorphism $\varphi_X : A \to \mathcal{L}(X)$. We will denote this by ${}_AX_B$. When $A = B$ we will simply refer to $X$ as C*-*correspondence over* $A$.

A subspace $Y$ of $X$ is a *subcorrespondence* of ${}_AX_B$, if it is a $C$-$D$-correspondence for some C*-subalgebras $C$ and $D$ of $A$ and $B$, respectively.

A C*-correspondence over $A$ is called *non-degenerate* if the closed linear span of $\varphi_X(A)X$ is dense in $X$. Moreover, $X$ is called *full* if $\langle X, X\rangle_X$ is dense in $A$. Also, $X$ is called *regular* if it is *injective*, i.e., $\varphi_X$ is injective, and $\varphi_X(A) \subseteq \mathcal{K}(X)$.

Two $A$-$B$-correspondences $X$ and $Y$ are called unitarily equivalent, if there is a unitary $u \in \mathcal{L}(X,Y)$ such that $u(\varphi_X(a)\xi b) = \varphi_Y(a)(u(\xi))b$, for all $a \in A, b \in B, \xi \in X$. In that case we write $X \approx Y$. We write $X \lesssim Y$ when $X \approx Y_0$ for a subcorrespondence $Y_0$ of $Y$.

**Example 1.3.** Every Hilbert $A$-module $X$ is a $\mathcal{K}(X)$-$A$-correspondence when endowed with the left multiplication $\varphi_X \equiv \mathrm{id}_{\mathcal{K}(X)} : \mathcal{K}(X) \to \mathcal{L}(X)$. A left inner product over $\mathcal{K}(X)$ can be defined by $[\xi, \eta]_X = \Theta_{\xi,\eta}$, for all $\xi, \eta \in X$. Also $X^*$ is an $A$-$\mathcal{K}(X)$-correspondence, when endowed with the following operations

$$\begin{aligned} \langle \xi^*, \eta^* \rangle_{X^*} &= [\xi, \eta]_X = \Theta_{\xi,\eta} && (\xi^*, \eta^* \in X^*), \\ \xi^* \cdot k &= (k^*\xi)^* && (\xi^* \in X^*, k \in \mathcal{K}(X)), \\ \varphi_{X^*}(a)\xi^* &= a \cdot \xi^* = (\xi \cdot a^*)^* && (\xi^* \in X^*, a \in A). \end{aligned}$$

**Example 1.4.** For Hilbert $A$-modules $X$ and $Y$, $\mathcal{L}(X,Y)$ becomes $\mathcal{L}(Y)$-$\mathcal{L}(X)$-correspondence by defining $\langle s, t\rangle := s^*t$, $t \cdot a := ta$ and $b \cdot t := bt$, for every $s, t \in \mathcal{L}(X,Y), a \in \mathcal{L}(X)$ and $b \in \mathcal{L}(Y)$.

Trivially, $\mathcal{K}(X,Y)$ is a $\mathcal{K}(Y)$-$\mathcal{K}(X)$-subcorrespondence of $\mathcal{L}(X,Y)$. Note that, when a c.a.i. in $\langle X, X\rangle_X$ is a right c.a.i. for $Y$, then $\mathcal{K}(Y)$ acts faithfully on $\mathcal{K}(X,Y)$. When $X = Y$ this is automatically true.

### 1.2. Interior tensor product

The interior tensor product of two Hilbert modules plays the role of a generalized multiplication, stabilized by the elements of a common C*-algebra (see [38] for more details). Let the C*-correspondences ${}_AX_B$ and ${}_BY_C$; the *interior* or *stabilized tensor product*, denoted by $X \otimes_B Y$ or simply by $X \otimes Y$, is the quotient of the vector space tensor product $X \otimes_{\mathrm{alg}} Y$ by the subspace generated by the elements of the form

$$\xi b \otimes y - \xi \otimes \varphi(b)y, \text{ for all } \xi \in X, y \in Y, b \in B.$$

It becomes a Hilbert $C$-module when equipped with

$$(\xi \otimes y)c := \xi \otimes (yc), \qquad (\xi \in X, y \in Y, c \in C),$$
$$\langle \xi_1 \otimes y_1, \xi_2 \otimes y_2 \rangle_{X \otimes Y} := \langle y_1, \varphi(\langle \xi_1, \xi_2 \rangle_X) y_2 \rangle_Y, \qquad (\xi_1, \xi_2 \in X, y_1, y_2 \in Y).$$

For $s \in \mathcal{L}(X)$ we define $s \otimes \mathrm{id}_Y \in \mathcal{L}(X \otimes Y)$ as the mapping $\xi \otimes y \mapsto s(\xi) \otimes y$. Clearly, $(s \otimes \mathrm{id}_Y)^* = s^* \otimes \mathrm{id}_Y$; hence $X \otimes Y$ becomes an $A$-$C$-correspondence by defining $\varphi_{X \otimes Y}(a) := \varphi_X(a) \otimes \mathrm{id}_Y$. One can prove that the interior tensor product is associative, that is if $Z$ is a $C$-$D$-correspondence, then $(X \otimes_B Y) \otimes_C Z = X \otimes_B (Y \otimes_C Z)$.

**Example 1.5.** When a Hilbert $A$-module $X$ is considered as the $\mathcal{K}(X)$-$A$-correspondence, then $X \otimes_A X^* \approx \mathcal{K}(X)$ as C*-correspondences over $\mathcal{K}(X)$, via the mapping

$$u_1 : X \otimes_A X^* \to \mathcal{K}(X) : \xi \otimes \zeta^* \mapsto \Theta_{\xi,\zeta},$$

and $X^* \otimes_{\mathcal{K}(X)} X \approx \langle X, X \rangle_X$, as C*-correspondences over $A$ via the mapping

$$u_2 : X^* \otimes_{\mathcal{K}(X)} X \to \langle X, X \rangle_X : \xi^* \otimes \zeta \mapsto \langle \xi, \zeta \rangle$$

In particular $X^* \otimes_{\mathcal{K}(X)} X \approx A$, when $X$ is full.

### 1.3. Hilbert bimodules

There are ${}_AX_B$ C*-correspondences that are both left and right Hilbert modules. If a compatibility relation is satisfied between the two inner products then ${}_AX_B$ is called *Hilbert bimodule.*

**Definition 1.6.** A *Hilbert $A$-$B$-bimodule* ${}_AX_B$ is a C*-correspondence $X$ together with a left inner product $[\cdot,\cdot]_X : X \times X \to A$, which satisfy:

$$[\varphi_X(a)\xi, \eta]_X = a\,[\xi, \eta]_X, \qquad (\xi, \eta \in X, a \in A),$$
$$[\xi, \eta]_X = [\eta, \xi]_X^*, \qquad (\xi, \eta \in X)$$
$$[\xi, \xi]_X \geq 0; \text{ if } [\xi, \xi]_X = 0 \text{ then } \xi = 0,$$
$$\varphi_X([\xi, \eta]_X)\zeta = \xi\,\langle \eta, \zeta \rangle_X, \qquad (\xi, \eta, \zeta \in X).$$

The last equation implies that $\varphi_X([\xi,\eta]_X) = \Theta^X_{\xi,\eta}$. It is clear that Hilbert bimodules are a special case of C*-correspondences. Let $I_X$ be the ideal,

$$I_X = \overline{\mathrm{span}}\{[\xi, \eta]_X : \xi, \eta \in X\},$$

in $A$. Using the very definitions, one can prove that $a \in \ker \varphi_X$, if and only if $a \in I_X^\perp$. Hence, $\varphi_X$ is $*$-injective, if and only if, the Hilbert $A$-bimodule $X$ is *essential*, i.e., when the ideal $I_X$ is essential in $A$.

**Definition 1.7.** An *$A$-$B$-imprimitivity bimodule* or *equivalence bimodule* is an $A$-$B$-bimodule $M$ which is simultaneously a full left and a full right Hilbert $A$-module. That is $[M, M]_M$ is dense in $A$ and $\langle M, M \rangle_M$ is dense in $B$.

It is easy to see that when $X$ is an $A$-$B$-imprimitivity bimodule then $A \simeq^{\varphi} \mathcal{K}(M)$. Thus imprimitivity bimodules are automatically non-degenerate and regular.

### 1.4. Matrix C*-correspondences

There is a number of ways of considering *a* direct sum of C*-correspondences. They are contained (as subcorrespondences) in the notion of the *matrix* C*-*correspondence* that is presented below. For the C*-correspondences ${}_AE_A$, ${}_BF_B$, ${}_AR_B$ and ${}_BS_A$ the *matrix* C*-*correspondence* $X = \left[\begin{array}{c|c} E & R \\ S & F \end{array}\right]$ over $A \oplus B$ is the Hilbert $(A \oplus B)$-module of the linear space of the matrices $\left[\begin{array}{c|c} e & r \\ s & f \end{array}\right]$, $e \in E, r \in R, s \in S, f \in F$, with

$$\left[\begin{array}{c|c} e & r \\ s & f \end{array}\right] \cdot (a,b) = \left[\begin{array}{c|c} ea & rb \\ sa & fb \end{array}\right],$$

$$\left\langle \left[\begin{array}{c|c} e_1 & r_1 \\ s_1 & f_1 \end{array}\right], \left[\begin{array}{c|c} e_2 & r_2 \\ s_2 & f_2 \end{array}\right] \right\rangle_X = \left( \langle e_1, e_2 \rangle_E + \langle s_1, s_2 \rangle_S , \langle r_1, r_2 \rangle_R + \langle f_1, f_2 \rangle_F \right),$$

such that the $*$-homomorphism $\varphi \colon A \oplus B \to \mathcal{L}\left( \left[\begin{array}{c|c} E & R \\ S & F \end{array}\right] \right)$ is defined as follows

$$\varphi(a,b) \left[\begin{array}{c|c} e & r \\ s & f \end{array}\right] = \left[\begin{array}{c|c} \varphi_E(a)e & \varphi_R(a)r \\ \varphi_S(b)s & \varphi_F(b)f \end{array}\right].$$

The careful reader can see that this is exactly the *exterior direct sum* C*-*correspondence* of the two *interior direct sum* C*-*correspondences* ${}_{A\oplus B}(E+S)_A$ and ${}_{A\oplus B}(R+F)_B$. Hence, the linear space of the matrices $\left[\begin{array}{c|c} e & r \\ s & f \end{array}\right]$, with $e \in E, r \in R, s \in S, f \in F$, is complete with respect to the induced norm, thus a Hilbert $(A \oplus B)$-module. Moreover $E, F, R, S$ imbed naturally as subcorrespondences in $\left[\begin{array}{c|c} E & R \\ S & F \end{array}\right]$. The following lemma explains the terminology *matrix* C*-*correspondence*, as tensoring corresponds to "multiplying" the matrices.

**Lemma 1.8.** *Let $E, F, R, S$ be* C*-*correspondences as above. Then*

$$\left[\begin{array}{c|c} E & R \\ S & F \end{array}\right] \otimes_{A\oplus B} \left[\begin{array}{c|c} E & R \\ S & F \end{array}\right] \approx \left[\begin{array}{c|c} (E\otimes_A E) + (R\otimes_B S) & (E\otimes_A R) + (R\otimes_B F) \\ (S\otimes_A E) + (F\otimes_B S) & (S\otimes_A R) + (F\otimes_B F) \end{array}\right],$$

*(unitarily equivalent) as* C*-*correspondences.*

### 1.5. Representations of C*-correspondences

Let us make a brief presentation on the representation theory of C*-correspondences. Let $X$ be a C*-correspondence over $A$. A (Toeplitz) representation $(\pi, t)$ of $X$ into a C*-algebra $B$, is a pair of a $*$-homomorphism $\pi \colon A \to B$ and a linear map $t \colon X \to B$, such that

(1) $\pi(a)t(\xi) = t(\varphi_X(a)(\xi))$,
(2) $t(\xi)^* t(\eta) = \pi(\langle \xi, \eta \rangle_X)$,

for $a \in A$ and $\xi, \eta \in X$. An easy application of the C*-identity shows that $t(\xi)\pi(a) = t(\xi a)$ is also valid. A representation $(\pi, t)$ is said to be *injective* if $\pi$ is injective; in that case $t$ is an isometry.

The C*-algebra generated by a representation $(\pi, t)$ equals the closed linear span of $t^n(\bar{\xi})t^m(\bar{\eta})^*$, where for simplicity we used the notation $\bar{\xi} \equiv \xi_1 \otimes \cdots \otimes \xi_n \in X^{\otimes n}$ and $t^n(\bar{\xi}) \equiv t(\xi_1)\dots t(\xi_n)$. For any representation $(\pi, t)$ there exists a $*$-homomorphism $\psi_t : \mathcal{K}(X) \to B$, such that $\psi_t(\Theta^X_{\xi,\eta}) = t(\xi)t(\eta)^*$.

Let $J$ be an ideal in $\varphi_X^{-1}(\mathcal{K}(X))$; we say that a representation $(\pi, t)$ is $J$-coisometric if

$$\psi_t(\varphi_X(a)) = \pi(a), \text{ for any } a \in J.$$

The representations $(\pi, t)$ that are $J_X$-coisometric, where

$$J_X = \ker \varphi_X^{\perp} \cap \varphi_X^{-1}(\mathcal{K}(X)),$$

are called *covariant representations* [33]. The ideal $J_X$ is the largest ideal on which the restriction of $\varphi_X$ is injective.

We define the Toeplitz–Cuntz–Pimsner algebra $\mathcal{T}_X$ as the universal C*-algebra for all Toeplitz representations of $X$. Similarly, the Cuntz–Pimsner algebra $\mathcal{O}_X$ is the universal C*-algebra for all covariant representations of $X$.

A concrete presentation of both $\mathcal{T}_X$ and $\mathcal{O}_X$ can be given in terms of the generalized Fock space $\mathcal{F}_X$ which we now describe. The *Fock space* $\mathcal{F}_X$ over the correspondence $X$ is the interior direct sum of the $X^{\otimes n}$ with the structure of a direct sum of C*-correspondences over $A$,

$$\mathcal{F}_X = A \oplus X \oplus X^{\otimes 2} \oplus \cdots .$$

Given $\xi \in X$, the (left) creation operator $t_\infty(\xi) \in \mathcal{L}(\mathcal{F}_X)$ is defined as

$$t_\infty(\xi)(\zeta_0, \zeta_1, \zeta_2, \dots) = (0, \xi\zeta_0, \xi \otimes \zeta_1, \xi \otimes \zeta_2, \dots),$$

where $\zeta_n \in X^{\otimes n}$, $n \geq 0$ and $\zeta_0 \in A$. (Here $X^{\otimes 0} \equiv A$, $X^{\otimes 1} \equiv X$ and $X^{\otimes n} = X \otimes X^{\otimes n-1}$, for $n \geq 2$.) For any $a \in A$, we define $\pi_\infty(a) \in \mathcal{L}(\mathcal{F}_X)$ to be the diagonal operator with $\varphi_X(c) \otimes \mathrm{id}_{n-1}$ at its $X^{\otimes n}$th entry. It is easy to verify that $(\pi_\infty, t_\infty)$ is a representation of $X$ which is called the *Fock representation* of $X$. Fowler and Raeburn [20] (resp. Katsura [33]) have shown that the C*-algebra $\mathrm{C}^*(\pi_\infty, t_\infty)$ (resp $\mathrm{C}^*(\pi_\infty, t_\infty)/\mathcal{K}(\mathcal{F}_X)J_X$) is $*$-isomorphic to $\mathcal{T}_X$ (resp. $\mathcal{O}_X$).

**Definition 1.9.** The *tensor algebra* $\mathcal{T}_X^+$ of a C*-correspondence ${}_AX_A$ is the norm-closed subalgebra of $\mathcal{T}_X$ generated by all elements of the form $\pi_\infty(a), t_\infty^n(\bar{\xi})$, $a \in A$, $\bar{\xi} \in X^n$, $n \in \mathbb{N}$.

The tensor algebras for C*-correspondences were pioneered by Muhly and Solel in [41]. They form a broad class of non-selfadjoint operator algebras which includes as special cases Peters' semicrossed products [46], Popescu's non-commutative disc algebras [48], the tensor algebras of graphs (introduced in [41] and further studied in [30]) and the tensor algebras for multivariable dynamics [13], to mention but a few. For more examples, see [32].

There is an important connection between $\mathcal{T}_X^+$ and $\mathcal{O}_X$ given in the following Theorem of Katsoulis and Kribs [31]. Recall that, for an operator algebra $\mathfrak{A}$ and a completely isometric representation $\iota\colon \mathfrak{A} \to A$, where $A = \mathrm{C}^*(\iota(\mathfrak{A}))$, the pair $(A, \iota)$ is called *a* C*-*cover for* $\mathfrak{A}$. The C*-*envelope of the operator algebra* $\mathfrak{A}$ is the universal C*-cover $(A, \iota)$ such that, if $(B, \iota')$ is any other C*-cover for $\mathfrak{A}$, then there exists a (unique) $*$-epimorphism $\Phi : B \to A$, such that $\Phi(\iota'(a)) = \iota(a)$, for any $a \in \mathfrak{A}$. For the existence of the C*-envelope see [21, 17, 2, 6, 25].

**Theorem 1.10** ([31, Theorem 3.7]). *The* C*-*envelope of the tensor algebra* $\mathcal{T}_X^+$ *is* $\mathcal{O}_X$.

As a consequence the Toeplitz–Cuntz–Pimsner algebra is the extension of the Cuntz–Pimsner algebra by the Šilov ideal. (Any ideal $\mathcal{J} \subseteq C$, for a C*-cover $(C, \iota)$ of an operator algebra $\mathfrak{A}$, with the property that the restriction of the natural projection $C \to C/\mathcal{J}$ on $\mathfrak{A}$ is a complete isometry, is called a *boundary ideal* and the Šilov ideal is the largest such ideal.)

Now, let us see how we can generalize the previous facts in the case of arbitrary C*-correspondences by using the notion of matrix C*-correspondences. A representation of an $A$-$B$-correspondence $X$ should be a triplet $(\pi_A, \pi_B, t)$ such that $\pi_A$ and $\pi_B$ are $*$-homomorphisms, $t$ is a linear mapping of $X$ and

(1) $\pi_A(a)t(\xi)\pi_B(b) = t(\varphi_X(a)(\xi)b)$,
(2) $t(\xi)^*t(\eta) = \pi_B(\langle \xi, \eta \rangle_X)$,

for all $\xi \in X, a \in A, b \in B$. If that is the case then one could define $\hat{t} : X \to \mathcal{B}(H^{(2)})$ such that $\hat{t}(\xi) = \begin{bmatrix} 0 & t(\xi) \\ 0 & 0 \end{bmatrix}$. Then $(\pi_A \oplus \pi_B, \hat{t})$ defines a representation of the $(A \oplus B)$-correspondence $\left[\begin{array}{c|c} 0 & X \\ 0 & 0 \end{array}\right]$. Conversely, if $(\pi, t)$ is a representation of $\left[\begin{array}{c|c} 0 & X \\ 0 & 0 \end{array}\right]$, then $(\pi|_A, \pi|_B, t)$ defines a representation of $X$. Hence, we can identify $X$ with $\left[\begin{array}{c|c} 0 & X \\ 0 & 0 \end{array}\right]$ and define the Toeplitz–Cuntz–Pimsner, the Cuntz–Pimsner and the tensor algebra of the $A$-$B$-correspondence $X$ as the corresponding algebras of the $(A \oplus B)$-correspondence $\left[\begin{array}{c|c} 0 & X \\ 0 & 0 \end{array}\right]$. However, note that most of the results known for C*-correspondences over the same C*-algebra must be verified, basically because $X^{\otimes n}$ is absurd for all $n \geq 2$.

**Remark 1.11.** We already gave a brief description of $X^*$ of the Hilbert $A$-module $X$. When $X$ is a correspondence over $A$ this can be simplified. Let $(\pi_u, t_u)$ be the universal representation of ${}_AX_A$; then $X^*$ is the closed linear span of $t_u(\xi)^*$, $\xi \in X$ with the left multiplication and inner product inherited by the trivial correspondence $\mathrm{C}^*(\pi_u, t_u)$. Via this identification one can produce a theory for the left analogue of C*-correspondences (that means also, to start with left Hilbert modules), but in most of the cases it can be recovered.

### 1.6. Examples

One of the fundamental examples in the theory of C*-correspondences are the C*-algebras of directed graphs. (For more details see [49].)

Let $\mathcal{G}$ be a countable directed graph with vertex set $\mathcal{G}^{(0)}$, edge set $\mathcal{G}^{(1)}$ and range and source maps $r$ and $s$ respectively. A family of partial isometries $\{L_e\}_{e\in\mathcal{G}^{(1)}}$ and projections $\{L_p\}_{p\in\mathcal{G}^{(0)}}$ is said to obey the Cuntz–Krieger relations associated with $\mathcal{G}$ if and only if they satisfy

$$(\dagger)\begin{cases} (1)\quad L_pL_q = 0 & \forall\, p,q \in \mathcal{G}^{(0)},\, p \neq q \\ (2)\quad L_e^*L_f = 0 & \forall\, e,f \in \mathcal{G}^{(1)},\, e \neq f \\ (3)\quad L_e^*L_e = L_{s(e)} & \forall\, e \in \mathcal{G}^{(1)} \\ (4)\quad L_eL_e^* \leq L_{r(e)} & \forall\, e \in \mathcal{G}^{(1)} \\ (5)\quad \sum_{r(e)=p} L_eL_e^* = L_p & \forall\, p \in \mathcal{G}^{(0)} \text{ with } |r^{-1}(p)| \neq 0,\infty. \end{cases}$$

The relations ($\dagger$) have been refined in a series of papers by the Australian school and reached the above form in [5, 50]. All refinements involved condition (5) and as it stands now, condition (5) gives the equality requirement for projections $L_p$ such that $p$ is not a source and receives finitely many edges. (Indeed, otherwise condition (5) would not be a C*-condition.)

It can been shown that there exists a universal C*-algebra, denoted as $\mathcal{O}_{\mathcal{G}}$, associated with the relations ($\dagger$). Indeed, one constructs a single family of partial isometries and projections obeying ($\dagger$). Then, $\mathcal{O}_{\mathcal{G}}$ is the C*-algebra generated by a 'maximal' direct sum of such families. It turns out that there $\mathcal{O}_{\mathcal{G}}$ is the Cuntz–Pimsner algebra of a certain C*-correspondence [41]. The associated Cuntz–Pimsner–Toeplitz algebra is the universal algebra for the first four relations in ($\dagger$) and is denoted as $\mathcal{T}_{\mathcal{G}}$.

**Example 1.12.** Let $G, G'$ be two graphs with adjacent matrices $A_G$ and $A_{G'}$. If $X_G$ and $X_{G'}$ are the corresponding C*-correspondences, then $X_G \otimes X_{G'}$ is unitarily equivalent to the C*-correspondence that comes from the adjacent matrix $A_G \cdot A_{G'}$.

Let $X$ be an imprimitivity bimodule that comes from a graph

$$\mathcal{G} = (\mathcal{G}^{(0)}, \mathcal{G}^{(1)}, r, s)$$

(we follow the notation in [49]); then $\mathcal{G}$ is either a cycle or a double infinite path. Pick your favorite completely isometric representation of $\mathcal{T}_X^+$; for us it is a Cuntz–Krieger family $\{P_v, S_e : v \in \mathcal{G}^{(0)}, e \in \mathcal{G}^{(1)}\}$ (because of [30]). Apart from the usual relations, due to the form of the graph we have the simplified relation $S_eS_e^* = P_{r(e)}$, for all $e \in \mathcal{G}^{(1)}$. Therefore $\{P_v, S_e^* : v \in \mathcal{G}^{(0)}, e \in \mathcal{G}^{(1)}\}$ defines a Cuntz–Krieger family of the graph $\mathcal{G}^* = (\mathcal{G}^{(0)}, \mathcal{G}^{(1)}, r^*, s^*)$ where the arrows are reversed, i.e., $r^* = s$ and $s^* = r$, thus $X^*$ is the C*-correspondence coming from this graph $\mathcal{G}^*$.

A second example comes from the class of dynamical systems. Let $\beta\colon A \to B$ be a $*$-homomorphism of C*-algebras. The trivial Hilbert module $B_B$ becomes a $A$-$B$-correspondence, denoted by $(X_\beta, A)$, when endowed with the left action $\varphi_B$ such that $\varphi_B(a)b = \beta(a)b$ for all $a \in A$ and $b \in B$.

**Example 1.13.** If $(X_\alpha, C)$ is a $C$-$A$-correspondence via a $*$-homomorphism $\alpha\colon C \to A$, and $(X_\beta, A)$ is a $A$-$B$-correspondence via a $*$-homomorphism $\beta\colon A \to B$, then

$A \otimes_A B$ is unitarily equivalent to the $C$-$B$-correspondence $(X_{\beta\circ\alpha}, C)$ associated to the $*$-homomorphism $\beta \circ \alpha \colon C \to B$.

Moreover, $(X_\beta, A)$ is an imprimitivity bimodule if and only if $\beta$ is a $*$-isomorphism. In this case $(X_\beta, A)^* \approx (X_{\alpha^{-1}}, B)$. In particular, when $A = B$, then $\mathcal{O}_{X_\beta}$ is the usual crossed product $B \rtimes_\beta \mathbb{Z}$ and the tensor algebra $\mathcal{T}^+_{X_\beta}$ is Peter's semicrossed product [46] (for this and various types of semicrossed products see also [23, 24, 27]).

## 2. Adding tails

In [28] the authors developed a method of "adding tails" that extends the one developed by Muhly and Tomforde in [44].

Let $\mathcal{G}$ be a connected, directed graph with a distinguished sink $p_0 \in \mathcal{G}^{(0)}$ and no sources. We assume that $\mathcal{G}$ is *contractible* at $p_0$, i.e., the subalgebra $\mathbb{C}L_{p_0}$ is a full corner of the Cuntz–Krieger algebra $\mathcal{O}_{\mathcal{G}}$. So there exists a unique infinite path $w_0 = e_1 e_2 e_3 \dots$ ending at $p_0$, i.e., $r(w_0) = p_0$. Let $p_n \equiv s(e_n)$, $n \geq 1$.

Let $(A_p)_{p\in\mathcal{G}^{(0)}}$ be a family of C*-algebras parameterized by the vertices of $\mathcal{G}$ so that $A_{p_0} = A$. For each $e \in \mathcal{G}^{(1)}$, we now consider a full, right Hilbert $A_{s(e)}$-module $X_e$ and a $*$-homomorphism

$$\varphi_e \colon A_{r(e)} \longrightarrow \mathcal{L}(X_e)$$

satisfying the following requirements.

For $e \neq e_1$, the homomorphism $\varphi_e$ are required to be injective and map onto $\mathcal{K}(X_e)$, i.e., $\varphi_e(A_{r(e)}) = \mathcal{K}(X_e)$. Therefore, each $X_e$, $e \neq e_1$, is an $A_{r(e)}$-$A_{s(e)}$-equivalence bimodule, in the language of Rieffel.

For $e = e_1$, we require $\mathcal{K}(X_{e_1}) \subseteq \varphi_{e_1}(A)$ and

$$J_X \subseteq \ker \varphi_{e_1} \subseteq (\ker \varphi_X)^\perp . \tag{1}$$

In addition, there is also a *linking condition*

$$\varphi_{e_1}^{-1}(\mathcal{K}(X_{e_1})) \subseteq \varphi_X^{-1}(\mathcal{K}(X)) \tag{2}$$

required between the maps $\varphi_X$ and $\varphi_{e_1}$.

Let $T_0 = c_0(\,(A_p)_{p\in\mathcal{G}_-^{(0)}})$ denote the $c_0$-sum of the family $(A_p)_{p\in\mathcal{G}_-^{(0)}}$, where $\mathcal{G}_-^{(0)} \equiv \mathcal{G}^{(0)}\backslash\{p_0\}$. Consider the set $c_{00}((X_e)_{e\in\mathcal{G}^{(1)}}) \subseteq c_0((X_e)_{e\in\mathcal{G}^{(1)}})$, consisting of sequences which are zero everywhere but on a finite number of entries. Equip $c_{00}((X_e)_{e\in\mathcal{G}^{(1)}})$ with a $T_0$-valued inner product defined by

$$\langle u, v\rangle (p) = \sum_{s(e)=p} u_e^* v_e, \quad p \in \mathcal{G}_-^{(0)},$$

for any $u, v \in c_{00}((X_e)_{e\in\mathcal{G}^{(1)}})$. Let $T_1$ be the completion of $c_{00}((X_e)_{e\in\mathcal{G}^{(1)}})$ with respect to that inner product. Equip now $T_1$ with a right $T_0$-action, so that

$$(ux)_e = u_e x_{s(e)}, \quad e \in \mathcal{G}^{(1)},$$

for any $x \in T_0$, so that the pair $(T_1, T_0)$ becomes a right $T_0$-Hilbert module. The pair $(T_0, T_1)$ is the *tail* for ${}_AX_A$.

To the C*-correspondence ${}_AX_A$ and the data

$$\tau \equiv \Big(\mathcal{G}, (X_e)_{e\in\mathcal{G}^{(1)}}, (A_p)_{p\in\mathcal{G}^{(0)}}, (\varphi_e)_{e\in\mathcal{G}^{(1)}}\Big),$$

we now associate

$$\begin{aligned} A_\tau &\equiv A \oplus T_0 \\ X_\tau &\equiv X \oplus T_1. \end{aligned} \tag{3}$$

Using the above, we view $X_\tau$ as a $A_\tau$-Hilbert module with the standard right action and inner product for direct sums of Hilbert modules. We also define a left $A_\tau$-action $\varphi_\tau : A_\tau \to \mathcal{L}(X_\tau)$ on $X_\tau$ by setting

$$\varphi_\tau(a, x)(\xi, u) = (\varphi_X(a)\xi, v),$$

where

$$v_e = \begin{cases} \varphi_{e_1}(a)(u_{e_1}), & \text{if } e = e_1 \\ \varphi_e(x_{r(e)})u_e, & \text{otherwise} \end{cases}$$

for $a \in A$, $\xi \in X$, $x \in T_0$ and $u \in T_1$.

**Theorem 2.1 ([28, Theorem 3.10]).** *Let ${}_AX_A$ be a non-injective* C*-*correspondence and let $X_\tau$ be the graph* C*-*correspondence over $A_\tau$ defined above. Then $X_\tau$ is an injective* C*-*correspondence and the Cuntz–Pimsner algebra $\mathcal{O}_X$ is a full corner of $\mathcal{O}_{X_\tau}$.*

Regarding Theorem 2.1 and the conditions imposed on the graph $\mathcal{G}$ and the maps $(\varphi_e)_{e\in\mathcal{G}^{(1)}}$, we have asked that the graph $\mathcal{G}$ be contractible. We cannot weaken this assumption to include more general graphs. Indeed, we want the tail associated with the data

$$\tau = \Big(\mathcal{G}, (X_e)_{e\in\mathcal{G}^{(1)}}, (A_p)_{p\in\mathcal{G}^{(0)}}, (\varphi_e)_{e\in\mathcal{G}^{(1)}}\Big),$$

to work with any possible Cuntz–Pimsner algebra $\mathcal{O}_X$ that can be "added on". This should apply in particular to the Cuntz–Krieger algebra $\mathcal{O}_{G_{p_0}}$ of the (trivial) graph $\mathcal{G}_{p_0}$ consisting only of one vertex $p_0$. By taking $\tau$ to be the "usual" tail associated with $\mathcal{G}$, i.e., $X_e = A_e = \mathbb{C}L_{p_0}$ and $\varphi_e$ left multiplication for all $e$, we see that $\mathcal{O}_{G_{p_0}}$ is a full corner of $O_{X_\tau}$ if and only if $\mathcal{G}$ is contractible at $p_0$.

Conditions (1) and (2) are also necessary, as the following result suggests.

**Proposition 2.2.** *Let ${}_AX_A$ be a non-injective* C*-*correspondence and let $X_\tau$ be the* C*-*correspondence over $A_\tau$ associated with the data*

$$\tau = \Big(\mathcal{G}, (X_e)_{e\in\mathcal{G}^{(1)}}, (A_p)_{p\in\mathcal{G}^{(0)}}, (\varphi_e)_{e\in\mathcal{G}^{(1)}}\Big),$$

*as defined at the beginning of the section. If $X_\tau$ is injective,*

$$(\varphi_1^{-1}(\mathcal{K}(X_1)) + J_X) \oplus 0 \subseteq J_{X_\tau},$$

*and the covariant representations of $X_\tau$ restrict to covariant representations of $\varphi_X$, then*

$$J_X \subseteq \ker \varphi_1 \subseteq (\ker \varphi_X)^\perp ,$$

*and the linking condition*

$$\varphi_1^{-1}(\mathcal{K}(X_1)) \subseteq \varphi_X^{-1}(\mathcal{K}(X))$$

*holds.*

Lets see now how the work of Muhly and Tomforde fits in our theory.

**Example 2.3 (The Muhly–Tomforde tail [44]).** Given a (non-injective) correspondence $(X, A, \varphi_X)$, Muhly and Tomforde construct in [44] the tail that results from the previous construction, with respect to data

$$\tau = \Big(\mathcal{G}, (X_e)_{e\in\mathcal{G}^{(1)}}, (A_p)_{p\in\mathcal{G}^{(0)}}, (\varphi_e)_{e\in\mathcal{G}^{(1)}}\Big)$$

defined as follows. The graph $\mathcal{G}$ is illustrated in the figure below.

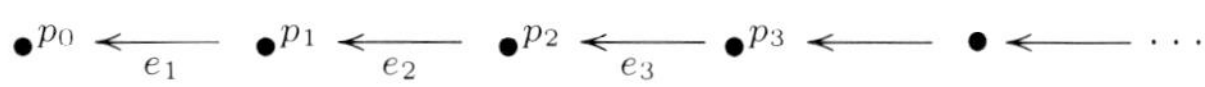


$A_p = X_e = \ker \varphi_X$, for all $p \in \mathcal{G}_-^{(0)}$ and $e \in \mathcal{G}^{(1)}$. Finally,

$$\varphi_e(a)u_e = au_e,$$

for all $e \in \mathcal{G}^{(1)}$, $u_e \in X_e$ and $a \in A_{r(e)}$.

## 2.1. Applications

**2.1.1. Semicrossed products.** The tail of Muhly and Tomforde has had significant applications in the theory of C*-correspondences, including a characterization for the C*-envelope of the tensor algebra of a non-injective correspondence [31]. However, it also has its limitations, as we are about to see.

**Example 2.4.** Let $(X_\alpha, A)$ be the C*-correspondence canonically associated with a dynamical system $(A, \alpha)$ and let $\mathcal{O}_{(A,\alpha)}$ be the associated Cuntz–Pimsner C*-algebra. If $\alpha$ is not injective, then by using the Muhly–Tomforde tail we obtain an injective C*-correspondence $(Y, B, \varphi_Y)$ so that $\mathcal{O}_{(A,\alpha)}$ is a full corner of $\mathcal{O}_Y$. Remarkably, $(Y, B, \varphi_Y)$ may not come from any dynamical system in general. Assume that $\ker \alpha \subseteq A$ is an essential ideal of $A$; then the Muhly–Tomforde tail produces an injective correspondence $(Y, B, \varphi_Y)$ with

$$Y = A \oplus c_0(\ker \alpha), \qquad B = A \oplus c_0(\ker \alpha)$$

and $\varphi_Y$ defined by

$$\varphi_Y\big(a, (c_i)_i\big)\big(a', (c'_i)_i\big) = \big(\alpha(a)a', \alpha(a)c'_1, c_1c'_2, c_2c'_3, \dots\big),$$

where $a, a' \in A$ and $(c_i)_i, (c'_i)_i \in c_0(\ker \alpha)$.

If there was a $*$-homomorphism $\beta$ satisfying

$$\varphi_Y(b)(b') = \beta(b)b', \tag{4}$$

then by equating second coordinates in the equation

$$\varphi_Y\big(1,(c_i)_i\big)\big(a',(c_i')_i\big) = \beta\big(1,(c_i)_i\big)\big(a',(c_i')_i\big)$$

we would obtain,

$$c_1' = \beta\big(1,(c_i)_i\big)_2 c_1',$$

for all $c_1' \in \ker\alpha$. Since $\ker\alpha$ is an essential ideal, we have $\ker\alpha \ni \beta\big(1,(c_i)_i\big)_2 = 1$, a contradiction.

Therefore, the Muhly–Tomforde tail produces an injective correspondence but not necessarily an injective dynamical system. Nevertheless, there exists a tail that can be added to $(X_\alpha, A)$ and produce an injective correspondence that comes from a dynamical system.

**Example 2.5.** If $(X_\alpha, A)$ is the C*-correspondence canonically associated with a dynamical system $(A,\alpha)$, then the appropriate tail for $(X_\alpha, A)$ comes from the data

$$\tau = \Big(\mathcal{G}, (X_e)_{e\in\mathcal{G}^{(1)}}, (A_p)_{p\in\mathcal{G}^{(0)}}, (\varphi_e)_{e\in\mathcal{G}^{(1)}}\Big)$$

where $\mathcal{G}$ is as in Example 2.4 *but* for any $p \in \mathcal{G}_-^{(0)}$ and $e \in \mathcal{G}^{(1)}$,

$$A_p = X_e = \theta(A),$$

where $\theta\colon A \longrightarrow M(\ker\alpha)$ is the map that extends the natural inclusion $\ker\alpha \subseteq M(\ker\alpha))$ in the multiplier algebra. Finally

$$\varphi_e(a)u_e = au_e,$$

for all $e \in \mathcal{G}^{(1)}$, $u_e \in X_e$ and $a \in A_{r(e)}$. Then the correspondence $X_\tau$ is canonically associated to the dynamical system $(B,\beta)$, where

$$B = A \oplus c_0(\theta(A) \text{ and } \beta(a,(c_i)_i) = (\alpha(a),\theta(a),(c_i)_i).$$

**2.1.2. Multivariable dynamical systems.** We now apply the method of adding tails to C*-dynamics. Apart from their own merit, this application will also address the necessity of using more elaborate tails than that of Muhly and Tomforde in the process of adding tails to C*-correspondences. This necessity has been already noted in the one-variable case.

A *multivariable* C*-*dynamical system* is a pair $(A,\alpha)$ consisting of a C*-algebra $A$ along with a tuple $\alpha = (\alpha_1,\alpha_2,\dots,\alpha_n)$, $n \in \mathbb{N}$, of $*$-endomorphisms of $A$. The dynamical system is called injective if $\cap_{i=1}^n \ker\alpha_i = \{0\}$.

In the C*-algebra literature, the algebras $\mathcal{O}_{(A,\alpha)}$ are denoted as $A \times_\alpha \mathcal{O}_n$ and go by the name "twisted tensor products by $\mathcal{O}_n$". They were first introduced and studied by Cuntz [8] in 1981. In the non-selfadjoint literature, these algebras are much more recent. In [13] Davidson and the second named author introduced the tensor algebra $\mathcal{T}_{(A,\alpha)}^+$ for a multivariable dynamical system $(A,\alpha)$. It turns out that $\mathcal{T}_{(A,\alpha)}^+$ is completely isometrically isomorphic to the tensor algebra for the C*-correspondence $(X_\alpha, A, \varphi_\alpha)$. As such, $\mathcal{O}_{(A,\alpha)}$ is the C*-envelope of $\mathcal{T}_{(A,\alpha)}^+$.

Therefore, $\mathcal{O}_{(A,\alpha)}$ provides a very important invariant for the study of isomorphisms between the tensor algebras $\mathcal{T}^+_{(A,\alpha)}$.

To the multivariable system $(A,\alpha)$ we associate a C*-correspondence $(X_\alpha, A, \varphi_\alpha)$ as follows. Let $X_\alpha = A^n = \oplus_{i=1}^n A$ be the usual right $A$-module. That is

(1) $(a_1,\dots,a_n)\cdot a = (a_1 a,\dots,a_n a)$,
(2) $\langle (a_1,\dots,a_n),(b_1,\dots,b_n)\rangle = \sum_{i=1}^n \langle a_i, b_i\rangle = \sum_{i=1}^n a_i^* b_i$.

Also, by defining the $*$-homomorphism

$$\varphi_\alpha \colon A \longrightarrow \mathcal{L}(X_\alpha)\colon a \longmapsto \oplus_{i=1}^n \alpha_i(a),$$

$X_\alpha$ becomes a C*-correspondence over $A$, with $\ker\varphi_\alpha = \cap_{i=1}^n \ker\alpha_i$ and $\varphi(A)\subseteq \mathcal{K}(X_\alpha)$. It is easy to check that in the case where $A$ and all $\alpha_i$ are unital, $X_\alpha$ is finitely generated as an $A$-module by the elements

$$e_1 := (1,0,\dots,0), e_2 := (0,1,\dots,0), \dots, e_n := (0,0,\dots,1),$$

where $1 \equiv 1_A$. In that case, $(\pi,t)$ is a representation of this C*-correspondence if and only if $t(\xi_i)$ are isometries with pairwise orthogonal ranges and

$$\pi(c)t(\xi) = t(\xi)\pi(\alpha_i(c)), \quad i = 1,\dots,n.$$

**Definition 2.6.** The Cuntz–Pimsner algebra $\mathcal{O}_{(A,\alpha)}$ of a multivariable C*-dynamical system $(A,\alpha)$ is the Cuntz–Pimsner algebra of the C*-correspondence $(X_\alpha, A, \varphi_\alpha)$ constructed as above.

The graph $\mathcal{G}$ that we associate with $(X_\alpha, A, \varphi_\alpha)$ has no loop edges and a single sink $p_0$. All vertices in $\mathcal{G}^{(0)}\backslash\{p_0\}$ emit $n$ edges, i.e., as many as the maps involved in the multivariable system, and receive exactly one. In the case where $n = 2$, the following figure illustrates $\mathcal{G}$.

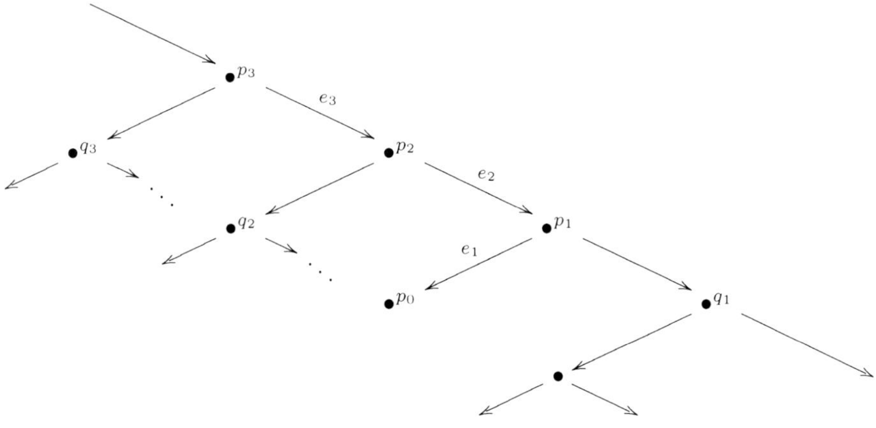


Clearly, $\mathcal{G}$ is $p_0$-accessible. There is also a unique infinite path $w$ ending at $p_0$ and so $\mathcal{G}$ is contractible at $p_0$.

Let $\mathcal{J} \equiv \cap_{i=1}^{n} \ker \alpha_i$ and let $M(\mathcal{J})$ be the multiplier algebra of $\mathcal{J}$. Let $\theta \colon A \longrightarrow M(\mathcal{J})$ the map that extends the natural inclusion $\mathcal{J} \subseteq M(\mathcal{J})$). Let $X_e = A_{s(e)} = \theta(A)$, for all $e \in \mathcal{G}^{(1)}$, and consider $(X_e, A_{s(e)})$ with the natural structure that makes it into a right Hilbert module.

For $e \in \mathcal{G}^{(1)} \backslash \{e_1\}$ we define $\varphi_e(a)$ as left multiplication by $a$. With that left action, clearly $X_e$ becomes an $A_{r(e)}$-$A_{s(e)}$-equivalence bimodule. For $e = e_1$, it is easy to see that

$$\varphi_{e_1}(a)(\theta(b)) \equiv \theta(ab), \quad a, b \in A$$

defines a left action on $X_{e_1} = \theta(A)$, which satisfies both (1) and (2).

**Theorem 2.7 ([28, Theorem 4.2]).** *If $(A, \alpha)$ is a non-injective multivariable* C**-dynamical system, then there exists an injective multivariable* C**-dynamical system $(B, \beta)$ so that the associated Cuntz–Pimsner algebras $\mathcal{O}_{(A,\alpha)}$ is a full corner of $\mathcal{O}_{(B,\beta)}$. Moreover, if $A$ belongs to a class $\mathcal{C}$ of* C**-algebras which is invariant under quotients and $c_0$-sums, then $B \in \mathcal{C}$ as well. Furthermore, if $(A, \alpha)$ is non-degenerate, then so is $(B, \beta)$.*

The reader familiar with the work of Davidson and Roydor may have noticed that the arguments in the proof of Theorem 2.7, when applied to multivariable systems over commutative C*-algebras produce a tail which is different from that of Davidson and Roydor in [14, Theorem 4.1]. It turns out that the proof of [14, Theorem 4.1] contains an error and the technique of Davidson and Roydor does not produce a full corner, as claimed in [14]. Nevertheless, [14, Theorem 4.1] is valid as Theorem 2.7 demonstrates (a fact also mentioned in [14, Corrigendum]). We illustrate this by examining their arguments in the following simple case.

**Example 2.8 (The Davidson–Roydor tail [14]).** Let $\mathcal{X} \equiv \{u, v\}$ and consider the maps $\sigma_i \colon \mathcal{X} \to \mathcal{X}$, $i = 1, 2$, with $\sigma_i(u) = v$ and $\sigma_i(v) = v$. Set $\sigma \equiv (\sigma_1, \sigma_2)$ and let $\mathcal{O}_{(\mathcal{X},\sigma)}$ be the Cuntz–Pimsner algebra associated with the multivariable system $(\mathcal{X}, \sigma)$, which by [13] is the C*-envelope of the associate tensor algebra.

We now follow the arguments of [14]. In order to obtain $\mathcal{O}_{(\mathcal{X},\sigma)}$ as a full corner of an injective Cuntz–Pimsner algebra, Davidson and Roydor add a tail to the multivariable system. They define

$$T = \{(u, k) \mid k < 0\} \text{ and } \mathcal{X}^T = \mathcal{X} \cup T.$$

For each $1 \leq i \leq 2$, they extend $\sigma_i$ to a map $\sigma_i^T \colon \mathcal{X}^T \to \mathcal{X}^T$ by

$$\sigma^T(u, k) = (u, k + 1) \text{ for } k < -1 \text{ and } \sigma_i^T(u, -1) = u.$$

They then consider the new multivariable system $(\mathcal{X}^T, \sigma^T)$ and its associated Cuntz–Pimsner algebra $\mathcal{O}(\mathcal{X}^T, \sigma^T)$.

It is easy to see that the Cuntz–Pimsner algebra $\mathcal{O}_{(\mathcal{X},\sigma)}$ for the multivariable system $(\mathcal{X}, \sigma)$ is the Cuntz–Krieger algebra $\mathcal{O}_{\mathcal{G}}$ of the graph $\mathcal{G}$ illustrated below, while the Cuntz–Pimsner algebra $\mathcal{O}(\mathcal{X}^T, \sigma^T)$ is isomorphic to the Cuntz–Krieger algebra $\mathcal{O}_{\mathcal{G}^T}$ of the following graph $\mathcal{G}^T$, where for simplicity we write $u_k$ instead

of $(u, k)$, $k < 0$,

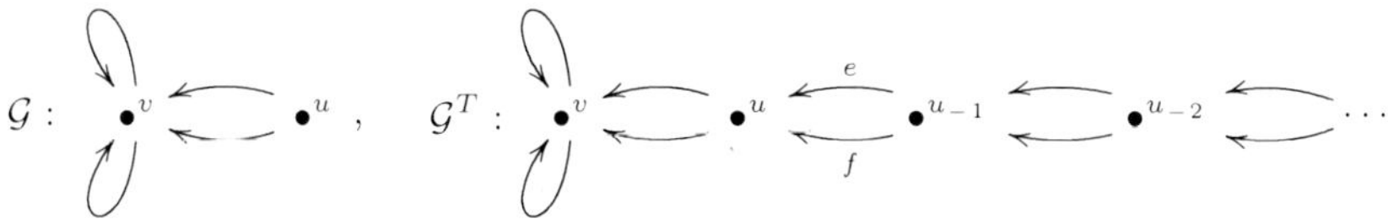

In [14, page 344], it is claimed that the projection P associated with the characteristic function of $\mathcal{X} \subseteq \mathcal{X}^T$ satisfies $P\mathcal{O}_{(\mathcal{X}^T,\sigma^T)}P = \mathcal{O}_{(\mathcal{X},\sigma)}$ and so $\mathcal{O}_{(\mathcal{X},\sigma)}$ is a corner of $\mathcal{O}_{(\mathcal{X}^T,\sigma^T)}$. In our setting, this claim translates as follows: if $P = L_u + L_v$, then $P\mathcal{O}_{\mathcal{G}^T}P = \mathcal{O}_{\mathcal{G}}$.

However this is not true. For instance, $P(L_f L_e^*)P = L_f L_e^* \notin \mathcal{O}_{\mathcal{G}}$.

**2.1.3. Multivariable dynamical systems and crossed products by endomorphism.** We can describe the Cuntz–Pimsner algebra of an injective and non-degenerate multivariable system as a crossed product of a C*-algebra $B$ by an endomorphism $\beta$. This idea appears first in [14] for a different crossed product than the one presented here. We start with the pertinent definitions.

**Definition 2.9.** Let $B$ be a (not necessary unital) C*-algebra and let $\beta$ be an injective endomorphism of $B$. A *covariant* representation $(\pi, v)$ of the dynamical system $(B, \beta)$ consists of a non-degenerate $*$-representation $\pi$ of $B$ and an isometry $v$ satisfying

(i) $\pi(\beta(b)) = v\pi(b)v^*, \forall\, b \in B$, i.e., $v$ *implements* $\beta$,
(ii) $v^*\pi(B)v \subseteq \pi(B)$, i.e., $v$ is *normalizing* for $\pi(B)$,
(iii) $v^k(v^*)^k\pi(B) \subseteq \pi(B), \forall\, k \in \mathbb{N}$.

The crossed product $B \times_\beta \mathbb{N}$ is the universal C*-algebra associated with this concept of a covariant representation for $(B, \beta)$. Specifically, $B \times_\beta \mathbb{N}$ is generated by $B$ and $BV$, where $V$ is an isometry satisfying (i), (ii) and (iii) in Definition 2.9 with $\pi = \mathrm{id}$. Furthermore, for any covariant representation $(\pi, v)$ of $(B, \beta)$, there exists a $*$-homomorphism $\hat{\pi}\colon B \times_\beta \mathbb{N} \to \mathcal{B}(H)$ extending $\pi$ and satisfying $\hat{\pi}(bV) = \pi(b)v$, for all $b \in B$.

In the case where $B$ is unital, condition (iii) is redundant and this version of a crossed product by an endomorphism was introduced by Paschke [45]; in the generality presented here, it is new. It has the advantage that for any covariant representation of $(\pi, v)$ of $(B, \beta)$ admitting a gauge action, the fixed point algebra of $(\pi, v)$ equals $\pi(B)$ (see Remark 2.10 below). This allows us to claim a gauge invariance uniqueness theorem for $B \times_\beta \mathbb{N}$: if $(\pi, v)$ is a faithful covariant representation of $(B, \beta)$ admitting a gauge action, then the C*-algebra generated by $\pi(B)$ and $\pi(B)v$ is isomorphic to $B \times_\beta \mathbb{N}$.

**Remark 2.10.** Observe that conditions (i), (ii) and (iii) in Definition 2.9 imply that $\hat{\pi}\,(B \times_\beta \mathbb{N})$ is generated by polynomials of the form $\pi(b_0) + \sum_k \pi(b_k)v^k + \sum_l (v^*)^l\pi(b_l)$, $\;b_0, b_k, b_l \in B$. Indeed, item (i) implies that $v\pi(b) = \pi(\beta(b))v$, thus $\pi(b)v^* = v^*\pi(\beta(b))$, for all $b \in B$ (since $B$ is self-adjoint). With this in hand and item (ii) of the definition we can see that the product $(v^*)^k\pi(b_k)\pi(b_l)v^l$ can be written as a trigonometric polynomial of the above form, for all $k, l \geq 0$. Item (iii) is used to show the same thing for the products $\pi(b_k)v^k(v^*)^l\pi(b_l)$, when $k \geq l$. Finally, for $k < l$ we use the fact that $v$ is an isometry and item (iii) so that

$$\begin{aligned}\pi(b_k)v^k(v^*)^l\pi(b_l) &= \pi(b_k)(v^*)^{l-k}v^l(v^*)^l\pi(b_l)\\ &= \pi(b_k)(v^*)^{l-k}\pi(b') = (v^*)^{l-k}\pi(b''),\end{aligned}$$

which completes the argument.

There is a related concept of a crossed product by an endomorphism which we now discuss. For a C*-algebra $B$ and an injective endomorphism $\beta$, Stacey [51] imposes on a covariant representation $(\pi, v)$ of $(B, \beta)$ only condition (i) from Definition 2.9. He then defines the crossed product $B \rtimes_\beta \mathbb{N}$ to be the universal C*-algebra associated with his concept of a covariant representation for $(B, \beta)$. Muhly and Solel have shown [42] that in the case where $B$ is unital, Stacey's crossed product is the Cuntz–Pimsner algebra of a certain correspondence. Using a gauge invariance uniqueness theorem one can prove that if the isometry V in $B \rtimes_\beta \mathbb{N}$ satisfies condition (ii) in Definition 2.9, then $B \rtimes_\beta \mathbb{N} \simeq B \times_\beta \mathbb{N}$.

**Theorem 2.11.** [28, Theorem 4.6] *If $(A, \alpha)$ is an injective multivariable system, then there exists a C*-algebra $B$ and an injective endomorphism $\beta$ of $B$ so that $\mathcal{O}_{(A,\alpha)}$ is isomorphic to the crossed product algebra $B \times_\beta \mathbb{N}$. Furthermore, if $A$ belongs to a class $\mathcal{C}$ which is invariant under direct limits and tensoring by $M_k(\mathbb{C})$, $k \in \mathbb{N}$, then $B$ also belongs to $\mathcal{C}$.*

Combining Theorem 2.11 with Theorem 2.7 we obtain the following.

**Corollary 2.12.** *If $(A, \alpha)$ is a multivariable system, then there exists a C*-algebra $B$ and an injective endomorphism $\beta$ of $B$ so that $\mathcal{O}_{(A,\alpha)}$ is isomorphic to a full corner of the crossed product algebra $B \times_\beta \mathbb{N}$. Furthermore, if $A$ belongs to a class $\mathcal{C}$ which is invariant under direct limits, quotients and tensoring by $M_k(\mathbb{C})$, $k \in \mathbb{N}$, then $B$ also belongs to $\mathcal{C}$.*

Finally, let us give a quick application of Theorem 2.11, that readily follows from Paschke's result [45] on the simplicity of $B \times_\beta \mathbb{N}$.

**Corollary 2.13.** *Let $A$ be a UHF C*-algebra and let $\alpha = (\alpha_1, \dots, \alpha_n)$ be a multivariable system with $n \geq 2$. If $\alpha_i(1) = 1$, for all $i = 1, 2, \dots, n$, then $\mathcal{O}_{(A,\alpha)}$ is simple.*

### 2.2. Graph C*-correspondences

Graph C*-correspondences can be used to summarize in a very delicate way properties and constructions of C*-correspondences [15].

Let $\mathcal{G} = (\mathcal{G}^{(0)}, \mathcal{G}^{(1)}, r, s)$ be a directed graph. Let $(A_p)_{p \in \mathcal{G}^{(0)}}$ be a family of C*-algebras parameterized by the vertices of $\mathcal{G}$ and for each $e \in \mathcal{G}^{(1)}$, we consider $X_e$ be a $A_{r(e)}$-$A_{s(e)}$-correspondences $X_e$.

Let $A_{\mathcal{G}} = c_0((A_p)_{p \in \mathcal{G}^{(0)}})$ denote the $c_0$-sum of the family $(A_p)_{p\in\mathcal{G}^{(0)}}$. Also, let $Y_0 = c_{00}((X_e)_{e\in\mathcal{G}^{(1)}})$ which is equipped with a $A_{\mathcal{G}}$-valued inner product

$$\langle u, v\rangle (p) = \sum_{s(e)=p} \langle u_e, v_e\rangle_{A_p}, \quad p \in \mathcal{G}^{(0)}.$$

If $X_{\mathcal{G}}$ is the completion of $Y_0$ with respect to the inner product, then $X_{\mathcal{G}}$ is a $A_{\mathcal{G}}$-Hilbert module when equipped with the right action

$$(ux)_e = u_e x_{s(e)}, \quad e \in \mathcal{G}^{(1)}.$$

It becomes a C*-correspondence, when equipped with the $*$-homomorphism $\varphi_{\mathcal{G}} : A_{\mathcal{G}} \to \mathcal{L}(X_{\mathcal{G}})$, such that

$$(\varphi_{\mathcal{G}}(x)u)_e = \varphi_e(x_{r(e)})(u_e), e \in \mathcal{G}^{(1)},$$

for $x \in A_{\mathcal{G}}$ and $u \in X_{\mathcal{G}}$. The C*-correspondence $X_{\mathcal{G}}$ over $A_{\mathcal{G}}$ is called *the graph* C*-*correspondence* with respect to the data

$$\{\mathcal{G}, \{A_p\}_{p\in\mathcal{G}^{(0)}}, \{X_e\}_{e\in\mathcal{G}^{(1)}}\}.$$

In other words, a graph C*-correspondence can be viewed as a graph $\mathcal{G}$ such that on every vertex sits a C*-algebra, on each edge sits a C*-correspondence and the actions and the inner product are defined via the information provided by the graph (they "remember" the form of the graph) as in the following image

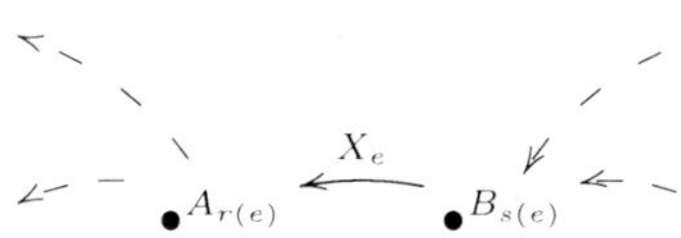


**Example 2.14.** Every C*-correspondence $X$ over $A$ can be visualized in the language of graph C*-correspondences trivially as

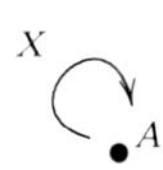


Moreover, every ${}_AX_B$ C*-correspondence can be visualized as

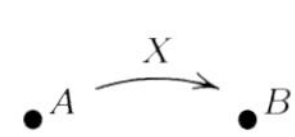


The Fock space of a graph C*-correspondence contains the *path* C*-*correspondences* $X_\mu$, for some path $\mu$. That is if $\mu = xe_n \dots e_1 y$ is a path of the graph $\mathcal{G}$ then $X_\mu := X_{e_n} \otimes_{A_{r(e_n)}} \cdots \otimes_{A_{r(e_2)}} X_{e_1}$ is a $A_x$-$A_y$-correspondence.

**Definition 2.15.** A graph C*-correspondence is called *commutative* if for any two paths $\mu = xe_n \dots e_1 y$ and $\nu = xf_k \dots f_1 y$, that have the same range and source, the corresponding *path* C*-*correspondences* $X_{e_n} \otimes \cdots \otimes X_{e_1}$ and $X_{f_n} \otimes \dots X_{f_1}$ are unitarily equivalent.

For example if $E \approx R \otimes_B S$ for some C*-correspondences ${}_AE_A$, ${}_AR_B$ and ${}_BS_A$, then one can form the commutative graph C*-correspondence

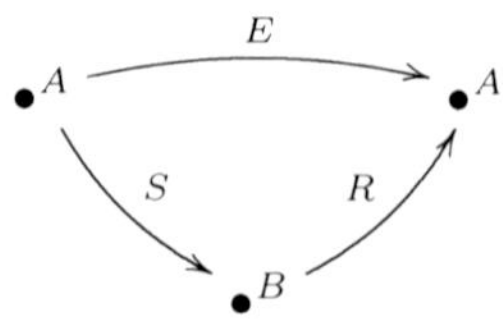


Note that the graph C*-correspondence that comes from this graph does not contain ${}_AE_A$ as a subcorrespondence, though. A graph C*-correspondence that has additionally that property is the following

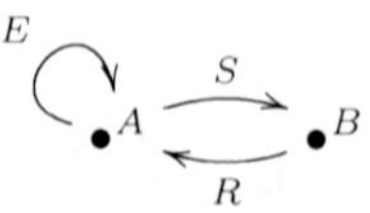


**Example 2.16.** Let $X_\tau$ be the correspondence that occurs from the adding-tail construction established in [28]. It can be visualized in the language of graph C*-correspondences as follows

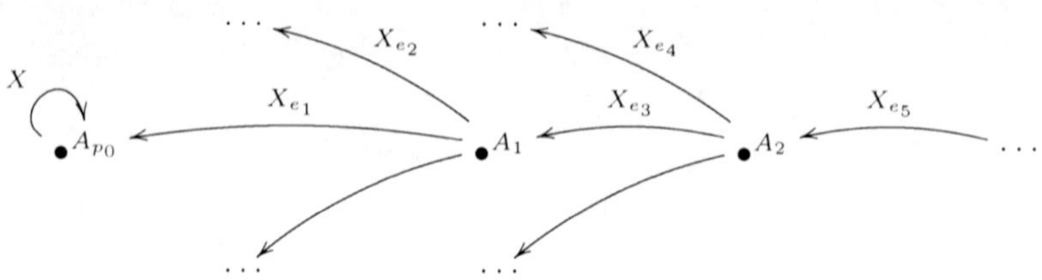

where $A_{p_0} \equiv A$.

**Example 2.17.** (Muhly–Tomforde tail [44]) When $X$ is visualized on a cycle graph, then Muhly–Tomforde tail produces the following graph C*-correspondence

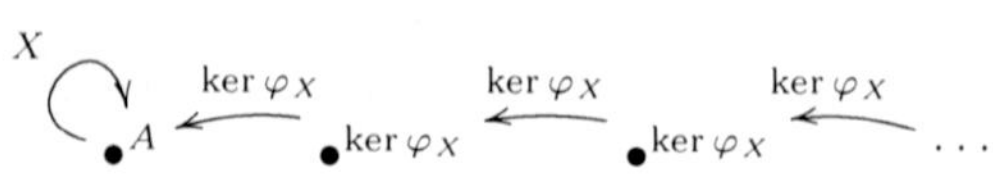

**Example 2.18.** If $(X_\alpha, A)$ is the C*-correspondence canonically associated with a dynamical system $(A, \alpha)$, then the tail produced by Muhly and Tomforde for $(X_\alpha, A)$ comes from the graph

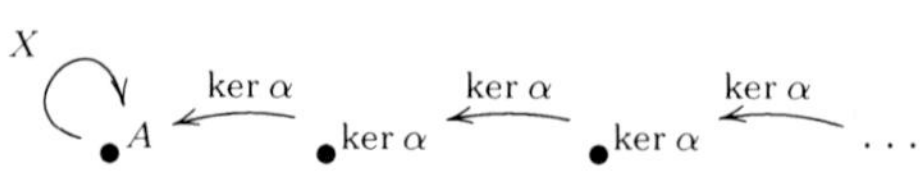

**Example 2.19.** If $(X_\alpha, A)$ is the C*-correspondence canonically associated with a dynamical system $(A, \alpha)$, then the appropriate tail for $(X_\alpha, A)$ comes from

the graph

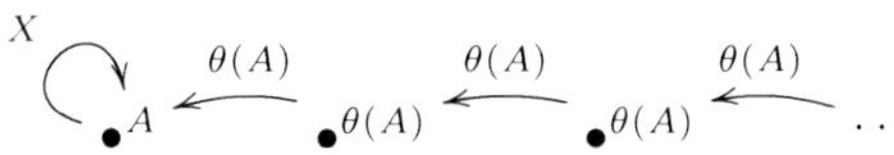


where $\theta\colon A \longrightarrow M(\ker\alpha)$ is the map that extends the natural inclusion $\ker\alpha \subseteq M(\ker\alpha))$ in the multiplier algebra. Therefore, the C*-correspondence $(Y, \varphi_Y, B)$ is the C*-correspondence canonically associated with a dynamical system $(B, \beta)$, where $B = A \oplus c_0(\theta(A))$ and $\beta(a, (x_n)) = (\alpha(a), \theta(a), (x_n))$.

**Example 2.20.** If $(X_\alpha, A)$ is the C*-correspondence canonically associated with a C*-dynamical system $(A, \alpha_1, \alpha_2)$, then the appropriate tail for $(X_\alpha, A)$ comes from the graph

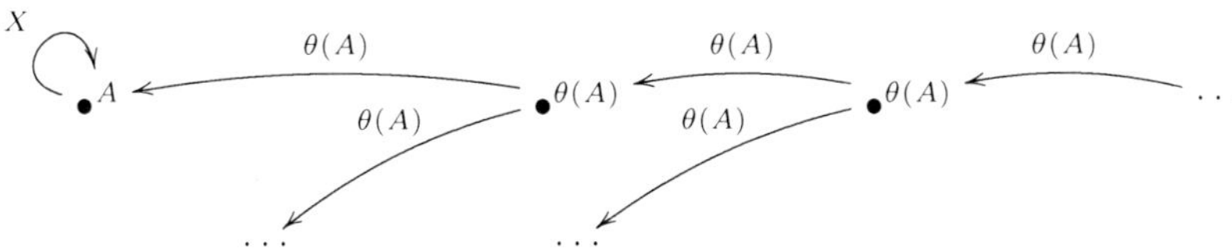


where $\mathcal{J} = \ker\alpha_1 \cap \ker\alpha_2$ and $\theta\colon A \longrightarrow M(\mathcal{J})$ is the map that extends the natural inclusion $\mathcal{J} \subseteq M(\mathcal{J})$ in the multiplier algebra.

## 3. Shift Equivalence Problem

In his pioneering paper [53] Williams studied certain notions of relations for the class of matrices with non-negative integer entries. We say that two such matrices $E$ and $F$ are *elementary strong shift equivalent*, and write $E \overset{\text{s}}{\sim} F$, if there are two matrices $R$ and $S$ such that $E = RS$ and $F = SR$. Whereas this relation is symmetric, it may not be transitive. In [53, Example 2] Williams gives the following counterexample; let

$$\begin{bmatrix} 10 & 2 \\ 2 & 1 \end{bmatrix} \overset{\text{s}}{\sim} \begin{bmatrix} 5 & 6 \\ 4 & 6 \end{bmatrix} \overset{\text{s}}{\sim} \begin{bmatrix} 9 & 4 \\ 3 & 2 \end{bmatrix},$$

but $\begin{bmatrix} 10 & 2 \\ 2 & 1 \end{bmatrix}$ is not elementary strong shift equivalent to $\begin{bmatrix} 9 & 4 \\ 3 & 2 \end{bmatrix}$. The transitive closure $\overset{\text{s}}{\sim}$, denoted by $\overset{\text{SSE}}{\sim}$, implies that $E \overset{\text{SSE}}{\sim} F$ if there is a sequence of matrices $T_i$, $i = 0, \dots, n$, such that $E = T_0$, $F = T_n$ and $T_i \overset{\text{SSE}}{\sim} T_{i+1}$. Williams also defined a third relation, which is proved to be transitive. We say that $E$ is *shift equivalent* to $F$, and write $E \overset{\text{SE}}{\sim} F$, if there are $R, S$ such that $E^n = RS$, $F^n = SR$ and $ER = SF$, $FR = SE$ for some $n \in \mathbb{N}$.

A purpose of [53] was to prove that the relations $\overset{\text{SSE}}{\sim}$ and $\overset{\text{SE}}{\sim}$ are equivalent. Unfortunately, an error in [53] made invalid the proof of a key lemma, and this task remained unsolved for over than 20 years, known as *Williams' Conjecture*. The research interest in this area contributed to the growth of symbolic dynamics and to the search of (complete) invariants for both the equivalence relations. A major change was made by Kim and Roush in [35], where they proved that Williams' Conjecture was false for the class of non-negative integral matrices. Their work suggests that Williams' Conjecture can be renamed as *Shift Equivalence Problem*; i.e., is $\stackrel{\text{SSE}}{\sim}$ equivalent to $\stackrel{\text{SE}}{\sim}$ for a class $\mathfrak{S}$. This formulation is a little vague as one has to extend the definition of the relations described above to *a* class $\mathfrak{S}$.

The notion of elementary and strong shift equivalence for C*-correspondences was studied by Muhly, Tomforde and Pask [40]. In addition they prove that strong shift equivalence of C*-correspondences implies the Morita equivalence of the associated Cuntz–Pimsner algebras, thus extending classical results of Cuntz and Krieger [9], Bates [4] and Drinen and Sieben [16] for graph C*-algebras. The concept of shift equivalence has been studied extensively from both the dynamical and the ring theoretic viewpoint. (See [52] for a comprehensive exposition.) In general, shift equivalence has been recognized to be a more manageable invariant than strong shift equivalence, as it is decidable over certain rings [34]. Unlike strong shift equivalence, the study of shift equivalence, from the viewpoint of C*-correspondences, has been met with limited success [39]. (Other operator theoretic viewpoints however have been quite successful [37].) The concept of strong Morita equivalence for C*-correspondences was first developed and studied by Abadie, Eilers and Exel [1] and Muhly and Solel [43], and plays the role of a generalized Conjugacy (see Example 3.2 below). Among others these authors show that if two C*-correspondences are strong Morita equivalent then the associated Cuntz–Pimsner algebras $\mathcal{O}_E$ and $\mathcal{O}_F$ are (strong) Morita equivalent as well. Let us give the definitions.

**Definition 3.1.** Let the C*-correspondences ${}_AE_A$ and ${}_BF_B$. Then we say that

(1) *E is Morita equivalent to F*, and we write $E \stackrel{\text{SME}}{\sim} F$, if there is an imprimitivity bimodule ${}_AM_B$ such that $E \otimes_A M = M \otimes_B F$.
(2) *E is elementary strong shift equivalent to F*, and we write $E \stackrel{\text{s}}{\sim} F$, if there are ${}_AR_B$ and ${}_BS_A$ such that $E = R \otimes_A S$ and $F = S \otimes_B R$,
(3) *E is strong shift equivalent to F*, and we write $E \stackrel{\text{SSE}}{\sim} F$, if there are $T_i$, $i = 0, \dots n$, such that $T_0 = E, T_n = F$ and $T_i \stackrel{\text{s}}{\sim} T_{i+1}$,
(4) *E is shift equivalent to F with lag m*, and we write $E \stackrel{\text{SE}}{\sim} F$, if there are ${}_AR_B$ and ${}_BS_B$ such that $E^{\otimes m} = R \otimes_B S$, $F^{\otimes m} = S \otimes_A R$ and $E \otimes_A R = R \otimes_B F$, $S \otimes_A E = F \otimes_B S$.

**Example 3.2.** Let ${}_AE_A$ and ${}_BF_B$ be C*-correspondences arising from two dynamical systems $(A, \alpha)$ and $(B, \beta)$ and assume that $E \stackrel{\text{SME}}{\sim} F$ via an imprimitivity bimodule $M$. If we wish $M$ to arise in a similar way then it should be the C*-correspondence associated to a $*$-isomorphism $\gamma \colon A \to B$. As showed in Example 1.13, $E \otimes_A M$ is the C*-correspondence associated to $\gamma \circ \alpha \colon A \to B$ and $M \otimes_B B$ is the C*-correspondence associated to $\beta \circ \gamma \colon A \to B$. Therefore the unitary equivalence $E \otimes_A M \approx M \otimes_B B$ induces a unitary $u \in \mathcal{L}(B) = B$, such that $a \cdot u(1_B) = u(a \cdot 1_B)$, for all $a \in A$; equivalently that $\beta \circ \gamma(a) u = u\gamma \circ \alpha(a)$, for all $a \in A$, hence that the systems $(A, \alpha)$ and $(B, \beta)$ are (outer) conjugate.

In [29] we studied these relations and the interaction between them. Note that $\overset{\text{SME}}{\sim}$, $\overset{\text{SSE}}{\sim}$ and $\overset{\text{SE}}{\sim}$ are equivalence relations for non-degenerate C*-correspondences. One of our main result is the following.

**Theorem 3.3.** [29] *The Shift Equivalence Problem Problem is true for the class of imprimitivity bimodules. In particular, the relations* $\overset{SME}{\sim}, \overset{s}{\sim}, \overset{SSE}{\sim}, \overset{SE}{\sim}$ *coincide.*

A weaker version of the previous theorem holds in general.

**Theorem 3.4.** [29] *Let the* C*-*correspondences* ${}_A E_A$ *and* ${}_B F_B$*; then*

$$E \overset{SME}{\sim} F \Longrightarrow E \overset{s}{\sim} F \Longrightarrow E \overset{SSE}{\sim} F \Longrightarrow E \overset{SE}{\sim} F \cdot$$

In [40] Muhly, Pask and Tomforde have provided a number of counterexamples showing that Morita equivalence of C*-correspondences differs from the elementary strong shift equivalence. The previous result shows that it is in fact stronger.

One of the basic tools we use in [29] is the *Pimsner dilation* of an injective C*-correspondence $X$ to a Hilbert bimodule $X_\infty$. This construction was first introduced by Pimsner in [47]. In [28, Appendix A] we revisited this construction by using direct limits. Assume first that $X$ is non-degenerate and consider the directed system

$$A \xrightarrow{\rho_0} \mathcal{L}(X) \xrightarrow{\rho_1} \mathcal{L}(X^{\otimes 2}) \xrightarrow{\rho_2} \cdots,$$

where

$$\begin{aligned} &\rho_0 = \varphi_X \colon A = \mathcal{L}(A) \longrightarrow \mathcal{L}(X), \\ &\rho_n \colon \mathcal{L}(X^{\otimes n}) \longrightarrow \mathcal{L}(X^{\otimes n+1}) \colon r \longmapsto r \otimes \mathrm{id}_X,\ n \geq 1, \end{aligned}$$

and let $A_\infty$ be the C*-subalgebra of $B = \varinjlim(\mathcal{L}(X^{\otimes n}), \rho_n)$ that is generated by the copies of $\mathcal{K}(X^{\otimes n})$, for $n \in \mathbb{Z}_+$. Consider also the directed system of Banach spaces

$$\mathcal{L}(A, X) \xrightarrow{\sigma_0} \mathcal{L}(X, X^{\otimes 2}) \xrightarrow{\sigma_1} \cdots,$$

where

$$\sigma_n \colon \mathcal{L}(X^{\otimes n}, X^{\otimes n+1}) \to \mathcal{L}(X^{\otimes n+1}, X^{\otimes n+2}) \colon s \mapsto s \otimes \mathrm{id}_X,\ n \geq 1,$$

and let $X_\infty$ be the Banach subalgebra of $Y = \varinjlim(\mathcal{L}(X^{\otimes n}, X^{\otimes n+1}), \sigma_n)$ generated by the copies of $\mathcal{K}(X^{\otimes n}, X^{\otimes n+1})$, for $n \in \mathbb{Z}_+$. Note that the map

$$\partial \colon X \longrightarrow \mathcal{L}(A, X) \colon \xi \longmapsto \partial_\xi,$$

where $\partial_\xi(a) = \xi a$, $\xi \in X$, maps a copy of $X$ isometrically into $\mathcal{K}(A, X) \subseteq X_\infty$. In particular, if $\varphi_X(A) \subseteq \mathcal{K}(X)$, then one can verify that

$$X_\infty = \varinjlim(\mathcal{K}(X^{\otimes n}, X^{\otimes n+1}), \sigma_n) \quad \text{and} \quad A_\infty = \varinjlim(\mathcal{K}(X^{\otimes n}), \rho_n).$$

Thus, in this case, $X_\infty$ is a full left Hilbert bimodule.

If $r \in \mathcal{L}(X^{\otimes n})$, $s \in \mathcal{L}(X^{\otimes n}, X^{\otimes n+1})$ and $[r], [s]$ are their equivalence classes in $B$ and $Y$ respectively, then we define $[s] \cdot [r] := [sr]$. From this, it is easy to define a right $B$-action on $Y$. Similarly, we may define a $B$-valued right inner product on $Y$ by setting

$$\langle [s'], [s] \rangle_Y \equiv [(s')^* s] \in B.$$

for $s, s' \in \mathcal{L}(X^{\otimes n}, X^{\otimes n+1})$, $n \in \mathbb{N}$, and then extending to $Y \times Y$. Finally we define a $*$-homomorphism $\varphi_Y \colon B \to \mathcal{L}(Y)$ by setting

$$\varphi_Y([r])([s]) \equiv [rs], \quad r \in \mathcal{L}(X^{\otimes n}), s \in \mathcal{L}(X^{\otimes n-1}, X^{\otimes n}), n \geq 0$$

and extending to all of $B$ by continuity. We therefore have a left $B$-action on $Y$ and thus $Y$ becomes a C*-correspondence over $B$.

The following diagrams depict the above construction in a heuristic way: the right action is "defined" through the diagram

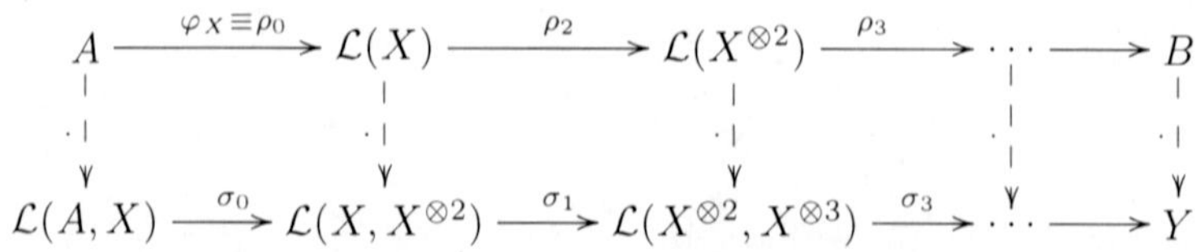


while the left action is "defined" through the diagram

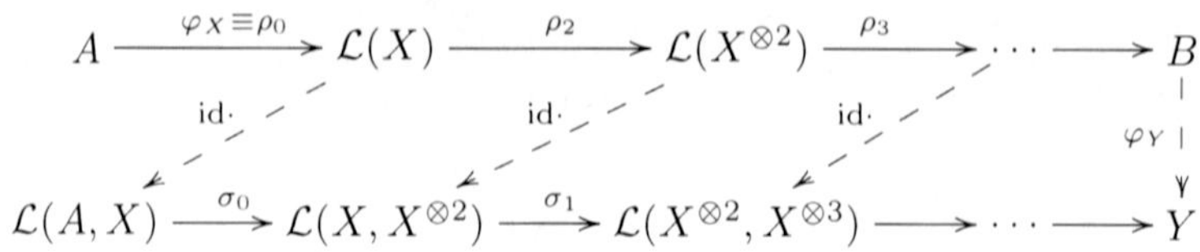


More generally one can use the mapping $\tau \colon X \to \mathcal{L}(X, X^{\otimes 2})$, with $\tau(\xi)\eta = \xi \otimes \eta$, instead of $\sigma_0$, so as to encounter degenerate cases.

For a proof of the following Theorem see [47, Theorem 2.5] or [28, Theorem 6.6], and [29, Section 3]. The main difference between the two approaches is that in [28] we have represented $X \otimes_A A_\infty$ as the direct limit $X_\infty$, thus having no concern in checking the form of the tensor product. Moreover, it appears that "tensoring" $X$ with $A_\infty$ is equivalent to multiplying $X$ with $A_\infty$ in $\varinjlim(\mathcal{L}(X^{\otimes n}, X^{\otimes n+1}), \sigma_n)$.

**Theorem 3.5.** [47, Theorem 2.5] [28, Theorem 6.6] *Let ${}_A X_A$ be an injective* C*-*correspondence and let $X_\infty$ be the $A_\infty$-correspondence defined above. Then $X_\infty$ is an essential Hilbert bimodule and its Cuntz–Pimsner algebra $\mathcal{O}_{X_\infty}$ is $*$-isomorphic to $\mathcal{O}_X$.*

The idea of using direct limits for Pimsner dilation is in complete analogy to the direct limit process for dynamical systems (see [27]).

**Example 3.6.** Let $(A, \alpha)$ denote a dynamical system where $\alpha$ is a $*$-injective endomorphism of $A$. We can define the direct limit dynamical system

$(A_\infty, \alpha_\infty)$ by

$$\begin{array}{ccccccc} A & \xrightarrow{\alpha} & A & \xrightarrow{\alpha} & A & \xrightarrow{\alpha} \cdots \longrightarrow & A_\infty \\ \downarrow{\scriptstyle \alpha} & & \downarrow{\scriptstyle \alpha} & & \downarrow{\scriptstyle \alpha} & & \downarrow{\scriptstyle \alpha_\infty} \\ A & \xrightarrow{\alpha} & A & \xrightarrow{\alpha} & A & \xrightarrow{\alpha} \cdots \longrightarrow & A_\infty \end{array}$$

The limit map $\alpha_\infty$ is an automorphism of $A_\infty$ and extends $\alpha$ (note that $A$ imbeds in $A_\infty$ since $\alpha$ is injective). Then the $A_\infty$-$A_\infty$-correspondence $X_{\alpha_\infty}$, is the Pimsner dilation of $X_\alpha$.

The main question examined in [29] was the following: is it true that $E_\infty \sim F_\infty$ (where $\sim$ may be $\overset{\text{SME}}{\sim}, \overset{\text{s}}{\sim}, \overset{\text{SSE}}{\sim}, \overset{\text{SE}}{\sim}$) if $E \sim F$ (in the same way)? When $E$ and $F$ are regular C*-correspondences we get the following result, by making use of the generalized notion of dilations.

**Theorem 3.7.** [29] *Let ${}_AE_A$ and ${}_BF_B$ be regular* C*-*correspondences. If $E \sim F$ (where $\sim$ may be $\overset{SME}{\sim}, \overset{s}{\sim}, \overset{SSE}{\sim}, \overset{SE}{\sim}$), then $E_\infty \sim F_\infty$ (in the same way).*

For full right, non-degenerate, regular C*-correspondences we have the following.

**Theorem 3.8.** [29] *Let ${}_AE_A$ and ${}_BF_B$ be full right, non-degenerate, regular* C*-*correspondences. Then:*

$$\begin{array}{ccccccc} E \overset{SME}{\sim} F & \Longrightarrow & E \overset{s}{\sim} F & \Longrightarrow & E \overset{SSE}{\sim} F & \Longrightarrow & E \overset{SE}{\sim} F \\ \Downarrow & & \Downarrow & & \Downarrow & & \Downarrow \\ E_\infty \overset{SME}{\sim} F_\infty & \Longleftrightarrow & E_\infty \overset{s}{\sim} F_\infty & \Longleftrightarrow & E_\infty \overset{SSE}{\sim} F_\infty & \Longleftrightarrow & E_\infty \overset{SE}{\sim} F_\infty \end{array}$$

The vertical arrows in Theorem 3.8 are not equivalences in general. Indeed, if $E_\infty \sim F_\infty$ implied $E \sim F$ then, in particular $E \overset{\text{SE}}{\sim} E_\infty$. The following Theorem shows that this happens only trivially, i.e., when $E = E_\infty$.

**Theorem 3.9.** [29] *Let ${}_AE_A$ be a full, non-degenerate and regular* C*-*correspondence. If $E \overset{SE}{\sim} E_\infty$ then $E$ is an imprimitivity bimodule.*

It would be interesting if we could prove the validity of the Shift Equivalence Problem for this class of C*-correspondences (as, after imprimitivity bimodules, it is the next best thing). An obstacle that prevents the construction of counterexamples that would give a negative answer to the Shift Equivalence Problem for non-degenerate and regular correspondences is that the theory of invariants of correspondences is rather poor. The best results (to our opinion) obtained so far are those appearing in [43, 40]. In particular [40, Theorem 3.14] holds for degenerate cases also, as proved in [29]. Note that the additional item (3) below is an immediate consequence of item (2) and item (4) below is an immediate consequence of Theorem 3.5 and Theorem 3.8.

**Theorem 3.10.**

(1) [43, Theorem 3.2] *Let* ${}_AE_A$ *and* ${}_BF_B$ *be non-degenerate, injective* C*-*correspondences. If* $E \overset{SME}{\sim} F$ *then the corresponding Toeplitz–Cuntz–Pimsner algebras and Cuntz–Pimsner algebras are strong Morita equivalent as* C*-*algebras, and the corresponding tensor algebras are strong Morita equivalent in the sense of* [7].

(2) [40, Theorem 3.14], [29] *Let* ${}_AE_A$ *and* ${}_BF_B$ *be regular* C*-*correspondences. If* $E \overset{s}{\sim} F$, *then* $\mathcal{O}_E \overset{SME}{\sim} \mathcal{O}_F$.

(3) [29] *Let* ${}_AE_A$ *and* ${}_BF_B$ *be regular* C*-*correspondences. If* $E \overset{SSE}{\sim} F$, *then* $\mathcal{O}_E \overset{SME}{\sim} \mathcal{O}_F$.

(4) [29] *Let* ${}_AE_A$ *and* ${}_BF_B$ *be full right, non-degenerate, regular* C*-*correspondences. If* $E \overset{SE}{\sim} F$, *then* $\mathcal{O}_E \overset{SME}{\sim} \mathcal{O}_F$.

In particular we obtain the following result for Cuntz–Krieger C*-algebras.

**Corollary 3.11.** *Let* $\mathcal{G}$ *and* $\mathcal{G}'$ *be finite graphs with no sinks or sources and let* $A_{\mathcal{G}}$ *and* $A_{\mathcal{G}'}$ *be their adjacent matrices. If* $A_{\mathcal{G}} \overset{SE}{\sim} A_{\mathcal{G}'}$, *in the sense of Williams, then the Cuntz–Krieger* C*-*algebras* $\mathcal{O}_{\mathcal{G}}$ *and* $\mathcal{O}_{\mathcal{G}'}$ *are strong Morita equivalent.*

There is also a direct application to unital injective dynamical systems.

**Corollary 3.12.** *Let* $(A,\alpha)$ *and* $(B,\beta)$ *be unital injective dynamical systems. If* $X_\alpha \overset{SE}{\sim} X_\beta$, *then* $X_{\alpha_\infty} \overset{SE}{\sim} Y_{\beta_\infty}$ *and the crossed products* $A_\infty \rtimes_{\alpha_\infty} \mathbb{Z}$ *and* $B \rtimes_{\beta_\infty} \mathbb{Z}$ *are strong Morita equivalent.*

Theorem 3.10 shows that Cuntz–Pimsner algebras is a rather coarse invariant. After all, Cuntz–Pimsner algebras are not a complete invariant for the restriction of the relations $\overset{\mathrm{SME}}{\sim}$, $\overset{\mathrm{s}}{\sim}$, $\overset{\mathrm{SSE}}{\sim}$ or $\overset{\mathrm{SE}}{\sim}$ to subclasses of C*-correspondences. For example, let $(A,\alpha)$ be the dynamical system constructed by Hoare and Parry in [22]. Then $\alpha$ is a $*$-isomorphism and $\alpha$ is not conjugate to its inverse. If $E$ is the C*-correspondence of $(A,\alpha)$ and $F$ is the C*-correspondence of $(A,\alpha^{-1})$, then there is not a C*-correspondence $M$ of a dynamical system $(B,\beta)$ such that $E \otimes M \approx M \otimes F$, because then the two dynamical systems would be conjugate. But $\mathcal{O}_A = A \rtimes_\alpha \mathbb{Z}$ is always $*$-isomorphic to $\mathcal{O}_B = A \rtimes_{\alpha^{-1}} \mathbb{Z}$.

On the other hand the tensor algebras of C*-correspondences may be more eligible. They were used in [40] to show that $\overset{\mathrm{s}}{\sim}$ does not imply $\overset{\mathrm{SME}}{\sim}$ and they provide a complete invariant for the conjugacy problem for dynamical systems in various cases, as shown by Davidson and Katsoulis [11, 12], and recently by Davidson and Kakariadis [10]. Moreover, for aperiodic C*-correspondences, Muhly and Solel [43] prove that $\overset{\mathrm{SME}}{\sim}$ is equivalent to strong Morita equivalence of the tensor algebras in the sense of [7]. Along with the further investigation of tensor algebras, it is natural to suggest to work on the development of other invariants such as periodicity, existence of cycles, saturated and/or hereditary submodules etc.

## 4. Hilbert bimodules

Recall that if $X$ is a Hilbert bimodule then

$$I_X = \overline{\operatorname{span}}\{[\xi, \eta] : \xi, \eta \in X\}.$$

If a Hilbert $A$-bimodule $X$ is considered as a C*-correspondence over $A$, then $J_X = I_X$. Hence, for $\xi, \eta \in X$, the element $[\xi, \eta] \in A$ is identified with the unique element $a \in J_X$ such that $\varphi_X(a) = \Theta^X_{\xi,\eta}$. The converse is also true.

The following result is well known.

**Proposition 4.1.** *Let $X$ be a C*-correspondence over $A$. Then the following are equivalent*

(1) *$X$ is a bimodule,*
(2) *$K(X) \subseteq \varphi_X(A)$ and $\varphi_X^{-1}(\mathcal{K}(X)) = \ker \varphi_X \oplus J_X$,*
(3) *the restriction of $\varphi_X$ to $J_X$ is a $*$-isomorphism onto $\mathcal{K}(X)$.*

*In particular, if $\varphi_X$ is injective then $X$ is a Hilbert bimodule if and only if $\mathcal{K}(X) \subseteq \varphi_X(A)$.*

It turns out that the property of a C*-correspondence being a Hilbert bimodule has an important non-selfadjoint operator algebra manifestation. We remind that an operator algebra $\mathfrak{A}$ is called *Dirichlet* if $\mathfrak{A} + \mathfrak{A}^*$ is dense (via a completely isometric homomorphism) in $\mathrm{C}^*_{\mathrm{env}}(\mathfrak{A})$.

**Theorem 4.2.** [26] *Let $X$ be a C*-correspondence over $A$. Then the following are equivalent:*

(1) *$X$ is a Hilbert bimodule,*
(2) *$\psi_t(\mathcal{K}(X)) \subseteq \pi(A)$, for any injective covariant representation $(\pi, t)$ that admits a gauge action,*
(3) *the tensor algebra $\mathcal{T}_X^+$ has the Dirichlet property.*

The above result allows us to correct a misconception regarding semicrossed products and Dirichlet algebras.

**Corollary 4.3.** *Let $(A, \alpha)$ be a dynamical system. Then the semicrossed product $A \times_\alpha \mathbb{Z}_+$ has the Dirichlet property if and only if $\alpha$ is surjective and* $\ker \alpha$ *is orthocomplemented in $A$.*

In particular, when $\alpha$ is injective we deduce that the semi-crossed product has the Dirichlet property if and only if $\alpha$ is onto (thus a $*$-isomorphism). Thus [18, Proposition 3] is false. Nevertheless the main results of [18] are correct since they do not require that Proposition. It remains of interest though to determine whether a semi-crossed product has the unique extension property, since this would allow us to extend Duncan's results to non-commutative dynamics.

Recall that an operator algebra $\mathfrak{A}$ is said to have the *the unique extension property* if the restriction of every faithful representation of $\mathrm{C}^*_{\mathrm{env}}(\mathfrak{A})$ to $\mathfrak{A}$ is maximal, i.e., it has no non-trivial dilations.

**Theorem 4.4 ([26]).** *Let* $(A, \alpha)$ *be a unital injective dynamical system of a* $\mathrm{C}^*$*-algebra. Then the semicrossed product* $A \times_\alpha \mathbb{Z}_+$ *has the unique extension property.*

The unique extension property of an operator algebra (or in general an operator space) implies the existence of the Choquet boundary in the sense of Arveson [3], i.e., the existence of sufficiently many irreducible representations such that their restriction is maximal. Indeed, let $PS(\mathrm{C}^*_{\mathrm{env}}(\mathfrak{A}))$ be the set of the pure states of $\mathrm{C}^*_{\mathrm{env}}(\mathfrak{A})$, and let $\Pi = \oplus_{\tau \in PS(\mathrm{C}^*_{\mathrm{env}}(\mathfrak{A}))} \pi_\tau$ be the free atomic representation of $\mathrm{C}^*_{\mathrm{env}}(\mathfrak{A})$. Then $\Pi$ is faithful on $\mathrm{C}^*_{\mathrm{env}}(\mathfrak{A})$, hence by the unique extension property its restriction to $\mathfrak{A}$ is maximal. Moreover, every $\pi_\tau$ is maximal as a direct summand of a maximal representation. Hence,

$$\|[x_{ij}]\| = \|[\Pi(x_{ij})]\| = \sup\{\|[\pi_\tau(x_{ij})]\| : \tau \in PS(\mathrm{C}^*_{\mathrm{env}}(\mathfrak{A}))\},$$

for all $[x_{ij}] \in M_\nu(\mathfrak{A})$ and $\nu \in \mathbb{N}$.

The existence of the Choquet boundary for separable operator systems (or operator spaces) was proved by Arveson in [3] and it is still an open problem for the non-separable cases. Recently it was proved by Kleski in [36] that the supremum above can be replaced by a maximum, at least for the separable cases, where Arveson's Theorem applies. Note that the semicrossed products can give examples of non-separable operator algebras that have a Choquet boundary.

Finally, we have a result that relates our adding of a tail process to the concept of a Hilbert bimodule.

**Theorem 4.5 ([26]).** *Let* $X$ *be a non-injective* $\mathrm{C}^*$*-correspondence. Then the graph* $\mathrm{C}^*$*-correspondence* $X_\tau$*, as defined in Section 2, is an (essential) Hilbert bimodule if and only if* $X$ *is a Hilbert bimodule and* $|s^{-1}(p)| = |r^{-1}(p)| = 1$ *for every* $p \neq p_0$.

## References


[1] B. Abadie, S. Eilers and R. Exel, *Morita equivalence for crossed products by Hilbert* $\mathrm{C}^*$*-bimodules*, Trans. Amer. Math. Soc. **350** (1998) 3043–3054.

[2] W. Arveson, *Notes on the unique extension property*, 2006, http://math.berkeley.edu /~arveson/Dvi/unExt.pdf.

[3] W. Arveson, *The noncommutative Choquet boundary*, J. Amer. Math. Soc. **21**(4) (2008), 1065–1084.

[4] T. Bates, *Applications of the Gauge-Invariant Uniqueness Theorem*, Bulletin of the Australian Mathematical Society **65** (2002), 57–67.

[5] T. Bates, J. Hong, I. Raeburn, W. Szymanski, *The ideal structure of the* $\mathrm{C}^*$*-algebras of infinite graphs*, Illinois J. Math. **46** (2002), 1159–1176.

[6] D.P. Blecher and C. Le Merdy, *Operator algebras and their modules – an operator space approach*, volume 30 of *London Mathematical Society Monographs, New Series*, The Clarendon Press, Oxford University Press, Oxford, 2004.

[7] D. Blecher, P. Muhly, V. Paulsen, *Categories of operator modules – Morita equivalence and projective modules*, Mem. Amer. Math. Soc. **143** (2000), no 681.

[8] J. Cuntz, *K-theory for certain* $\mathrm{C}^*$*-algebras II*, J. Operator Theory **5** (1981), 101–108.

[9] J. Cuntz and W. Krieger, *A class of* C*-*algebras and topological Markov chains*, Inventiones Math. **56** (1980), 251–268.

[10] K.R. Davidson, E.T.A. Kakariadis, *Conjugate Dynamical Systems on* C*-*algebras*, Int. Math. Res. Not., to appear.

[11] K.R. Davidson and E.G. Katsoulis, *Isomorphisms between topological conjugacy algebras*, J. Reine Angew. Math. (Crelle), **621** (2008), 29–51.

[12] K.R. Davidson and E.G. Katsoulis, *Semicrossed Products of Simple* C*-*algebras*, Math. Ann. **342** (2008), 515–525.

[13] K.R. Davidson, E.G. Katsoulis, *Operator algebras for multivariable dynamics*, Mem. Amer. Math. Soc. **209**, (2011), no. 983.

[14] K.R. Davidson, J. Roydor, C*-*envelopes of tensor algebras for multivariable dynamics*, Proc. Edinb. Math. Soc. (2) **53** (2010), 333–351; corrigendum Proc. Edinb. Math. Soc. (2) **54** (2011) 643–644.

[15] V. Deaconu, A. Kumjian, D. Pask, A. Sims, *Graphs of* C*-*correspondences and Fell bundles*, Indiana Univ. Math. J. **59** (2010), 1687–1736.

[16] D. Drinen and N. Sieben, C*-*equivalences of graphs*, J. Operator Theory **45** (2001), 209–229.

[17] M.A. Dritschel, S.A. McCullough, *Boundary representations for families of representations of operator algebras and spaces*, J. Operator Theory, **53**(1) (2005), 159–167.

[18] B.L. Duncan, C*-*envelopes of universal free products and semicrossed products for multivariable dynamics*, Indiana Univ. Math. J. **57**(4) (2008), 1781–1788.

[19] N. Fowler, P. Muhly, I. Raeburn, *Representations of Cuntz–Pimsner algebras* Indiana Univ. Math. J. **52** (2003), 569–605.

[20] N. Fowler, I. Raeburn, *The Toeplitz algebra of a Hilbert bimodule*, Indiana Univ. Math. J. **48** (1999), 155–181.

[21] M. Hamana, *Injective envelopes of operator systems*, Publ. RIMS Kyoto Univ. **15**(1979), 773–785.

[22] H. Hoare, W. Parry, *Affine transformations with quasi-discrete spectrum. I*, J. London Math. Soc. **41** (1966), 88–96.

[23] E.T.A. Kakariadis, *Semicrossed products and reflexivity*, J. Operator Theory **67**(2) (2012), 379–395.

[24] E.T.A. Kakariadis, *Semicrossed products of* C*-*algebras and their* C*-*envelopes*, preprint.

[25] E.T.A. Kakariadis, *The Šilov Boundary for Operator Spaces*, Int. Eq. Op. Th., to appear.

[26] E.T.A. Kakariadis, *The Dirichlet Property for tensor algebras*, Bulletin of London Math. Soc., to appear.

[27] E.T.A. Kakariadis, E.G. Katsoulis *Semicrossed products of operator algebras and their* C*-*envelopes*, J. Funct. Anal., **262**(7) (2012), 3108–3124.

[28] E.T.A. Kakariadis, E.G. Katsoulis, *Contributions to the theory of* C*-*correspondences with applications to multivariable dynamics*, Trans. Amer. Math. Soc. **364** (2012), 6605–6630.

[29] E.T.A. Kakariadis, E.G. Katsoulis, C*-*algebras and Equivalences for* C*-*correspondences*, preprint.

[30] E. Katsoulis and D. Kribs, *Isomorphisms of algebras associated with directed graphs*, Math. Ann. **330**, (2004), 709–728.

[31] E.G. Katsoulis, D. Kribs, *Tensor algebras of* C**-correspondences and their* C**-envelopes*, J. Funct. Anal. **234**(1) (2006), 226–233.

[32] T. Katsura, *On $C^*$-algebras associated with* C**-correspondences*, Contemp. Math. **335** (2003), 173–182.

[33] T. Katsura, *On $C^*$-algebras associated with* C**-correspondences*, J. Funct. Anal. **217**(2) (2004), 366–401.

[34] K.H. Kim, F.W. Roush, *Decidability of shift equivalence*, Dynamical systems (College Park, MD, 1986–87), 374–424, Lecture Notes in Math. 1342, Springer, Berlin, 1988.

[35] K.H. Kim, F.W. Roush *The Williams conjecture is false for irreducible subshifts*, Ann. of Math **149**(2) (1999), 545–558.

[36] C. Kleski, *Boundary representations and pure completely positive maps*, J. Operator Theory, to appear.

[37] W. Krieger, *On dimension functions and topological Markov chains*, Invent. Math. **56** (1980), 239–250.

[38] C. Lance, *Hilbert* C**-modules. A toolkit for operator algebraists*, London Mathematical Society Lecture Note Series, **210** Cambridge University Press, Cambridge, 1995. x+130 pp. ISBN: 0-521-47910-X.

[39] K. Matsumoto, *Actions of symbolic dynamical systems on* C**-algebras*, J. Reine Angew. Math. **605** (2007), 23–49.

[40] P.S. Muhly, D. Pask, M. Tomforde, *Strong Shift Equivalence of* C**-correspondences*, Israel J. Math. **167** (2008), 315–345.

[41] P.S. Muhly, B. Solel, *Tensor algebras over* C**-correspondences: representations, dilations and* C**-envelopes* J. Funct. Anal. **158** (1998), 389–457.

[42] P.S. Muhly, B. Solel, *On the simplicity of some Cuntz–Pimsner algebras*, Math. Scand. **83** (1998), 53–73.

[43] P.S. Muhly, B. Solel, *On the Morita Equivalence of Tensor Algebras*, Proc. London Math. Soc. **3** (2000), 113–168.

[44] P.S. Muhly, M. Tomforde, *Adding tails to* C**-correspondences*, Doc. Math. **9** (2004), 79–106.

[45] W. Paschke, *The crossed product of a* C**-algebra by an endomorphism* Proc. Amer. Math. Soc. **80** (1980), 113–118.

[46] J. Peters, *Semicrossed products of* C**-algebras*, J. Funct. Anal. **59** (1984), 498–534.

[47] M. Pimsner, *A class of $C^*$-algebras generalizing both Cuntz–Krieger algebras and crossed products by* $\mathbb{Z}$, Free probability theory (Waterloo, ON, 1995), 189–212, Fields Inst. Commun., **12**, Amer. Math. Soc., Providence, RI, 1997.

[48] G. Popescu, *Non-commutative disc algebras and their representations*, Proc. Amer. Math. Soc. **124**, (1996), 2137–2148.

[49] I. Raeburn, *Graph algebras*, CBMS Regional Conference Series in Mathematics, 103, 2005.

[50] I. Raeburn, W. Szymanski, *Cuntz–Krieger algebras of infinite graphs and matrices*, Trans. Amer. Math. Soc. **356** (2004) 39–59.

[51] P. Stacey, *Crossed products of* $C^*$*-algebras by endomorphisms*, J. Austr. Math. Soc., Ser. A **56** (1993), 204–212.

[52] J.B. Wagoner, *Strong shift equivalence theory and the shift equivalence problem*, Bull. Amer. Math. Soc. (N.S.) **36** (1999), 271–296.

[53] R. Williams, *Classification of subshifts of finite type*, Annals of Math. 98 (1973), 120–153; erratum, Annals of Math. 99 (1974), 380–381.

Evgenios T.A. Kakariadis
Pure Mathematics Department
University of Waterloo
ON N2L-3G1, Canada
e-mail: ekakaria@uwaterloo.ca

Elias G. Katsoulis
Department of Mathematics
University of Athens
15784 Athens, Greece

*Alternate address*:

Department of Mathematics
East Carolina University
Greenville, NC 27858, USA
e-mail: katsoulise@ecu.edu

Operator Theory:
Advances and Applications, Vol. 233, 75–84


# Some Operator Algebras from Semigroups

A. Katavolos

**Abstract.** This is a study of reflexivity and structure properties of operator algebras generated by representations of the Heisenberg semigroup. We briefly revise earlier joint work with S.C. Power [14] on the continuous Heisenberg semigroup. We then show that the (restricted) left regular representation of the discrete Heisenberg semigroup gives rise to a reflexive operator algebra, which is semisimple. An example of a representation giving rise to a non-reflexive algebra is presented.

Report on joint work with M. Anoussis (Aegean) and I.G. Todorov (Belfast) [2].

**Mathematics Subject Classification (2010).** Primary 47L75; Secondary 43A65, 47L99.

**Keywords.** Reflexive algebras; Heisenberg semigroup; Weyl relations; Fourier binest algebra; irrational rotation algebra.

## 1. Introduction: From semigroups to operator algebras

The theory of unitary representations of groups on Hilbert space has been a motivating force in the development of operator algebra theory since the early years of its history.

Unitary representations of groups very often arise from the dynamics of physical systems; for example, the time evolution might be represented by a one-parameter group $\{U_t : t \in \mathbb{R}\}$ of unitary operators on a Hilbert space $H$. In case the dynamical system is irreversible, often the operators $U_t$ are only defined for $t \geq 0$; this leads to a unitary representation of the semigroup $\mathbb{R}_+$. On the other hand, often there are subspaces $M \subseteq H$ which are invariant under the semigroup $\{U_t : t \geq 0\}$, but not under the whole group $\{U_t : t \in \mathbb{R}\}$. In such cases, it is of interest to analyse the semigroup $\{U_t|_M : t \geq 0\}$. The operators $U_t|_M$ are usually non-unitary isometries; as is well known, these are very different objects.

Presented at the conference *"Operator Theory and its Applications" in honour of Victor Shulman*, Gothenburg, April 2011.

We are thus led to the following

**Definition 1.1.** Let $\mathbb{G}$ be a group and $\mathbb{S} \subseteq \mathbb{G}$ a (unital) semigroup. A *representation* of $\mathbb{S}$ on a Hilbert space $H$ is a unital morphism

$$\sigma : \mathbb{S} \to B(H) : g \to S_g,$$

such that each $S_g$ is an isometry.

Unitary representations of *groups* give rise to operator algebras that are selfadjoint: C*-algebras or von Neumann algebras. By contrast, representations of *semigroups* typically (but not always, see case (3) in Section 5) generate nonselfadjoint operator algebras. Both the norm closed and the weak-* closed algebras are of interest. In the present work, we concentrate mainly on the weak-* closed case.

**Notation** We denote by $W(\sigma, \mathbb{S}) \subseteq B(H)$ the w*-closed algebra generated by $\{S_g : g \in \mathbb{S}\}$.

The symbol $\mathrm{Lat}(W(\sigma, \mathbb{S}))$ denotes the set of all closed *invariant subspaces* of the representation, namely, all closed subspaces $M \subseteq H$ such that $S_g(M) \subseteq M$ for all $g \in \mathbb{S}$.

## 2. Some examples

### 2.1. The analytic Toeplitz algebra

The simplest semigroup is $\mathbb{Z}_+$; it acts on $\ell^2(\mathbb{Z}_+)$ by the left regular representation:

$$n \to \lambda_n = S^n, \quad \text{where} \quad S : e_k \to e_{k+1},$$

the unilateral shift. In this case, we have an explicit characterisation of the invariant subspace lattice: if we identify $\ell^2(\mathbb{Z}_+)$ with the usual Hardy space $H^2 \subseteq L^2(\mathbb{T})$ by Fourier transform, $S$ corresponds to the operator $T_\zeta$ given by $(T_\zeta f)(z) = zf(z)\,(f \in H^2)$. The classical result of Beurling provides a complete characterisation of the invariant subspaces of $T_\zeta$ in terms of unimodular inner functions:

**Theorem 2.1 (Beurling [8], 1949).** *If $M \in \mathrm{Lat}(W(\lambda, \mathbb{Z}_+))$ (i.e., $S(M) \subseteq M$), there exists $\phi \in H^2$ with $|\phi(z)| = 1$ for almost all $z \in \mathbb{T}$ so that (after Fourier transform)*

$$M = \phi H^2.$$

In the case of the semigroup $\mathbb{Z}_+^2$, although there are partial results for the invariant subspaces of two commuting shifts, an analogous parametrisation of *all* invariant subspaces is not available: In particular, the two-variable analogue of Beurling's theorem does not hold [21, Theorem 4.4.1]. Some partial results may be found in [1, 10, 11, 13, 15, 18, 19] and their references.

Let us move on to the dual question: Which operators on $H^2$ leave invariant *all* shift-invariant subspaces? Obviously, all analytic Toeplitz operators $T_f$ ($f \in H^\infty$) have this property (they are w*-limits of polynomials in the shift $T_\zeta$). The fact that there are no others was settled in a paper that could be considered the first publication in non-selfadjoint operator algebra theory:

**Theorem 2.2 (Sarason [22], 1966).** *If $T \in B(\ell^2(\mathbb{Z}_+))$ satisfies $T(M) \subseteq M$ for all $M \in \mathrm{Lat}(\lambda(\mathbb{Z}_+))$ then $T$ is in the $w^*$-closed algebra $W(\lambda, \mathbb{Z}_+)$ generated by $\{\lambda_n : n \in \mathbb{Z}_+\}$.*

In the terminology introduced by Halmos [12], Sarason's result says that the algebra $W(\lambda, \mathbb{Z}_+)$ acting on $\ell^2(\mathbb{Z}_+)$ is *reflexive*. Halmos' definition was extended by Loginov and Shulman as follows.

**Definition 2.1 (Loginov–Shulman [17], 1975).** The *reflexive cover* of a subspace $\mathcal{W}$ of $B(H)$ is defined to be

$$\mathrm{Ref}\,\mathcal{W} := \{T \in B(H) : Tx \in \overline{\mathcal{W}x} \quad \text{for all} \quad x \in H\}.$$

A subspace $\mathcal{W}$ of $B(H)$ is called *reflexive* if $\mathcal{W} = \mathrm{Ref}\,\mathcal{W}$.

Contrary to the failure of a 'multidimensional Beurling theorem', Sarason's result has wide-ranging generalisations. In the abelian case, not only the regular representation, *any* representation gives a reflexive algebra:

**Theorem 2.3 (Commuting isometries; Bercovici [5], 1994/Li–McCarthy [16], 1997).** *For all $d \in \mathbb{N}$ and any representation $\pi$ of $\mathbb{Z}_+^d$, the algebra $W(\pi, \mathbb{Z}_+^d)$ is reflexive.*

### 2.2. The free semigroup algebra

Let us now move to the 'extreme non-abelian situation': consider the *free semigroup* $\mathbb{F}_n$ *in $n$ generators* $a_1, \dots, a_n$ and its semigroup $\mathbb{F}_n^+$ consisting of all *words* $w$ in the generators (and the identity). The semigroup $\mathbb{F}_n^+$ acts naturally on itself by left multiplication (concatenation) and hence on the Hilbert space $\ell^2(\mathbb{F}_n^+)$ by the *left regular representation*: $\ell^2(\mathbb{F}_n^+)$ has an orthonormal basis $\{e_w : w \in \mathbb{F}_n^+\}$ indexed by the elements of $\mathbb{F}_n^+$ and we define

$$\lambda_w : \ell^2(\mathbb{F}_n^+) \to \ell^2(\mathbb{F}_n^+) : \quad \lambda_w : e_{w'} \to e_{ww'}.$$

With this terminology, the generalisation of Beurling's theorem is

**Proposition 2.4 (Popescu [20], 1989).** *Every $M \in \mathrm{Lat}(W(\lambda, \mathbb{F}_n^+))$ is a direct sum of cyclic subspaces, each of the form $M_w = U(\ell^2(\mathbb{F}_n^+))$ where $U$ is an isometry commuting with all $\lambda_w$.*

To see that this is the natural generalisation of Beurling's theorem, notice that each subspace of $H^2$ of the form $\phi H^2$ is cyclic ($\phi$ is the cyclic vector) and is the range of the isometry $T_\phi$; moreover these $T_\phi$ are the only isometries on $H^2$ that commute with the shift.

On the operator algebra side (which is nearer to the topic of our work) we have a generalisation of Sarason's result:

**Proposition 2.5 (Arias–Popescu, [3], 1995).** *The algebra $W(\lambda, \mathbb{S})$ is reflexive.*

In fact $W(\lambda, \mathbb{S})$ is hyper-reflexive (Davidson–Pitts [9]) with constant at most 3 (Bercovici [6]). Hyper-reflexivity is a property much stronger and rarer than reflexivity, which need not concern us here.

## 3. The Heisenberg semigroup

We shall now focus on an intermediate situation between abelian groups and free groups: the Heisenberg group. This is important both for its historical roots in quantum mechanics (recall the 'Heisenberg commutation relations' or its integrated form the 'Weyl relations' – see below), but also for producing interesting operator algebras. We will consider the discrete case (based on $\mathbb{Z}$) as well as the continuous case (based on $\mathbb{R}$).

**Definition 3.1.** The Heisenberg group $\mathbb{H}$ consists of all matrices

$$[x, y, z] \equiv \begin{bmatrix} 1 & x & z \\ 0 & 1 & y \\ 0 & 0 & 1 \end{bmatrix}$$

where if $x, y, z$ are allowed to vary over $\mathbb{R}$ we call $\mathbb{H}$ *continuous*, while if $x, y, z \in \mathbb{Z}$ we call $\mathbb{H}$ *discrete.*

The *semigroup* $\mathbb{H}_+$ consists of all $[x, y, z]$ with $x, y \geq 0$ (but $z$ free).

Note that $[0, 0, 1]$ generates the centre of $\mathbb{H}$.

**The continuous Heisenberg semigroup**

For $t, s \in \mathbb{R}_+$ consider the operators on $L^2(\mathbb{R})$ given by:

$$(D_t f)(x) = f(x - t) \qquad \text{Translations}$$

$$(M_s f)(x) = e^{isx} f(x) \qquad \text{Multiplications} \quad (f \in L^2(\mathbb{R})).$$

These give rise to the representation $\sigma$ of the continuous semigroup $\mathbb{H}_+ = \{[s, t, \mu] : s, t \in \mathbb{R}_+, \mu \in \mathbb{R}\}$ on $L^2(\mathbb{R})$ as follows:

$$\sigma : [s, t, \mu] = \begin{bmatrix} 1 & s & \mu \\ 0 & 1 & t \\ 0 & 0 & 1 \end{bmatrix} \longrightarrow e^{i\mu} D_t M_s \,.$$

The operators $M_s$ and $D_t$ satisfy the famous

$$\text{Weyl relations :} \quad M_s D_t = e^{ist} D_t M_s.$$

The weak*-closed algebra $W(\sigma, \mathbb{H}_+)$ is the weak-* closed linear span of $\{M_s D_t : s, t \geq 0\}$. When looking for common invariant subspaces of these two one-parameter families, one immediately notices two chains of subspaces: $(i)$ the *Volterra nest* $N_v$ consisting of all subspaces of $L^2(\mathbb{R})$ of the form $L^2([a, \infty)) = D_a(L^2(\mathbb{R}_+))$, $a \in \mathbb{R}$ together with the 'endpoints' 0 and $L^2(\mathbb{R})$ and $(ii)$ its Fourier transform, the *analytic nest* $N_a = \{M_b(H^2(\mathbb{R})) : b \in \mathbb{R}\} \cup \{0, L^2(\mathbb{R})\}$. Let us picture these as two semicircles meeting only at the endpoints: the South pole 0 and the North pole $L^2(\mathbb{R})$. Are there any other invariant subspaces? Yes! For any $c > 0$, if we let $\phi_c(x) = e^{-icx^2}$, the subspace $\phi_c H^2(\mathbb{R})$ is also invariant under both families; thus we get another 'intermediate' chain $N_c = \{\phi_c K : K \in N_a\}$ of common invariant subspaces. We may picture $N_c$ as a semicircle between $N_v$ and $N_a$ meeting them only at the poles. Now, we have found all common invariant

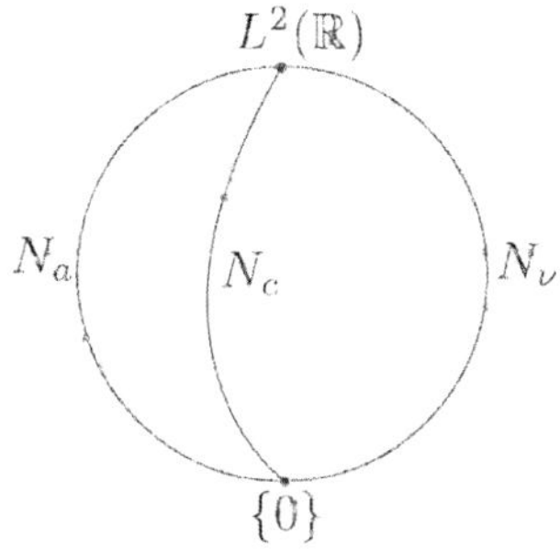


FIGURE 1. The onion (picture several $N_c$'s)

subspaces; and not only that, but the picture of the two 'boundary' semicircles $N_v$ and $N_a$ together with the intermediate longitudinal semicircles $N_c$ (*the onion!*) turns out to be 'topologically correct':

**Theorem 3.1 (A.K. & S.C. Power, [14]).**

(i) *The set* $\mathrm{Lat}\{M_s, D_t : s, t \geq 0\}$ *equals the union* $\bigcup_{c>0} N_c \cup N_v \cup N_a$. *Considered as a set of projections in the strong operator topology, it is a topological manifold homeomorphic to the closed disc.*

(ii) *The* $w^*$*-closed algebra* $W(\sigma, \mathbb{H}_+)$ *generated by* $\{M_s, D_t : s, t \geq 0\}$ *is reflexive. In fact, it consists of all operators leaving invariant both chains (or nests)* $N_v$ *and* $N_a$*: the* Fourier binest algebra.

For part (ii), one should notice that $W(\sigma, \mathbb{H}_+)$ contains no non-trivial self-adjoint or finite rank operators. In order to prove reflexivity, one first obtains a complete characterisation of Hilbert Schmidt operators in $W(\sigma, \mathbb{H}_+)$ as having 'bianalytic' kernels in a certain sense, and then one constructs an approximate identity consisting of such bi-analytic integral operators. Thus $W(\sigma, \mathbb{H}_+)$ is generated, as a w*-closed linear space, by two non-commuting copies of $H^\infty(\mathbb{R})$.

## 4. The discrete Heisenberg semigroup

Henceforth the symbol $\mathbb{H}_+$ will stand for the discrete Heisenberg semigroup; it has three generators $u = [1, 0, 0], v = [0, 1, 0]$ and $w = [0, 0, 1]$ satisfying the relation $uv = wvu$.

### 4.1. The restricted left regular representation $\mathcal{T}_L(\mathbb{H}_+)$: Algebraic properties.

This representation is defined by the left action of $\mathbb{H}_+$ on $\ell^2(\mathbb{H}_+) \simeq \ell^2(\mathbb{Z}) \otimes \ell^2(\mathbb{Z}_+) \otimes \ell^2(\mathbb{Z}_+)$ given by

$$\begin{aligned} L_w(w^n \otimes u^k \otimes v^m) &= w^{n+1} \otimes u^k \otimes v^m \\ L_u(w^n \otimes u^k \otimes v^m) &= w^n \otimes u^{k+1} \otimes v^m \\ L_v(w^n \otimes u^k \otimes v^m) &= w^{n-k} \otimes u^k \otimes v^{m+1} \qquad (n, k, m) \in \mathbb{Z} \times \mathbb{Z}_+ \times \mathbb{Z}_+. \end{aligned}$$

Here we have identified the elements of the group with the elements of the canonical basis of $\ell^2(\mathbb{Z}) \otimes \ell^2(\mathbb{Z}_+) \otimes \ell^2(\mathbb{Z}_+)$; thus $w^n u^k v^m \in \mathbb{H}^+$ is identified with the elementary tensor $w^n \otimes u^k \otimes v^m$.

It is convenient, in this subsection, to make a Fourier transform in the first variable (corresponding to $w$). Thus $\mathbb{H}_+$ is considered as acting on $L^2(\mathbb{T}) \otimes \ell^2(\mathbb{Z}_+) \otimes \ell^2(\mathbb{Z}_+)$ and the operator $L_w$ is identified with $M_\xi \otimes 1 \otimes 1$, where $M_\xi h(z) = zh(z)$, $h \in L^2(\mathbb{T})$. More generally, for $f \in L^\infty(\mathbb{T})$, we write $L_f(w^n \otimes u^k \otimes v^m) = fw^n \otimes u^k \otimes v^m$.

We define $\mathcal{T}_L(\mathbb{H}_+)$ to be the weak-* closed algebra of bounded operators on $L^2(\mathbb{T}) \otimes \ell^2(\mathbb{Z}_+) \otimes \ell^2(\mathbb{Z}_+)$ generated by $L_w, L_u$ and $L_v$; it is unitarily equivalent to the algebra $W(\lambda, \mathbb{H}_+)$.

The main properties of the algebra $\mathcal{T}_L(\mathbb{H}_+)$ are the following:

**Fourier expansion** *A basic tool in our analysis is a 'two-dimensional Fourier expansion' of each element* $A \in \mathcal{T}_L(\mathbb{H}_+)$*:*

$$A \sim \sum_{k \geq 0, m \geq 0} L_{f_{k,m}} L_u^k L_v^m$$

*with centre-valued coefficients* $L_{f_{k,m}}$. *This expansion is Abel – summable to* $A$ *in the weak-** *topology.*

**Proposition 4.1.** *The diagonal* $\mathcal{T}_L(\mathbb{H}_+) \cap \mathcal{T}_L(\mathbb{H}_+)^*$ *and the centre of* $\mathcal{T}_L(\mathbb{H}_+)$ *are both equal to* $\mathcal{M} \otimes \mathbf{1} \otimes \mathbf{1}$ *where* $\mathcal{M}$ *is the multiplication algebra of* $L^\infty(\mathbb{T})$ *acting on* $L^2(\mathbb{T})$.

By comparison, the Fourier binest algebra defined in the previous section has trivial diagonal.

**Proposition 4.2.** *The algebra* $\mathcal{T}_L(\mathbb{H}_+)$ *contains no compact operators.*

Compare this with the fact that the Fourier binest algebra has an approximate identity consisting of Hilbert-Schmidt operators.

**Theorem 4.3.** *The algebra* $\mathcal{T}_L(\mathbb{H}^+)$ *contains no nonzero quasinilpotent operators. In particular, it is semi-simple.*

**Theorem 4.4.** *The algebra* $\mathcal{T}_L(\mathbb{H}^+)$ *has the bicommutant property* $\mathcal{T}_L(\mathbb{H}^+)'' = \mathcal{T}_L(\mathbb{H}^+)$.

It follows that $\mathcal{T}_L(\mathbb{H}^+)$ is inverse closed: if $A \in \mathcal{T}_L(\mathbb{H}^+)$ is a bijective operator, then $A^{-1} \in \mathcal{T}_L(\mathbb{H}^+)$.

It also follows that $\mathcal{T}_L(\mathbb{H}^+)$ is closed in the weak operator topology (a necessary – but far from sufficient – condition for reflexivity).

### 4.2. The restricted left regular representation $\mathcal{T}_L(\mathbb{H}_+)$: Reflexivity

Let us now make a Fourier transform in all three variables;

$$\ell^2(\mathbb{H}_+) \simeq \ell^2(\mathbb{Z}) \otimes \ell^2(\mathbb{Z}_+) \otimes \ell^2(\mathbb{Z}_+) \sim L^2(\mathbb{T}) \otimes H^2 \otimes H^2 \simeq L^2(\mathbb{T}, \mathcal{K}),$$

where $\mathcal{K} = H^2 \otimes H^2 \simeq H^2(\mathbb{T} \times \mathbb{T})$. We consider $L^2(\mathbb{T}, \mathcal{K})$ as a direct integral over $(\mathbb{T}, \mu)$ of the constant field $\xi \to \mathcal{K}(\xi) = \mathcal{K}$ of Hilbert spaces.

Let $W, U, V \in \mathcal{B}(L^2(\mathbb{T}, \mathcal{K}))$ be the images of the generators $L_w, L_u, L_v$ of $\mathcal{T}_L(\mathbb{H}_+)$ after this unitary equivalence. Thus $W$ is diagonalised as $M_\xi \otimes I \otimes I$ and so on each "fiber" $\mathcal{K}(\xi)$, $\xi \in \mathbb{T}$ the three generators become:

$$\begin{aligned} (W_\xi f)(z_1, z_2) &= \xi f(z_1, z_2), & & \\ (U_\xi f)(z_1, z_2) &= z_1 f(z_1, z_2), & \qquad (z_1, z_2) &\in \mathbb{T} \times \mathbb{T} \\ (V_\xi f)(z_1, z_2) &= z_2 f(\bar{\xi} z_1, z_2), & \qquad f &\in H^2(\mathbb{T} \times \mathbb{T}) \end{aligned}$$

**Theorem 4.5.** *The $w^*$-closed algebra $\mathcal{W}_\xi$ generated by $U_\xi, V_\xi \in \mathcal{B}(\mathcal{K})$ is reflexive for each $\xi$.*

Using this, together with the fact that the direct integral of reflexive algebras is reflexive [4], we are able to show

**Theorem 4.6.** *The $w^*$-closed algebra $\mathcal{T}_L(\mathbb{H}_+)$ is reflexive.*

### 4.3. Reflexivity: proof of Theorem 4.5 (sketch)

It is convenient to use the (unitarily equivalent) representation $\mathcal{K} \simeq H^2 \otimes H^2$. Thus the algebra $\mathcal{W}_\xi$ is generated by

$$U_\xi(u^k \otimes v^m) = u^{k+1} \otimes v^m \quad \text{and} \quad V_\xi(u^k \otimes v^m) = (\bar{\xi} u)^k \otimes v^{m+1}$$

(here $u(z_1) = z_1$, $v(z_2) = z_2$). Thus

$$U_\xi = S \otimes I \quad \text{and} \quad V_\xi = D \otimes S$$

where $S$ is the unilateral shift and $(Df)(z_1) = f(\bar{\xi} z_1)$ acting on $H^2$.

To prove that $\mathcal{W}_\xi$ is reflexive, let $T$ be in $\operatorname{Ref} \mathcal{W}_\xi$. This has a formal 'Fourier series'

$$T \sim \sum_{n \geq 0} \widehat{T}_n \otimes S^n$$

where the 'Fourier coefficients' $\widehat{T}_n$ are in $B(H^2)$. Averaging over the dual action $\{R_s : e^{is} \in \mathbb{T}\}$ given by $R_s(u^k \otimes v^m) = u^k \otimes (e^{is} v)^m$, we show that $\widehat{T}_n \otimes S^n \in \operatorname{Ref} \mathcal{W}$.

The next step is to show that $D^{-n} \widehat{T}_n$ is in fact an analytic Toeplitz operator $T_{f_n} = f_n(S)$ (where $f_n \in H^\infty$). Thus we may write $\widehat{T}_n \otimes S^n = D^n f_n(S) \otimes S^n = V^n f_n(U)$, a function of $U$ and $V$, and so $\widehat{T}_n \otimes S^n \in \mathcal{W}$.

But now as in the classical Féjer theorem, the Fourier series of $T$ is Cesàro – summable to $T$ in the weak-* topology. Therefore $T \in \mathcal{W}$.

## 5. A class of representations for the discrete Heisenberg semigroup

We now proceed to study a class of representations of $\mathbb{H}^+$ which arise from representations of the irrational rotation algebra studied by Brenken [7].

Fix $\lambda = e^{2\pi i\theta}$ where $\theta$ is irrational. Given a Borel probability measure $\nu$ on $\mathbb{T}$ which is quasi-invariant and ergodic under 'irrational rotations' $z \to \lambda z$ we define a representation $\pi_\nu$ of $\mathbb{H}_+$ on $L^2(\mathbb{T}, \nu)$ by

$$[k, m, n] = \begin{bmatrix} 1 & k & n \\ 0 & 1 & m \\ 0 & 0 & 1 \end{bmatrix} \xrightarrow{\pi_\nu} \lambda^n V^m U^k$$

where

$$\begin{aligned} (Vf)(z) &= \sqrt{r_\lambda(z)} f(\lambda z) && \text{Rotation} \\ (Uf)(z) &= z f(z) && \text{Multiplication} \qquad (f \in L^2(\mathbb{T}, \nu)) \end{aligned}$$

where $r_\lambda$ is a Radon–Nikodym factor that makes $V$ isometric. The generators $U, V$ satisfy the Weyl relations $UV = \lambda VU$.

We show that the weak-* closed algebras $W(\pi_\nu, \mathbb{H}^+)$ generated by these representations are always reflexive and we describe them completely. Because of ergodicity, there are three cases for the measure $\nu$:

(1) $\nu$ is equivalent to Lebesgue measure. In this case $\pi_\nu(u)$ generates (in the weak-* topology) an algebra unitarily equivalent to the diagonal operators on $\ell^2(\mathbb{Z})$, while $\pi_\nu(v)$ decomes the bilateral shift; thus $\mathcal{W}(\pi_\nu, \mathbb{H}^+)$ is unitarily equivalent to the lower triangular operators on $\ell^2(\mathbb{Z})$.

(2) $\nu$ is singular to Lebesgue measure and not continuous. Now $\pi_\nu(u)$ turns out equivalent to the bilateral shift while $\pi_\nu(v)$ generates the diagonal operators; thus again one obtains the algebra of lower triangular operators, but with the generators reversed.

(3) $\nu$ is singular to Lebesgue measure and is continuous. In this case $\mathcal{W}(\pi_\nu, \mathbb{H}^+)$ is, surprisingly, selfadjoint: it equals all of $\mathcal{B}(L^2(\mathbb{T}, \nu))$.

It is interesting to observe that, although the representations of $\mathbb{H}_+$ in cases (1) and (2) generate unitarily equivalent weak-* closed algebras, the corresponding representations of the irrational rotation C*-algebra are inequivalent, since the measures are inequivalent [7].

Recall that any family of *commuting* isometries generates a reflexive algebra. We conclude with an example showing that even a 'small' departure from commutativity may destroy reflexivity:

**Example 5.1.** A non-reflexive representation: For a fixed irrational rotation $\lambda$, consider the representation of $\mathbb{H}_+$ on $H^2(\mathbb{T})$ generated by

$$(Uf)(z) = z f(z) \quad \text{and} \quad (Vf)(z) = z f(\lambda z) \quad (f \in H^2(\mathbb{T})).$$

Here we have two isometries which 'almost commute', in the sense that $UV = \lambda VU$. However, they do not generate a reflexive algebra, even in the weak operator topology. Indeed, we show that the weakly closed algebra they generate is lower triangular, but with *finite-dimensional* diagonals (with respect to the basis $\{\zeta_k : k \in \mathbb{Z}_+\}$ of $H^2$), while its reflexive hull contains *all* lower triangular matrices.

## References


[1] P. Ahern, D. Clark, *Invariant subspaces and analytic continuation in several variables,* J. Math. Mech 19 No. 11 (1970), 963–969.

[2] M. Anoussis, A. Katavolos and I.G. Todorov, *Operator algebras from the discrete Heisenberg semigroup,* Proc. Edinburgh Math. Soc. 55 (2012), 1–22. *arXiv:1001.2755.*

[3] A. Arias, G. Popescu, *Factorization and reflexivity on Fock spaces*, Integral Equations Operator Theory 23 (1995), 268–286.

[4] E.A. Azoff, C.K. Fong, F. Gilfeather, *A reduction theory for non-selfadjoint operator algebras*, Trans. Amer. Math. Soc. 224 no. 2 (1976), 351–366.

[5] H. Bercovici, *A factorization theorem with applications to invariant subspaces and the reflexivity of isometries*, Math. Res. Lett. 1 (4) (1994), 511–518.

[6] H. Bercovici, *Hyper-reflexivity and the factorization of linear functionals*, J. Funct. Anal. 158 No. 1 (1998), 242–252.

[7] B. Brenken, *Representations and automorphisms of the irrational rotation algebras*, Pacific J. Math. 111 (1984), 257–282.

[8] A. Beurling, *On two problems concerning linear transformations in Hilbert space*, Acta Math. 81 (1949), 239–255.

[9] K.R. Davidson, D.R. Pitts, *Invariant subspaces and hyper-reflexivity for free semigroup algebras*, Proc. London Math. Soc. 78 (1999), 401–430.

[10] D. Gaspar, N. Suciu, *On invariant subspaces in the bitorus*, J. Operator Theory 30 (1993), 227–241.

[11] P. Ghatage, V. Mandrekar, *On Beurling type invariant subspaces of $L^2(T^2)$ and their equivalence*, J. Operator Theory 20 (1988), 83–89.

[12] P.R. Halmos, Ten problems in Hilbert space. *Bull. Amer. Math. Soc.* 76 (1970), 887–933.

[13] A. Hasegawa, *The invariant subspace structure of $L^2(\mathbb{T}^2)$ for certain von Neumann algebras*, Hokkaido Math. Journal 35 (2006), 601–611.

[14] A. Katavolos, S.C. Power, *The Fourier binest Algebra*, Math. Proc. Cambridge Philos. Soc. 122 No 3 (1997), 525–539.

[15] M. Kosiek, *"On common invariant subspaces for commuting contractions with rich spectrum" – erratum*, Indiana Univ. Math. J. 59 No. 4 (2010), 1541–1546.

[16] W.-S. Li, J.E. McCarthy, *Reflexivity of isometries*, Studia Math. 124 (2) (1997), 101–105.

[17] A.I. Loginov, V.S. Shul'man, *Hereditary and intermediate reflexivity of W*-algebras*, Izv. Akad. Nauk. SSSR, 39 (1975), 1260–1273; Math. USSR-Izv. 9 (1975), 1189–1201.

[18] V. Mandrekar, *The validity of Beurling theorems in polydiscs*, Proc. Amer. Math. Soc. 103 (1988), 145–148.

[19] T. Nakazi, *Invariant Subspaces in the Bidisc and Wandering Subspaces*, J. Aust. Math. Soc. 84 No. 3 (2008), 367–374.

[20] G. Popescu, *A generalization of Beurling's Theorem and a class of reflexive algebras*, J. Operator Theory 41 (1999), 391–420.

[21] W. Rudin, *Function theory in polydiscs*, W.A. Benjamin, Inc., New York-Amsterdam, (1969).

[22] D. Sarason, *Invariant subspaces and unstarred operator algebras*, Pacific J. Math. 17 no. 3 (1966), 511–517.

A. Katavolos
Department of Mathematics
University of Athens
Greece
e-mail: akatavol@math.uoa.gr

Operator Theory:
Advances and Applications, Vol. 233, 85–92


# Lifting Algebraic Contractions in C*-algebras

Terry Loring and Tatiana Shulman

**Abstract.** Let $p$ be a polynomial in one variable. It is shown that the universal $C^*$-algebra of the relation $p(x) = 0$, $\|x\| \leq C$ is semiprojective, residually finite dimensional and has trivial extension group.

**Mathematics Subject Classification (2010).** 46 L05; 46L35.

**Keywords.** Projective and semiprojective $C^*$-algebras, stable relation, lifting problem.

## Introduction

The notions of projectivity and semiprojectivity for $C^*$-algebras were introduced by Effros and Kaminker [3] and, in its modern form, by Blackadar [1] as non-commutative analogues of absolute retract and absolute neighborhood retract in topology.

A $C^*$-algebra $D$ is *projective* if for any $C^*$-algebra $A$, its ideal $I$ and every $*$-homomorphism $\phi : D \to A/I$, there exists a $*$-homomorphism $\tilde{\phi}$ such that the diagram

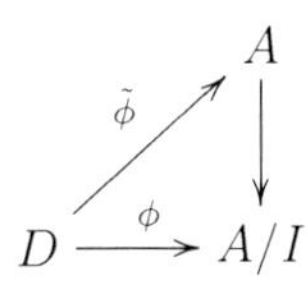


commutes.

A $C^*$-algebra $D$ is *semiprojective* if for any $C^*$-algebra $A$, any increasing chain of ideals $I_1 \subseteq I_2 \subseteq \cdots$ in $A$ and for every $*$-homomorphism $\phi : D \to A/\overline{\bigcup_k I_k}$,

This work was partially supported by a grant from the Simons Foundation (208723 to Loring) and by the NordForsk Research Network "Operator Algebras and Dynamics" (grant 11580).

there exist $n$ and a $*$-homomorphism $\tilde{\phi} : D \to A/I_n$ such that the diagram

$$\begin{array}{ccc} & & A/I_n \\ & \overset{\tilde{\phi}}{\nearrow} & \downarrow \\ D & \xrightarrow{\ \phi\ } & A/\overline{\bigcup_k I_k} \end{array}$$

commutes.

The notion of weak semiprojectiivty was introduced by Eilers and Loring [4].

A $C^*$-algebra $D$ is *weakly semiprojective* if for any sequence of $C^*$-algebras $A_i$ and any $*$-homomorphism $\phi : D \to \prod A_i / \bigoplus A_i$, there exists a $*$-homomorphism $\tilde{\phi} : D \to \prod A_i$ such that the diagram

$$\begin{array}{ccc} & & \prod A_i \\ & \overset{\tilde{\phi}}{\nearrow} & \downarrow \\ D & \xrightarrow{\ \phi\ } & \prod A_i / \bigoplus A_i \end{array}$$

commutes.

Examples and basic properties of projective and (weakly) semiprojective $C^*$-algebras can be found in [7].

The notion of (weak) (semi)projectivity provides an algebraic setting of lifting and perturbation problems for relations in $C^*$-algebras. Namely a relation is liftable (which means that in any quotient $C^*$-algebra $A/I$ any elements satisfying the relation have preimages in $A$ also satisfying the relation) if and only if the universal $C^*$-algebra of the relation is projective.

Similarly a relation is stable under small perturbations if and only if the universal $C^*$-algebra of the relation is weakly semiprojective.

The only problem is that not all relations have the universal $C^*$-algebras. However if relations are noncommutative $*$-polynomial equations combined with norm restrictions on generators

$$\|x_i\| \le c_i$$

then such system of relations defines a universal $C^*$-algebra. That is why in (weak) (semi)projectivity questions it is important to solve lifting problems for relations combined with norm restrictions on generators.

It is not easy for a polynomial relation to be liftable. For example, let us consider a polynomial in one variable. Suppose in a $C^*$-quotient we have an element $x$ satisfying $p(x) = 0$. If $p$ has a non-zero root, the spectral idempotent of $x$ on this root also belongs to the quotient. But idempotents are not liftable: consider for example the unit of $\mathbb{C} = C_0(0,1]/C_0(0,1)$. It has no idempotent lift because $C_0(0,1]$ contains no non-zero idempotents. Thus only monomials $x^n = 0$ have a chance to be liftable. And they indeed are liftable by deep result of Olsen and Pedersen [10]. Immediately there arises a question ([7]) of whether the universal $C^*$-algebra of $x^n = 0$, $\|x\| \le C$ is projective. It was answered positively in [12] and in this paper we will give a short proof of that (Corollary 3).

As to non-monomial relations $p(x) = 0$, they are known to be stable under small perturbations [5]. Moreover the proof in [5] can be generalized to show that relations $p(x) = 0$ are liftable from quotients of the form $A/\overline{\bigcup I_n}$ arising in the definition of semiprojectivity, and one is led to ask if the universal $C^*$-algebra of $p(x) = 0$, $\|x\| \le C$ is semiprojective. In [13] this was proved for the case where all roots of the polynomial have multiplicity more than 1. In this paper we prove it for arbitrary polynomial (Theorem 9).

We also show that these universal $C^*$-algebras are RFD (Theorem 10) and that $*$-homomorphisms from these $C^*$-algebras to the Calkin algebra lift to $*$-homomorphisms to $B(H)$ (Theorem 11). The last result is a generalization of Olsen's structure theorem for polynomially compact operators [11].

Our main technical tool is a generalized spectral radius formula we introduced in [9] in connection with question of Olsen about best approximation of operators by compacts. It turns out to be a useful tool also for lifting polynomial relations combined with restrictions on norms of generators.

## A generalized spectral radius formula

For $x \in A$, we denote by $\dot{x}$ its image in $A/I$ and by $\rho(x)$ its spectral radius.

The following theorem is a generalization of spectral radius formula of Murphy and West [8]. The spectral radius formula is a particular case of the generalized spectral radius formula when $I = A$.

**Theorem 1 ([9]).** *Let $A$ be a $C^*$-algebra, $I$ its ideal, $x \in A$. Then*

$$\max\{\rho(x), \|\dot{x}\|\} = \inf \|(1+i)x(1+i)^{-1}\|$$

*(here* inf *is taken over all $i \in I$ such that $1+i$ is invertible). If $\|\dot{x}\| > \rho(x)$ then the infimum in the right-hand side is attained.*

**Lemma 2.** *Let $p$ be a polynomial in one variable and $t_1, \dots, t_k$ its roots. Let $A$ be a $C^*$-algebra, $I$ its ideal, $x \in A/I$, $p(x) = 0$ and $\|x\| > \max\{t_i\}$. Suppose $x$ has a lift $X \in A$ such that $p(X) = 0$. Then there is a lift $\tilde{X} \in A$ of $x$ such that $p(\tilde{X}) = 0$ and $\|\tilde{X}\| = \|x\|$.*

*Proof.* Since $p(X) = 0$,

$$\rho(X) = \max\{t_i\} < \|x\|.$$

By the generalized spectral radius formula there exists $i \in I$ such that

$$\|(1+i)X(1+i)^{-1}\| = \|x\|.$$

Let $\tilde{X} = (1+i)X(1+i)^{-1}$. Then $\tilde{X}$ is a lift of $x$, $\|\tilde{X}\| = \|x\|$ and

$$p(\tilde{X}) = p((1+i)X(1+i)^{-1}) = (1+i)p(X)(1+i)^{-1} = 0. \qquad \square$$

**Corollary 3 ([12]).** *The universal $C^*$-algebra*

$$C^*\langle x \mid x^n = 0, \|x\| \le C\rangle$$

*is projective.*

*Proof.* For $C = 0$ the statement obviously holds. So let $C > 0$ and let $x \in A/I$, $x^n = 0$, $\|x\| \le C$. We need to show that there is a lift of $x$ with the same properties. If $x = 0$ then it is obvious, so let us assume $x \neq 0$. By [10] there is a lift $X$ of $x$ such that $X^n = 0$. By Lemma 2 there is a lift $\tilde{X}$ of $x$ such that $(\tilde{X})^n = 0$ and $\|\tilde{X}\| = \|x\| \le C$. □

## Semiprojectivity of the universal $C^*$-algebra $C^*\langle x \mid p(x) = 0, \|x\| \le C\rangle$

**Lemma 4.** *Let $T \in B(H)$ and $(T - t_N)^{k_N}(T - t_{N-1})^{k_{N-1}} \dots (T - t_1)^{k_1} = 0$. Let*

$$\begin{aligned} H_1 &= \ker(T - t_1) \\ H_2 &= \ker(T - t_1)^2 \ominus H_1 \\ &\dots \\ H_{k_1} &= \ker(T - t_1)^{k_1} \ominus H_{k_1 - 1} \\ H_{k_1+1} &= \ker(T - t_2)(T - t_1)^{k_1} \ominus H_{k_1} \\ &\dots \\ H_{k_1+\dots+k_N} &= \ker(T - t_N)^{k_N - 1}(T - t_{N-1})^{k_{N-1}} \dots (T - t_1)^{k_1} \ominus H_{k_1+\dots+k_N - 1}. \end{aligned}$$

*Then with respect to the decomposition $H = H_1 \oplus \dots \oplus H_{k_1+\dots+k_N}$ the operator $T$ is uppertriangular with $t_1 1, \dots, t_1 1, \dots, t_N 1, \dots, t_N 1$ on the diagonal, where each $t_i 1$ is repeated $k_i$ times.*

*Proof.* If $x \in H_1$, then $Tx = t_1 x$. If $x \in H_2$, then $Tx = (T - t_1)x + t_1 x$, where $(T - t_1)x \in H_1$. And so on. □

**Lemma 5.** *Let $B \subseteq B(H)$ be a $C^*$-algebra, $b \in B$ an idempotent. Then the projection onto the range of $b$ also belongs to $B$.*

*Proof.* By Lemma 4, $b$ can be written as

$$b = \begin{pmatrix} 1 & X \\ 0 & 0 \end{pmatrix}. \tag{1}$$

Hence

$$\begin{pmatrix} 1 + XX^* & 0 \\ 0 & 0 \end{pmatrix} = \begin{pmatrix} 1 & X \\ 0 & 0 \end{pmatrix} \begin{pmatrix} 1 & 0 \\ X^* & 0 \end{pmatrix} = bb^* \in B.$$

Let $f$ be a continuous function on $\mathbb{R}$ which vanishes at 0 and equal 1 at $[1, \infty)$. Then

$$\begin{pmatrix} 1 & 0 \\ 0 & 0 \end{pmatrix} = f\left(\begin{pmatrix} 1 + XX^* & 0 \\ 0 & 0 \end{pmatrix}\right) \in B.$$

From (1) it is seen that $\begin{pmatrix} 1 & 0 \\ 0 & 0 \end{pmatrix}$ is exactly the projection onto the range of $b$. □

Let $L_1$ be the direct sum of first $k_1$ summands in the decomposition $H = H_1 \oplus \cdots \oplus H_{k_1+\cdots+k_N-1}$ (that is $\ker(T-t_1)^{k_1}$), $L_2$ be the direct sum of next $k_2$ summands, and so on. For any $m$, let $M_m = H \ominus (L_1 \oplus \cdots \oplus L_m)$.

**Corollary 6.** *Let $T \in B(H)$,*

$$(T-t_N)^{k_N}(T-t_{N-1})^{k_{N-1}} \ldots (T-t_1)^{k_1} = 0$$

*and subspaces $L_i$ be as above. Then the projections onto $L_i$ belong to $C^*(T,1)$.*

*Proof.* By transposing factors in the product $(T-t_N)^{k_N}(T-t_{N-1})^{k_{N-1}} \ldots (T-t_1)^{k_1}$ the general case can be reduced to the case $i=1$. So let us prove that the projection onto $L_1$ belongs to $C^*(T,1)$. Since $t_1$ is an isolated point of $\sigma(T)$, there exists the spectral idempotent $Q$ corresponding to $t_1$, that is

$$Q = \chi(T),$$

where $\chi$ is equal to 1 in a neighborhood of $t_1$ and is equal to zero in a neighborhood of $\sigma(T) \setminus \{t_1\}$. By Lemma 5, it is sufficient to prove that $\operatorname{Ran} Q = L_1$.

With respect to the decomposition $H = L_1 \oplus M_1$ the operator $T$ is of the form

$$T = \begin{pmatrix} A & B \\ 0 & C \end{pmatrix},$$

where $\sigma(A) = \{t_1\}$ , $\sigma(C) = \sigma(T) \setminus \{t_1\}$. Hence

$$Q = \chi(T) = \begin{pmatrix} \chi(A) & * \\ 0 & \chi(C) \end{pmatrix} = \begin{pmatrix} 1 & * \\ 0 & 0 \end{pmatrix}.$$

Hence $\operatorname{Ran} Q = L_1$. □

**Lemma 7.** *Let $A \in B(H)$ be given by $A = (A_{ij})$ with respect to some orthonormal basis $\{e_i\}$ in $H$. If $|A_{11}| = \|A\|$, then $A_{1j} = A_{j1} = 0$, when $j \neq 1$.*

*Proof.* We have

$$(AA^*)_{11} = \sum_{j \geq 1} |A_{1j}|^2 = \|A\|^2 + \sum_{j>1} |A_{1j}|^2.$$

Since

$$(AA^*)_{11} = (AA^*e_1, e_1) \leq \|AA^*\| = \|A\|^2,$$

we get $A_{1j}$=0, when $j > 1$. Applying this to $A^*$, we get $A_{j1} = 0$, when $j > 1$. □

In what follows we will assume $|t_1| \geq |t_2| \geq \cdots$ which always can be done by transposition of factors in the product.

**Corollary 8.** *Suppose $T \in B(H)$ and*

$$(T-t_N)^{k_N}(T-t_{N-1})^{k_{N-1}} \ldots (T-t_1)^{k_1} = 0.$$

*Then there exists $0 \leq m \leq N$ such that with respect to the decomposition $H = L_1 \oplus \cdots \oplus L_m \oplus M_m$*

$$T = t_1 1 \oplus \cdots \oplus t_m 1 \oplus S,$$

*where $S$ is such that $(S-t_{m+1})^{k_{m+1}} \ldots (S-t_N)^{k_N} = 0$ and $\|S\| > |t_i|$, for all $i \geq m+1$.*

*Proof.* We write $T$ in upper-triangular form as in Lemma 4 and then use Lemma 7. □

Let $p$ be a polynomial in one variable, $C \geq 0$. Below the universal $C^*$-algebra

$$\mathcal{A} = C^* \langle p(x) = 0, \ \|x\| \leq C \rangle$$

is denoted by $\mathcal{A}$.

**Theorem 9.** *$\mathcal{A}$ is semiprojective.*

*Proof.* Write $p$ as $p(x) = (x - t_N)^{k_N} \dots (x - t_1)^{k_1} = 0$. Let $b \in A/I$, $I = \overline{\bigcup I_n}$,

$$(b - \ t_N \ ) \ ^{k_N} \dots (b - t_1)^{k_1} = 0, \ \|b\| \leq C.$$

Embed $A/I$ into $B(H)$ and write $b$ as in Corollary 8. Let $p_1, \dots, p_m$ be the projections onto $L_1, \dots, L_m$ and $p_{m+1}$ be the projection onto $M_m$. By Corollary 6, they all belong to $A/I$. Then

$$b = \sum_{i=1}^{m} t_i p_i + s,$$

where $s \in p_{m+1} A/I p_{m+1}$ satisfies the equation

$$(s - t_{m+1})^{k_{m+1}} \dots (s - t_N)^{k_N} = 0 \tag{2}$$

and

$$\|s\| > \max_{i \geq m+1} |t_i|. \tag{3}$$

By [1] there exists $n$ such that $p_i$'s can be lifted to projections $P_i$'s in $A/I_n$ with $\sum_{i=1}^{m+1} \ P_i \ = 1$. Since $A/I = (A/I_n)/(I/I_n)$, we have

$$p_{m+1} \ A/I \ p_{m+1} = (P_{m+1} \ A/I_n \ P_{m+1})/(P_{m+1} \ I/I_n \ P_{m+1})$$

and, by (2), (3) and Lemma 2 we can lift $s$ to $S \in P_{m+1} \ A/I_n \ P_{m+1}$ with

$$(S - t_{m+1})^{k_{m+1}} \dots (S - t_N)^{k_N} = 0, \ \ \|S\| = \|s\| \leq C.$$

Let

$$a = \sum_{i=1}^{m} t_i P_i + S. \tag{4}$$

It is a lift of $b$, $\|a\| \leq C$ and $(a - t_N)^{k_N} \dots (a - t_1)^{k_1} = 0$. The last equality can be checked by direct calculations, but it is easier to say that (4) corresponds to upper-triangular form of $a$ as in Corollary 8 and then the last equality follows instantly. □

**Theorem 10.** *$\mathcal{A}$ is RFD.*

*Proof.* Let $H = l^2(\mathbb{N})$. We will identify the algebra $M_n$ of $n$-by-$n$ matrices with

$$B(l^2\{1, \dots, n\}) \subseteq B(H).$$

Let $\mathcal{B} \subseteq \prod M_n$ be the $C^*$-algebra of all $*$-strongly convergent sequences and let $\mathcal{I}$ be the ideal of all sequences $*$-strongly convergent to zero. Then we can identify $\mathcal{B}/\mathcal{I}$ with $B(H)$ by sending each sequence to its $*$-strong limit.

We claim that any family $p_1, \ldots, p_n$ of projections with sum 1 in $B(H)$ lifts to a family of projections $P_1, \ldots, P_n$ with sum 1 in $\mathcal{B}$. One way to prove this is to modify the argument of Choi, used in the proof of Theorem 7 of [2]. A more modern approach is to use Hadwin's [6] characterization of separable RFD $C^*$-algebras: in the unital separable case, $D$ is RFD if and only if we can lift all elements of $\hom_1(D, B(H))$ to $\hom_1(D, \mathcal{B})$. Clearly $\mathbb{C}^n$ is RFD and this lifting problem for $\mathbb{C}^n$ is equivalent to the needed lift of $n$ projections that sum to the identity.

Let $\pi : \mathcal{A} \to B(H)$ be the universal representation. Arguments from the proof of Theorem 9 can be repeated without any change to show that $\pi$ lifts to a $*$-homomorphism $\tilde{\pi} : \mathcal{A} \to \mathcal{B}$. This lift gives a separating family of finite-dimensional representations. □

**Theorem 11.** *Any $*$-homomorphism from $\mathcal{A}$ to the Calkin algebra lifts to a $*$-homomorphism to $B(H)$. In particular,* $\mathrm{Ext}(\mathcal{A}) = 0$.

*Proof.* It is well known, orthogonal projections (with sum 1) in the Calkin algebra can be lifted to orthogonal projections (with sum 1) in $B(H)$. Now we can repeat the arguments from the proof of Theorem 9. □

**Remark 12.** To each $x$ in any $C^*$-algebra with $p(x) = 0$ we can assign in a canonical and functorial way a collection of projections that are orthogonal and sum to one. See Corollary 6, or [5]. If $x \in A/I$ and these projections lift, then $x$ lifts, preserving the relation $p(x) = 0$ and the norm. In formal terms,

$$\mathbb{C}^{N-1} \to C^* \langle x \mid p(x) = 0, \ \|x\| \leq C \rangle$$

is conditionally projective. Thus we have improved upon Theorem 2 in [5] by incorporating the norm condition.

## References


[1] B. Blackadar, Shape theory for C*-algebras, Math. Scand., 56 (1985), 249–275.

[2] M.-D. Choi, The full $C^*$-algebra of the free group on two generators, 8, no. 1, (1980), 41–48.

[3] E.G. Effros and J. Kaminker, Homotopy continuity and shape theory for $C^*$-algebras. Geometric methods in operator algebras, Proc. US-Jap. Semin., Kyoto/Jap. 1983, Pitman Res. Notes Math. Ser. 123, 152–180 (1986).

[4] S. Eilers and T.A. Loring, Computing contingencies for stable relations. International J. of Math. 10 (1999), 301–326.

[5] D. Hadwin, Lifting algebraic elements in C*-algebras, J. Funct. Anal., 127 (1995), 431–437.

[6] D. Hadwin, Free Products of Generalized RFD C*-algebras, arXiv:1108.0049.

[7] T.A. Loring, Lifting solutions to perturbing problems in $C^*$-algebras, volume 8 of Fields Institute Monographs. American Mathematical Society, Providence, RI, 1997.

[8] Murphy, G.J., West, T.T., Spectral radius formulae. Proc. Edinburgh Math. Soc. (2) 22 (1979), no. 3, 271–275.

[9] T.A. Loring and T. Shulman, A generalized spectral radius formula and Olsen's question. J. Funct. Anal., 262, no. 2 (2012), 719–731.

[10] C.L. Olsen and G.K. Pedersen, Corona-algebras and their applications to lifting problems, Math. Scand. 64 (1989), 63–86.

[11] C.L. Olsen, A structure theorem for polynomially compact operators, Amer. J. Math. 93 (1971), 686–698.

[12] T. Shulman, Lifting of nilpotent contractions, Bull. London Math. Soc., 40, 6 (2008), 1002–1006.

[13] T. Shulman, Semiprojectivity of universal C*-algebras generated by algebraic elements. Proceedings of Amer. Math. Soc., to appear.

Terry Loring
Department of Mathematics and Statistics
University of New Mexico
Albuquerque, NM 87131, USA
e-mail: loring@math.unm.edu

Tatiana Shulman
Department of Mathematics
Siena College, 515 Loudon Road
Loudonville, NY 12211, USA
e-mail: shulman@math.ku.dk

Operator Theory:
Advances and Applications, Vol. 233, 93–102


# The Second Local Multiplier Algebra of a Separable $C^*$-algebra

Martin Mathieu

*This paper is dedicated to Victor Shulman on his 65th birthday.*

**Abstract.** Several examples of (separable) $C^*$-algebras with the property that their second (iterated) local multiplier algebra is strictly larger than the first have been found by various groups of authors over the past few years, thus answering a question originally posed by G.K. Pedersen in 1978. This survey discusses a systematic approach by P. Ara and the author to produce such examples on the one hand; on the other hand, we present new criteria guaranteeing that the second and the first local multiplier algebra of a separable $C^*$-algebra agree. For this class of $C^*$-algebras, each derivation of the local multiplier algebra is inner.

**Mathematics Subject Classification (2010).** Primary 46L05. Secondary 46L06, 46M20.

**Keywords.** Local multiplier algebra, injective envelope, $C^*$-algebra, sheaf theory.

## 1. Introduction

In 1978 the late G.K. Pedersen introduced an algebra extension of a general $C^*$-algebra $A$ which he called the "$C^*$-algebra of essential multipliers". His interest in this $C^*$-algebra was stipulated by possible applications in operator theory on $C^*$-algebras as was the interest of the authors of the few papers that appeared on this topic around the same time; see [15] and [16]. Having lain dormant for a good while, this theme was taken up again by P. Ara and the present author who rediscovered the same $C^*$-algebra independently of Pedersen's work in the late 1980s and coined the terminology "local multiplier algebra", introducing, accordingly, the notation $M_{\mathrm{loc}}(A)$. An important connection with the symmetric algebra of quotients in noncommutative ring theory was subsequently made, which led to a fuller understanding of the structure of $M_{\mathrm{loc}}(A)$. For a discussion of this interplay, see [19]. A comprehensive account can be found in our monograph [2], which also contains a wealth of applications to a number of classes of operators between $C^*$-algebras, thus continuing Pedersen's ideas.

In his seminal paper [21], Pedersen asked two questions which spurred quite some research in the past decades. Iterating the construction of the local multiplier algebra (see the definition in the subsequent section) one obtains the following tower of $C^*$-algebras which, a priori, does not have a largest element.

$$A \subseteq M_{\mathrm{loc}}(A) \subseteq M_{\mathrm{loc}}(M_{\mathrm{loc}}(A)) \subseteq \cdots . \tag{1.1}$$

This led Pedersen to ask

**Question 1.** *Is $M_{\mathrm{loc}}(M_{\mathrm{loc}}(A)) = M_{\mathrm{loc}}(A)$ for every $C^*$-algebra $A$?*

The main result in [21] is the following.

**Theorem 1.1.** *Let $A$ be a separable $C^*$-algebra. Every derivation $d\colon A \to A$ extends uniquely to a derivation $d\colon M_{\mathrm{loc}}(A) \to M_{\mathrm{loc}}(A)$ and there is $y \in M_{\mathrm{loc}}(A)$ such that $d = \operatorname{ad} y$ (that is, $dx = [x, y] = xy - yx$ for all $x \in M_{\mathrm{loc}}(A)$).*

A derivation of $M_{\mathrm{loc}}(A)$ may or may not leave $A$ invariant; thus Pedersen's second question reads:

**Question 2.** *Is every derivation $d\colon M_{\mathrm{loc}}(A) \to M_{\mathrm{loc}}(A)$ inner in $M_{\mathrm{loc}}(A)$, provided $A$ is a separable $C^*$-algebra?*

It appears that the answer to Question 2 is still not known in full generality. However, a negative answer to Question 1 was provided in [3]. Since then, several other examples of $C^*$-algebras with the property that the second local multiplier algebra is strictly larger than the first have surfaced [4], [9]. Some of them are separable $C^*$-algebras, others are not. On the other hand, Somerset, in [24], gave fairly general conditions on a separable $C^*$-algebra $A$ implying that both Questions 1 and 2 have positive answers. Until recently, it was not understood how these two different directions could fit into a more comprehensive framework. This is now achieved in our paper [6], and the aim of the present survey is to explain the historical development that led to this more detailed analysis of the differences between the first and the second local multiplier algebra of a separable $C^*$-algebra.

## 2. The history

Let $I$ be a closed, two-sided ideal of a $C^*$-algebra $A$ which is essential (that is, $aI = 0$ for some $a \in A$ implies that $a = 0$). If $J$ is another such essential ideal of $A$ which is contained in $I$, then the multiplier algebra $M(I)$ is canonically embedded as a $C^*$-subalgebra into $M(J)$ by restriction of multipliers to the smaller ideal. In this way, we obtain a directed system of $C^*$-algebras with isometric connecting morphisms, where $I$ runs through the directed set $\mathscr{I}_{ce}(A)$ of all closed, two-sided, essential ideals of $A$. The direct limit of this system is $M_{\mathrm{loc}}(A) = \varinjlim_{\mathscr{I}_{ce}(A)} M(I)$, the *local multiplier algebra* of $A$.

There are several other very useful descriptions of the local multiplier algebra, which are all discussed in detail in our monograph [2]. These lead, e.g., to a representation of the centre $Z = Z(M_{\mathrm{loc}}(A))$ of $M_{\mathrm{loc}}(A)$ as $Z = \varinjlim_{\mathscr{I}_{ce}(A)} Z(M(I))$ [2, Section 3.1] which cannot be deduced directly from the defining formula for $M_{\mathrm{loc}}(A)$ above.

Another important characterisation of $M_{\mathrm{loc}}(A)$, which we will rely on heavily in the following, was first obtained by Frank and Paulsen in [17]; see also [4, Section 4.3]. For a $C^*$-algebra $A$, let us denote by $I(A)$ its *injective envelope* as introduced by Hamana in [18]; see also [20]. We emphasise that $I(A)$ is not an injective object in the category of $C^*$-algebras and $^*$-homomorphisms but in the category of operator spaces and complete contractions. However, it turns out that, nevertheless, $I(A)$ is a $C^*$-algebra canonically containing $A$ as a $C^*$-subalgebra. For a concise discussion of these facts suited for our purposes, see [4].

Under this embedding of $A$ into $I(A)$, the local multiplier algebra $M_{\mathrm{loc}}(A)$ is nothing but the completion of the set of all $y \in I(A)$ which act as a multiplier on some $I \in \mathscr{I}_{ce}(A)$. Since $I(M_{\mathrm{loc}}(A)) = I(A)$ [4, Proposition 2.14], we see that all higher local multiplier algebras of $A$ are contained in $I(A)$ wherefore (1.1) improves to

$$A \subseteq M_{\mathrm{loc}}(A) \subseteq M_{\mathrm{loc}}(M_{\mathrm{loc}}(A)) \subseteq \cdots \subseteq I(A). \tag{2.1}$$

As a result, we can use the injective envelope to study Question 1. For instance, if $A$ is commutative, then $M_{\mathrm{loc}}(A)$ is a commutative $AW^*$-algebra, hence injective. It follows that $M_{\mathrm{loc}}(A) = I(A)$ and, since the local multiplier algebra of every $AW^*$-algebra is the algebra itself [2, Theorem 2.3.8], we find that $M_{\mathrm{loc}}(M_{\mathrm{loc}}(A)) = M_{\mathrm{loc}}(I(A)) = I(A) = M_{\mathrm{loc}}(A)$, one of the possible ways to affirm Pedersen's question in the commutative case.

Evidently, $M_{\mathrm{loc}}(A) = M(A)$ for each simple $C^*$-algebra $A$. Since

$$M_{\mathrm{loc}}(M(A)) = M_{\mathrm{loc}}(A),$$

it follows that $M_{\mathrm{loc}}(M_{\mathrm{loc}}(A)) = M_{\mathrm{loc}}(A)$ in this case too. (We note in passing that $M_{\mathrm{loc}}(A)$ itself can be simple without $A$ being simple unital, see [1], in which case we also have $M_{\mathrm{loc}}(M_{\mathrm{loc}}(A)) = M_{\mathrm{loc}}(A)$.) On the basis of these two positive answers, it appears to be close at hand to investigate a $C^*$-algebra $A$ which is the tensor product of a commutative and a simple one: $A = C \otimes B$ with $C$ commutative (and, without loss of generality, unital) and $B$ simple. The surprise comes in Section 3 below where we shall pin down the properties of $C$ and $B$ determining whether the answer to Question 1 is negative or positive.

Another important contribution is due to Somerset who proved in [24] that the answer to both Question 1 and Question 2 is positive for every unital, separable $C^*$-algebra $A$ which contains sufficiently many maximal ideals; to be precise, he assumed that the primitive ideal space $\mathrm{Prim}(A)$ contains a dense $G_\delta$ subset consisting of closed points. This topological condition will feature again in Section 3 below.

The first class of examples of $C^*$-algebras for which Question 1 has a negative answer was found in [3]. These are certain unital separable approximately finite-dimensional $C^*$-algebras which are primitive and (necessarily) anti-liminal. We employed non-stable $K$-theory to describe these $C^*$-algebras. A very different method, the theory of Hilbert modules over commutative $AW^*$-algebras, was applied in [4] to prove that algebras of the form $C(X)\otimes B(H)$, where $H$ is a separable Hilbert space and $X$ is a Stonean space with additional properties, also provide a negative answer to Question 1. From this it is easy to see that some separable, liminal $C^*$-algebras such as $C[0,1]\otimes K(H)$ have the same property. The latter example was independently found by Argerami, Farenick and Massey in [9]. Both approaches make use of the injective envelope as well as of formulas for $M_{\mathrm{loc}}(A)$ and $I(A)$ which are (fortunately) available in these special cases. The same three authors recently studied the local multiplier algebra of certain continuous trace $C^*$-algebras with similar techniques [10], [11] leading to the general result that the second local multiplier algebra of those $C^*$-algebras is injective [11, Theorem 6.7].

## 3. The results

In this section, we shall explain how the following dichotomy arises. Let $X$ be a compact metric space which is perfect (that is, contains no isolated points). Let $B$ be any separable simple $C^*$-algebra. Then we have the following alternative.

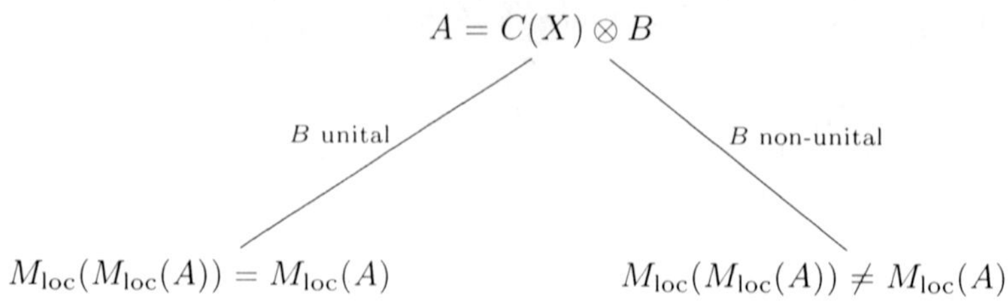


Consequently, there is a plethora of examples of $C^*$-algebras for which Question 1 has a negative answer!

The first of the two main results below explains the non-unital case.

**Theorem 3.1 ([6, Corollary 3.8]).** *Let $B$ and $C$ be separable $C^*$-algebras and suppose that at least one of them is nuclear. Suppose further that $B$ is simple and non-unital and that* $\mathrm{Prim}(C)$ *contains a dense $G_\delta$ subset consisting of closed points. Let $A = C\otimes B$. Then $M_{\mathrm{loc}}(A) = M_{\mathrm{loc}}(M_{\mathrm{loc}}(A))$ if and only if* $\mathrm{Prim}(A)$ *contains a dense subset of isolated points.*

The sufficient condition in this theorem applies to any $C^*$-algebra $A$: Let $X = \mathrm{Prim}(A)$, $X_1$ the set of isolated points in $X$ and $X_2 = X\setminus\overline{X_1}$. Then $X_1$ and $X_2$ are disjoint open subsets of $X$ with corresponding orthogonal closed ideals $I_1 = A(X_1)$ and $I_2 = A(X_2)$ of $A$. If $X_1$ is dense, $I_1$ is the minimal essential closed ideal of $A$ so $M_{\rm loc}(A) = M(I_1)$. It follows that

$$M_{\rm loc}(M_{\rm loc}(A)) = M_{\rm loc}(M(I_1)) = M_{\rm loc}(I_1) = M_{\rm loc}(A).$$

It is thus surprising that this very general condition is indeed necessary for $C^*$-algebras of the above type – and therefore provides us with an easy, systematic way of producing counterexamples. Note that, since $B$ or $C$ is nuclear, the tensor product is unambiguous and $\mathrm{Prim}(C \otimes B) = \mathrm{Prim}(C) \times \mathrm{Prim}(B)$ [23, Theorem B.45]. We shall indicate in Section 4 how to obtain Theorem 3.1.

The unital case in the above alternative fits into the following more general result which applies to $C(X)\otimes B$ since this $C^*$-algebra is quasicentral if and only if $B$ is unital (the following formula for the centres applies: $Z(C\otimes B) = Z(C)\otimes Z(B)$ [8, Theorem 3]).

**Theorem 3.2 ([6, Theorem 4.7]).** *Let $A$ be a quasicentral separable $C^*$-algebra such that* $\mathrm{Prim}(A)$ *contains a dense $G_\delta$ subset consisting of closed points. If $B$ is a $C^*$-subalgebra of $M_{\rm loc}(A)$ containing $A$ then $M_{\rm loc}(B) \subseteq M_{\rm loc}(A)$. In particular, $M_{\rm loc}(M_{\rm loc}(A)) = M_{\rm loc}(A)$.*

Therefore, for every unital separable $C^*$-algebra with Hausdorff primitive spectrum, the first and the second local multiplier algebras agree; this was already observed in [24].

Recall that $A$ is said to be *quasicentral* if no primitive ideal of $A$ contains the centre $Z(A)$ of $A$. This class of $C^*$-algebras was introduced and studied initially by Delaroche in the late 1960s [12], [13] and has turned out to be useful on various occasions. Another "classical" notion that arises when dealing with topological spaces associated to a $C^*$-algebra is the one of a separated point. A point $t$ in a topological space $X$ is called *separated* if $t$ and every point $t'$ outside the closure of $\{t\}$ can be separated by disjoint open neighbourhoods. The set $\mathrm{Sep}(A)$ of all separated points in a separable $C^*$-algebra $A$ is a dense $G_\delta$ subset of $\mathrm{Prim}(A)$ and consists precisely of those $t \in \mathrm{Prim}(A)$ for which the norm function $t \mapsto \|a+t\|$ is continuous for every $a \in A$ [14]. We shall soon make good use of $\mathrm{Sep}(A)$ when we outline the arguments for Theorem 3.2 in Section 4 below.

But first let us draw an immediate consequence of Theorem 3.2 for Question 2.

**Corollary 3.3 ([6, Corollary 4.9]).** *Let $A$ be a quasicentral separable $C^*$-algebra such that* $\mathrm{Prim}(A)$ *contains a dense $G_\delta$ subset consisting of closed points. Then every derivation of $M_{\rm loc}(A)$ is inner.*

The argument is in fact a reduction to Pedersen's theorem, Theorem 1.1 above. Starting with a derivation $d\colon M_{\rm loc}(A) \to M_{\rm loc}(A)$ one can construct a separable $C^*$-subalgebra $B$ of $M_{\rm loc}(A)$ containing $A$ which is $d$-invariant. By Theorem 1.1, $d$ is inner in $M_{\rm loc}(B)$ which, however, is contained in $M_{\rm loc}(A)$ by Theorem 3.2. For the details of the proof, see [6]. Note that the same argument applies to every separable $C^*$-algebra $A$ with the property that any separable $C^*$-subalgebra $B$ of $M_{\rm loc}(A)$ that contains $A$ has its local multiplier algebra $M_{\rm loc}(B)$ contained in $M_{\rm loc}(A)$.

Corollary 3.3 was obtained in the unital case in [24].

## 4. Proofs

We will first attend to the proof of Theorem 3.2. The key tool are two formulas, valid for an arbitrary $C^*$-algebra $A$, that describe both $M_{\text{loc}}(A)$ and $I(A)$ in a compatible way. They rest on our sheaf theory for $C^*$-algebras which was developed in [5].

Rather than giving the formal definition of a sheaf of a $C^*$-algebra [5, Definition 3.1], we shall content ourselves here with two examples. It is straightforward that both constitute contravariant functors from the category $\mathcal{O}_{\text{Prim}(A)}$ of open subsets of the primitive ideal space of the $C^*$-algebra $A$ with inclusions as arrows into the category $\mathcal{C}_1^*$ of unital $C^*$-algebras with unital $*$-homomorphisms as arrows, that is, *presheaves* of unital $C^*$-algebras over the base space $\mathcal{O}_{\text{Prim}(A)}$. It takes a bit more work to verify the unique gluing property for the multiplier sheaf [5, Proposition 3.4] whereas it is fairly easy to establish this in the case of the injective envelope sheaf.

**Example 4.1 (The multiplier sheaf).** Let $A$ be a $C^*$-algebra with primitive ideal space $\text{Prim}(A)$. We define

$$\mathfrak{M}_A\colon \mathcal{O}_{\text{Prim}(A)} \to \mathcal{C}_1^*, \quad \mathfrak{M}_A(U) = M(A(U)),$$

where $M(A(U))$ denotes the multiplier algebra of the closed ideal $A(U)$ of $A$ associated to the open subset $U \subseteq \text{Prim}(A)$ and $M(A(U)) \to M(A(V))$, $V \subseteq U$, are the restriction homomorphisms.

This is the *multiplier sheaf* of $A$ over $\mathcal{O}_{\text{Prim}(A)}$.

**Example 4.2 (The injective envelope sheaf).** Let $I(B)$ denote the injective envelope of a $C^*$-algebra $B$. We define

$$\mathfrak{I}_A\colon \mathcal{O}_{\text{Prim}(A)} \to \mathcal{C}_1^*, \quad \mathfrak{I}_A(U) = p_U I(A) = I(A(U)),$$

where $p_U$ denotes the unique central open projection in $I(A)$ such that $p_U I(A)$ is the injective envelope of $A(U)$. The mappings $I(A(U)) \to I(A(V))$, $V \subseteq U$, are given by multiplication by $p_V$.

This is the *injective envelope sheaf* of $A$ over $\mathcal{O}_{\text{Prim}(A)}$.

To every presheaf we can associate in a canonical way an upper semicontinuous $C^*$-bundle [5, Theorem 5.6]. Let $\mathfrak{A}$ be a presheaf of $C^*$-algebras over the topological space $X$. For $t \in X$, we define the stalk at $t$ by $\mathsf{A}_t := \varinjlim_{t\in U} \mathfrak{A}(U)$ as the direct limit of $C^*$-algebras of the directed family $\{\mathfrak{A}(U)\}$, where $U$ ranges over the family of all open neighbourhoods of $t$ in $X$. Take $\mathsf{A} := \bigsqcup_{t\in X} \mathsf{A}_t$ and let $\pi(a) = t$ if $a \in \mathsf{A}_t$. For $s \in \mathfrak{A}(U)$ and $t \in U$, we have a canonical $*$-homomorphism $\mathfrak{A}(U) \to \mathsf{A}_t$ and we denote by $s(t)$ the image of $s$ under this mapping. There is a canonical topology on the total space $\mathsf{A}$ (which is uniquely determined in case $X$ is Hausdorff) turning $(\mathsf{A}, \pi, X)$ into an upper semicontinuous $C^*$-bundle over $X$.

Let us denote by $\mathsf{A}_{\mathfrak{M}_A}$ and $\mathsf{A}_{\mathfrak{I}_A}$, respectively, the $C^*$-bundles associated in this way to $\mathfrak{M}_A$ and $\mathfrak{I}_A$, respectively.

For any upper semicontinuous $C^*$-bundle $(\mathsf{A}, \pi, X)$ and a subset $Y \subseteq X$, we shall write $\Gamma_b(Y, \mathsf{A})$ for the $C^*$-algebra of bounded continuous local sections on $Y$ [5, Lemma 5.2]. Moreover, $\mathcal{T}$ stands for the downwards directed family of dense $G_\delta$ subsets of $X$.

The following two formulas are the main results, Theorem 7.6 and Theorem 7.7, in [5]. They require the additional concept of the *derived sheaf* of a sheaf of $C^*$-algebras [5, Proposition 7.4]:

$$\begin{aligned} M_{\text{loc}}(A) &= \operatorname{alg}\varinjlim_{T \in \mathcal{T}} \Gamma_b(T, \mathsf{A}_{\mathfrak{M}_A}) \\ I(A) &= \operatorname{alg}\varinjlim_{T \in \mathcal{T}} \Gamma_b(T, \mathsf{A}_{\mathfrak{I}_A}). \end{aligned} \tag{4.1}$$

The algebraic direct limit, that is, the limit before completion to a $C^*$-algebra, is already complete since $\operatorname{Prim}(A)$ is a Baire space and hence countable intersections of dense $G_\delta$ subsets remain within $\mathcal{T}$. As $\mathsf{A}_{\mathfrak{M}_A}$ is a sub-bundle of $\mathsf{A}_{\mathfrak{I}_A}$, the above formulas (4.1) open up a new way to compare the second local multiplier algebra with the first. Indeed, if we take $y \in M_{\text{loc}}(M_{\text{loc}}(A)) \subseteq I(A)$, by (4.1), $y$ is contained in some $C^*$-subalgebra $\Gamma_b(T, \mathsf{A}_{\mathfrak{I}_A})$ and will belong to $M_{\text{loc}}(A)$ once we find $T' \subseteq T$, $T' \in \mathcal{T}$ such that $y \in \Gamma_b(T', \mathsf{A}_{\mathfrak{M}_A})$.

We shall now outline how to use this approach to prove Theorem 3.2 for the case $B = M_{\text{loc}}(A)$ (the general case only requires small modifications). Take $y \in M(J)$ for some closed essential ideal $J$ of $M_{\text{loc}}(A)$. There is $T \in \mathcal{T}$ such that $y \in \Gamma_b(T, \mathsf{A}_{\mathfrak{I}_A})$. By our assumption on $\operatorname{Prim}(A)$ and the remarks before Corollary 3.3 we can assume, without restricting the generality, that $T$ consists of closed separated points of $\operatorname{Prim}(A)$. By the results in [6], there is an element $h \in J$ such that $h(t) \neq 0$ for all $t \in T$ when viewed as a section on $T$, and there is a separable $C^*$-subalgebra $B \subseteq J$ with the properties $AhA \subseteq B$ and $y \in M(B)$. Now take a countable dense subset $\{b_n \mid n \in \mathbb{N}\}$ in $B$ and $T_n \in \mathcal{T}$ such that $b_n \in \Gamma_b(T_n, \mathsf{A}_{\mathfrak{M}_A})$ for each $n \in \mathbb{N}$. In order to simplify the notation, let us put $\mathsf{A} = \mathsf{A}_{\mathfrak{M}_A}$ for the rest of the proof. Setting $T' = \bigcap_n T_n \cap T \in \mathcal{T}$, we have $B \subseteq \Gamma_b(T', \mathsf{A})$ and hence

$$B_t = \{b(t) \mid b \in B\} \subseteq \mathsf{A}_t \qquad (t \in T').$$

In a next step, we aim to describe the fibres $\mathsf{A}_t$ in more detail. In general, there is a homomorphism $\varphi_t \colon \mathsf{A}_t \to M_{\text{loc}}(A/t)$ [5, Section 6]; however, this need neither be injective nor surjective. From our hypotheses, we can conclude for each $t \in T'$:

$$\left.\begin{aligned} A \text{ quasicentral } &\Rightarrow A/t \text{ unital} \\ t \text{ closed } &\Rightarrow A/t \text{ simple} \end{aligned}\right\} \Rightarrow \quad M_{\text{loc}}(A/t) = A/t$$

and, moreover, that $\varphi_t$ is an isomorphism. The surjectivity of $\varphi_t$ rests on the existence of local identities in quasicentral $C^*$-algebras:

$$\begin{aligned} \forall\, t \in \operatorname{Prim}(A) \quad & \exists\, U_1 \subseteq \operatorname{Prim}(A) \text{ open}, t \in U_1, \\ & \exists\, z \in Z(A)_+,\ \|z\| = 1\colon\ z + A(U_2) = 1_{A/A(U_2)}, \end{aligned}$$

where $U_2 = \mathrm{Prim}(A) \setminus \overline{U_1}$ [6, Lemma 4.3]. As a result, $\mathsf{A}_t$ is a simple, unital $C^*$-algebra for each $t \in T'$ and thus,

$$\mathsf{A}_t = \mathsf{A}_t\, h(t)\, \mathsf{A}_t = (A/t)h(t)(A/t) = (AhA)_t \subseteq B_t \subseteq \mathsf{A}_t \qquad (t \in T').$$

Taking $b_t \in B$ with $b_t(t) = 1_{\mathsf{A}_t}$ we obtain $y(t) = y(t)\, 1_{\mathsf{A}_t} = (yb_t)(t) \in \mathsf{A}_t$ for all $t \in T'$. This entails that $y \in \Gamma_b(T', \mathsf{A}_{\mathfrak{M}_A})$ with $T' \subseteq T$, proving that $y \in M_{\mathrm{loc}}(A)$, as desired.

In the proof of the "only if"-part in Theorem 3.1, the sheaf theoretic concepts can be pushed to the background, and we will merely roughly indicate the main argument in the following. It is easy to reduce the general case to the case when $\mathrm{Prim}(A)$ is perfect (i.e., has no isolated points) using the direct sum decomposition $M_{\mathrm{loc}}(A) = M_{\mathrm{loc}}(I_1) \oplus M_{\mathrm{loc}}(I_2)$ (in the notation of the argument for the "if"-part directly after the statement of Theorem 3.1 above).

The strategy now is the following: to show that $M_{\mathrm{loc}}(M_{\mathrm{loc}}(A)) \neq M_{\mathrm{loc}}(A)$ under the hypothesis that $\mathrm{Prim}(A)$ is perfect (and all the other assumptions in Theorem 3.1) we aim to identify a closed essential ideal $K$ of $M_{\mathrm{loc}}(A)$ with the property that $M_{\mathrm{loc}}(A) \subsetneq M(K)$. The following $C^*$-subalgebra $K_A$ of $M_{\mathrm{loc}}(A)$ was already introduced by Somerset in [24]: $K_A$ is the closure of the set of all elements of the form $\sum_{n\in\mathbb{N}} a_n z_n$, where $\{a_n\} \subseteq A$ is a bounded family and $\{z_n\} \subseteq Z(M_{\mathrm{loc}}(A))$ consists of mutually orthogonal projections. (These infinite sums exist in $M_{\mathrm{loc}}(A)$ by [2, Lemma 3.3.6], for example. Note also that $Z(M_{\mathrm{loc}}(A))$ is countably decomposable since $A$ is separable.)

For a separable $C^*$-algebra $A$ with the property that $\mathrm{Prim}(A)$ contains a dense $G_\delta$ subset of closed points, the following statements hold (see [6, Section 2]).

(i) $K_A$ is an essential ideal in $M_{\mathrm{loc}}(A)$.
(ii) If $K_I = K_A$ for all $I \in \mathscr{I}_{ce}(A)$ then $M_{\mathrm{loc}}(K_A) = M(K_A)$.
(iii) Let $y \in I(A)$. If $ya \in K_A$ for all $a \in A$ then $y \in M(K_A)$.

Combining these we obtain the following result.

**Proposition 4.3 ([6, Theorem 3.2]).** $M_{\mathrm{loc}}^{(3)}(A) = M_{\mathrm{loc}}^{(2)}(A) = M(K_A)$, *where* $M_{\mathrm{loc}}^{(n)}(A) = M_{\mathrm{loc}}(M_{\mathrm{loc}}^{(n-1)}(A))$, $n \geq 2$ *denotes the* $n$*-fold iterated local multiplier algebra of* $A$.

The main work in the proof of Theorem 3.1 consists in constructing an element $q \in M(K_A) \setminus M_{\mathrm{loc}}(A)$. This uses the special tensor product structure $A = C \otimes B$. The topological assumptions on $X = \mathrm{Prim}(A) = \mathrm{Prim}(C)$ lead us to a dense $G_\delta$ subset $S$ of $X$ consisting of closed separated points which is a Polish space (using [14]). Since $S$ itself is perfect, every non-empty open subset of $S$ contains an open subset which has non-empty boundary. This allows us to choose a suitable sequence $(z_n)_{n\in\mathbb{N}}$ of projections in $Z(M(K_n \otimes B)) \subseteq M_{\mathrm{loc}}(A)$, where $K_n \in \mathscr{I}_{ce}(C)$. As $B$ is simple and non-unital, there is a strictly increasing approximate identity $(e_n)_{n\in\mathbb{N}}$ of $B$ with $e_n e_{n+1} = e_n$ and $\|e_{n+1} - e_n\| = 1$ for all $n$. Put $p_1 = e_1$, $p_n = e_n - e_{n-1}$ for $n \geq 2$ and set $q_n = \sum_{j=1}^{n} z_j \otimes p_{2j}$, $n \in \mathbb{N}$. We thus obtain an increasing sequence $(q_n)_{n\in\mathbb{N}}$ in $M_{\mathrm{loc}}(A)_+$ bounded by 1. Since $I(A)$ is monotone complete, $q = \sup_n q_n = \sum_{n=1}^{\infty} z_n \otimes p_{2n}$ exists in $I(A)_+$ and has norm 1. It remains to show

(a) $q \in M(K_A)$;
(b) $q \notin M_{\text{loc}}(A)$.

The first assertion is established by using that the bounded central closure ${}^cA = \overline{AZ}$ is $\sigma$-unital (as $A$ is separable) together with approximation in the strict topology of $M(K_A)$. For (b), we use special properties of the sequence $(z_n)_{n\in\mathbb{N}}$ chosen above relying on the topological properties of $S$ indicated above. It is shown that, if $q$ were in $M_{\text{loc}}(A)$, this would lead to a contradiction.

This completes the proof of Theorem 3.1.

It is clear from Proposition 4.3, which holds for every separable $C^*$-algebra $A$ with the property that Prim$(A)$ contains a dense $G_\delta$ subset of closed points, that the methods discussed above cannot create an example of a $C^*$-algebra $A$ with the property that $M_{\text{loc}}^{(2)}(A) \neq M_{\text{loc}}^{(3)}(A)$; in fact, it appears that no such concrete example is known at present.

## References

[1] P. Ara and M. Mathieu, *A simple local multiplier algebra*, Math. Proc. Cambridge Phil. Soc. **126** (1999), 555–564.

[2] P. Ara and M. Mathieu, Local multipliers of $C^*$-algebras, Springer-Verlag, London, 2003.

[3] P. Ara and M. Mathieu, *A not so simple local multiplier algebra*, J. Funct. Anal. **237** (2006), 721–737.

[4] P. Ara and M. Mathieu, *Maximal $C^*$-algebras of quotients and injective envelopes of $C^*$-algebras*, Houston J. Math. **34** (2008), 827–872.

[5] P. Ara and M. Mathieu, *Sheaves of $C^*$-algebras*, Math. Nachr. **283** (2010), 21–39.

[6] P. Ara and M. Mathieu, *When is the second local multiplier algebra of a $C^*$-algebra equal to the first?*, Bull. London Math. Soc. **43** (2011), 1167–1180.

[7] R.J. Archbold, *Density theorems for the centre of a $C^*$-algebra*, J. London Math. Soc. (2) **10** (1975), 189–197.

[8] R.J. Archbold, *On the centre of a tensor product of $C^*$-algebras*, J. London Math. Soc. (2) **10** (1975), 257–262.

[9] M. Argerami, D.R. Farenick and P. Massey, *The gap between local multiplier algebras of $C^*$-algebras*, Quart. J. Math. **60** (2009), 273–281.

[10] M. Argerami, D.R. Farenick and P. Massey, *Injective envelopes and local multiplier algebras of some spatial continuous trace $C^*$-algebras*, Quart. J. Math. **63** (2012), 1–20.

[11] M. Argerami, D.R. Farenick and P. Massey, *Second-order local multiplier algebras of continuous trace $C^*$-algebras*, J. Math. Anal. Appl., to appear.

[12] C. Delaroche, *Sur les centres des $C^*$-algèbres*, Bull. Sc. math. **91** (1967), 105–112.

[13] C. Delaroche, *Sur les centres des $C^*$-algèbres*, II, Bull. Sc. math. **92** (1968), 111–128.

[14] J. Dixmier, *Sur les espaces localement quasi-compacts*, Canad. J. Math. **20** (1968), 1093–1100.

[15] G.A. Elliott, *Derivations determined by multipliers on ideals of a $C^*$-algebra*, Publ. Res. Inst. Math. Sci. **10** (1974/75), 721–728.

[16] G.A. Elliott, *Automorphisms determined by multipliers on ideals of a $C^*$-algebra*, J. Funct. Anal. **23** (1976), 1–10.

[17] M. Frank and V.I. Paulsen, *Injective envelopes of $C^*$-algebras as operator modules*, Pacific J. Math. **212** (2003), 57–69.

[18] M. Hamana, *Injective envelopes of $C^*$-algebras*, J. Math. Soc. Japan **31** (1979), 181–197.

[19] M. Mathieu, *The local multiplier algebra: blending noncommutative ring theory and functional analysis*, in: *Proc. Conf. in honour of Robert Wisbauer's 65th birthday*, (Porto, 8–10 September 2006), Trends in Math., Birkhäuser, Basel, 2008, 301–312.

[20] V.I. Paulsen, Completely bounded maps and operator algebras, Cambridge Studies in Adv. Math. **78**, Cambridge Univ. Press, Cambridge, 2002.

[21] G.K. Pedersen, *Approximating derivations on ideals of $C^*$-algebras*, Invent. Math. **45** (1978), 299–305.

[22] G.K. Pedersen, $C^*$-algebras and their automorphism groups, Academic Press, London, 1979.

[23] I. Raeburn and D.P. Williams, Morita equivalence and continuous-trace $C^*$-algebras, Math. Surveys and Monographs **60**, Amer. Math. Soc., Providence, RI, 1998.

[24] D.W.B. Somerset, *The local multiplier algebra of a $C^*$-algebra*. II, J. Funct. Anal. **171** (2000), 308–330.

Martin Mathieu
Department of Pure Mathematics
Queen's University Belfast
Belfast BT7 1NN
Northern Ireland
e-mail: `m.m@qub.ac.uk`

Operator Theory:
Advances and Applications, Vol. 233, 103–126


# Test-space Characterizations of Some Classes of Banach Spaces

M.I. Ostrovskii

**Abstract.** Let $\mathcal{P}$ be a class of Banach spaces and let $T = \{T_\alpha\}_{\alpha\in A}$ be a set of metric spaces. We say that $T$ is a set of *test-spaces* for $\mathcal{P}$ if the following two conditions are equivalent: (1) $X \notin \mathcal{P}$; (2) The spaces $\{T_\alpha\}_{\alpha\in A}$ admit uniformly bilipschitz embeddings into $X$.

The first part of the paper is devoted to a simplification of the proof of the following test-space characterization obtained in M.I. Ostrovskii [Different forms of metric characterizations of classes of Banach spaces, *Houston J. Math.*, **39** (2013), 889–906]:

For each sequence $\{X_m\}_{m=1}^\infty$ of finite-dimensional Banach spaces there is a sequence $\{H_n\}_{n=1}^\infty$ of finite connected unweighted graphs with maximum degree 3 such that the following conditions on a Banach space $Y$ are equivalent: (A) $Y$ admits uniformly isomorphic embeddings of $\{X_m\}_{m=1}^\infty$; (B) $Y$ admits uniformly bilipschitz embeddings of $\{H_n\}_{n=1}^\infty$.

The second part of the paper is devoted to the case where $\{X_m\}_{m=1}^\infty$ is an increasing sequence of spaces. It is shown that in this case the class of spaces given by (A) can be characterized using one test-space, which can be chosen to be an infinite graph with maximum degree 3.

**Mathematics Subject Classification (2010).** Primary: 46B07; Secondary: 05C12, 46B85, 54E35.

**Keywords.** Banach space, Bourgain discretization theorem, Lipschitz map, metric characterization, test-space.

## 1. Introduction

Embeddings of metric spaces into Banach spaces play an important role in Computer Science (see, for example, [WS11, Chapter 15]) and in Topology (see [Yu06]). In connection with problems of embeddability of metric spaces into Banach spaces it would be interesting to find metric characterizations of well-known classes of Banach spaces. By *metric characterizations* we mean characterizations which can be tested on an arbitrary metric space. So, in a metric characterization only distances between elements of the space are involved, and no linear combinations of any kind can be used. At this point it seems to be unclear: how to define the notion of a metric characterization which would be the most useful for applications in the theory of metric embeddings? One can try to define a metric characterization in the following way: a *metric characterization* is a set of *formulas* understood as in logic (see [Han77, p. 19] for a definition of a first-order formula). Some of the variables in the formulas are elements in an unknown metric space $X$ (so the formulas make sense for an arbitrary metric space $X$). We say that such set of formulas *characterizes* a class $\mathcal{P}$ of Banach spaces if the following conditions are equivalent for a Banach space $X$:

- $X \in \mathcal{P}$
- All of the formulas of the set hold for $X$.

Metric characterizations which we are going to study in this paper use the following definition.

**Definition 1.1.** Let $0 \leq C < \infty$. A map $f : (A, d_A) \to (Y, d_Y)$ between two metric spaces is called *C-Lipschitz* if

$$\forall u, v \in A \quad d_Y(f(u), f(v)) \leq C d_A(u, v).$$

A map $f$ is called *Lipschitz* if it is $C$-Lipschitz for some $0 \leq C < \infty$. For a Lipschitz map $f$ we define its *Lipschitz constant* by

$$\mathrm{Lip} f := \sup_{d_A(u,v) \neq 0} \frac{d_Y(f(u), f(v))}{d_A(u, v)}.$$

Let $1 \leq C < \infty$. A map $f : A \to Y$ is called a *C-bilipschitz embedding* if there exists $r > 0$ such that

$$\forall u, v \in A \quad r d_A(u, v) \leq d_Y(f(u), f(v)) \leq r C d_A(u, v). \tag{1}$$

A *bilipschitz embedding* is an embedding which is $C$-bilipschitz for some $1 \leq C < \infty$. The smallest constant $C$ for which there exist $r > 0$ such that (1) is satisfied is called the *distortion* of $f$. A family of embeddings is called *uniformly bilipschitz* if they have uniformly bounded distortions.

**Remark 1.2.** Linear embeddings $T_i : A_i \to Y$ of Banach (or normed) spaces into a Banach (normed) space are uniformly bilipschitz if and only if

$$\sup_i \left( ||T_i|| \cdot ||T_i^{-1}|_{T_i(A_i)}|| \right) < \infty.$$

Such embeddings $T_i$ are called *uniformly isomorphic*.

**Remark 1.3.** The definition of a metric characterization suggested above does not seem to be completely satisfactory. It includes trivial (in a certain sense) characterizations of the type: A Banach space is nonreflexive if and only if it contains a (separable) subset which is bilipschitz equivalent to a nonreflexive separable Banach space. (The validity of this characterization is a consequence of the following well-known facts: (1) Each nonreflexive Banach space contains a separable nonreflexive subspace. This fact follows, for example, from the Eberlein–Smulian theorem [DS58, Theorem V.6.1]; (2) If a Banach space $Y$ admits a bilipschitz embedding of a (separable) nonreflexive Banach space $X$, then $Y$ is nonreflexive, see [BL00, Theorem 6.42].)

At this point it is not clear how to define a metric characterization in such a way that, on one hand, all interesting examples are included, but, on the other hand, characterizations like the trivial characterization of nonreflexivity mentioned in Remark 1.3 are excluded. We shall focus on one of the classes of metric characterizations which is known to be interesting (see [Bau07, Bau12, Bou86, BMW86, JS09, MN08, Ost11a, Ost11c, Pis86]). We mean metric characterizations of the following type.

**Definition 1.4.** Let $\mathcal{P}$ be a class of Banach spaces and let $T = \{T_\alpha\}_{\alpha\in A}$ be a set of metric spaces. We say that $T$ is a set of *test-spaces* for $\mathcal{P}$ if the following two conditions are equivalent

1. $X \notin \mathcal{P}$.
2. The spaces $\{T_\alpha\}_{\alpha\in A}$ admit uniformly bilipschitz embeddings into $X$.

**Remark 1.5.** We use $X \notin \mathcal{P}$ in condition 1 of Definition 1.4 rather than $X \in \mathcal{P}$ for terminological reasons: we would like to use terms "test-spaces for reflexivity, superreflexivity, etc." rather than "test-spaces for **non**reflexivity, **non**superreflexivity, etc."

**Remark 1.6.** One can introduce the notion of test-spaces differently, requiring, for example: "at least one of the spaces $\{T_\alpha\}_{\alpha\in A}$ admits a bilipschitz embedding into $X$". However, this version of test-space characterizations includes the trivial characterization of reflexivity mentioned in Remark 1.3. Another reason why we have chosen the introduced in Definition 1.4 notion of test-spaces is: many important known characterizations are of this form (see [Bau07, Bau12, Bou86, BMW86, JS09, MN08, Ost11a, Ost11c, Pis86]).

## 2. Simplification of the proof of metric characterizations of classes with excluded subspaces

The main purpose of this section is to give a simpler proof of the following result of [Ost11c]:

**Theorem 2.1 ([Ost11c]).** *For each sequence $\{X_m\}_{m=1}^\infty$ of finite-dimensional Banach spaces there exists a sequence $\{H_n\}_{n=1}^\infty$ of finite connected unweighted graphs with maximum degree 3 such that the following conditions on a Banach space $Y$ are equivalent:*

- *$Y$ admits uniformly isomorphic embeddings of $\{X_m\}_{m=1}^\infty$.*
- *$Y$ admits uniformly bilipschitz embeddings of $\{H_n\}_{n=1}^\infty$.*

Everywhere in this paper we consider graphs as metric spaces with their *shortest path metric*: the distance between two vertices is the length of the shortest path between them. In some cases we consider weighted graphs with some positive weights assigned to their edges. In such a case the length of the path is the sum of weights of edges included in it. For graphs with no weights (sometimes we emphasize this by calling them *unweighted graphs*) the length of a path is the number of edges in it (this corresponds to the case when all weights are equal to 1).

**Remark 2.2.** For the reasons explained in [Ost11c] we restrict our attention to the case $\sup_m \dim X_m = \infty$.

Our purpose is to simplify the proof of the step which has the longest proof in [Ost11c]. We shall also present other steps of the proof, in a more general form than in [Ost11c]. The reason for doing so is that we need these steps in the more general form later in the paper. Recall some standard definitions.

**Definition 2.3.** Let $\alpha > 0$. We say that a subset $A$ in a metric space $(X, d)$ is $\alpha$*-dense* in a subset $B \subset X$ if $A \subset B$ and

$$\forall x \in B\ \exists y \in A\ d(x, y) \leq \alpha.$$

A subset $D$ in a metric space $(X, d)$ is called $\alpha$*-separated* if $d(x, y) \geq \alpha$ for each $x, y \in D$, $x \neq y$.

If $A$ and $B$ are subsets in a metric space $(X, d)$, we let

$$d(A, B) = \inf\{d(x, y) : \ x \in A, \ y \in B\}.$$

First we introduce an approximate description of convex sets in Banach spaces using unweighted graphs. Let $C$ be a convex set in a Banach space $X$, $\alpha, \beta > 0$, and let $V$ be an $\alpha$-separated $\beta$-dense subset of $C$.

**Remark 2.4.** It is easy to see that such subset $V$ does not exist if $\beta < \frac{\alpha}{2}$ and $\mathrm{diam} C > \beta$. In this paper only the case where $\beta \geq \alpha$ is considered. In this case the existence is immediate from Zorn's lemma.

**Definition 2.5.** Let $G$ be the graph whose vertex set is $V$ and whose edge set is defined in the following way: vertices $u, v \in V$ are joined by an edge if and only if $||u - v|| \leq 3\beta$. The graph $G$ is called a $(C, \alpha, \beta)$*-graph*. If $\alpha = \beta$, $G$ is called a $(C, \alpha)$*-graph*.

It is easy to check that $G$ is not uniquely determined by $C$, $\alpha$, and $\beta$, but for our purposes it does not matter which of $(C, \alpha, \beta)$-graphs we pick. We endow the vertex set $V$ of $G$ with its shortest path metric $d_G$.

**Lemma 2.6.** *The natural embedding* $f : (V, d_G) \to (X, ||\cdot||)$ *is bilipschitz with distortion* $\leq \max\left\{3, \frac{3\beta}{\alpha}\right\}$. *More precisely,*

$$\mathrm{Lip}(f) \leq 3\beta \text{ and } \mathrm{Lip}(f^{-1}|_{f(V)}) \leq \max\left\{\frac{1}{\beta}, \frac{1}{\alpha}\right\}.$$

*Proof.* The inequality $\mathrm{Lip}(f) \leq 3\beta$ follows immediately from the fact that adjacent vertices in $G$ are at distance $\leq 3\beta$ in $X$, and the definition of the shortest path metric.

To prove the inequality for $\mathrm{Lip}(f^{-1})$ we consider two distinct vertices $u, v \subset V$, write $||u - v|| = d\beta$ for some $d > 0$, and consider two cases:

**Case A.** $d \leq 3$. In such a case $d_G(u, v) = 1$. Since $||u - v|| \geq \alpha$, we have

$$\frac{d_G(u,v)}{||u-v||} \leq \frac{1}{\alpha}.$$

**Case B.** $d > 3$. In this case, and even in a wider case $d > 2$, we show that

$$d_G(u,v) \leq \lfloor d \rfloor - 1 \tag{2}$$

and therefore

$$\frac{d_G(u,v)}{||u-v||} \leq \frac{\lfloor d \rfloor - 1}{d\beta} \leq \frac{1}{\beta}.$$

We prove the inequality (2) by induction starting with $2 < d \leq 3$. In this case $d_G(u, v) = 1$ and so it is clear that (2) is satisfied.

Suppose that we have proved the inequality (2) for $2 < d \leq n$. Let us show that this implies the inequality for $n < d \leq n + 1$. We do this as follows:

Consider the vertex $\tilde{u}$ lying on the line segment joining $u$ and $v$ at distance $2\beta$ from $u$. Since $\tilde{u} \in C$ (this is the point where we use the convexity of $C$), there is $w \in V$ satisfying $||w - \tilde{u}|| \leq \beta$.

By the triangle inequality, we have $||w - u|| \leq 3\beta$ and $||w - v|| \leq (d - 1)\beta$. The first inequality implies $d_G(w, u) = 1$. Applying the triangle inequality and the Induction Hypothesis, we get $d_G(u, v) \leq d_G(w, v) + 1 \leq (\lfloor d - 1 \rfloor - 1) + 1 = \lfloor d \rfloor - 1$. (We cannot apply the Induction Hypothesis if $||w - v|| \leq 2\beta$, but in this case the result is also easy to verify.) □

If $X$ is a Banach space, we use the notation $B_X(r), r > 0$, for $\{x \in X : ||x|| \leq r\}$. The unit ball $B_X(1)$ is also denoted by $B_X$. Observe that if $X$ is finite dimensional (and $\beta \geq \alpha$), $(B_X(r), \alpha, \beta)$-graphs are finite.

The next step in the proof of Theorem 2.1 in [Ost11c] is the following lemma (in [Ost11c] the lemma is stated in slightly less general form).

**Lemma 2.7.** *If $\{X_m\}_{m=1}^{\infty}$ are finite dimensional Banach spaces, and a Banach space $Y$ admits uniformly bilipschitz embeddings of a collection of $\big(B_{X_m}(n), \alpha(n), \beta(n)\big)$-graphs $m, n \in \mathbb{N}$, where $\alpha(n) \leq \beta(n)$, $\lim_{n\to\infty} \alpha(n) = 0$, $\lim_{n\to\infty} \beta(n) = 0$, and $\sup_n \frac{\beta(n)}{\alpha(n)} < \infty$, then $\{X_m\}_{m=1}^{\infty}$ are uniformly isomorphic to subspaces of $Y$.*

Lemma 2.7 can be derived from the following discretization result.

**Theorem 2.8.** *For each finite-dimensional Banach space $X$ and each $\gamma > 0$ there exists $\varepsilon > 0$ such that for each bilipschitz embedding $L$ of an $\varepsilon$-dense subset of $B_X$, with the metric inherited from $X$, into a Banach space $Y$ there is a linear embedding $T : X \to Y$ such that $(1 - \gamma)||T|| \cdot ||T^{-1}|| \leq \mathrm{Lip}(L) \cdot \mathrm{Lip}(L^{-1})$.*

This theorem goes back to Ribe [Rib76], a new proof of the essential ingredients was found by Heinrich–Mankiewicz [HM82], versions of these proofs are presented in [BL00]. The first explicit bound on $\varepsilon$ was found by Bourgain [Bou87]. Bourgain's proof was simplified and explained by Begun [Beg99] and Giladi–Naor–Schechtman [GNS11]. We recommend everyone who would like to study this result to start by reading [GNS11].

It is clear that the fact that the ball in Theorem 2.8 has radius 1 plays no role. To derive Lemma 2.7 from Theorem 2.8 we observe that the vertex set of a $(B_{X_m}(n), \alpha(n), \beta(n))$-graph is a $\beta(n)$-dense subset of $B_{X_m}(n)$. Furthermore, by Lemma 2.6, the graph distance on this set is $\max\left\{3, \frac{3\beta(n)}{\alpha(n)}\right\}$-equivalent to the metric inherited from $X$. This proves Lemma 2.7.

Lemmas 2.6 and 2.7 show that a Banach space $Y$ admits uniformly isomorphic embeddings of $X_m$ if and only if $Y$ admits uniformly bilipschitz embeddings of some (or any) collection of $(B_{X_m}(n), \alpha(n), \beta(n))$-graphs $m, n \in \mathbb{N}$, where $\alpha(n) \le \beta(n)$, $\lim_{n\to\infty} \alpha(n) = 0$, $\lim_{n\to\infty} \beta(n) = 0$, and $\sup_n \frac{\beta(n)}{\alpha(n)} < \infty$. This statement does not complete the proof of Theorem 2.1 (even of a weaker version of it, with 3 replaced by any other uniform bound on degrees of $\{H_n\}_{n=1}^{\infty}$). In fact, it is easy to see that if $\sup_m \dim X_m = \infty$, the degrees of any collection of $(B_{X_m}(n), \alpha(n), \beta(n))$-graphs, $m, n \in \mathbb{N}$, where $\alpha(n) \le \beta(n)$, $\lim_{n\to\infty} \alpha(n) = 0$, $\lim_{n\to\infty} \beta(n) = 0$, and $\sup_n \frac{\beta(n)}{\alpha(n)} < \infty$, are unbounded.

**Observation 2.9.** *Since we assumed* $\sup_m \dim X_m = \infty$, *each Banach space admitting uniformly isomorphic embeddings of* $X_m$ *has to be infinite-dimensional. For this reason, in order to prove Theorem* 2.1 *it suffices to show that for each finite-dimensional Banach space* $X$, *each* $(B_X(r), \delta)$*-graph* $G$ $(0 < \delta < r < \infty)$, *and each infinite-dimensional Banach space* $Z$ *containing* $X$ *as a subspace, there exist a graph* $H$ *and bilipschitz embeddings* $\psi : G \to H$ *and* $\varphi : H \to Z$, *such that distortions of* $\psi$ *and* $\varphi$ *are bounded from above by absolute constants and the maximum degree of* $H$ *is* 3. *It is clear that it is enough to consider the case* $\delta = 1$.

**Remark 2.10.** Observation 2.9 is the main step towards simplification of the proof of Theorem 2.1 given in [Ost11c]: in [Ost11c] the graph $H$ was embedded into $X$ (if $\dim X \ge 3$). This is substantially more difficult and, as we see from Observation 2.9, is not needed to prove Theorem 2.1. However, the proof of [Ost11c, Section 4] could be of independent interest, see Remark 2.12 below.

The construction of $H$ which we use is the same as in [Ost11c]: We introduce the graph $MG$ as the following "expansion" of $G$: we replace each edge in $G$ by a path of length $M$. It is clear that the graph $MG$ is well defined for each $M \in \mathbb{N}$. In our construction of $H$ the number $M$ will be chosen to be much larger than the number of edges of $G$. We use the term *long paths* for the paths of length $M$ which replace edges of $G$. Next step in the construction of $H$: For each vertex $v$ of $G$, we introduce a path $p_v$ in the graph $H$ whose length is equal to the number of edges of $G$, we call each such path a *short path*. At the moment these paths do not interact. We continue our construction of $H$ in the following way. We label vertices of short paths in a monotone way by long paths. "In a monotone way" means that the first vertex of **each** short path corresponds to the long path $p_1$, the second vertex of **each** short path corresponds to the long path $p_2$ etc. We complete our construction of $H$ introducing, for a long path $p$ in $MG$ corresponding to an edge $uv$ in $G$, a path of the same length in $H$ (we also call it *long*) which joins those vertices of the short paths $p_u$ and $p_v$ which have label $p$. There is no further interaction between short and long paths in $H$. It is obvious that the maximum degree of $H$ is 3.

It remains to define embeddings $\psi$ and $\varphi$ and to estimate their distortions.

To define $\psi$ we pick a long path $p$ in $MG$ (in an arbitrary way) and map each vertex $u$ of $G$ onto the vertex in $H$ having label $p$ in the short path $p_u$ corresponding to $u$. The estimates for $\mathrm{Lip}(\psi)$ and $\mathrm{Lip}(\psi^{-1})$ given below are taken from [Ost11c]. We reproduce them because they do not take much space.

We have $\mathrm{Lip}(\psi) \le 2e(G) + M$, where $e(G)$ is the number of edges of $G$. In fact, to estimate the Lipschitz constant it suffices to find an estimate from above for the distances in $H$ between $\psi(u)$ and $\psi(v)$ where $u$ and $v$ are adjacent vertices of $G$. To see that $2e(G)+M$ provides the desired estimate we consider the following three-stage walk from $\psi(u)$ from $\psi(v)$:

- We walk from $\psi(u)$ along the short path $p_u$ to the vertex labelled by the long path corresponding to the edge $uv$ in $G$.
- Then we walk along the corresponding long path to its end in $p_v$.
- We conclude the walk with the piece of the short path $p_v$ which we need to traverse in order to reach $\psi(v)$.

We claim that $\mathrm{Lip}(\psi^{-1}) \le M^{-1}$. This gives an absolute upper bound for the distortion of $\psi$ provided the quantity $e(G)$ is controlled by $M$, we need $M$ to be much larger than $e(G)$ only if we would like to make the distortion close to 1. To prove $\mathrm{Lip}(\psi^{-1}) \le M^{-1}$ we let $\psi(u)$ and $\psi(v)$ be two vertices of $\psi(V(G))$. We need to estimate $d_G(u,v)$ from below in terms of $d_H(\psi(u),\psi(v))$. Let

$$P = \psi(u), w_1, \dots, w_{n-1}, \psi(v)$$

be one of the shortest $\psi(u)\psi(v)$-paths in $H$. Let $u, u_1, \dots, u_{k-1}, v$ be those vertices of $G$ for which the path $P$ visits the corresponding short paths $p_u$, $p_{u_1}, \dots, p_{u_{k-1}}$, $p_v$. We list $u_1, \dots, u_{k-1}$ in the order of visits. It is clear that in such a case the sequence $u, u_1, \dots, u_{k-1}, v$ is a $uv$-walk in $G$. Therefore $d_G(u,v) \le k$. On the other hand, in $H$, to move from one short path to another, one has to traverse at least $M$ edges, therefore $d_H(\psi(u),\psi(v)) \ge kM$. This implies $\mathrm{Lip}(\psi^{-1}) \le M^{-1}$.

Our next purpose is to introduce $\varphi : H \to Z$. First we prove (Lemma 2.11 below) that there is a bilipschitz embedding of $MG$ into some finite-dimensional subspace $W$ of $Z$ with distortion bounded by an absolute constant.

It is convenient to handle all $M \in \mathbb{N}$ simultaneously by considering the following thickening of the graph $G$ (see [Gro93, Section 1.B] for the general notion of thickening). For each edge $uv$ in $G$ we join $u$ and $v$ with a set isometric to $[0,1]$, we denote this set $t(uv)$. The *thickening* $TG$ of $G$ is the union of all sets $t(uv)$ (such sets can intersect at their ends only) with the distance between two points defined as the length of the shortest curve joining the points.

**Lemma 2.11.** *If a finite unweighted graph $G$ endowed with its graph distance admits a bilipschitz embedding $\tau$ into a finite-dimensional Banach space $X$, then the graph $TG$ admits a bilipschitz embedding $f$ into any infinite-dimensional Banach space $Z$ containing $X$ as a subspace, and the distortion of $f$ is bounded in terms of the distortion of $\tau$ and some absolute constants.*

**Remark 2.12.** In [Ost11c, Section 4] a stronger result was proved, namely, it was proved that in the case where $\dim X \geq 3$, the bilipschitz embedding $f$ whose existence is claimed in Lemma 2.11 can be required to map $TG$ into $X$. As we shall see, this result is not needed for Theorem 2.1. However, it could be of independent interest.

*Proof of Lemma* 2.11. We may assume without loss of generality that $\mathrm{Lip}(\tau^{-1}) = 1$, that is, $||\tau(u) - \tau(v)|| \geq d_G(u, v)$. We construct a bilipschitz embedding of $TG$ into an (arbitrary) Banach space $W$ containing $X$ as a subspace and satisfying $\dim(W/X) = e(G)$, where $e(G)$ is the number of edges in $G$ and $W/X$ is the quotient space.

We find an Auerbach basis in $W/X$. Recall the definition. Let $\{x_i\}_{i=1}^n$ be a basis in an $n$-dimensional Banach space $Y$, its *biorthogonal functionals* are defined by $x_i^*(x_j) = \delta_{ij}$ (Kronecker delta). The basis $\{x_i\}_{i=1}^n$ is called an *Auerbach basis* if $||x_i|| = ||x_i^*|| = 1$ for all $i \in \{1, \dots, n\}$. This notion goes back to [Aue30]. See [Ost11b, Section 2] and [Pli95] for historical comments and proofs.

Since the cardinality of the Auerbach basis is equal to the number of edges in $G$, we label its elements by edges. Also we lift the elements of this Auerbach basis into $W$. Since $W$ is finite dimensional, we may assume that the norms of the lifted elements are also equal to 1. We use the notation $\{e_{uv}\}_{uv\in E(G)}$ for the lifted elements of the Auerbach basis and the notation $\{e_{uv}^*\}_{uv\in E(G)}$ for its biorthogonal system. It is clear that $\{e_{uv}^*\}_{uv\in E(G)}$ may be regarded as elements of $W^*$.

If $uv$ is an edge in $G$, we map $t(uv)$ onto the concatenation of two line segments in $W \subset Z$, namely, onto the concatenation of $\left[\tau(u), \frac{\tau(u)+\tau(v)}{2} + e_{uv}\right]$ and $\left[\frac{\tau(u)+\tau(v)}{2} + e_{uv}, \tau(v)\right]$. More precisely, we map the point in $t(uv)$ which is at distance $\alpha \in [0, 1]$ from $u$ onto the point

$$\alpha\tau(v) + (1 - \alpha)\tau(u) + \min\{2\alpha, 2(1 - \alpha)\}e_{uv}. \tag{3}$$

It is easy to check that such points cover the concatenation of the line segments $\left[\tau(u), \frac{\tau(u)+\tau(v)}{2} + e_{uv}\right]$ and $\left[\frac{\tau(u)+\tau(v)}{2} + e_{uv}, \tau(v)\right]$. We denote this map by $f : TG \to W$.

We claim that $f$ is a bilipschitz embedding and its distortion is bounded in terms of distortion of $\tau$ and some absolute constant. To estimate $\mathrm{Lip}(f)$ observe that the derivative of the function in (3) with respect to $\alpha$ is $\tau(v) - \tau(u) \pm 2e_{uv}$ (at points where the derivative is defined). Hence $\mathrm{Lip}(f) \le \mathrm{Lip}(\tau) + 2$.

To estimate the Lipschitz constant of $f^{-1}$, we need to estimate from above the quotient

$$\frac{d_{TG}(x,y)}{||f(x) - f(y)||},$$

where $x, y \in TG$. Let $\alpha$ be the distance in $TG$ from $x \in t(uv)$ to $u$ and let $\beta$ be the distance from $y \in t(wz)$ to $w$. We may choose our notation in such a way that $\alpha, \beta \le \frac{1}{2}$. Let $D = d_{TG}(x, y)$.

First we consider the case when the edges $uv$ and $wz$ are different. We have $d_{TG}(u,w) \ge D - \alpha - \beta$. Thus $||f(u) - f(w)|| = ||\tau(u) - \tau(w)|| \ge D - \alpha - \beta$ and $||f(x) - f(y)|| \ge D - (\mathrm{Lip}(\tau) + 3)(\alpha + \beta)$. Here we use the fact that, since $\mathrm{Lip}(f) \le \mathrm{Lip}(\tau) + 2$, we have $||f(x) - f(u)|| \le (\mathrm{Lip}(\tau) + 2)\alpha$ and $||f(y) - f(w)|| \le (\mathrm{Lip}(\tau) + 2)\beta$. On the other hand $||f(x) - f(y)|| \ge e^*_{uv}(f(x) - f(y)) = e^*_{uv}(f(x) - f(u)) + e^*_{uv}(f(u) - f(y)) = e^*_{uv}(f(x) - f(u)) = 2\alpha$ (we use $uv \ne wz$, the definition of $\alpha$ and the fact that $\{e_{uv}\}$ is a lifted Auerbach basis in $W/X$). Similarly we get $||f(x) - f(y)|| \ge 2\beta$. Therefore

$$\frac{d_{TG}(x,y)}{||f(x) - f(y)||} \le \min\left\{\frac{D}{\max\{0, D - (\mathrm{Lip}(\tau) + 3)(\alpha + \beta)\}}, \frac{D}{2\alpha}, \frac{D}{2\beta}\right\}.$$

It is easy to see that the minimum in this inequality is bounded from above in terms of $\mathrm{Lip}(\tau)$ and an absolute constant. If $\alpha = 0$, or $\beta = 0$, or both, we modify this argument in a straightforward way.

It remains to consider the case where $x, y \in t(uv)$. Let $d_{TG}(x, u) = \alpha$ and $d_{TG}(y, u) = \beta$, so $d_{TG}(x, y) = |\alpha - \beta|$. It is easy to see that

$$|e^*_{uv}(f(x) - f(y))| = \begin{cases} 2\,|\alpha - \beta| & \text{if } \alpha \text{ and } \beta \text{ are on the same side of } \frac{1}{2} \\ 2\,|1 - \alpha - \beta| & \text{otherwise.} \end{cases}$$

In the former case we get

$$\frac{d_{TG}(x,y)}{||f(x) - f(y)||} \le \frac{1}{2}.$$

If the latter case we use

$$f(x) - f(y) = (\alpha - \beta)(\tau(v) - \tau(u)) \pm 2(1 - \alpha - \beta)e_{uv}.$$

We get from here that

$$||f(x) - f(y)|| \ge \max\{2|1 - \alpha - \beta|, |\alpha - \beta|||\tau(u) - \tau(v)|| - 2|1 - \alpha - \beta|\}.$$

The desired estimate follows. □

This proof shows that there exists an embedding $\varphi_0 : MG \to W$ such that $\mathrm{Lip}(\varphi_0) \le 1$ and $\mathrm{Lip}(\varphi_0^{-1})$ is bounded from above by an absolute constant. The rest of the proof is quite similar to the proof in [Ost11c], we only replace the embedding $\varphi_0 : MG \to X$ constructed in [Ost11c] by the embedding $\varphi_0 : MG \to W$ constructed in Lemma 2.11. For convenience of the reader we reproduce this part of the proof with necessary modifications. We number vertices along short paths using numbers from 1 to $e(G)$ in such a way that vertices numbered 1 correspond to the same long path in the correspondence described above.

At this point we are ready to describe the action of the map $\varphi$ on vertices of short paths. We construct the map $\varphi$ as a map into the Banach space $W \oplus_1 \mathbb{R}$. This is enough because $W \oplus_1 \mathbb{R}$ admits a linear bilipschitz embedding into $Z$ with distortion bounded by an absolute constant.

For vertex $w$ of $H$ having number $i$ on the short path $p_u$ the image of $w$ in $W \oplus_1 \mathbb{R}$ is $\varphi(w) = \varphi_0(u) \oplus i$ (here we use the same notation $u$ both for a vertex of $G$ and the corresponding vertex in $MG$).

To map vertices of long paths of $H$ into $W \oplus_1 \mathbb{R}$ we observe that the numbering of vertices of short paths leads to a one-to-one correspondence between long paths and numbers $\{1, \dots, e(G)\}$. We define the map $\varphi$ on a long path corresponding to $i$ by $\varphi(w) = \varphi_0(w') \oplus i$, where $w'$ is the uniquely determined vertex in a long path of $MG$ corresponding to a vertex $w$ in a long path of $H$.

The fact that $\mathrm{Lip}(\varphi) \le 1$ follows immediately from the easily verified claim that the distance between $\varphi$-images of adjacent vertices of $H$ is at most 1 (here we use $\mathrm{Lip}(\varphi_0) \le 1$).

We turn to an estimate of $\mathrm{Lip}(\varphi^{-1})$. In this part of the proof we assume that $M > 2e(G)$. Let $w$ and $z$ be two vertices of $H$. As we have already mentioned our construction implies that there are uniquely determined corresponding vertices $w'$ and $z'$ in $MG$.

Obviously there are two possibilities:

(1) $d_{MG}(w', z') \ge \frac{1}{2} d_H(w, z)$. In this case we observe that the definitions of $\varphi$ and of the norm on $W \oplus_1 \mathbb{R}$ imply that
$$\begin{aligned} ||\varphi(w) - \varphi(z)|| \ge ||\varphi_0(w') - \varphi_0(z')|| &\ge d_{MG}(w', z')/\mathrm{Lip}(\varphi_0^{-1}) \\ &\ge \frac{1}{2} d_H(w, z)/\mathrm{Lip}(\varphi_0^{-1}). \end{aligned}$$

(2) $d_{MG}(w', z') < \frac{1}{2} d_H(w, z)$. This inequality implies that there is a path joining $w$ and $z$ for which the naturally defined *short-paths-portion* is longer than the *long-paths-portion*. The inequality $M > 2e(G)$ implies that the short-paths-portion of this path consists of one path of length $> \frac{1}{2} d_H(w, z)$. This implies that the difference between the second coordinates of $w$ and $z$ in the decomposition $W \oplus_1 \mathbb{R}$ is $> \frac{1}{2} d_H(w, z)$. Thus $||\varphi(w) - \varphi(z)|| > \frac{1}{2} d_H(w, z)$.

Since $\mathrm{Lip}(\varphi_0^{-1}) \ge 1$ (this follows from the assumption $\mathrm{Lip}(\varphi_0) \le 1$), we get $\mathrm{Lip}(\varphi^{-1}) \le 2\mathrm{Lip}(\varphi_0^{-1})$ in each of the cases (1) and (2). The proof of Theorem 2.1 is completed.

## 3. Metric characterization with one test-space

Bourgain [Bou86] proved that a Banach space is nonsuperreflexive if and only if it admits uniformly bilipschitz embeddings of binary trees of all finite depths (see [BL00, pp. 412, 436] for the definition and equivalent characterizations of superreflexivity). Baudier [Bau07] strengthened the "only if" part of this result by proving that each nonsuperreflexive Banach space admits a bilipschitz embedding of an infinite binary tree.

Our purpose in this section is to find similar one-test-space-characterizations for classes of Banach spaces defined in terms of excluded finite-dimensional subspaces. At this moment we do not know how to do this for an arbitrary sequence of finite-dimensional subspaces, we found such a characterization only for increasing sequences of finite-dimensional subspaces.

**Theorem 3.1.** *Let $\{X_n\}_{n=1}^{\infty}$ be an increasing sequence of finite-dimensional Banach spaces with dimensions going to $\infty$. Then there exists an infinite graph $G$ such that the following conditions are equivalent:*

- *$G$ admits a bilipschitz embedding into a Banach space $X$.*
- *The spaces $\{X_n\}$ admit uniformly isomorphic embeddings into $X$.*

**Remark 3.2.** Theorem 3.1 is obviously weaker that Theorem 4.1 proved in Section 4. We give an independent proof of Theorem 3.1 because it is substantially simpler.

*Proof of Theorem* 3.1. Our proof is based on the construction of graphs providing approximate descriptions of convex sets, see Definition 2.5. We use the following immediate consequence of Theorem 2.8:

**Lemma 3.3.** *Let $X$ be a finite-dimensional Banach space, $Y$ be a Banach space admitting uniformly bilipschitz embeddings of some $(B_X(n), 1)$-graphs, and $C \in [1, \infty)$ be an upper bound for distortions of these embeddings. Then for each $\varepsilon > 0$ there is a linear embedding $T : X \to Y$ satisfying $||T|| \cdot ||T^{-1}|| \leq 3(1+\varepsilon)C$.*

Let $L$ be the inductive limit of the sequence $\{X_n\}$, that is, $L = \bigcup_{n=1}^{\infty} X_n$ with its natural vector operations and the norm whose restriction to each of $X_n$ is the norm of $X_n$. So $L$ is an incomplete normed space (we can, of course, consider its completion, but for our purposes completeness is not needed). We construct $G$ as a graph whose vertex set $V$ is a countable infinite subset of $L \oplus_1 \mathbb{R}$. (We fix the $\ell_1$-sum because it is convenient to have a precise formula for the norm, of course in the bilipschitz category all direct sums are equivalent.) The main features of the construction are:

(1) The graph $G$ with its shortest path metric is a locally finite metric space. (Recall that a metric space is called *locally finite* if all balls of finite radius in it have finite cardinality.)
(2) We have $V = \bigcup_{n=1}^{\infty} V_n$, where $V_n$ are finite and there exist uniformly bilipschitz embeddings $f_n : V_n \to (X_n \oplus_1 \mathbb{R})$, where we assume that the distance in $V_n$ is inherited from $G$.
(3) The set $V$ endowed with its shortest path metric $d_G$ contains images of bilipschitz embeddings of some $(B_{X_n}(m), 1, 2)$-graphs (with their shortest path metrics), $m, n \in \mathbb{N}$, with uniformly bounded distortions.

First let us explain why such graph $G$ satisfies the conclusion of Theorem 3.1. Suppose that $X$ is such that the spaces $\{X_n\}$ admit uniformly isomorphic embeddings into $X$. Then condition (2) implies that $V_n$ admit uniformly bilipschitz embeddings into $X$. Since $G$ is locally finite, by the main result of [Ost12], we get that $G$ admits a bilipschitz embedding into $X$.

Now suppose that $G$ admits a bilipschitz embedding into $X$. By (3) we get that $X$ admits uniformly bilipschitz embeddings of some $(B_{X_n}(m), 1)$-graphs, $m, n \in \mathbb{N}$. Applying Lemma 3.3, we get that the spaces $\{X_n\}$ admit uniformly isomorphic embeddings into $X$.

We construct $V$ as an infinite union $\bigcup_{n=1}^{\infty} V_n$, where each $V_n$ is a finite subset of

$$C_n := \operatorname{conv}\left(\bigcup_{k=1}^{n} (B_{X_k}(k), s_k)\right) \subset L \oplus_1 \mathbb{R},$$

where $s_k \in \mathbb{R}$, the pairs $(z, s_k)$ are in the sense of the decomposition $L \oplus_1 \mathbb{R}$, and $(B_{X_k}(k), s_k) = \{(z, s_k) : \ z \in B_{X_k}(k)\}$. Recall that $B_{X_k}(k)$ is the centered at 0 ball of $X_k$ of radius $k$.

Now we describe our choice of $V_n$ and $s_n$ such that the conditions (1)–(3) above are satisfied. One of the requirements is

$$s_{n+1} - s_n > 1. \tag{4}$$

We let $s_1 = 0$ and let $V_1$ to be a 1-separated 1-dense subset of $(B_{X_1}(1), s_1)$ (see Definition 2.3).

The choice of $s_2$ is less restrictive than further choices. We let $s_2 = 2$ and let $V_2$ be the extension of $V_1$ to a 1-separated 1-dense subset of

$$C_2 := \operatorname{conv}\left(\bigcup_{n=1}^{2} (B_{X_n}(n), s_n)\right)$$

satisfying the condition: some part of $V_2$ is a 1-dense in $(B_{X_1}(2), s_2)$, and some part of it is a 1-dense in $(B_{X_2}(2), s_2)$ (we use $s_2 - s_1 > 1$).

We use the following notation for a subsets $A$ of $L \oplus_1 \mathbb{R}$:

$$A[a, b] := A \cap (L \oplus_1 [a, b]), \tag{5}$$

where $[a, b]$ is an interval in $\mathbb{R}$. We use similar notation for open and half-open intervals.

Now we turn to the choice of $s_3$. We choose $s_3$ to satisfy (4) and to be so large that $V_2$ is $\left(1 + \frac{1}{2}\right)$-dense in $C_3[s_1, s_2]$. It is easy to see that sufficiently large $s_3$ satisfy these conditions. Then we extend $V_2$ to a 1-separated $\left(1 + \frac{1}{2}\right)$-dense in $C_3$ subset $V_3$ in such a way that

- $(V_3 \backslash V_2) \subset C_3(s_2, s_3]$
- Some parts of $V_3$ are $\left(1 + \frac{1}{2}\right)$-dense in the sets $(B_{X_1}(3), s_3)$, $(B_{X_2}(3), s_3)$, and $(B_{X_3}(3), s_3)$, respectively (here we use (4)).

We continue in the following way. We pick $s_4$ in such a way that

- $V_2$ is $\left(1 + \frac{1}{2} + \frac{1}{4}\right)$-dense in $C_4[s_1, s_2]$.
- $V_3$ is $\left(1 + \frac{1}{2} + \frac{1}{4}\right)$-dense in $C_4[s_1, s_3]$.

Now we extend $V_3$ to a 1-separated $\left(1+\frac{1}{2}+\frac{1}{4}\right)$-dense subset $V_4$ of $C_4$ in such a way that

- $(V_4 \backslash V_3) \subset C_4(s_3, s_4]$
- Some parts of $V_4$ are $\left(1+\frac{1}{2}+\frac{1}{4}\right)$-dense subsets in $(B_{X_1}(4), s_4)$, $(B_{X_2}(4), s_4)$, $(B_{X_3}(4), s_4)$, and $(B_{X_4}(4), s_4)$, respectively (here we use (4)).

We continue in an obvious way: In step $n$ we pick $s_n$, $s_n - s_{n-1} > 1$, in such a way that for each $m = 2, \ldots, n-1$ we have:

$V_m$ is a $\left(1+\frac{1}{2}+\frac{1}{4}+\cdots+\left(\frac{1}{2}\right)^{n-2}\right)$-dense subset of $C_n[s_1, s_m]$.

We extend $V_{n-1}$ to a 1-separated $\left(1+\frac{1}{2}+\cdots+\left(\frac{1}{2}\right)^{n-2}\right)$-dense subset $V_n$ of $C_n$ in such a way that

- $(V_n \backslash V_{n-1}) \subset C_n(s_{n-1}, s_n]$
- Some parts of $V_n$ are $\left(1+\frac{1}{2}+\cdots+\left(\frac{1}{2}\right)^{n-2}\right)$-dense subsets in
$$(B_{X_1}(n), s_n), \ldots, (B_{X_n}(n), s_n),$$
respectively (here we use (4)).

Let $V = \bigcup_{n=1}^{\infty} V_n$ and

$$C = \operatorname{conv}\left(\bigcup_{n=1}^{\infty} (B_{X_n}(n), s_n)\right) \subset L \oplus_1 \mathbb{R}.$$

Our construction implies that $V$ is a 1-separated 2-dense subset of $C$. We let $G$ be the corresponding $(C, 1, 2)$-graph (see Definition 2.5). It remains to verify that $G$ satisfies the conditions (1)–(3) above.

Condition (1). The set $V$ is a locally finite subset of $L \oplus_1 \mathbb{R}$ because it is contained in $L \oplus_1 [s_1, \infty)$ and its intersection with each subset of the form $L \oplus_1 [s_1, s_n]$ is a finite set $V_n$. The graph $G$ is locally finite because, by Lemma 2.6, its natural embedding into $L \oplus_1 \mathbb{R}$ is bilipschitz.

Condition (2). We apply Lemma 2.6 to $V$, the corresponding $(C, 1, 2)$-graph, and $L \oplus \mathbb{R}$. We get that the natural embedding of $V$ with the metric inherited from $G$ into $L \oplus \mathbb{R}$ is bilipschitz. Hence its restrictions to $V_n$ are uniformly bilipschitz. The fact that the restriction of this map to $V_n$ maps $V_n$ into $X_n \oplus \mathbb{R}$ follows from the definitions.

Condition (3). Our construction of set $V$ is such that it contains subsets which are 1-separated 2-dense subsets in shifted $B_{X_m}(n)$. We apply Lemma 2.6 twice. First time to $V$, the corresponding $(C, 1, 2)$-graph, and $L \oplus \mathbb{R}$. Second time we apply it to the set $(B_{X_m}(n), s_n)$, the corresponding $((B_{X_m}(n), s_n), 1, 2)$-graph (vertex set of this graph is the intersection of $V$ with $(B_{X_m}(n), s_n)$). We get that embeddings of all of these graphs into $L \oplus_1 \mathbb{R}$ have uniformly bounded distortions. Therefore the metrics of these $((B_{X_m}(n), s_n), 1, 2)$-graphs are bilipschitz equivalent to the metrics inherited from $G$. Hence the condition (3) is also satisfied. □

## 4. Characterization in terms of an infinite graph with maximum degree 3

Our next purpose is to show that the test-space for Theorem 3.1 can be chosen to have maximum degree 3:

**Theorem 4.1.** *Let $\{X_n\}_{n=1}^{\infty}$ be an increasing sequence of finite-dimensional Banach spaces with dimensions going to $\infty$. Then there exists an infinite graph $H$ with maximum degree 3 such that the following conditions are equivalent:*

- *$H$ admits a bilipschitz embedding into a Banach space $X$.*
- *The spaces $\{X_n\}$ admit uniformly isomorphic embeddings into $X$.*

*Proof.* Our proof uses some of the ideas of the proof of Theorem 3.1. For this reason we keep the same notation for some of the objects, although now they are somewhat different. We use Definition 2.3 and the notation introduced in formula (5).

We may assume without loss of generality that $\dim X_n = n$. We introduce convex sets $C_n$ and $C$ in $L \oplus \mathbb{R}$ of the form

$$C_n := \operatorname{conv}\left(\bigcup_{k=1}^{n}(B_{X_{i(k)}}(4^k), s_k)\right) \quad \text{and} \quad C = \bigcup_{n=1}^{\infty} C_n,$$

where $\{i(k)\}_{k=1}^{\infty}$ is a sequence of natural numbers satisfying $i(1) = 1$, $i(k) \leq i(k+1) \leq i(k)+1$, and such that the equality $i(k+1) = i(k)+1$ holds **rarely** (the exact condition will be described later), so the dimension of the sets $C_n$ increases slowly.

Let $\{s_n\}_{n=1}^{\infty}$ be a sequence of real numbers, such that $s_1 = 0$ and the following two conditions are satisfied:

**Gap condition:**

$$s_n - s_{n-1} > 6^{n+1}. \tag{6}$$

**Density condition:**

$$C_{n-1}[a, s_{n-1}] \text{ is } \left(\frac{1}{2}\right)^{n-2} - \text{dense in } C_n[a, s_{n-1}] \text{ for every } a \in [s_1, s_{n-1}]. \tag{7}$$

**Remark 4.2.** It is easy to verify that condition (7) is satisfied for each sufficiently rapidly increasing sequence $\{s_n\}_{n=1}^{\infty}$.

We construct two sequences of finite subsets in $L \oplus \mathbb{R}$, $\{A_n\}_{n=1}^{\infty}$ and $\{B_n\}_{n=2}^{\infty}$. The desired properties of these sequences of sets are the following:

1. $A_n$ is $2^n$-separated $2^n$-dense subset in $C_n[s_{n-1}+2^n, s_n]$. If $n = 1$, this condition is replaced by: $A_1$ is a 2-separated 2-dense set in $C_1$.
2. $A_n$ contains $2^n$-separated $2^n$-dense subsets in $\{(B_{X_{i(k)}}(4^n), s_n)\}_{k=1}^{n}$.
3. $B_n$, $n \geq 2$ is a $2^n$-separated subset of $C_n(s_{n-1}, s_{n-1}+2^n)$ such $d(A_n, B_n) \geq 2^n$, $d(A_{n-1}, B_n) \geq 2^n$ and $A_{n-1} \cup B_n \cup A_n$ is a $2^{n-1}$-separated $2^n$-dense subset in $C_n[s_{n-2}+2^{n-1}, s_n]$. If $n = 2$, the last condition is replaced by: 2-separated and $2^2$-dense in $C_2[s_1, s_2] = C_2$.

We construct such sets in steps. First we construct $A_n$, then $B_n$ (for $n \geq 2$). We start by letting $A_1$ to be any 2-separated 2-dense subset of $(B_{X_1}(4), s_1)$.

The construction of $A_n$ ($n \geq 2$) starts with picking a $2^n$-separated $2^n$-dense subset of

$$(B_{X_1}(4^n), s_n).$$

Then we gradually extend this subset to $2^n$-separated $2^n$-dense subsets of

$$(B_{X_{i(2)}}(4^n), s_n), \dots (B_{X_{i(n)}}(4^n), s_n). \tag{8}$$

Observe that our description of the sequence $\{i(n)\}$ implies that many sets in the sequence (8) are the same.

We complete the construction of $A_n$ extending the obtained set to a $2^n$-separated $2^n$-dense subset of $C_n[s_{n-1} + 2^n, s_n]$.

To construct $B_n$ we remove from $C_n(s_{n-1}, s_{n-1} + 2^n)$ all elements which are covered by $2^n$-balls centered in $A_{n-1} \cup A_n$. If the obtained set $R$ is empty, we let $B_n = \emptyset$. Otherwise we let $B_n$ be a $2^n$-separated $2^n$-dense subset of $R$.

The only condition which has to be verified is the condition that $A_{n-1} \cup B_n \cup A_n$ is $2^n$-dense in $C_n[s_{n-2} + 2^{n-1}, s_n]$ (and its version for $n = 2$). Here we use the condition (7). By this condition, since $A_{n-1}$ is $2^{n-1}$-dense in $C_{n-1}[s_{n-2} + 2^{n-1}, s_{n-1}]$, it is $\left(2^{n-1} + \left(\frac{1}{2}\right)^{n-2}\right)$-dense in $C_n[s_{n-2} + 2^{n-1}, s_{n-1}]$. The conclusion follows from the construction of $A_n$ and $B_n$ since $\left(2^{n-1} + \left(\frac{1}{2}\right)^{n-2}\right) < 2^n$.

Let

$$V = \left(\bigcup_{n=1}^{\infty} A_n\right) \cup \left(\bigcup_{n=2}^{\infty} B_n\right).$$

We create a weighted graph with the vertex set $V$ by joining a vertex $v \in (A_n \cup B_n)$, $n \geq 2$, to all vertices of

$$\left(\bigcup_{k=1}^{n} A_k\right) \cup \left(\bigcup_{k=2}^{n} B_k\right)$$

which are within distance $3 \cdot 2^n$ to $v$ in the normed space $L \oplus_1 \mathbb{R}$. (Also we join each vertex $v \in A_1$ to all vertices of $A_1$ which are within distance 6 to $v$.) The inequality (6) implies that in this way vertices of $A_n \cup B_n$ are joined only to some of the vertices in $A_{n-1} \cup B_{n-1}$, $A_n \cup B_n$, and $A_{n+1} \cup B_{n+1}$. The vertex $v$ is joined to some vertices in $A_{n+1} \cup B_{n+1}$ if $v$ is within distance $3 \cdot 2^{n+1}$ in $L \oplus_1 \mathbb{R}$ to those vertices.

We assign weight (length) $2^n$ to all edges joining $v \in (A_n \cup B_n)$ with vertices of $(A_{n-1} \cup B_{n-1}) \bigcup (A_n \cup B_n)$. Thus we assign weight (length) $2^{n+1}$ to the edges joining $v \in (A_n \cup B_n)$ with vertices of $(A_{n+1} \cup B_{n+1})$. It is clear that we get a well-defined weighted graph. We denote the obtained weighted graph by $W$ and endow it with its (weighted) shortest path metric.

We estimate the number of edges incident to $v \in (A_n \cup B_n)$ in the following way. All vertices joined to $v$ by edges are in a ball of radius $3 \cdot 2^{n+1}$ centered at $v$.

The distance between two vertices joined to $v$ is at least $2^{n-1}$ because all such vertices are in the set $A_{n-1} \cup B_{n-1} \cup A_n \cup B_n \cup A_{n+1} \cup B_{n+1}$, and it is clear from the conditions on $A_n$ and $B_n$ that the set $A_{n-1} \cup B_{n-1} \cup A_n \cup B_n \cup A_{n+1} \cup B_{n+1}$ is $2^{n-1}$-separated. All of the elements of this set are in $X_{i(n+1)} \oplus \mathbb{R}$, and the dimension of this space is $d(n) = i(n+1)+1$. Therefore $X_{i(n+1)} \oplus \mathbb{R}$-balls of radii $2^{n-2}$ centered at points joined to $v$ with an edge have disjoint interiors and are contained in a ball of radius $3 \cdot 2^{n+1} + 2^{n-2}$ centered at $v$. Comparing the volumes of the union of the $2^{n-2}$-balls and the $(3 \cdot 2^{n+1} + 2^{n-2})$-ball containing them (it is the standard volumetric argument, see, e.g., [MS86, Lemma 2.6]), we get that the number of vertices adjacent to $v$ is at most

$$\left(\frac{3 \cdot 2^{n+1} + 2^{n-2}}{2^{n-2}}\right)^{d(n)} = 25^{d(n)}.$$

**Lemma 4.3.** *The natural embedding of $W$ into the normed space $L \oplus_1 \mathbb{R}$ has distortion $\leq 3$. More precisely, its Lipschitz constant is $\leq 3$, and the Lipschitz constant of the inverse map is $\leq 1$.*

*Proof.* The statement about the Lipschitz constant of the natural embedding is immediate. In fact, ends of an edge of length $2^n$ in $W$ correspond to a vector in $L \oplus_1 \mathbb{R}$ whose length is $\leq 3 \cdot 2^n$.

The fact that the Lipschitz constant of the inverse map is $\leq 1$ can be proved as follows:

Let $a$ and $b$ be vertices of $W$. We may assume that $b \in (A_n \cup B_n)$ and $a \in (A_k \cup B_k)$ for some $k \leq n$. We use double induction. This means the following: First we prove the result for $n = 1$ using induction on $\lfloor ||a-b||/2 \rfloor$ (here we use Lemma 2.6).

Next, we assume that the result holds for $n = m$ and any $k \leq m$. We show that this assumption can be used to prove the result for $n = m+1$ using the induction on $\lfloor ||a-b||/(2^{m+1}) \rfloor$.

So let us follow the described program. If we divide all distances by 2, the desired inequality for $a, b \in A_1$ is a special case of Lemma 2.6 for $\alpha = \beta = 1$.

**Assumption:** Now we assume that we have proved the statement for all pairs $b \in (A_m \cup B_m)$, $a \in (A_k \cup B_k)$, $k \leq m$.

We show that this assumption can be used to prove the statement for $b \in (A_{m+1} \cup B_{m+1})$, $a \in (A_k \cup B_k)$, $k \leq m+1$, using induction on $\lfloor ||a-b||/(2^{m+1}) \rfloor$.

If $\lfloor ||a-b||/(2^{m+1}) \rfloor \leq 2$, then $a$ and $b$ are joined by an edge of length $2^{m+1}$. In addition, $||a-b|| \geq 2^{m+1}$ (see the conditions 1 and 3 in the description of $A_n$ and $B_n$). So in this case we get the desired $d_W(a,b) \leq ||a-b||$.

**Induction Hypothesis:** Suppose that the statement has been proved for all pairs $b \in (A_{m+1} \cup B_{m+1})$, $a \in (A_k \cup B_k)$, $k \leq m+1$, and

$$\lfloor ||a-b||/(2^{m+1}) \rfloor \leq D. \tag{9}$$

We show that this implies the same conclusion for pairs $b \in (A_{m+1} \cup B_{m+1})$, $a \in (A_k \cup B_k)$, $k \leq m+1$, satisfying $\lfloor ||a-b||/(2^{m+1}) \rfloor \leq D+1$.

We need to consider the case where the inequality $(D+2) \cdot 2^{m+1} > ||a-b|| \geq (D+1) \cdot 2^{m+1}$ is satisfied, $D \geq 2$, $D \in \mathbb{N}$. In such a case let $\widetilde{b}$ be the vector on the line segment joining $b$ to $a$ at distance $2 \cdot 2^{m+1}$ from $b$.

It is clear from the construction that $\widetilde{b} \in C_{m+1}$. Therefore (see condition 3 in the list of conditions on $A_n$ and $B_n$) there is a point $\widehat{b} \in V$ such that $||\widehat{b}-\widetilde{b}|| \leq 2^{m+1}$. We have $||\widehat{b}-b|| \leq 3 \cdot 2^{m+1}$ and

$$||\widehat{b}-a|| \leq ||b-a|| - 2^{m+1}. \tag{10}$$

Also we have $\widehat{b} \in A_m \cup B_m \cup A_{m+1} \cup B_{m+1}$. Therefore $d_W(\widehat{b}, b) \leq 2^{m+1}$.

In the case when $\widehat{b} \in A_{m+1} \cup B_{m+1}$, we get the desired conclusion using the Induction Hypothesis as follows: The inequality (10) implies that the pair $a, \widehat{b}$ satisfies the inequality (9). By the Induction Hypothesis,

$$||a-\widehat{b}|| \geq d_W(a, \widehat{b}).$$

Therefore

$$||a-b|| \geq 2^{m+1} + ||a-\widehat{b}|| \geq 2^{m+1} + d_W(a, \widehat{b}) \geq d_W(a, b).$$

It remains to consider the case $\widehat{b} \in A_m \cup B_m$. This case is to be divided into two subcases: $k = m+1$ and $k \leq m$. In the latter case we use the assumption that we have proved the statement for points in $(\bigcup_{i=1}^m A_i) \cup (\bigcup_{i=2}^m B_i)$. In the former case we use the Induction Hypothesis. □

The graph $W$ contains $(B_{X_k}(2^n), 1)$-graphs (in the sense of Definition 2.5) as subgraphs for all $k, n \in \mathbb{N}$, with all edges having the same weight of $2^n$. This follows from the choice of $4^n$ as the diameter of the ball used in the construction of $C_n$ and the fact that $A_n$ contains subsets which are $2^n$-separated and $2^n$-dense in the ball $(B(X_k(4^n)), s_n)$ for $k \leq i(n)$ (by combining these facts with our definitions). We claim that these $(B(X_k(2^n)), 1)$-graphs embed in a uniformly bilipschitz way into $W$. This claim can be proved in the following way: Applying Lemma 2.6 we get that the natural embeddings of the $(B(X_k(4^n)), 2^n)$-graphs (constructed using the $2^n$-separated and $2^n$-dense in the ball $(B(X_k(4^n)), s_n)$) into $L \oplus_1 \mathbb{R}$ are uniformly bilipschitz. By Lemma 4.3, the natural embedding of $W$ into $L \oplus_1 \mathbb{R}$ is also bilipschitz. The conclusion on uniformity of bilipschitz embeddings of the $(B(X_k(2^n)), 1)$-graphs into $W$ follows. Thus, by Lemma 2.7, bilipschitz embeddability of the graph $W$ into a Banach space $X$ implies that $X$ admits uniformly isomorphic embeddings of the spaces $\{X_i\}_{i=1}^\infty$.

Now we construct an unweighted graph $H$ with maximum degree 3 whose existence is claimed in Theorem 4.1. The graph $H$ will admit a bilipschitz embedding into $(L \oplus_1 \mathbb{R}) \oplus_1 \mathbb{R}$. The graph $H$ will be a modification of the weighted graph $W$. As in the construction of Section 2, the graph $H$ consists of long paths and short paths. Short paths correspond to vertices of $W$, long paths correspond to (weighted) edges of $W$. The length of a short path corresponding to a vertex in

$A_n \cup B_n$ is $2 \cdot 25^{d(n+1)}$. These lengths of short paths are chosen because they provide a sufficient number of colors for the coloring introduced in the next paragraph.

We are going to color edges of $W$. For our purposes we need a proper edge coloring (that is, edges having a common end should have different colors). Of course, since the degrees of $W$ are unbounded, we need infinitely many colors. Our purpose is to bound from above the number of colors used for edges incident with a vertex $v \in A_n \cup B_n$ by $2 \cdot 25^{d(n+1)}$. To achieve this goal we order vertices in $V$ according to the magnitude of their $\mathbb{R}$-coordinate in the decomposition $L \oplus \mathbb{R}$ (starting with those for which the $\mathbb{R}$-coordinate is 0), resolving ties in an arbitrary way.

We color edges incident with the first vertex arbitrarily (there are at most $25^{d(1)}$ of such edges). For each of the further vertices in our list we need to color all uncolored edges incident with them. We do this according to the following procedure. Let $v \in A_n \cup B_n$ be the next vertex in our list. We pick an uncolored edge incident with $v$, let $u$ be the other end of this edge. We cannot use for the edge $vu$ the colors which have already been used for other edges incident to $v$ and to $u$. There are at most $25^{d(n)} - 1$ colors already used for edges incident with $v$. As for $u$, we know that (see the construction of $W$) $u \in A_{n-1} \cup B_{n-1} \cup A_n \cup B_n \cup A_{n+1} \cup B_{n+1}$, therefore the degree of $u$ is $\leq 25^{d(n+1)}$. Therefore among $2 \cdot 25^{d(n+1)}$ colors there should be an available color for the edge $vu$.

Now we create a graph as in Section 2. The only difference is that we paste a long path to the level corresponding to its color (rather than to the level corresponding to the path itself as we did in Section 2). The length of the long path corresponding to an edge of $W$ of weight $2^n$ is $M \cdot 2^n$, where $M$ is a positive integer which we are going to select now, together with the sequence $\{i(n)\}_{n=1}^{\infty}$ (which, as we have already mentioned, is a slowly increasing sequence). The main condition describing our choice of both objects is

$$M \cdot 2^n > 4 \cdot 25^{d(n+1)}. \tag{11}$$

(Recall that $d(n) = i(n+1) + 1$.) This condition ensures that the length of a long path is larger than the sum of the lengths of the short paths at its ends.

We get an unweighted graph, let us denote it $H$. The maximum degree of $H$ is 3 because we use a proper edge coloring. The graph $W$ admits a bilipschitz embedding into $H$: consider the map which maps each vertex of $W$ to the vertex of level 1 on the corresponding short path. The Lipschitz constant of this embedding is $\leq 2 \cdot M$ by (11). The Lipschitz constant of the inverse map is $\leq M^{-1}$. In fact, if we consider two vertices in the image of $W$, and join them by a shortest path in $H$, the path goes through some collection of short paths (possibly at the end vertices only). The vertices in $W$ corresponding to these short paths form a path in $W$. Each time the weight of the edge in $W$ is $M^{-1} \times$ (the length of the corresponding path in $H$). The conclusion about the Lipschitz constant of the inverse map follows.

Observe that the graph $H$ is locally finite because its maximum degree is 3. It remains to show that the graph $H$ admits a bilipschitz embedding into any Banach space $X$ containing uniformly isomorphic $\{X_i\}$. We do this by proving the fact that the graph $H$ admits a bilipschitz embedding into $(L \oplus_1 \mathbb{R}) \oplus_1 \mathbb{R}$. By the finite determination result of [Ost12] this is enough because it is easy to see that finite-dimensional subspaces of the space $(L \oplus_1 \mathbb{R}) \oplus_1 \mathbb{R}$ are uniformly isomorphic to subspaces in $\{X_i\}$.

**Lemma 4.4.** *The graph $H$ admits a bilipschitz embedding into $(L \oplus_1 \mathbb{R}) \oplus_1 \mathbb{R}$.*

*Proof.* Vertices of $W$ are in $L \oplus_1 \mathbb{R}$. We map the short path corresponding to a vertex $v \in A_n \cup B_n$ onto those points of the line segment joining $(Mv, 1)$ with $(Mv, 2 \cdot 25^{d(n+1)})$ whose second coordinate is an integer. (The number $M$ is introduced by (11). In a pair $(Mv, a)$ the first component is in $L \oplus_1 \mathbb{R}$ and the second component is in the second $\mathbb{R}$-summand.)

Now we describe our map for long paths of $H$. For each long path of $H$, or, what is the same, for each edge $uv$ of $W$ we pick a vector $x_{uv} \in L$ (the space $L$ is identified with the corresponding summand in $(L \oplus_1 \mathbb{R}) \oplus_1 \mathbb{R}$). Suppose that the edge $uv$ has weight $2^n$ and color $i$ in the coloring above. We number vertices of the long path corresponding to $uv$ as $u_0, u_1, \ldots, u_N$, where $N = M \cdot 2^n$, $u$ corresponds to $u_0$ and $v$ corresponds to $u_N$. The image of $u_m$ $(m = 0, 1, \ldots, N)$ under the map which we are constructing is given by

$$Tu_m = \begin{cases} \left(\left(1 - \frac{m}{N}\right) Mu + \frac{m}{N} Mv + mx_{uv}, i\right) & \text{if } m \leq \frac{N}{2} \\ \left(\left(1 - \frac{m}{N}\right) Mu + \frac{m}{N} Mv + (N-m) x_{uv}, i\right) & \text{if } m \geq \frac{N}{2}. \end{cases} \tag{12}$$

So we map vertices of the long path in $H$ corresponding to an edge $uv$ of $W$ onto a sequence of evenly distributed points in the union of two line segments joining $(Mu, i)$ and $(Mv, i)$. We introduce also a map $O$ given by $Ou_m = \left(1 - \frac{m}{N}\right) Mu + \frac{m}{N} Mv$.

The map $T$ introduced by (12) is a Lipschitz map of the vertex set of $H$ into $(L \oplus_1 \mathbb{R}) \oplus_1 \mathbb{R}$ for an arbitrary uniformly bounded set of vectors $\{x_{uv}\}$ in $L$. To show this it suffices to find a bound for the distances between images of ends of an edge of $H$. For short-path-edges the distances are equal to 1 because their ends are mapped onto pairs of the form $(Mv, i)$, $(Mv, i+1)$. For a long-path-edge, the distance between the ends is

$$\left\| \frac{1}{N} Mv - \frac{1}{N} Mu \pm x_{uv} \right\|$$

This norm can be estimated from above by $3 + \sup_{uv} ||x_{uv}||$ (we use the estimate for the Lipschitz constant of Lemma 4.3).

Therefore the purpose of a suitable selection of the set $\{x_{uv}\}$ is to ensure that $T^{-1}$ is Lipschitz. In a similar situation in Section 2 we used Auerbach bases. For this construction we use a somewhat different type of biorthogonal sequences. We use systems whose existence was shown by Ovsepian-Pełczyński [OP75]. We mean the following result proved in [OP75] (see also [LT77, p. 44]):

**Theorem 4.5.** *There is an absolute constant $C > 0$ such that for each separable Banach space $Z$, each sequence $\{f_i^*\} \subset Z^*$, and each sequence $\{f_i\} \subset Z$ there exists a biorthogonal sequence $\{z_i, z_i^*\}_{i=1}^{\infty}$ in $Z$ such that $||z_i|| \leq C$, $||z_i^*|| \leq C$, the* *linear span of* $\{z_i\}$ *contains the sequence* $\{f_i\}$, *and the linear span of* $\{z_i^*\}$ *contains the sequence* $\{f_i^*\}$.

**Remark 4.6.** Pełczyński [Pel76] and Plichko [Pli76] proved that the constant $C$ can be chosen to be an arbitrary number $> 1$.

We apply Theorem 4.5 to $Z = L$ in the following situation. We form the sequence $\{f_i^*\}$ in the following way. We denote by $P_L : L \oplus \mathbb{R} \to L$ the natural projection. For any two edges $uv$ and $wz$ in $W$ we consider all vectors of the form

$$P_L(Ou_m - Ow_p) \tag{13}$$

for all admissible values of $m$ and $p$. The map $O$ was defined after formula (12) and $Ow_p$ is the image of a vertex $w_p$ of a long path corresponding to $wz$.

For each vector of the form (13) we pick a supporting functional $f^* \in L^*$, that is, a functional $f^*$ satisfying

$$f^*(P_L(Ou_m - Ow_p)) = ||P_L(Ou_m - Ow_p)||$$

and $||f^*|| = 1$. There are countably many such functionals, so we can form a sequence $\{f_i^*\}$ containing all of them. Also we form a sequence $\{f_i\}$ containing all of the vectors of the form (13).

Now we describe a suitable choice of the vectors $x_{uv}$ for (12). We enumerate edges of $W$ in the non-decreasing order of the larger $\mathbb{R}$-coordinates of their ends, resolving ties arbitrarily. Let $uv$ be the first edge in the ordering. We pick as $x_{uv}$ an element of the sequence $\{z_i\}$ satisfying the following two conditions:

- $z_i$ is annihilated by all functionals $f_j^*$ of the sequence $\{f_i^*\}$ supporting vectors $P_L(Ou_m - Ow_p)$, where $u_m$ is in the long path corresponding to $uv$ and $w_p$ is in the long path corresponding to an edge $wz$ for which the smaller $\mathbb{R}$-coordinate of its ends is $\le s_3$.
- $x_{uv}^* := z_i^*$ annihilates all vectors $f_j$ of the sequence $\{f_i\}$ of the form $P_L(Ou_m - Ow_p)$, where $u_m$ is in the long path corresponding to $uv$ and $w_p$ is in the long path corresponding to an edge $wz$ for which the smaller $\mathbb{R}$-coordinate of its ends is $\le s_3$.

Such pair $z_i, z_i^*$ exists because there are finitely many $f_j$ and $f_j^*$ satisfying the conditions. (Here we use the following conditions of Theorem 4.5: the linear span of $\{z_i\}$ contains the sequence $\{f_i\}$, and the linear span of $\{z_i^*\}$ contains the sequence $\{f_i^*\}$.)

We make a similar choice of $x_{uv}$ for all further edges in the ordering. More details: If we consider an edge for which the larger $\mathbb{R}$-coordinate of an end is in the interval $(s_{n-1}, s_n]$, we replace $s_3$ by $s_{n+2}$ in the conditions above. Also we pick different pairs $z_i, z_i^*$ for different edges $uv$.

With this choice of vectors $x_{uv}$, let us estimate from above the Lipschitz constant of the inverse of $T$. Let $x, y$ be two vertices of $H$. Let $x$ be on a long path joining $u$ and $v$ and $y$ be on a long path joining $w$ and $z$ (this is a generic description because $x$ and $y$ are allowed to be the end vertices of the long paths). We need to estimate from above the quotient

$$\frac{d_H(x,y)}{||Tx-Ty||}. \tag{14}$$

We consider a shortest $xy$-path in $H$. It has naturally defined *short-path portion* and *long-path portion*. There are two cases: (1) The length of the short-path portion of this path has length $\geq \frac{1}{2}d_H(x,y)$; The length of the long-path portion of this path has length $> \frac{1}{2}d_H(x,y)$.

The construction of the graph $H$ (see inequality (11)) is such that in the case (1) the short-path portion consists of just one piece. Let the short path portion start at level (color) $i$ and end at level $j$. Then $|i-j| \geq \frac{1}{2}d_H(x,y)$. On the other hand, since the sum $(L\oplus_1\mathbb{R})\oplus_1\mathbb{R}$ is direct, we have $||Tx-Ty|| \geq |i-j| \geq \frac{1}{2}d_H(x,y)$.

In the case (2) we ignore the difference in the second $\mathbb{R}$-coordinate (caused by the different colors of the edges $uv$ and $wz$). There are two subcases: Subcase (A): The vertices $x$ and $y$ are on the same long path; Subcase (B): The vertices $x$ and $y$ are on different long paths.

**Subcase (A):** Observe that our construction is such that there is a functional supporting $P_L(u-v)$ (let us denote it $f^*_{uv}$) which annihilates $x_{uv}$ (because $u-v$ is of the form $Ou_m - Ow_p$). Let $x = u_p$, $y = u_s$, we have $d_H(x,y) = |s-p|$ and

$$||Tu_p - Tu_s|| \geq |f^*_{uv}P_L(Ou_p - Ou_s)| = \frac{M|s-p|}{N}||P_L(u-v)||.$$

If $||P_L(u-v)|| = ||P_Lu - P_Lv||$ is a nontrivially large part of $||u-v|| \geq \frac{N}{M}$, we get the desired estimate.

If $||P_Lu - P_Lv||$ is only a small part of $||u-v||$, then the difference between the $\mathbb{R}$-coordinates of $u$ and $v$ is the large part of $||u-v||$. Denoting the projection of $L\oplus\mathbb{R}$ to $\mathbb{R}$ by $P_{\mathbb{R}}$, we use $P_{\mathbb{R}}(x_{uv}) = 0$ and get

$$||Tu_p - Tu_s|| \geq \frac{M|s-p|}{N}|P_{\mathbb{R}}u - P_{\mathbb{R}}v|,$$

so we get the estimate in this case, too.

**Subcase (B):** Let $x$ be on a long path corresponding to an edge $uv$ in $W$, and $y$ be on a long path corresponding to $wz$. Then there are two possibilities:

(i) One of the edges $uv$ and $wz$ has the largest $\mathbb{R}$-coordinate of its ends in the interval $(s_{n-1}, s_n]$, and the other edge has the least $\mathbb{R}$-coordinate of its ends in the interval $[s_m, s_{m+1})$, where $m \geq n+2$

(ii) It is not the case.

**Subsubcase (i):** Let $x = u_m$ and $y = w_p$. Ignoring the second $\mathbb{R}$-coordinate (in the sum $L\oplus\mathbb{R}\oplus\mathbb{R}$) and using $P_{\mathbb{R}}x_{uv} = P_{\mathbb{R}}x_{wz} = 0$ we get

$$||Tx - Ty|| \geq M|P_{\mathbb{R}}u - P_{\mathbb{R}}w| \geq M(s_m - s_n), \tag{15}$$

where $u$ and $w$ are the corresponding vertices picked in such a way that $u$ has larger $\mathbb{R}$-coordinate than $v$ and $w$ has smaller $\mathbb{R}$-coordinate than $w$. On the other hand

$$\begin{aligned} d_H(x,y) &\le d_H(u,w) + M2^n + M2^{m+1} \\ &\le M||u-w|| + M2^n + M2^{m+1} \\ &\le M|P_{\mathbb{R}}u - P_{\mathbb{R}}w| + M2^n + M2^{m+1} + M4^n + M4^{m+1} \\ &\le M|P_{\mathbb{R}}u - P_{\mathbb{R}}w| + M(s_m - s_n) \\ &\le 2M|P_{\mathbb{R}}u - P_{\mathbb{R}}w|. \end{aligned}$$

To get these inequalities we use

- The triangle inequality in $H$ for the first inequality.
- Lemma 4.3 for the second inequality.
- The observation that $||P_L z|| \le 4^n$ if $z \in C_n$ for the third inequality (see the definition of $C_n$).
- The gap condition (6) for the fourth inequality.
- The second inequality in (15) for the fifth inequality.

The conclusion follows.

**Subsubcase (ii):** We may assume without loss of generality that $x$ is closer (in $H$) to the short path corresponding to $u$ than to the short path corresponding to $v$ and that $y$ is closer to the short path corresponding to $w$ rather than to the short path corresponding to $z$. We have

$$||Tx-Ty|| \ge \left\|\left(1-\frac{m}{N}\right)Mu + \frac{m}{N}Mv + mx_{uv} - \left(1-\frac{p}{N'}\right)Mw - \frac{p}{N'}Mz - px_{wz}\right\|, \tag{16}$$

where $N'$ is the length of the long path corresponding to $wz$. Let us denote the vector whose norm is taken in the right-hand side of (16) by $B$. We get

$$||Tx - Ty|| \ge |x^*_{uv}(B)| = m.$$
$$||Tx - Ty|| \ge |x^*_{wz}(B)| = p.$$

Writing $x^*_{uv}(B), x^*_{wz}(B)$ we mean that the functionals $x^*_{uv}, x^*_{wz} \in L^*$ act on a vector $t \in L \oplus \mathbb{R}$ by acting on $P_L t$. We use the fact that the difference $\left(1-\frac{m}{N}\right)Mu + \frac{m}{N}Mv - \left(1-\frac{p}{N'}\right)Mw - \frac{p}{N'}Mz$ is of the from $Ou_m - Ow_p$ for edges satisfying the conditions above. Thus this difference is annihilated by $x^*_{uv}$ and $x^*_{wz}$.

We apply the triangle inequality to (16) and get

$$\begin{aligned} ||Tx - Ty|| &\ge M||u-w|| - \frac{mM}{N}||u-v|| - \frac{pM}{N'}||w-z|| - mC - pC \\ &= M||u-w|| - m\left(\frac{M}{N}||u-v|| + C\right) - p\left(\frac{M}{N'}||w-z|| + C\right) \\ &\ge d_H(u,w) - m\left(\frac{M}{N}||u-v|| + C\right) - p\left(\frac{M}{N'}||w-z|| + C\right), \end{aligned}$$

where $C = \sup_i ||z_i|| = \sup_{u,v} ||x_{uv}||$. (We used Lemma 4.3 to get the last inequality.) Observe that the numbers in brackets in the last line are bounded by an absolute constant, let us denote it by $D$. We also have $d_H(x,y) \leq d_H(u,w)+m+p$. Therefore

$$\frac{d_H(x,y)}{||Tx-Ty||} \leq \min\left\{\frac{d_H(u,w)+m+p}{d_H(u,w)-D(m+p)}, \frac{d_H(u,w)+m+p}{m}, \frac{d_H(u,w)+m+p}{p}\right\}.$$

It is easy to see that the minimum in the last formula is bounded from above by an absolute constant. □□

## References


[Aue30] H. Auerbach, *On the area of convex curves with conjugate diameters* (Polish), Ph. D. thesis, University of Lwów, 1930.

[Bau07] F. Baudier, Metrical characterization of super-reflexivity and linear type of Banach spaces, *Archiv Math.*, **89** (2007), no. 5, 419–429.

[Bau12] F. Baudier, Embeddings of proper metric spaces into Banach spaces, *Houston J. Math.*, **38** (2012), 209–223.

[Beg99] B. Begun, A remark on almost extensions of Lipschitz functions, *Israel J. Math.*, **109** (1999), 151–155.

[BL00] Y. Benyamini, J. Lindenstrauss, *Geometric Nonlinear Functional Analysis*, volume 1, American Mathematical Society, Providence, R.I., 2000.

[Bou86] J. Bourgain, The metrical interpretation of superreflexivity in Banach spaces, *Israel J. Math.*, **56** (1986), no. 2, 222–230.

[Bou87] J. Bourgain, Remarks on the extension of Lipschitz maps defined on discrete sets and uniform homeomorphisms, in: *Geometrical aspects of functional analysis* (1985/86), 157–167, *Lecture Notes in Math.*, **1267**, Springer, Berlin, 1987.

[BMW86] J. Bourgain, V. Milman, H. Wolfson, On type of metric spaces, *Trans. Amer. Math. Soc.*, **294** (1986), no. 1, 295–317.

[DS58] N. Dunford, J.T. Schwartz, *Linear Operators.* Part **I**: General Theory, New York, Interscience Publishers, 1958.

[GNS11] O. Giladi, A. Naor, G. Schechtman, Bourgain's discretization theorem, *Ann. Fac. Sci. Toulouse Math.* (6) **21** (2012), no. 4, 817–837; `arXiv:1110.5368`

[Gro93] M. Gromov, Asymptotic invariants of infinite groups, in: A. Niblo, M. Roller (eds.) *Geometric group theory*, London Math. Soc. Lecture Notes, **182**, 1–295, Cambridge University Press, 1993.

[Han77] *Handbook of mathematical logic.* Edited by Jon Barwise. With the cooperation of H.J. Keisler, K. Kunen, Y.N. Moschovakis, A.S. Troelstra. Studies in Logic and the Foundations of Mathematics, Vol. **90**, North-Holland Publishing Co., Amsterdam-New York-Oxford, 1977.

[HM82] S. Heinrich, P. Mankiewicz, Applications of ultrapowers to the uniform and Lipschitz classification of Banach spaces, *Studia Math.*, **73** (1982), no. 3, 225–251.

[JS09] W.B. Johnson, G. Schechtman, Diamond graphs and super-reflexivity, *J. Topol. Anal.*, **1** (2009), no. 2, 177–189.

[LT77] J. Lindenstrauss, L. Tzafriri, *Classical Banach spaces*, v. **I**, Berlin, Springer-Verlag, 1977.

[MN08] M. Mendel, A. Naor, Metric cotype, *Ann. Math.*, **168** (2008), 247–298.

[MS86] V.D. Milman, G. Schechtman, *Asymptotic theory of finite dimensional normed spaces*, Berlin, Springer-Verlag, 1986.

[Ost11a] M.I. Ostrovskii, On metric characterizations of some classes of Banach spaces, *C. R. Acad. Bulgare Sci.*, **64** (2011), no. 6, 775–784.

[Ost11b] M.I. Ostrovskii, Auerbach bases and minimal volume sufficient enlargements, *Canadian Math. Bull.*, **54** (2011), 726–738.

[Ost11c] M.I. Ostrovskii, Different forms of metric characterizations of classes of Banach spaces, *Houston J. Math.*, **39** (2013), 889–906.

[Ost12] M.I. Ostrovskii, Embeddability of locally finite metric spaces into Banach spaces is finitely determined, *Proc. Amer. Math. Soc.*, **140** (2012), 2721–2730.

[OP75] R.I. Ovsepian, A. Pełczyński, On the existence of a fundamental total and bounded biorthogonal sequence in every separable Banach space, and related constructions of uniformly bounded orthonormal systems in $L^2$, *Studia Math.*, **54** (1975), no. 2, 149–159.

[Pel76] A. Pełczyński, All separable Banach spaces admit for every $\varepsilon > 0$ fundamental total and bounded by $1 + \varepsilon$ biorthogonal sequences, *Studia Math.*, **55** (1976), no. 3, 295–304.

[Pis86] G. Pisier, Probabilistic methods in the geometry of Banach spaces, in: *Probability and analysis* (*Varenna,* 1985), 167–241, *Lecture Notes in Math.*, **1206**, Springer, Berlin, 1986.

[Pli76] A. Plichko, Existence of a complete $\varepsilon$-orthonormal system in a separable normed space (Russian), *Dokl. Akad. Nauk Ukrain. SSR*, Ser. A, **1976**, no. 1, 22–23.

[Pli95] A. Plichko, On the volume method in the study of Auerbach bases of finite-dimensional normed spaces, *Colloq. Math.*, **69** (1995), 267–270.

[Rib76] M. Ribe, On uniformly homeomorphic normed spaces, *Ark. Mat.*, **14** (1976), no. 2, 237–244.

[WS11] D.P. Williamson, D.B. Shmoys, *The Design of Approximation Algorithms*, Cambridge University Press, 2011.

[Yu06] G. Yu, Higher index theory of elliptic operators and geometry of groups, in: *International Congress of Mathematicians*, Vol. **II**, 1623–1639, Eur. Math. Soc., Zürich, 2006.

M.I. Ostrovskii
Department of Mathematics and Computer Science
St. John's University
8000 Utopia Parkway
Queens, NY 11439, USA
e-mail: ostrovsm@stjohns.edu

Operator Theory:
Advances and Applications, Vol. 233, 127–134


# Splittings of Masa-bimodules

S. Papapanayides and I.G. Todorov

**Abstract.** We study masa-bimodules whose support is an amplification of a given $\omega$-closed set $\kappa$, showing that its corresponding minimal and maximal masa-bimodules are amplifications of the minimal and the maximal masa-bimodules of $\kappa$. We provide a description of the convex hull of an amplification of a commutative bilattice.

**Mathematics Subject Classification (2010).** Primary 47L05; Secondary 47L35.

**Keywords.** Masa, bimodule, bilattice.

## 1. Introduction and preliminaries

Tomita's Commutation Theorem states that if $\mathcal{M}$ and $\mathcal{N}$ are von Neumann algebras then the commutant $(\mathcal{M}\bar{\otimes}\mathcal{N})'$ of the (spatial weak* closed) tensor product of $\mathcal{M}$ and $\mathcal{N}$ coincides with the tensor product $\mathcal{M}'\bar{\otimes}\mathcal{N}'$ of the commutants of $\mathcal{M}$ and $\mathcal{N}$; in particular, it splits as a tensor product of two von Neumann algebras. In the Theory of Non-selfadjoint Operator Algebras, the role of the commutant is often played by the invariant subspace lattice $\operatorname{Lat}\mathcal{A}$ of the corresponding algebra $\mathcal{A}$. It is thus natural to consider the following question: given subspace lattices $\mathcal{L}_1$ and $\mathcal{L}_2$ and a weak* closed unital operator algebra with $\operatorname{Lat}\mathcal{A} = \mathcal{L}_1 \otimes \mathcal{L}_2$, when can one conclude that $\mathcal{A}$ splits as a tensor product of the form $\mathcal{A}_1\bar{\otimes}\mathcal{A}_2$, where $\mathcal{A}_1$ and $\mathcal{A}_2$ are operator algebras with $\operatorname{Lat}\mathcal{A}_1 = \mathcal{L}_1$ and $\operatorname{Lat}\mathcal{A}_2 = \mathcal{L}_2$? In this note, we initiate the study of this general question by considering the simplest case, namely the case where one of the subspace lattices is trivial. More precisely, we obtain an affirmative answer to this question in case $\mathcal{L}_1$ is a commutative subspace lattice, while $\mathcal{L}_2 = \{0, I\}$ is the subspace lattice with two elements. It is convenient to work in the more general setting of operator subspaces rather than algebras; the absence of the notion of a unit imposes in this case additional restrictions on the results we obtain. Our splitting result for amplifications of masa-bimodules, Theorem 2.6, relies on a characterisation of splitting for weak* closed subspaces (Theorem 2.1), which can be viewed as a subspace version of a special case of, and can be obtained from, a result due to Kraus [8, Theorem 3]. Here we provide an alternative proof which, we believe, may be interesting in its own right. We show that the minimal (resp. maximal) masa-bimodule corresponding to the amplification of a set $\kappa$ is the amplification of the minimal (resp. maximal) masa-bimodule corresponding to $\kappa$. We also obtain characterisations of the convex hull of an amplification of a commutative bilattice (see the second part of this section for the definitions), an object which plays an important role in the theory of masa-bimodules [11].

In the rest of this section, we collect some preliminary notions, notation and results. Throughout the note, all Hilbert spaces will be assumed to be complex and separable. If $H$, $H_1$ and $H_2$ are Hilbert spaces, we denote as usual by $\mathcal{B}(H_1, H_2)$ the space of all bounded linear operators from $H_1$ into $H_2$, and set $\mathcal{B}(H) = \mathcal{B}(H, H)$. The identity operator on a Hilbert space will be denoted by $I$. We denote by $\mathcal{B}(H)_1^+$ the positive part of the unit ball of $\mathcal{B}(H)$. For a set of operators $\mathcal{M}$, we let $\mathcal{P}_{\mathcal{M}}$ be the set of all projections contained in $\mathcal{M}$, and we write $\mathcal{P}(H) = \mathcal{P}_{\mathcal{B}(H)}$.

Let $(X_i, \mu_i)$ be a standard measure space, set $H_i = L^2(X_i, \mu_i)$ and let $\mathcal{D}_i$ be the multiplication masa of $L^\infty(X_i, \mu_i)$, $i = 1, 2$. A $\mathcal{D}_2, \mathcal{D}_1$-bimodule (or, simply, a masa-bimodule) is a subspace $\mathcal{U} \subseteq \mathcal{B}(H_1, H_2)$ such that $\mathcal{D}_2 \mathcal{U} \mathcal{D}_1 \subseteq \mathcal{U}$.

We will need several facts and notions from the theory of masa-bimodules [1], [5], [11]. Given a masa-bimodule $\mathcal{U} \subseteq \mathcal{B}(H_1, H_2)$, we let

$$\mathrm{Bil}\,\mathcal{U} = \{(P, Q) \in \mathcal{D}_1 \times \mathcal{D}_2 : P, Q \text{ projections with } Q\mathcal{U}P = \{0\}\}$$

be its *bilattice* and define, following [9], the *reflexive hull of* $\mathcal{U}$ by

$$\mathrm{Ref}\,\mathcal{U} = \{T \in \mathcal{B}(H_1, H_2) : T\xi \in \overline{\mathcal{U}\xi}, \text{ for all } \xi \in H_1\}. \tag{1}$$

We note that $\mathrm{Ref}\,\mathcal{U} = \{T \in \mathcal{B}(H_1, H_2) : QTP = 0, \text{ for all } (P, Q) \in \mathrm{Bil}\,\mathcal{U}\}$. The space $\mathcal{U}$ is called *reflexive* [9] if $\mathcal{U} = \mathrm{Ref}\,\mathcal{U}$.

We equip the direct product $X_1 \times X_2$ with the product measure (defined on the $\sigma$-algebra generated by the rectangles $\alpha \times \beta$, where $\alpha \subseteq X_1$ and $\beta \subseteq X_2$ are measurable). A subset $E \subseteq X_1 \times X_2$ is called *marginally null* if $E \subseteq (M \times X_2) \cup (X_1 \times N)$, where $\mu_1(M) = \mu_2(N) = 0$. We call two subsets $E, F \subseteq X_1 \times X_2$ *marginally equivalent* (and write $E \simeq F$) if the symmetric difference of $E$ and $F$ is marginally null. A set $\kappa \subseteq X_1 \times X_2$ is called $\omega$-*open* if it is marginally equivalent to a (countable) union of the form $\cup_{i=1}^\infty \alpha_i \times \beta_i$, where $\alpha_i \subseteq X_1$ and $\beta_i \subseteq X_2$ are measurable, $i \in \mathbb{N}$. The complements of $\omega$-open sets are called $\omega$-*closed.* An operator $T \in \mathcal{B}(H_1, H_2)$ is said to be *supported on* $\kappa$ if, for all rectangles $\alpha \times \beta$ with $(\alpha \times \beta) \cap \kappa \simeq \emptyset$, we have that $P(\beta)TP(\alpha) = 0$. (Here $P(\gamma)$ stands for the projection of multiplication by the characteristic function of the measurable subset $\gamma$.) If $\kappa$ is an $\omega$-closed set, let

$$\mathfrak{M}_{\max}(\kappa) = \{T \in \mathcal{B}(H_1, H_2) : T \text{ is supported on } \kappa\}.$$

The space $\mathfrak{M}_{\max}(\kappa)$ is a reflexive masa-bimodule; conversely, every reflexive masa-bimodule is of this form, for a unique, up to marginal equivalence, $\omega$-closed set $\kappa$ [5]. Given a masa-bimodule $\mathcal{U}$, its *support* is the $\omega$-closed set $\kappa$ such that $\mathrm{Ref}\,\mathcal{U} = \mathfrak{M}_{\max}(\kappa)$. Given an $\omega$-closed subset $\kappa \subseteq X_1 \times X_2$, there exists a weak* closed masa-bimodule $\mathfrak{M}_{\min}(\kappa)$ such that $\operatorname{supp}\mathfrak{M}_{\min}(\kappa) = \kappa$ and that if $\mathcal{U}$ is any weak* closed masa-bimodule whose support is $\kappa$ then

$$\mathfrak{M}_{\min}(\kappa) \subseteq \mathcal{U} \subseteq \mathfrak{M}_{\max}(\kappa).$$

For all such $\mathcal{U}$, we have that

$$\operatorname{Bil}\mathcal{U} = \mathcal{S}(\kappa) \stackrel{def}{=} \{(P(\alpha), P(\beta)) : (\alpha \times \beta) \cap \kappa \simeq \emptyset\}.$$

The reflexive masa-bimodules in $\mathcal{B}(H)$ which are at the same time subalgebras of $\mathcal{B}(H)$ are precisely the CSL algebras introduced by Arveson in [1]. We recall that a commutative subspace lattice (CSL) is a lattice of projections $\mathcal{L}$ acting on a Hilbert space $H$, closed in the strong operator topology and such that $LM = ML$ for all $L, M \in \mathcal{L}$. A CSL algebra is the algebra $\mathcal{A}_{\max}(\mathcal{L})$ of all operators which leave invariant the range of every projection belonging to a given CSL $\mathcal{L}$; we have that $\operatorname{Lat}\mathcal{A}_{\max}(\mathcal{L}) = \mathcal{L}$. Given a CSL $\mathcal{L}$, Arveson showed in [1] that there exist smallest weak* closed algebra $\mathcal{A}_{\min}(\mathcal{L})$ containing a masa such that $\operatorname{Lat}\mathcal{A}_{\min}(\mathcal{L}) = \mathcal{L}$.

Given Hilbert spaces $H$ and $K$, we denote by $H \otimes K$ their Hilbertian tensor product. If $H_1$, $H_2$, $K_1$, $K_2$ are Hilbert spaces and $\mathcal{U}_i \subseteq \mathcal{B}(H_i, K_i)$, $i = 1, 2$, we denote by $\mathcal{U}_1 \bar{\otimes} \mathcal{U}_2$ the weak* closed subspace of $\mathcal{B}(H_1 \otimes K_1, H_2 \otimes K_2)$ generated by the elementary tensors $T_1 \otimes T_2$, where $T_1 \in \mathcal{U}_1$ and $T_2 \in \mathcal{U}_2$.

## 2. The results

Throughout this note, we will make the natural identification

$$\mathcal{B}(H_1 \otimes K_1, H_2 \otimes K_2) \equiv \mathcal{B}(H_1, H_2) \bar{\otimes} \mathcal{B}(K_1, K_2).$$

The following fact can be obtained from a special case of a result of Kraus' [8, Theorem 3]. Below we provide an alternative, operator space theoretic, proof.

**Theorem 2.1.** *Let $H_1$, $H_2$, $K_1$, $K_2$ be Hilbert spaces and $\mathcal{V} \subseteq \mathcal{B}(H_1 \otimes K_1, H_2 \otimes K_2)$ be a weak* closed subspace. The following are equivalent:*

(i) *there exists a weak* closed subspace $\mathcal{U} \subseteq \mathcal{B}(H_1, H_2)$ such that $\mathcal{V} = \mathcal{U} \bar{\otimes} \mathcal{B}(K_1, K_2)$;*

(ii) *$\mathcal{V}$ is a $1 \otimes \mathcal{B}(K_2), 1 \otimes \mathcal{B}(K_1)$-bimodule.*

*Proof.* (i)$\Rightarrow$(ii) is trivial.

(ii)$\Rightarrow$(i) For each $\omega \in \mathcal{B}(K_1, K_2)_*$, let $R_\omega : \mathcal{B}(H_1 \otimes K_1, H_2 \otimes K_2) \to \mathcal{B}(H_1, H_2)$ be the corresponding right Tomiyama slice map, i.e., the (unique) weak* continuous linear mapping from $\mathcal{B}(H_1 \otimes K_1, H_2 \otimes K_2)$ into $\mathcal{B}(H_1, H_2)$ satisfying $R_\omega(X \otimes Y) = \omega(Y)X$, $X \in \mathcal{B}(H_1, H_2)$, $Y \in \mathcal{B}(K_1, K_2)$. Fix a non-zero operator $S \in \mathcal{B}(K_1, K_2)$ and let $\tilde{\omega} : \mathcal{B}(K_1, K_2) \to \mathcal{B}(K_1, K_2)$ be the map $\tilde{\omega}(Y) = \omega(Y)S$, $Y \in \mathcal{B}(K_1, K_2)$. Since bounded functionals on $\mathcal{B}(K_1, K_2)$ are completely bounded [4, p.26], the map $\tilde{\omega}$ is completely bounded. Let $\widetilde{R}_\omega$ be the (linear) map on $\mathcal{B}(H_1 \otimes K_1, H_2 \otimes K_2)$ given by $\widetilde{R}_\omega(T) = R_\omega(T) \otimes S$; it is easy to see that $\widetilde{R}_\omega$ is weak* continuous. Note that, if $T = X \otimes Y$ then

$$\widetilde{R}_\omega(T) = R_\omega(X \otimes Y) \otimes S = \omega(Y)X \otimes S = X \otimes \omega(Y)S = (\operatorname{id} \otimes \tilde{\omega})(T).$$

The maps id and $\widetilde{\omega}$ are weak* continuous. It follows from [2, p. 40] that $\mathrm{id}\otimes\widetilde{\omega}$ is weak* continuous and thus we have that $\widetilde{R}_\omega = \mathrm{id}\otimes\widetilde{\omega}$.

Since $\widetilde{\omega}$ is a weak* continuous completely bounded map, it follows from [3, Theorem 2.2] that there exist sequences $(A_k)_{k\in\mathbb{N}} \subseteq \mathcal{B}(K_1)$ and $(B_k)_{k\in\mathbb{N}} \subseteq \mathcal{B}(K_2)$ such that $\sum_{k=1}^\infty B_k B_k^*$ and $\sum_{k=1}^\infty A_k^* A_k$ are weak* convergent and

$$\widetilde{\omega}(Y) = \sum_{k=1}^\infty B_k Y A_k, \quad Y \in \mathcal{B}(K_1, K_2).$$

Hence

$$\widetilde{R}_\omega(X\otimes Y) = \sum_{k=1}^\infty X\otimes(B_k Y A_k) = \sum_{k=1}^\infty (I\otimes B_k)(X\otimes Y)(I\otimes A_k).$$

Since the maps $\widetilde{R}_\omega$ and $\sum_{k=1}^\infty (I\otimes B_k)(\cdot)(I\otimes A_k)$ agree on elementary tensors and they are linear and weak* continuous, it follows that

$$R_\omega(T)\otimes S = \widetilde{R}_\omega(T) = \sum_{k=1}^\infty (I\otimes B_k)T(I\otimes A_k), \quad T\in\mathcal{B}(H_1\otimes K_1, H_2\otimes K_2). \quad (2)$$

Let

$$\mathcal{U} = \{X\in\mathcal{B}(H_1,H_2) : X\otimes S\in\mathcal{V}\};$$

clearly, $\mathcal{U}\bar{\otimes}S \subseteq \mathcal{V}$. Since $\mathcal{V}$ is a $1\otimes\mathcal{B}(K_2), 1\otimes\mathcal{B}(K_1)$-bimodule, we have that $\mathcal{U}\bar{\otimes}\mathcal{B}(K_1,K_2)\subseteq\mathcal{V}$. Let $T\in\mathcal{V}$ and $\omega\in\mathcal{B}(K_1,K_2)_*$. Since $\mathcal{V}$ is weak* closed and is a $1\otimes\mathcal{B}(K_2), 1\otimes\mathcal{B}(K_1)$-bimodule, it follows by (2) that $R_\omega(T)\otimes S\in\mathcal{V}$. Hence, $R_\omega(T)\in\mathcal{U}$ for every $\omega\in\mathcal{B}(K_1,K_2)_*$. Since $\mathcal{B}(K_1,K_2)$ possesses (the subspace version of) Kraus' property $S_\sigma$ [7], we conclude that $T\in\mathcal{U}\bar{\otimes}\mathcal{B}(K_1,K_2)$. □

**Corollary 2.2.** *Let $H$ and $K$ be Hilbert spaces and $\mathcal{B}\subseteq\mathcal{B}(H\otimes K)$ be a weak* closed unital algebra. The following are equivalent:*

(i) *there exists a weak* closed unital algebra $\mathcal{A}\subseteq\mathcal{B}(H)$ such that $\mathcal{B}=\mathcal{A}\bar{\otimes}\mathcal{B}(K)$;*
(ii) $1\otimes\mathcal{B}(K)\subseteq\mathcal{B}$.

*Proof.* (i)⇒(ii) is trivial. Assuming (ii), the fact that $\mathcal{B}$ is an algebra implies that it is a $1\otimes\mathcal{B}(K)$-bimodule. By Theorem 2.1, there exists a weak* closed subspace $\mathcal{A}\subseteq\mathcal{B}(H)$ with $\mathcal{B}=\mathcal{A}\bar{\otimes}\mathcal{B}(H)$. The fact that $\mathcal{B}$ is a unital algebra now easily implies that $\mathcal{A}$ is a unital algebra. □

In the remainder of this note, we fix standard measure spaces $(X_i,\mu_i)$, $(Y_i,\nu_i)$, and let $H_i = L^2(X_i,\mu_i)$, $K_i = L^2(Y_i,\nu_i)$, $i=1,2$. We let $\mathcal{D}_i$ (resp. $\mathcal{C}_i$) be the multiplication masa of $L^\infty(X_i,\mu_i)$ (resp. $L^\infty(Y_i,\nu_i)$) acting on $H_i$ (resp. $K_i$), $i=1,2$. It will be convenient to introduce the notation $\mathcal{E}_i=\mathcal{D}_i\bar{\otimes}\mathcal{C}_i$; we have that $\mathcal{E}_i$ is the multiplication masa of $L^\infty(X_i\times Y_i,\mu_i\times\nu_i)$, $i=1,2$.

If $\kappa\subseteq X_1\times X_2$ is an $\omega$-closed set, we let

$$\widetilde{\kappa} = \{(x_1,y_1,x_2,y_2) : (x_1,x_2)\in\kappa, (y_1,y_2)\in Y_1\times Y_2\}.$$

If $H'$ and $H''$ are Hilbert spaces and $\mathcal{S} \subseteq \mathcal{B}(H') \times \mathcal{B}(H'')$ then, following [11], we denote by $\widetilde{\mathcal{S}}$ the set

$$\widetilde{\mathcal{S}} = \{(P, Q) \in \mathcal{P}(\ell^2 \otimes H') \times \mathcal{P}(\ell^2 \otimes H'') : (L_\phi(P), L_\phi(Q)) \in \operatorname{Conv} \mathcal{S},$$
$$\text{for every state } \phi \in \mathcal{B}(\ell^2)_*\}.$$

Here, $\operatorname{Conv} \mathcal{S}$ is the weakly closed convex hull of the set $\mathcal{S}$ and $L_\phi$ is the left Tomiyama slice map corresponding to the functional $\phi$. In the sequel, we will use the above construction in the cases where $H' = H_1$, $H'' = H_2$, or $H' = H_1 \otimes K_1$, $H'' = H_2 \otimes K_2$. We note that if $\mathcal{S} = \mathcal{S}(\kappa) \subseteq \mathcal{D}_1 \times \mathcal{D}_2$ is the bilattice corresponding to an $\omega$-closed set $\kappa \subseteq X_1 \times X_2$ then, as was shown in [11],

$$\mathfrak{M}_{\min}(\kappa) = \{T \in \mathcal{B}(L^2(X_1), L^2(X_2)) : Q(I \otimes T)P = 0, \ \forall\ (P, Q) \in \widetilde{\mathcal{S}}\}.$$

Given a weakly closed set $\mathcal{S} \subseteq \mathcal{P}_{\mathcal{D}_1} \times \mathcal{P}_{\mathcal{D}_2}$, we let

$$\mathcal{S} \otimes 1 = \{(P, Q) \in \mathcal{P}_{\mathcal{E}_1} \times \mathcal{P}_{\mathcal{E}_2} : \ \exists\ (p, q) \in \mathcal{S} \text{ with } P \leq p \otimes I \text{ and } Q \leq q \otimes I\}.$$

**Lemma 2.3.** *Let $\kappa \subseteq X_1 \times X_2$ be an $\omega$-closed set. The following hold:*

(i) $\mathfrak{M}_{\max}(\widetilde{\kappa}) = \mathfrak{M}_{\max}(\kappa) \bar{\otimes} \mathcal{B}(K_1, K_2)$;

(ii) $\mathcal{S}(\widetilde{\kappa}) = \mathcal{S}(\kappa) \otimes 1$.

*Proof.* (i) By [10], $\operatorname{supp}(\mathfrak{M}_{\max}(\kappa) \bar{\otimes} \mathcal{B}(K_1, K_2)) = \widetilde{\kappa}$. Since $\mathfrak{M}_{\max}(\widetilde{\kappa})$ is the unique reflexive masa-bimodule with support $\widetilde{\kappa}$, the proof of part (i) will be completed if we show that $\mathfrak{M}_{\max}(\kappa) \bar{\otimes} \mathcal{B}(K_1, K_2)$ is reflexive. To this end, we will establish the identity

$$\mathfrak{M}_{\max}(\kappa) \bar{\otimes} \mathcal{B}(K_1, K_2) = \{T \in \mathcal{B}(H_1 \otimes K_1, H_2 \otimes K_2) :$$
$$(q \otimes I)T(p \otimes I) = 0, \ \forall\ (p, q) \in \mathcal{S}(\kappa)\}.$$

It is clear that the left-hand side is a subset of the right-hand side; let, therefore, $T$ be an operator with $(q \otimes I)T(p \otimes I) = 0$ whenever $(p, q) \in \mathcal{S}(\kappa)$. Then, for each $\omega \in \mathcal{B}(K_1, K_2)_*$, we have that $qR_\omega(T)p = R_\omega((q \otimes I)T(p \otimes I)) = 0$. It follows that $R_\omega(T) \in \mathfrak{M}_{\max}(\kappa)$; since this holds for all normal functionals $\omega$, we conclude by [7] that $T \in \mathfrak{M}_{\max}(\kappa) \bar{\otimes} \mathcal{B}(K_1, K_2)$.

(ii) By (i), $\mathcal{S}(\widetilde{\kappa})$ is the bilattice of the amplification $\mathfrak{M}_{\max}(\kappa) \bar{\otimes} \mathcal{B}(K_1, K_2)$. It is trivial to check that it contains $\mathcal{S}(\kappa) \otimes 1$. Conversely, suppose that $(P, Q) \in \mathcal{S}(\widetilde{\kappa})$, and let $p$ and $q$ be the projections defined by

$$pH_1 \otimes K_1 = \overline{(\mathcal{D}_1 \bar{\otimes} \mathcal{B}(K_1))P(H_1 \otimes K_1)}$$

and

$$qH_2 \otimes K_2 = \overline{(\mathcal{D}_2 \bar{\otimes} \mathcal{B}(K_2))Q(H_2 \otimes K_2)}.$$

Since $\mathfrak{M}_{\max}(\kappa)$ is a $\mathcal{D}_2, \mathcal{D}_1$-bimodule, we can easily verify that $p \in \mathcal{P}_{\mathcal{D}_1}$ and $q \in \mathcal{P}_{\mathcal{D}_2}$. Moreover, $(p \otimes I, q \otimes I)$ is by its definition an element of the bilattice of $\mathfrak{M}_{\max}(\kappa) \bar{\otimes} \mathcal{B}(K_1, K_2)$ and $P \leq p \otimes I$, $Q \leq q \otimes I$; thus, $\mathcal{S}(\widetilde{\kappa}) \subseteq \mathcal{S}(\kappa) \otimes 1$ and the proof is complete. □

**Lemma 2.4.** *If $\mathcal{S} \subseteq \mathcal{D}_1 \times \mathcal{D}_2$ is a bilattice then*

$$\operatorname{Conv}(\mathcal{S} \otimes 1) = \{(A, B) \in \mathcal{E}_1^+ \times \mathcal{E}_2^+ : \text{there exists } (a, b) \in \operatorname{Conv} \mathcal{S} \text{ with } A \leq a \otimes 1, B \leq b \otimes 1\}.$$

*Proof.* Let $V$ denote the right-hand side of the above equality. Since $\operatorname{Conv} \mathcal{S}$ is convex, $V$ is convex, too. Suppose that $((P_i, Q_i))_{i \in \mathbb{I}} \subseteq V$ is a net, weakly convergent to the pair $(P, Q)$. Let $(p_i, q_i) \in \operatorname{Conv} \mathcal{S}$ be such that $P_i \leq p_i \otimes I$ and $Q_i \leq q_i \otimes I$. By the uniform boundedness of $((p_i, q_i))_{i \in \mathbb{I}}$, there exists a subnet $((p_j, q_j))_{j \in \mathbb{J}}$ of $((p_i, q_i))_{i \in \mathbb{I}}$ such that $p_j \to p$ and $q_j \to q$ in the weak operator topology; since $\operatorname{Conv} \mathcal{S}$ is weakly closed, we have that $(p, q) \in \operatorname{Conv} \mathcal{S}$. Passing to a limit along $J$, we conclude that $P \leq p \otimes I$ and $Q \leq q \otimes I$, and so $(P, Q) \in V$. We thus showed that $V$ is weakly closed and convex. Since it contains $\mathcal{S} \otimes 1$, we have that $\operatorname{Conv}(\mathcal{S} \otimes 1) \subseteq V$.

We next note that $\operatorname{Conv}(\mathcal{S} \otimes 1)$ is hereditary in the sense that if $(A, B) \in \operatorname{Conv}(\mathcal{S} \otimes 1)$ and $A_0 \in \mathcal{E}_1$ and $B_0 \in \mathcal{E}_2$ are positive operators with $A_0 \leq A$ and $B_0 \leq B$ then $(A_0, B_0) \in \operatorname{Conv}(\mathcal{S} \otimes 1)$. Fix $(A, B) \in \operatorname{Conv}(\mathcal{S} \otimes 1)$. By [11, Lemma 3.2], $(E_A[s, 1], E_B[t, 1]) \in \mathcal{S} \otimes 1$ whenever $s + t > 1$. Since $A_0 \leq A$ and $B_0 \leq B$, we have that $E_{A_0}[s, 1] \leq E_A[s, 1]$ and $E_{B_0}[t, 1] \leq E_B[t, 1]$; thus, $(E_{A_0}[s, 1], E_{B_0}[t, 1]) \in \mathcal{S} \otimes 1$ whenever $s + t > 1$ and hence $(A_0, B_0) \in \operatorname{Conv}(\mathcal{S} \otimes 1)$.

If $(a, b) \in \operatorname{Conv} \mathcal{S}$ then clearly $(a \otimes I, b \otimes I) \in \operatorname{Conv}(\mathcal{S} \otimes 1)$; by the previous paragraph, $V \subseteq \operatorname{Conv}(\mathcal{S} \otimes 1)$ and the proof is complete. □

**Lemma 2.5.** *Let $\mathcal{S} \subseteq \mathcal{D}_1 \times \mathcal{D}_2$ be a bilattice. Then*

$$\widetilde{\mathcal{S} \otimes 1} = \{(E, F) \in (\mathcal{P}_{\ell^2} \otimes \mathcal{P}_{\mathcal{E}_1}) \times (\mathcal{P}_{\ell^2} \otimes \mathcal{P}_{\mathcal{E}_2}) : \text{there exists } (P, Q) \in \widetilde{\mathcal{S}} \text{ with } E \leq P \otimes I \text{ and } F \leq Q \otimes I\}.$$

*Proof.* Let $\kappa \subseteq X_1 \times X_2$ be the $\omega$-closed set with $\mathcal{S} = \mathcal{S}(\kappa)$. Denote by $\widetilde{\mathcal{S}} \otimes 1$ the right-hand side of the displayed equality in the statement of Lemma 2.5, and by $V$ the right-hand side of the displayed equality in the statement of Lemma 2.4. Let $(P, Q) \in \widetilde{\mathcal{S}}$. By definition, $(L_\phi(P), L_\phi(Q)) \in \operatorname{Conv} \mathcal{S}$ for every state $\phi \in \mathcal{B}(\ell^2)_*$ and thus

$$(L_\phi(P \otimes I), L_\phi(Q \otimes I)) = (L_\phi(P) \otimes I, L_\phi(Q) \otimes I) \in V,$$

for every state $\phi \in \mathcal{B}(\ell^2)_*$. It follows from Lemma 2.4 that $(L_\phi(P \otimes I), L_\phi(Q \otimes I)) \in \operatorname{Conv}(\mathcal{S} \otimes 1)$ for every state $\phi \in \mathcal{B}(\ell^2)_*$. Hence, $(P \otimes I, Q \otimes I) \in \widetilde{\mathcal{S} \otimes 1}$. Now, if $(E, F) \in \widetilde{\mathcal{S}} \otimes 1$ then $E \leq P \otimes I$ and $F \leq Q \otimes I$ for some $(P, Q) \in \widetilde{\mathcal{S}}$. By Lemma 2.4, $\widetilde{\mathcal{S} \otimes 1}$ is hereditary, and the previous arguments imply that $(E, F) \in \widetilde{\mathcal{S} \otimes 1}$.

Conversely, suppose that $(E, F) \in \widetilde{\mathcal{S} \otimes 1}$. By [11, Lemma 4.1],

$$E(x_1, y_1) F(x_2, y_2) = 0 \quad \text{whenever } (x_1, x_2) \in \kappa \text{ and } (y_1, y_2) \in Y_1 \times Y_2.$$

(Here we identify a projection in $\mathcal{B}(\ell^2) \bar{\otimes} \mathcal{E}_i$ with a measurable function $X_i \times Y_i \to \mathcal{P}(\ell^2)$ in the natural way.) Let $P(x_1) = \vee_{y_1 \in Y_1} E(x_1, y_1)$, $x_1 \in X_1$, and $Q(x_2) = \vee_{y_2 \in Y_2} F(x_2, y_2)$, $x_2 \in X_2$. We have that $P(x_1) Q(x_2) = 0$ whenever $(x_1, x_2) \in \kappa$, and so, by [11, Lemma 4.1], $(P, Q) \in \widetilde{\mathcal{S}}$; moreover, $E \leq P \otimes I$ and $F \leq Q \otimes I$. It follows that $(E, F) \in \widetilde{\mathcal{S}} \otimes 1$, and the proof is complete. □

**Theorem 2.6.** *Let $\kappa \subseteq X_1 \times X_2$ be an $\omega$-closed set. Then*

(i) $\mathfrak{M}_{\min}(\widetilde{\kappa}) = \mathfrak{M}_{\min}(\kappa)\bar{\otimes}\mathcal{B}(K_1, K_2)$, *and*

(ii) $\mathfrak{M}_{\max}(\widetilde{\kappa}) = \mathfrak{M}_{\max}(\kappa)\bar{\otimes}\mathcal{B}(K_1, K_2)$.

*Moreover, the following are equivalent for a weak* closed masa-bimodule $\mathcal{V} \subseteq \mathcal{B}(H_1 \otimes K_1, H_2 \otimes K_2)$ with $\operatorname{supp} \mathcal{V} = \widetilde{\kappa}$:*

(a) *there exists a weak* closed masa-bimodule $\mathcal{U} \subseteq \mathcal{B}(H_1, H_2)$ with $\mathfrak{M}_{\min}(\kappa) \subseteq \mathcal{U} \subseteq \mathfrak{M}_{\max}(\kappa)$ such that $\mathcal{V} = \mathcal{U}\bar{\otimes}\mathcal{B}(K_1, K_2)$;*

(b) *$\mathcal{V}$ is a $1 \otimes \mathcal{B}(K_2), 1 \otimes \mathcal{B}(K_1)$-bimodule.*

*Proof.* (i) Let $\mathcal{S} = \mathcal{S}(\kappa)$. By [10], the masa-bimodule $\mathfrak{M}_{\min}(\kappa)\bar{\otimes}\mathcal{B}(K_1, K_2)$ has support $\widetilde{\kappa}$, and since $\mathfrak{M}_{\min}(\widetilde{\kappa})$ is the smallest weak* closed masa-bimodule with support $\widetilde{\kappa}$, we have that $\mathfrak{M}_{\min}(\widetilde{\kappa}) \subseteq \mathfrak{M}_{\min}(\kappa)\bar{\otimes}\mathcal{B}(K_1, K_2)$. To prove the converse inclusion, let $X \in \mathfrak{M}_{\min}(\kappa)$ and $Y \in \mathcal{B}(K_1, K_2)$. Suppose that $(P, Q)$ is a pair of projections lying in $\widetilde{\mathcal{S}(\widetilde{\kappa})}$. By Lemma 2.3, $(P, Q) \in \widetilde{\mathcal{S} \otimes 1}$, and by Lemma 2.5, $P \leq P_0 \otimes I$ and $Q \leq Q_0 \otimes I$, for some pair $(P_0, Q_0)$ of projections lying in $\widetilde{\mathcal{S}}$. Then

$$(Q_0 \otimes I)(I \otimes X \otimes Y)(P_0 \otimes I) = (Q_0(I \otimes X)P_0) \otimes Y = 0.$$

It follows that $Q(I \otimes X \otimes Y)P = 0$. By [11], $X \otimes Y \in \mathfrak{M}_{\min}(\widetilde{\kappa})$, and (i) is proved.

(ii) is Lemma 2.3 (i).

(a)⇒(b) is trivial.

(b)⇒(a) By Theorem 2.1, $\mathcal{V} = \mathcal{U}\bar{\otimes}\mathcal{B}(K_1, K_2)$, and (i) and (ii) imply that

$$\mathfrak{M}_{\min}(\kappa)\bar{\otimes}\mathcal{B}(K_1, K_2) \subseteq \mathcal{U}\bar{\otimes}\mathcal{B}(K_1, K_2) \subseteq \mathfrak{M}_{\max}(\kappa)\bar{\otimes}\mathcal{B}(K_1, K_2).$$

Now a simple slice map argument implies that $\mathfrak{M}_{\min}(\kappa) \subseteq \mathcal{U} \subseteq \mathfrak{M}_{\max}(\kappa)$, and the proof is complete. □

Using Corollary 2.2 and arguments similar to the ones used in the proof of Theorem 2.6, we can establish the following operator algebraic version of Theorem 2.6, whose proof we omit. For a CSL $\mathcal{L}$, we let below $\mathcal{L} \otimes 1 = \{L \otimes I : L \in \mathcal{L}\}$.

**Theorem 2.7.** *Let $\mathcal{L}$ be a CSL acting on a Hilbert space $H$ and $K$ be a Hilbert space. Suppose that $\mathcal{B}$ is a weak* closed algebra containing a masa such that* $\operatorname{Lat} \mathcal{B} = \mathcal{L} \otimes 1$. *Then there exists a weak* closed algebra containing a masa such that* $\operatorname{Lat} \mathcal{A} = \mathcal{L}$ *and $\mathcal{B} = \mathcal{A}\bar{\otimes}\mathcal{B}(K)$.*

For the last corollary, we recall [1], [11] that an $\omega$-closed set $\kappa$ is said to satisfy operator synthesis if $\mathfrak{M}_{\min}(\kappa) = \mathfrak{M}_{\max}(\kappa)$.

**Corollary 2.8.** *An $\omega$-closed set $\kappa \subseteq X_1 \times X_2$ satisfies operator synthesis if and only if the set $\widetilde{\kappa}$ satisfies operator synthesis.*

*Proof.* Immediate from Theorem 2.6. □

**Acknowledgement**

We would like to thank G. Eleftherakis and L. Turowska for the helpful discussions on the topic of this paper.

## References

[1] W. ARVESON, *Operator Algebras and Invariant Subspaces*, Ann. Math. (2) 100 (1974), 433–532

[2] D.P. BLECHER AND C. LE MERDY , *Operator Algebras and Their Modules – An operator space approach*, Oxford University Press (2004)

[3] D.P. BLECHER AND R.R. SMITH, *The dual of the Haagerup tensor product*, J. London Math. Soc. (2) 45 (1992), 126–144

[4] E.G. EFFROS AND Z. RUAN, *Operator spaces*, Oxford University Press (2000)

[5] J.A. ERDŐS, A. KATAVOLOS AND V.S. SHULMAN, *Rank one subspaces of bimodules over maximal abelian selfadjoint algebras*, J. Funct. Anal. **157** (1998), no. 2, 554–587

[6] L. GE AND R. KADISON, *On tensor products of von Neumann algebras*, Invent. Math. 123 (1996), 453–466

[7] J. KRAUS, *The slice map problem for σ-weakly closed subspaces of von Neumann algebras*, Trans. Amer. Math. Soc. 279 (1983), no. 1, 357–376

[8] J. KRAUS, *The splitting problem for subspaces of tensor products of operator algebras*, Proc. Amer. Math. Soc. 132 (2003), no. 4, 1125–1131

[9] A.I. LOGINOV AND V.S. SHUL'MAN, *Hereditary and intermediate reflexivity of W*-algebras*, Izv. Akad. Nauk. SSSR, 39 (1975), 1260–1273; Math. USSR-Izv. 9 (1975), 1189–1201

[10] M. MCGARVEY, L. OLIVEIRA AND I. G. TODOROV, *Normalisers and tensor product formulas*, Rev. Math. Iber., to appear

[11] V.S. SHULMAN AND L. TUROWSKA, *Operator synthesis. I. Synthetic sets, bilattices and tensor algebras*, Journal of Functional Analysis 209 (2004), no. 2, 293–331

S. Papapanayides and I.G. Todorov
Pure Mathematics Research Centre
Queen's University Belfast
Belfast BT7 1NN, UK
e-mail: spapapanayides01@qub.ac.uk
i.todorov@qub.ac.uk

Operator Theory:
Advances and Applications, Vol. 233, 135–153


# Noncommutative Analogues of Stein Spaces of Finite Embedding Dimension

A.Yu. Pirkovskii

**Abstract.** We introduce and study *holomorphically finitely generated* (HFG) Fréchet algebras, which are analytic counterparts of affine (i.e., finitely generated) $\mathbb{C}$-algebras. Using a theorem of O. Forster, we prove that the category of commutative HFG algebras is anti-equivalent to the category of Stein spaces of finite embedding dimension. We also show that the class of HFG algebras is stable under some standard constructions. This enables us to give a series of concrete examples of HFG algebras, including Arens–Michael envelopes of affine algebras (such as the algebras of holomorphic functions on the quantum affine space and on the quantum torus), the algebras of holomorphic functions on the free polydisk, on the quantum polydisk, and on the quantum ball. We further concentrate on the algebras of holomorphic functions on the quantum polydisk and on the quantum ball and show that they are isomorphic, in contrast to the classical case. Finally, we interpret our algebras as Fréchet algebra deformations of the classical algebras of holomorphic functions on the polydisk and on the ball in $\mathbb{C}^n$.

**Mathematics Subject Classification (2010).** 46L89, 46L52, 46H30, 14A22, 16S38, 46L65, 16S80.

**Keywords.** Fréchet algebra, noncommutative geometry, Stein space.

## 1. Introduction

To motivate our constructions, let us start by recalling some basic principles of noncommutative geometry. For more details, see, e.g., [20].

Noncommutative geometry is based on the observation that, very often, all essential information about a geometric space (e.g., an affine algebraic variety, or a smooth manifold, or a topological space, or a measure space) is contained in a suitably chosen algebra of functions on the space. Sometimes this observation leads to a theorem which establishes an anti-equivalence between a certain category of spaces and the respective category of algebras of functions on such spaces. Moreover, the resulting category of algebras often admits an abstract characterization not involving any functions on any spaces. So the ideal starting point for any kind of noncommutative geometry is roughly as follows. Let $\mathscr{A}$ be a category of associative algebras (maybe endowed with an additional structure), and suppose that the full subcategory of $\mathscr{A}$ consisting of commutative algebras is anti-equivalent to a certain category $\mathscr{C}$ of "spaces". In such a situation we may think of algebras belonging to $\mathscr{A}$ as noncommutative analogues of spaces belonging to $\mathscr{C}$. Let us give some concrete illustrations of this phenomenon.

**Example 1.1.** By the Gelfand–Naimark Theorem, the category of compact Hausdorff topological spaces is anti-equivalent to the category of commutative unital $C^*$-algebras via the functor taking a compact topological space $X$ to the algebra $C(X)$ of continuous functions on $X$. This leads to the idea that arbitrary (i.e., not necessarily commutative) unital $C^*$-algebras may be viewed as "noncommutative compact topological spaces". The above point of view has proved to be very productive. It gave birth to noncommutative topology, a deep and important subject having many applications (see, e.g., [5, 6, 15, 20]).

**Example 1.2.** Hilbert's Nullstellensatz implies that the category of affine algebraic varieties (over $\mathbb{C}$) is anti-equivalent to the category of commutative unital finitely generated algebras without nilpotents. As above, the equivalence takes an affine variety $V$ to the algebra $\mathscr{O}^{\mathrm{reg}}(V)$ of regular functions on $V$. More generally, the same functor yields an anti-equivalence between the category of affine schemes of finite type over $\mathbb{C}$ and the category of commutative unital finitely generated algebras (with nilpotents allowed). Thus unital finitely generated algebras may be viewed as "noncommutative affine algebraic varieties", or "noncommutative affine schemes of finite type". This point of view leads to noncommutative affine algebraic geometry, also a challenging and rapidly growing subject (see, e.g., [3, 23, 33, 41, 42]).

**Example 1.3.** One more illustration is a theorem by O. Forster. Recall (see, e.g., [16]) that for each complex space $X = (X, \mathscr{O}_X^{\mathrm{hol}})$ the algebra $\mathscr{O}^{\mathrm{hol}}(X)$ has a canonical topology making it into a nuclear Fréchet algebra. If $X$ is reduced (e.g., if $X$ is a complex manifold), then the elements of $\mathscr{O}^{\mathrm{hol}}(X)$ are holomorphic functions on $X$, and the canonical topology on $\mathscr{O}^{\mathrm{hol}}(X)$ is the topology of compact convergence. By definition, a Fréchet algebra $A$ is a *Stein algebra* if it is topologically isomorphic to $\mathscr{O}^{\mathrm{hol}}(X)$ for some Stein space $X$. Forster's theorem [13] states that the functor $(X, \mathscr{O}_X^{\mathrm{hol}}) \mapsto \mathscr{O}^{\mathrm{hol}}(X)$ is an anti-equivalence between the category of Stein spaces and the category of Stein algebras.

By comparing the above examples, we see that Example 1.3 does not entirely match the above "ideal" setup for noncommutative geometry. Indeed, in contrast to Examples 1.1 and 1.2, Forster's theorem does not give an abstract characterization of Stein algebras, and so it does not give us any hint of what noncommutative Stein spaces are.

Our goal here is to introduce a category $\mathscr{A}$ of Fréchet algebras such that the commutative part of $\mathscr{A}$ is anti-equivalent to the category of Stein spaces of finite embedding dimension. We hope that such a category may be useful for developing noncommutative complex analytic geometry, a field much less investigated than other types of noncommutative geometry (like, for example, noncommutative algebraic geometry, noncommutative differential geometry, noncommutative topology, and noncommutative measure theory).

This paper is mostly a survey. The proofs are either sketched or omitted. We plan to present the details elsewhere.

## 2. Holomorphically finitely generated algebras

We shall work over the field $\mathbb{C}$ of complex numbers. All algebras are assumed to be associative and unital, and all algebra homomorphisms are assumed to be unital (i.e., to preserve identity elements). By a *Fréchet algebra* we mean a complete metrizable locally convex algebra (i.e., a topological algebra whose underlying space is a Fréchet space). A *locally $m$-convex algebra* [25] is a topological algebra $A$ whose topology can be defined by a family of submultiplicative seminorms (i.e., seminorms $\|\cdot\|$ satisfying $\|ab\| \le \|a\|\|b\|$ for all $a, b \in A$). A complete locally $m$-convex algebra is called an *Arens–Michael algebra* [18].

Recall that for each Arens–Michael algebra $A$ and for each commuting $n$-tuple $a = (a_1, \dots, a_n) \in A^n$ there exists an *entire functional calculus*, i.e., a unique continuous homomorphism $\gamma_a$ from the algebra $\mathscr{O}(\mathbb{C}^n)$ of holomorphic functions on $\mathbb{C}^n$ to $A$ taking the complex coordinates $z_1, \dots, z_n$ to $a_1, \dots, a_n$, respectively. Explicitly, $\gamma_a$ is given by

$$\gamma_a(f) = f(a) = \sum_{\alpha \in \mathbb{Z}_+^n} c_\alpha a^\alpha \quad \text{for } f = \sum_{\alpha \in \mathbb{Z}_+^n} c_\alpha z^\alpha \in \mathscr{O}(\mathbb{C}^n),$$

where we use the standard notation $a^\alpha = a_1^{\alpha_1} \cdots a_n^{\alpha_n}$ for $\alpha = (\alpha_1, \dots, \alpha_n) \in \mathbb{Z}_+^n$. Below we will need a noncommutative (or, more exactly, free) version of the entire functional calculus due to J.L. Taylor [35].

Let $F_n = \mathbb{C}\langle \zeta_1, \dots, \zeta_n \rangle$ be the free algebra with generators $\zeta_1, \dots, \zeta_n$. For each $k \in \mathbb{Z}_+$, let $W_{n,k} = \{1, \dots, n\}^k$, and let $W_n = \bigsqcup_{k \in \mathbb{Z}_+} W_{n,k}$. Thus a typical element of $W_n$ is a $k$-tuple $\alpha = (\alpha_1, \dots, \alpha_k)$ of arbitrary length $k \in \mathbb{Z}_+$, where $\alpha_j \in \{1, \dots, n\}$ for all $j$. The only element of $W_{n,0}$ will be denoted by $\emptyset$. For each $\alpha = (\alpha_1, \dots, \alpha_k) \in W_n$ we let $\zeta_\alpha = \zeta_{\alpha_1} \cdots \zeta_{\alpha_k} \in F_n$ if $k > 0$; it is also convenient to set $\zeta_\emptyset = 1 \in F_n$. The set $\{\zeta_\alpha : \alpha \in W_n\}$ of all words in $\zeta_1, \dots, \zeta_n$ is the standard vector space basis of $F_n$.

For each $\alpha \in W_{n,k} \subset W_n$, let $|\alpha| = k$. The algebra of *free entire functions* [35, 36] is defined to be

$$\mathscr{F}_n = \Big\{ a = \sum_{\alpha \in W_n} c_\alpha \zeta_\alpha : \|a\|_\rho = \sum_{\alpha \in W_n} |c_\alpha| \rho^{|\alpha|} < \infty \ \forall \rho > 0 \Big\}. \tag{1}$$

The topology on $\mathscr{F}_n$ is given by the seminorms $\|\cdot\|_\rho$ ($\rho > 0$), and the product is given by concatenation (like on $F_n$). Each seminorm $\|\cdot\|_\rho$ is easily seen to be submultiplicative, so $\mathscr{F}_n$ is a Fréchet–Arens–Michael algebra containing $F_n$ as a dense subalgebra. As was observed by D. Luminet [24], $\mathscr{F}_n$ is nuclear. Note also that $\mathscr{F}_1$ is topologically isomorphic to $\mathscr{O}(\mathbb{C})$.

Taylor [35] observed that for each Arens–Michael algebra $A$ and for each $n$-tuple $a = (a_1, \ldots, a_n) \in A^n$ there exists a *free entire functional calculus*, i.e., a unique continuous homomorphism $\gamma_a^{\text{free}} \colon \mathscr{F}_n \to A$ taking the free generators $\zeta_1, \ldots, \zeta_n$ to $a_1, \ldots, a_n$, respectively. Explicitly, $\gamma_a^{\text{free}}$ is given by

$$\gamma_a^{\text{free}}(f) = f(a) = \sum_{\alpha \in W_n} c_\alpha a_\alpha \quad \text{for } f = \sum_{\alpha \in W_n} c_\alpha \zeta_\alpha \in \mathscr{F}_n, \tag{2}$$

where $a_\alpha = a_{\alpha_1} \cdots a_{\alpha_k} \in A$ for $\alpha = (\alpha_1, \ldots, \alpha_k) \in W_n$, and $a_\emptyset = 1 \in A$.

From now on, we assume that $A$ is a Fréchet–Arens–Michael algebra.

**Definition 2.1.** We say that a subalgebra $B \subseteq A$ is *holomorphically closed* if for each $n \in \mathbb{N}$, each $b \in B^n$, and each $f \in \mathscr{F}_n$ we have $f(b) \in B$.

In the commutative case, we may use $\mathscr{O}(\mathbb{C}^n)$ instead of $\mathscr{F}_n$:

**Proposition 2.2.** *Let $A$ be a commutative Fréchet–Arens–Michael algebra. Then a subalgebra $B \subseteq A$ is holomorphically closed if and only if for each $n \in \mathbb{N}$, each $b \in B^n$, and each $f \in \mathscr{O}(\mathbb{C}^n)$ we have $f(b) \in B$.*

**Examples.** Of course, if $B$ is closed in $A$, then it is holomorphically closed. More generally, if $B$ is an Arens–Michael algebra under a topology stronger than the topology inherited from $A$, then $B$ is holomorphically closed in $A$. For example, the algebra $C^\infty(M)$ of smooth functions on a compact manifold $M$ is holomorphically closed in $C(M)$.

**Remark 2.3.** Our notion of a holomorphically closed subalgebra should not be confused with the more common notion of a subalgebra stable under the holomorphic functional calculus (see, e.g., [5, 15]). Recall that a subalgebra $B \subseteq A$ is stable under the holomorphic functional calculus if for each $b \in B$, each neighbourhood $U$ of the spectrum $\sigma_A(b)$, and each $f \in \mathscr{O}(U)$ we have $f(b) \in B$. Such a subalgebra is necessarily spectrally invariant in $A$, i.e., $\sigma_B(b) = \sigma_A(b)$ for all $b \in B$. In contrast, a holomorphically closed subalgebra need not be spectrally invariant. For example, for each domain $D \subseteq \mathbb{C}$ the algebra $\mathscr{O}(\mathbb{C})$ may be viewed (via the restriction map) as a holomorphically closed subalgebra of $\mathscr{O}(D)$, but it is not spectrally invariant in $\mathscr{O}(D)$ unless $D = \mathbb{C}$.

It is clear from the definition that the intersection of any family of holomorphically closed subalgebras of $A$ is holomorphically closed. This leads naturally to the following definition.

**Definition 2.4.** The *holomorphic closure*, $\operatorname{Hol}(S)$, of a subset $S \subseteq A$ is the intersection of all holomorphically closed subalgebras of $A$ containing $S$.

Clearly, $\operatorname{Hol}(S)$ is the smallest holomorphically closed subalgebra of $A$ containing $S$. It can also be described more explicitly as follows.

**Proposition 2.5.** *For each subset $S \subseteq A$ we have*

$$\operatorname{Hol}(S) = \{f(a) : f \in \mathscr{F}_n,\ a \in S^n,\ n \in \mathbb{Z}_+\}.$$

*If, in addition, $A$ is commutative, then*

$$\operatorname{Hol}(S) = \{f(a) : f \in \mathscr{O}(\mathbb{C}^n),\ a \in S^n,\ n \in \mathbb{Z}_+\}.$$

Now we are ready to introduce our main objects of study.

**Definition 2.6.** We say that $A$ is *holomorphically generated* by a subset $S \subseteq A$ if $\operatorname{Hol}(S) = A$. We say that $A$ is *holomorphically finitely generated* (HFG for short) if $A$ is holomorphically generated by a finite subset.

Proposition 2.5, together with the Open Mapping Theorem, implies the following characterization of HFG algebras.

**Proposition 2.7.** *Let $A$ be a Fréchet–Arens–Michael algebra.*

(i) *$A$ is holomorphically finitely generated if and only if $A$ is topologically isomorphic to $\mathscr{F}_n/I$ for some $n \in \mathbb{Z}_+$ and for some closed two-sided ideal $I \subseteq \mathscr{F}_n$.*
(ii) *If $A$ is commutative, then $A$ is holomorphically finitely generated if and only if $A$ is topologically isomorphic to $\mathscr{O}(\mathbb{C}^n)/I$ for some $n \in \mathbb{Z}_+$ and for some closed two-sided ideal $I \subseteq \mathscr{O}(\mathbb{C}^n)$.*

**Corollary 2.8.** *Each HFG algebra is nuclear.*

Combining Proposition 2.7 with the Remmert–Bishop–Narasimhan–Wiegmann Embedding Theorem for Stein spaces [44] and with Forster's theorem (see Example 1.3), we obtain the following.

**Theorem 2.9.** *A commutative Fréchet–Arens–Michael algebra is holomorphically finitely generated if and only if it is topologically isomorphic to $\mathscr{O}(X)$ for some Stein space $(X, \mathscr{O}_X)$ of finite embedding dimension. Moreover, the functor $(X, \mathscr{O}_X) \mapsto \mathscr{O}(X)$ is an anti-equivalence between the category of Stein spaces of finite embedding dimension and the category of commutative HFG algebras.*

Theorem 2.9 looks similar to the Gelfand–Naimark Theorem (Example 1.1) and to the categorical consequence of the Nullstellensatz (Example 1.2). Thus HFG algebras may be considered as candidates for "noncommutative Stein spaces of finite embedding dimension". Of course, this naive point of view needs a solid justification, and it is perhaps too early to say whether it will lead to an interesting theory. As a first step towards this goal, we will give some concrete examples of HFG algebras below, showing that the class of HFG algebras is rather large.

## 3. Arens–Michael envelopes

Many natural examples of HFG algebras can be obtained from the following general construction.

**Definition 3.1.** Let $A$ be an algebra. A pair $(\widehat{A}, i_A)$ consisting of an Arens–Michael algebra $\widehat{A}$ and a homomorphism $i_A\colon A \to \widehat{A}$ is called the *Arens–Michael envelope* of $A$ if for each Arens–Michael algebra $B$ and for each homomorphism $\varphi\colon A \to B$ there exists a unique continuous homomorphism $\widehat{\varphi}\colon \widehat{A} \to B$ making the following diagram commute:

$$\begin{array}{ccc} \widehat{A} & \overset{\widehat{\varphi}}{\dashrightarrow} & B \, . \\ {\scriptstyle i_A}\uparrow & \nearrow{\scriptstyle \varphi} & \\ A & & \end{array}$$

Arens–Michael envelopes were introduced by J.L. Taylor [34] under the name of "completed locally $m$-convex envelopes". Now it is customary to call them "Arens–Michael envelopes", following the terminology suggested by A.Ya. Helemskii [18]. For a more detailed study of Arens–Michael envelopes we refer to [8–12, 27–29].

It is clear from the definition that if $\widehat{A}$ exists, then it is unique up to a unique topological isomorphism over $A$. In fact, $\widehat{A}$ always exists, and it can be obtained as the completion of $A$ with respect to the family of all submultiplicative seminorms on $A$. Obviously, the correspondence $A \mapsto \widehat{A}$ is a functor from the category of algebras to the category of Arens–Michael algebras, and this functor is the left adjoint to the forgetful functor acting in the opposite direction. In what follows, the functor $A \mapsto \widehat{A}$ will be called the *Arens–Michael functor.*

Here are some basic examples of Arens–Michael envelopes.

**Example 3.2 ([35]).** If $A = \mathbb{C}[x_1, \dots, x_n]$ is the polynomial algebra, then $\widehat{A} = \mathscr{O}(\mathbb{C}^n)$, the Fréchet algebra of entire functions on $\mathbb{C}^n$.

**Example 3.3 ([29]).** If $X$ is an affine algebraic variety and $A$ is the algebra $\mathscr{O}^{\mathrm{reg}}(X)$ of regular functions on $X$, then $\widehat{A}$ is the algebra $\mathscr{O}^{\mathrm{hol}}(X)$ of holomorphic functions on $X$. More generally, if $(X, \mathscr{O}_X^{\mathrm{reg}})$ is an affine scheme of finite type over $\mathbb{C}$, and if $A = \mathscr{O}^{\mathrm{reg}}(X)$, then $\widehat{A} = \mathscr{O}^{\mathrm{hol}}(X)$, where $(X, \mathscr{O}_X^{\mathrm{hol}})$ is the Stein space associated to $(X, \mathscr{O}_X^{\mathrm{reg}})$ (cf. [17, Appendix B]).

**Example 3.4 ([35]).** The Arens–Michael envelope of the free algebra $F_n$ is the algebra $\mathscr{F}_n$ of free entire functions (1).

By combining Example 3.4 with Proposition 2.7, and by using the fact that the Arens–Michael functor commutes with quotients [29], we obtain the following.

**Corollary 3.5.** *If $A$ is a finitely generated algebra, then $\widehat{A}$ is holomorphically finitely generated.*

Thus we may interpret the Arens–Michael functor as a "noncommutative analytization functor" acting from the category of noncommutative affine schemes of finite type to the category of noncommutative Stein spaces of finite embedding dimension. Example 3.3 shows that, in the commutative case, the Arens–Michael functor reduces to the classical analytization functor $(X, \mathscr{O}_X^{\mathrm{reg}}) \mapsto (X, \mathscr{O}_X^{\mathrm{hol}})$.

Here are some more examples of Arens–Michael envelopes.

### 3.1. Quantum affine space

Let $\mathbb{C}^\times = \mathbb{C}\setminus\{0\}$, and let $q \in \mathbb{C}^\times$. Recall that the algebra $\mathscr{O}_q^{\mathrm{reg}}(\mathbb{C}^n)$ *of regular functions on the quantum affine n-space* is generated by $n$ elements $x_1, \dots, x_n$ subject to the relations $x_i x_j = q x_j x_i$ for all $i < j$ (see, e.g., [2]). If $q = 1$, then $\mathscr{O}_q^{\mathrm{reg}}(\mathbb{C}^n)$ is nothing but the polynomial algebra $\mathbb{C}[x_1, \dots, x_n] = \mathscr{O}^{\mathrm{reg}}(\mathbb{C}^n)$. Of course, $\mathscr{O}_q^{\mathrm{reg}}(\mathbb{C}^n)$ is noncommutative unless $q = 1$, but the monomials $x^\alpha = x_1^{\alpha_1} \cdots x_n^{\alpha_n}$ $(\alpha \in \mathbb{Z}_+^n)$ still form a basis of $\mathscr{O}_q^{\mathrm{reg}}(\mathbb{C}^n)$. Thus $\mathscr{O}_q^{\mathrm{reg}}(\mathbb{C}^n)$ may be viewed as a "deformed" polynomial algebra.

Taking into account Example 3.2, it is natural to give the following definition.

**Definition 3.6.** The Arens–Michael envelope of $\mathscr{O}_q^{\mathrm{reg}}(\mathbb{C}^n)$ is denoted by $\mathscr{O}_q^{\mathrm{hol}}(\mathbb{C}^n)$ and is called the *algebra of holomorphic functions on the quantum affine n-space.*

The algebra $\mathscr{O}_q^{\mathrm{hol}}(\mathbb{C}^n)$ has the following explicit description. Define a weight function $w_q \colon \mathbb{Z}_+^n \to \mathbb{R}_+$ by

$$w_q(\alpha) = \begin{cases} 1 & \text{if } |q| \geq 1, \\ |q|^{\sum_{i<j} \alpha_i \alpha_j} & \text{if } |q| < 1. \end{cases} \tag{3}$$

**Theorem 3.7** ([29]). *For each $q \in \mathbb{C}^\times$ we have*

$$\mathscr{O}_q^{\mathrm{hol}}(\mathbb{C}^n) = \Big\{ a = \sum_{\alpha \in \mathbb{Z}_+^n} c_\alpha x^\alpha : \|a\|_\rho = \sum_{\alpha \in \mathbb{Z}_+^n} |c_\alpha| w_q(\alpha) \rho^{|\alpha|} < \infty \ \forall \rho > 0 \Big\}$$

*(where $|\alpha| = \alpha_1 + \cdots + \alpha_n$ for $\alpha = (\alpha_1, \dots, \alpha_n) \in \mathbb{Z}_+^n$). The topology on $\mathscr{O}_q^{\mathrm{hol}}(\mathbb{C}^n)$ is given by the seminorms $\|\cdot\|_\rho$ $(\rho > 0)$.*

### 3.2. Quantum torus

Let $q \in \mathbb{C}^\times$. Recall that the algebra $\mathscr{O}_q^{\mathrm{reg}}((\mathbb{C}^\times)^n)$ *of regular functions on the quantum n-torus* is generated by $2n$ elements $x_1, \dots, x_n, x_1^{-1}, \dots, x_n^{-1}$ subject to the relations $x_i^{-1} x_i = x_i x_i^{-1} = 1$ for all $i$, and $x_i x_j = q x_j x_i$ for all $i < j$ (see, e.g., [2]). If $q = 1$, then $\mathscr{O}_q^{\mathrm{reg}}((\mathbb{C}^\times)^n)$ is nothing but the Laurent polynomial algebra $\mathbb{C}[x_1^{\pm 1}, \dots, x_n^{\pm 1}] = \mathscr{O}^{\mathrm{reg}}((\mathbb{C}^\times)^n)$. Of course, $\mathscr{O}_q^{\mathrm{reg}}((\mathbb{C}^\times)^n)$ is noncommutative unless $q = 1$, but the monomials $x^\alpha = x_1^{\alpha_1} \cdots x_n^{\alpha_n}$ $(\alpha \in \mathbb{Z}^n)$ still form a basis of $\mathscr{O}_q^{\mathrm{reg}}(\mathbb{C}^n)$. Thus $\mathscr{O}_q^{\mathrm{reg}}((\mathbb{C}^\times)^n)$ may be viewed as a "deformed" Laurent polynomial algebra.

**Definition 3.8.** If $|q| = 1$, then the Arens–Michael envelope of $\mathscr{O}_q^{\mathrm{reg}}((\mathbb{C}^\times)^n)$ is denoted by $\mathscr{O}_q^{\mathrm{hol}}((\mathbb{C}^\times)^n)$ and is called the *algebra of holomorphic functions on the quantum affine n-torus.*

**Theorem 3.9 ([29]).** *If* $|q| = 1$, *then*

$$\mathscr{O}_q^{\mathrm{hol}}((\mathbb{C}^\times)^n) = \Big\{ a = \sum_{\alpha \in \mathbb{Z}_+^n} c_\alpha x^\alpha : \|a\|_\rho = \sum_{\alpha \in \mathbb{Z}^n} |c_\alpha| \rho^{|\alpha|} < \infty \ \forall \rho > 0 \Big\}.$$

*The topology on* $\mathscr{O}_q^{\mathrm{hol}}((\mathbb{C}^\times)^n)$ *is given by the seminorms* $\|\cdot\|_\rho$ $(\rho > 0)$.

**Remark 3.10.** If $|q| \neq 1$, then the only submultiplicative seminorm on $\mathscr{O}_q^{\mathrm{reg}}((\mathbb{C}^\times)^n)$ is identically zero [29], and so the Arens–Michael envelope of $\mathscr{O}_q^{\mathrm{reg}}((\mathbb{C}^\times)^n)$ is trivial. Another example of an algebra with trivial Arens–Michael envelope is the Weyl algebra generated by two elements $x, y$ subject to the relation $[x, y] = 1$.

We refer to [10, 12, 29] for explicit descriptions of Arens–Michael envelopes of some other algebras, including quantum Weyl algebras, the algebra of quantum $2 \times 2$-matrices, and universal enveloping algebras.

## 4. Free products and free polydisk

To construct more examples of HFG algebras (not necessarily related to Arens–Michael envelopes), we need the simple fact that the category of HFG algebras is stable under the analytic free product.

**Definition 4.1 (cf. [7]).** Let $A_1$ and $A_2$ be Arens–Michael algebras. The *analytic free product* of $A_1$ and $A_2$ is the coproduct of $A_1$ and $A_2$ in the category of Arens–Michael algebras. More explicitly, the analytic free product of $A_1$ and $A_2$ is a triple $(A_1 \mathbin{\widehat{*}} A_2, j_1, j_2)$ consisting of an Arens–Michael algebra $A_1 \mathbin{\widehat{*}} A_2$ and continuous homomorphisms $j_k \colon A_k \to A_1 \mathbin{\widehat{*}} A_2$ $(k = 1, 2)$ such that for each Arens–Michael algebra $B$ and each pair $\varphi_k \colon A_k \to B$ $(k = 1, 2)$ of continuous homomorphisms there exists a unique continuous homomorphism $A_1 \mathbin{\widehat{*}} A_2 \to B$ making the following diagram commute:

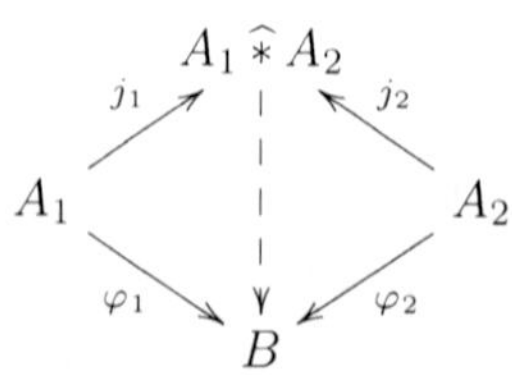


Clearly, if $A_1 \mathbin{\widehat{*}} A_2$ exists, then it is unique up to a unique topological algebra isomorphism over $A_1$ and $A_2$. To show that $A_1 \mathbin{\widehat{*}} A_2$ exists, recall the definition (due to J. Cuntz [7]; cf. also [29]) of an analytic tensor algebra. Let $E$ be a complete locally convex space, and let $\{\|\cdot\|_\lambda : \lambda \in \Lambda\}$ be a directed family of seminorms generating the topology on $E$. For each $\lambda \in \Lambda$ and each $n \in \mathbb{Z}_+$, let $\|\cdot\|_\lambda^{(n)}$ denote the $n$th projective tensor power of $\|\cdot\|_\lambda$ (we let $\|\cdot\|_\lambda^{(0)} = |\cdot|$ by definition). The *analytic tensor algebra* $\widehat{T}(E)$ is defined by

$$\widehat{T}(E) = \Big\{ a = \sum_{n=0}^{\infty} a_n : a_n \in E^{\widehat{\otimes} n},\ \|a\|_{\lambda,\rho} = \sum_n \|a_n\|_\lambda^{(n)} \rho^n < \infty \ \forall \lambda \in \Lambda,\ \forall \rho > 0 \Big\}.$$

The topology on $\widehat{T}(E)$ is defined by the seminorms $\|\cdot\|_{\lambda,\rho}$ $(\lambda \in \Lambda,\ \rho > 0)$, and the product on $\widehat{T}(E)$ is given by concatenation, like on the usual tensor algebra $T(E)$. Each seminorm $\|\cdot\|_{\lambda,\rho}$ is easily seen to be submultiplicative, and so $\widehat{T}(E)$ is an Arens–Michael algebra containing $T(E)$ as a dense subalgebra. As was observed by Cuntz [7], $\widehat{T}(E)$ has the universal property that, for every Arens–Michael algebra $A$, each continuous linear map $E \to A$ uniquely extends to a continuous homomorphism $\widehat{T}(E) \to A$. Note that $\widehat{T}(\mathbb{C}^n) \cong \mathscr{F}_n$, and that the free entire functional calculus (2) is a special case of the universal property of $\widehat{T}(E)$.

Now let $A_1 \mathbin{\widehat{*}} A_2$ be the completion of the quotient of $\widehat{T} = \widehat{T}(A_1 \oplus A_2)$ by the two-sided closed ideal generated by all elements of the form

$$a_i \otimes b_i - a_i b_i, \quad 1_{\widehat{T}} - 1_{A_i} \quad (a_i, b_i \in A_i,\ i = 1, 2).$$

The canonical homomorphisms $j_k \colon A_k \to A_1 \mathbin{\widehat{*}} A_2$ are defined in the obvious way. Using the universal property of $\widehat{T}$, it is easy to show that $A_1 \mathbin{\widehat{*}} A_2$ is indeed the analytic free product of $A_1$ and $A_2$.

A more explicit construction of $A_1 \mathbin{\widehat{*}} A_2$ (in the nonunital category) was given by Cuntz [7]. We adapt the construction to the unital case below, assuming that for each $i = 1, 2$ there exists a closed two-sided ideal $\overline{A}_i \subset A_i$ such that $A_i = \overline{A}_i \oplus \mathbb{C}1_{A_i}$ as locally convex spaces.

For each $k \geq 2$, let

$$W'_{2,k} = \bigl\{\alpha = (\alpha_1, \ldots, \alpha_k) \in W_{2,k} : \alpha_i \neq \alpha_{i+1}\ \forall i = 1, \ldots, k-1\bigr\}.$$

Let also

$$W'_{2,1} = W_{2,1}, \quad W'_{2,0} = W_{2,0} \quad \text{and} \quad W'_2 = \bigsqcup_{k \in \mathbb{Z}_+} W'_{2,k}.$$

Given $\alpha = (\alpha_1, \ldots, \alpha_k) \in W'_2$, let $\overline{A}_\alpha = \overline{A}_{\alpha_1} \mathbin{\widehat{\otimes}} \cdots \mathbin{\widehat{\otimes}} \overline{A}_{\alpha_k}$. For $k = 0$, we let $\overline{A}_\emptyset = \mathbb{C}$. Choose directed defining families $\{\|\cdot\|_{\lambda_i} : \lambda_i \in \Lambda_i\}$ $(i = 1, 2)$ of submultiplicative seminorms on $\overline{A}_1$ and $\overline{A}_2$, respectively. Let $\Lambda = \Lambda_1 \times \Lambda_2$, and, for each $\lambda = (\lambda_1, \lambda_2) \in \Lambda$ and each $\alpha = (\alpha_1, \ldots, \alpha_k) \in W'_2$, let $\|\cdot\|_\lambda^{(\alpha)}$ denote the projective tensor product of the seminorms $\|\cdot\|_{\lambda_{\alpha_1}}, \ldots, \|\cdot\|_{\lambda_{\alpha_k}}$. For $k = 0$, it is convenient to set $\|\cdot\|_\lambda^{(\emptyset)} = |\cdot|$. We have

$$A_1 \mathbin{\widehat{*}} A_2 = \left\{ \begin{array}{c} a = \displaystyle\sum_{\alpha \in W'_2} a_\alpha : a_\alpha \in \overline{A}_\alpha, \\ \|a\|_{\lambda,\rho} = \displaystyle\sum_\alpha \|a_\alpha\|_\lambda^{(\alpha)} \rho^{|\alpha|} < \infty\ \forall \lambda \in \Lambda,\ \forall \rho > 0 \end{array} \right\}. \tag{4}$$

The topology on $A_1 \mathbin{\widehat{*}} A_2$ is defined by the seminorms $\|\cdot\|_{\lambda,\rho}$ $(\lambda \in \Lambda, \rho > 0)$. The product on $A_1 \mathbin{\widehat{*}} A_2$ is given by concatenation composed (if necessary) with the product maps $\overline{A}_i \mathbin{\widehat{\otimes}} \overline{A}_i \to \overline{A}_i$ $(i = 1, 2)$.

**Example 4.2.** If $E$ and $F$ are complete locally convex spaces, then we have a topological algebra isomorphism $\widehat{T}(E) \mathbin{\widehat{*}} \widehat{T}(F) \cong \widehat{T}(E \oplus F)$. In particular, $\mathscr{F}_m \mathbin{\widehat{*}} \mathscr{F}_n \cong$

$\mathscr{F}_{m+n}$, and

$$\mathscr{F}_n \cong \mathscr{O}(\mathbb{C}) \widehat{*} \cdots \widehat{*} \mathscr{O}(\mathbb{C}). \tag{5}$$

The next result easily follows from Example 4.2 and Proposition 2.7.

**Proposition 4.3.** *If $A_1$ and $A_2$ are HFG algebras, then so is $A_1 \widehat{*} A_2$.*

Using Proposition 4.3, we can construct more examples of HFG algebras. Let $r > 0$, and let $\mathbb{D}_r = \{z \in \mathbb{C} : |z| < r\}$ denote the open disk of radius $r$. The following definition is motivated by (5).

**Definition 4.4.** We define the *algebra of holomorphic functions on the free $n$-dimensional polydisk of radius $r$* to be

$$\mathscr{F}(\mathbb{D}_r^n) = \mathscr{O}(\mathbb{D}_r) \widehat{*} \cdots \widehat{*} \mathscr{O}(\mathbb{D}_r). \tag{6}$$

By Proposition 4.3, $\mathscr{F}(\mathbb{D}_r^n)$ is an HFG algebra. Note that replacing in (5) and (6) the analytic free product $\widehat{*}$ by the projective tensor product $\widehat{\otimes}$ yields the algebras of holomorphic functions on $\mathbb{C}^n$ and $\mathbb{D}_r^n$, respectively.

We have a canonical "restriction" map $\mathscr{F}_n \to \mathscr{F}(\mathbb{D}_r^n)$ defined to be the $n$th free power of the restriction map $\mathscr{O}(\mathbb{C}) \to \mathscr{O}(\mathbb{D}_r)$. For each $i = 1, \dots, n$, the canonical image of the free generator $\zeta_i \in \mathscr{F}_n$ in $\mathscr{F}(\mathbb{D}_r^n)$ will also be denoted by $\zeta_i$. It is easy to see that $\mathscr{F}(\mathbb{D}_r^n)$ has the following universal property.

**Proposition 4.5.** *Let $A$ be an Arens–Michael algebra, and let $a = (a_1, \dots, a_n)$ be an $n$-tuple in $A^n$ such that $\sigma_A(a_i) \subseteq \mathbb{D}_r$ for all $i = 1, \dots, n$. Then there exists a unique continuous homomorphism $\gamma_a^{\mathrm{free}} \colon \mathscr{F}(\mathbb{D}_r^n) \to A$ such that $\gamma_a^{\mathrm{free}}(\zeta_i) = a_i$ for all $i = 1, \dots, n$.*

The algebra $\mathscr{F}(\mathbb{D}_r^n)$ can also be described more explicitly as follows. Given $k \geq 2$ and $\alpha = (\alpha_1, \dots, \alpha_k) \in W_n$, let $d(\alpha)$ denote the cardinality of the set

$$\{i \in \{1, \dots, k-1\} : \alpha_i \neq \alpha_{i+1}\}.$$

If $|\alpha| \in \{0, 1\}$, we set $d(\alpha) = |\alpha| - 1$. The next result follows from (4).

**Proposition 4.6.** *We have*

$$\mathscr{F}(\mathbb{D}_r^n) = \left\{ \begin{array}{l} a = \sum\limits_{\alpha \in W_n} c_\alpha \zeta_\alpha : \|a\|_{\rho_1,\rho_2} = \sum\limits_{\alpha \in W_n} |c_\alpha| \rho_1^{|\alpha|} \rho_2^{d(\alpha)+1} \\ \qquad < \infty \ \forall 0 < \rho_1 < r, \ \forall \rho_2 > 0 \end{array} \right\}.$$

*The topology on $\mathscr{F}(\mathbb{D}_r^n)$ is given by the seminorms $\|\cdot\|_{\rho_1,\rho_2}$ $(0 < \rho_1 < r,\ \rho_2 > 0)$, and the product is given by concatenation.*

**Remark 4.7.** It is instructive to compare $\mathscr{F}(\mathbb{D}_r^n)$ with Taylor's "free power series algebras" $\mathscr{F}_n(r)$ [35, 36]. By definition,

$$\mathscr{F}_n(r) = \Big\{ a = \sum_{\alpha \in W_n} c_\alpha \zeta_\alpha : \|a\|_\rho = \sum_{\alpha \in W_n} |c_\alpha| \rho^{|\alpha|} < \infty \ \forall 0 < \rho < r \Big\}.$$

Comparing (1), (5), and (6), we see that $\mathscr{F}_n(r) = \mathscr{F}_n = \mathscr{F}(\mathbb{D}_r^n)$ if $r = \infty$. However, if $r < \infty$, then $\mathscr{F}(\mathbb{D}_r^n)$ differs from $\mathscr{F}_n(r)$. Indeed, $\mathscr{F}_n(r)$ is not nuclear [24], and so it is not even an HFG algebra. Although such algebras fall outside the scope of this paper, we would like to make a short digression and to show that $\mathscr{F}_n(r)$ has a remarkable universal property, similar in spirit to Proposition 4.5.

Recall that, for a Banach algebra $A$ and an $n$-tuple $a = (a_1, \dots, a_n) \in A^n$, the *joint spectral radius* $r_\infty(a)$ is given by

$$r_\infty(a) = \limsup_{k\to\infty} \Bigl( \sup_{\alpha\in W_{n,k}} \|a_\alpha\| \Bigr)^{1/k}.$$

If now $A$ is an Arens–Michael algebra, we say that an $n$-tuple $a \in A^n$ is *strictly $r$-contractive* if, for each Banach algebra $B$ and each continuous homomorphism $\varphi\colon A \to B$, we have $r_\infty(\varphi^{\times n}(a)) < r$. An equivalent, but more handy definition is as follows. Let $\{\|\cdot\|_\lambda : \lambda \in \Lambda\}$ be a directed defining family of submultiplicative seminorms on $A$. For each $\lambda \in \Lambda$, let $A_\lambda$ denote the completion of $A$ with respect to $\|\cdot\|_\lambda$, and let $a_\lambda$ denote the canonical image of $a$ in $A_\lambda^n$. Then it is easy to show that $a$ is strictly $r$-contractive if and only if $r_\infty(a_\lambda) < r$ for all $\lambda \in \Lambda$. For example, if $\zeta = (\zeta_1, \dots, \zeta_n) \in \mathscr{F}_n(r)^n$ is the canonical system of free generators, then $\zeta$ is strictly $r$-contractive, but is not strictly $\rho$-contractive for $\rho < r$.

Now it is easy to show that $\mathscr{F}_n(r)$ has the following universal property. *Given an Arens–Michael algebra $A$ and a strictly $r$-contractive $n$-tuple $a = (a_1, \dots, a_n) \in A^n$, there exists a unique continuous homomorphism $\gamma_a^{\mathrm{free}}\colon \mathscr{F}_n(r) \to A$ such that $\gamma_a^{\mathrm{free}}(\zeta_i) = a_i$ for all $i = 1, \dots, n$.*

To complete our discussion of $\mathscr{F}_n(r)$ and $\mathscr{F}(\mathbb{D}_r^n)$, let us note that the $n$-tuple $(\zeta_1, \dots, \zeta_n)$ is not strictly $R$-contractive in $\mathscr{F}(\mathbb{D}_r^n)^n$ for any $R > 0$.

## 5. Quantum polydisk and quantum ball

**Definition 5.1.** Let $q \in \mathbb{C}^\times$, and let $r > 0$. We define the *algebra of holomorphic functions on the quantum $n$-polydisk of radius $r$* by

$$\mathscr{O}_q^{\mathrm{hol}}(\mathbb{D}_r^n) = \Bigl\{ a = \sum_{\alpha\in\mathbb{Z}_+^n} c_\alpha x^\alpha : \|a\|_\rho = \sum_{\alpha\in\mathbb{Z}_+^n} |c_\alpha| w_q(\alpha) \rho^{|\alpha|} < \infty\ \forall 0 < \rho < r \Bigr\},$$

where the function $w_q\colon \mathbb{Z}_+^n \to \mathbb{R}_+$ is given by (3). The topology on $\mathscr{O}_q^{\mathrm{hol}}(\mathbb{D}_r^n)$ is defined by the seminorms $\|\cdot\|_\rho$ $(0 < \rho < r)$, and the product on $\mathscr{O}_q^{\mathrm{hol}}(\mathbb{D}_r^n)$ is uniquely determined by $x_i x_j = q x_j x_i$ for all $i < j$.

In other words, $\mathscr{O}_q^{\mathrm{hol}}(\mathbb{D}_r^n)$ is the completion of $\mathscr{O}_q^{\mathrm{reg}}(\mathbb{C}^n)$ (see Section 3) with respect to the family $\{\|\cdot\|_\rho : 0 < \rho < r\}$ of submultiplicative seminorms. Clearly, if $q = 1$, then $\mathscr{O}_q^{\mathrm{hol}}(\mathbb{D}_r^n)$ is topologically isomorphic to the algebra $\mathscr{O}^{\mathrm{hol}}(\mathbb{D}_r^n)$ of holomorphic functions on the polydisk $\mathbb{D}_r^n$.

**Proposition 5.2.** *For each $q \in \mathbb{C}^\times$, $\mathscr{O}_q^{\mathrm{hol}}(\mathbb{D}_r^n)$ is topologically isomorphic to the quotient of $\mathscr{F}(\mathbb{D}_r^n)$ by the two-sided closed ideal generated by the elements $\zeta_i\zeta_j - q\zeta_j\zeta_i$ for all $i < j$. As a corollary, $\mathscr{O}_q^{\mathrm{hol}}(\mathbb{D}_r^n)$ is an HFG algebra.*

Note that $\mathscr{F}(\mathbb{D}_r^n)$ can be replaced by $\mathscr{F}_n(r)$ in the above proposition.

Our last example is the algebra of holomorphic functions on the quantum ball. The construction below is a slight modification of L.L. Vaksman's $q$-analogue of $A(\bar{\mathbb{B}}^n)$, the algebra of functions continuous on the closed ball and holomorphic on the open ball [39]. In order to motivate the construction, let us start from the classical situation. Let $z_1, \dots, z_n$ be the complex coordinates on $\mathbb{C}^n$, and let $\mathbb{B}_r^n = \{z \in \mathbb{C}^n : \sum_i |z_i|^2 < r^2\}$ denote the open ball of radius $r$. For brevity, we write $\mathbb{B}^n$ for $\mathbb{B}_1^n$. Denote by $\mathrm{Pol}(\mathbb{C}^n)$ the algebra of polynomials in $z_1, \dots, z_n$ and their complex conjugates $\bar{z}_1, \dots, \bar{z}_n$. There is a natural involution on $\mathrm{Pol}(\mathbb{C}^n)$ uniquely determined by $z_i^* = \bar{z}_i$ $(i = 1, \dots, n)$. By the Stone–Weierstrass Theorem, the completion of $\mathrm{Pol}(\mathbb{C}^n)$ with respect to the uniform norm $\|f\| = \sup_{z \in \bar{\mathbb{B}}^n} |f(z)|$ is the algebra $C(\bar{\mathbb{B}}^n)$ of continuous functions on the closed ball $\bar{\mathbb{B}}^n$. For each $r \in (0,1)$ and each $f \in \mathbb{C}[z_1, \dots, z_n]$, let

$$\|f\|_r = \sup_{z \in \bar{\mathbb{B}}_r^n} |f(z)| = \|\gamma_r(f)\|,$$

where $\gamma_r$ is the automorphism of $\mathbb{C}[z_1, \dots, z_n]$ uniquely determined by $\gamma_r(z_i) = rz_i$ $(i = 1, \dots, n)$. It is easy to see that the completion of $\mathbb{C}[z_1, \dots, z_n]$ with respect to the family $\{\|\cdot\|_r : 0 < r < 1\}$ of seminorms is topologically isomorphic to the algebra $\mathscr{O}(\mathbb{B}^n)$ of holomorphic functions on $\mathbb{B}^n$.

Now let us "quantize" the above data. Fix $q \in (0,1)$, and denote by $\mathrm{Pol}_q(\mathbb{C}^n)$ the $*$-algebra generated (as a $*$-algebra) by $n$ elements $z_1, \dots, z_n$ subject to the relations

$$\begin{aligned} z_i z_j &= q z_j z_i \quad (i < j); \\ z_i^* z_j &= q z_j z_i^* \quad (i \neq j); \\ z_i^* z_i &= q^2 z_i z_i^* + (1 - q^2)\Big(1 - \sum_{k>i} z_k z_k^*\Big). \end{aligned} \tag{7}$$

Clearly, for $q = 1$ we have $\mathrm{Pol}_q(\mathbb{C}^n) = \mathrm{Pol}(\mathbb{C}^n)$. The algebra $\mathrm{Pol}_q(\mathbb{C}^n)$ was introduced by W. Pusz and S.L. Woronowicz [32], although they used different $*$-generators $a_1, \dots, a_n$ given by $a_i = (1 - q^2)^{-1/2} z_i^*$. Relations (7) divided by $1 - q^2$ and written in terms of the $a_i$'s are called the "twisted canonical commutation relations", and the algebra $A_q = \mathrm{Pol}_q(\mathbb{C}^n)$ defined in terms of the $a_i$'s is sometimes called the "quantum Weyl algebra" (see, e.g., [1, 19, 22, 43]). Note that, while $\mathrm{Pol}_q(\mathbb{C}^n)$ becomes $\mathrm{Pol}(\mathbb{C}^n)$ for $q = 1$, $A_q$ becomes the Weyl algebra. The idea to use the generators $z_i$ instead of the $a_i$'s and to consider $\mathrm{Pol}_q(\mathbb{C}^n)$ as a $q$-analogue of $\mathrm{Pol}(\mathbb{C}^n)$ is probably due to Vaksman [37]; the one-dimensional case was considered in [21]. The algebra $\mathrm{Pol}_q(\mathbb{C}^n)$ serves as a basic example in the general theory of "quantum bounded symmetric domains" developed by Vaksman and his collaborators (see [38, 40] and references therein).

The $q$-analogue of the algebra $C(\bar{\mathbb{B}}^n)$ is defined as follows. For each $k \in \mathbb{N}$, let $[k]_q = \sum_{i=0}^{k-1} q^{2i}$. Fix a separable Hilbert space $H$ with an orthonormal basis $\{e_\alpha : \alpha \in \mathbb{Z}_+^n\}$. Following [32], for each $\alpha = (\alpha_1, \dots, \alpha_n) \in \mathbb{Z}_+^n$ we will write $|\alpha_1, \dots, \alpha_n\rangle$ for $e_\alpha$. As was proved by Pusz and Woronowicz [32], there exists a faithful irreducible $*$-representation $\pi\colon \operatorname{Pol}_q(\mathbb{C}^n) \to \mathscr{B}(H)$ uniquely determined by

$$\pi(z_j)e_\alpha = \sqrt{1-q^2}\sqrt{[\alpha_j+1]}q^{\sum_{k>j}\alpha_k}|\alpha_1,\ldots,\alpha_j+1,\ldots,\alpha_n\rangle$$
$$(j=1,\ldots,n,\ \alpha=(\alpha_1,\ldots,\alpha_n)\in\mathbb{Z}_+^n).$$

The completion of $\operatorname{Pol}_q(\mathbb{C}^n)$ with respect to the operator norm $\|a\|_{\mathrm{op}} = \|\pi(a)\|$ is denoted by $C_q(\bar{\mathbb{B}}^n)$ and is called the *algebra of continuous functions on the closed quantum ball* [39]; see also [31, 32].

Now we are ready to define the algebra of holomorphic functions on the quantum ball. Observe that the subalgebra of $\operatorname{Pol}_q(\mathbb{C}^n)$ generated by $z_1,\ldots,z_n$ is exactly $\mathscr{O}_q^{\mathrm{reg}}(\mathbb{C}^n)$. For each $r\in(0,1)$, let $\gamma_r$ be the automorphism of $\mathscr{O}_q^{\mathrm{reg}}(\mathbb{C}^n)$ uniquely determined by $\gamma_r(z_i)=rz_i$ $(i=1,\ldots,n)$. Define a norm $\|\cdot\|_r$ on $\mathscr{O}_q^{\mathrm{reg}}(\mathbb{C}^n)$ by $\|a\|_r = \|\gamma_r(a)\|_{\mathrm{op}}$.

**Definition 5.3.** The completion of $\mathscr{O}_q^{\mathrm{reg}}(\mathbb{C}^n)$ with respect to the family $\{\|\cdot\|_r : 0<r<1\}$ of seminorms is denoted by $\mathscr{O}_q^{\mathrm{hol}}(\mathbb{B}^n)$ and is called the *algebra of holomorphic functions on the quantum ball.*

It follows from the above discussion that, if we replace $C_q(\bar{\mathbb{B}}^n)$ by $C(\bar{\mathbb{B}}^n)$ in the above construction, then the result will be the algebra $\mathscr{O}^{\mathrm{hol}}(\mathbb{B}^n)$ of holomorphic functions on $\mathbb{B}^n$. In Section 6, we will study relations between $\mathscr{O}_q^{\mathrm{hol}}(\mathbb{B}^n)$ and $\mathscr{O}^{\mathrm{hol}}(\mathbb{B}^n)$ in more detail.

At the moment, it is not obvious whether $\mathscr{O}_q^{\mathrm{hol}}(\mathbb{B}^n)$ is an HFG algebra. The positive answer follows from the next result. Let $\mathbb{D}^n=\mathbb{D}_1^n$ denote the open unit polydisk in $\mathbb{C}^n$.

**Theorem 5.4.** *For each $0<q<1$ the algebras $\mathscr{O}_q^{\mathrm{hol}}(\mathbb{B}^n)$ and $\mathscr{O}_q^{\mathrm{hol}}(\mathbb{D}^n)$ are topologically isomorphic.*

Of course, Theorem 5.4 is a "purely quantum" phenomenon. Indeed, the classical function algebras $\mathscr{O}^{\mathrm{hol}}(\mathbb{B}^n)$ and $\mathscr{O}^{\mathrm{hol}}(\mathbb{D}^n)$ are not isomorphic (unless $n=1$) because $\mathbb{B}^n$ and $\mathbb{D}^n$ are not biholomorphically equivalent.

Let us briefly discuss the idea of the proof. Consider the following families of seminorms on $\mathscr{O}_q^{\mathrm{reg}}(\mathbb{C}^n)$:

$$\|a\|_r^{(1)} = \sum_{\alpha\in\mathbb{Z}_+^n}|c_\alpha|w_q(\alpha)r^{|\alpha|}\quad (0<r<1); \tag{8}$$

$$\|a\|_r^{(2)} = \Big(\sum_{\alpha\in\mathbb{Z}_+^n}|c_\alpha|^2w_q^2(\alpha)r^{2|\alpha|}\Big)^{1/2}\quad (0<r<1); \tag{9}$$

$$\|a\|_r^{(\infty)} = \sup_{\alpha\in\mathbb{Z}_+^n}|c_\alpha|w_q(\alpha)r^{|\alpha|}\quad (0<r<1); \tag{10}$$

$$\|a\|_r = \|\gamma_r(a)\|_{\mathrm{op}}\quad (0<r<1). \tag{11}$$

Recall that the completion of $\mathscr{O}_q^{\mathrm{reg}}(\mathbb{C}^n)$ w.r.t. the seminorms (8) is $\mathscr{O}_q^{\mathrm{hol}}(\mathbb{D}^n)$, while the completion of $\mathscr{O}_q^{\mathrm{reg}}(\mathbb{C}^n)$ w.r.t. the seminorms (11) is $\mathscr{O}_q^{\mathrm{hol}}(\mathbb{B}^n)$. A standard argument shows that the families (8)–(10) are equivalent. The following lemma gives a less obvious estimate.

**Lemma 5.5.** *We have*

$$\Big(\prod_{j=1}^{\infty}(1-q^{2j})\Big)^{\frac{n}{2}}\|\cdot\|_1^{(2)} \le \|\cdot\|_{\mathrm{op}} \le \|\cdot\|_1^{(1)}. \tag{12}$$

The first inequality in (12) follows from the estimate $\|a\|_{\mathrm{op}} \ge \|\pi(a)e_0\|$, where $e_0 = |0,\dots,0\rangle \in H$ is the "vacuum vector". The second inequality in (12) follows from the fact that $\|\cdot\|_1^{(1)}$ is the largest submultiplicative seminorm on $\mathscr{O}_q^{\mathrm{reg}}(\mathbb{C}^n)$ such that $\|z_i\|_1^{(1)} = 1$ for all $i = 1,\dots,n$ [29, 5.10].

Combining Lemma 5.5 with the equivalence of the families (8) and (9), we obtain the following.

**Corollary 5.6.** *For each* $0 < \rho < r < 1$ *we have*

$$\Big(\frac{r^2-\rho^2}{r^2}\prod_{j=1}^{\infty}(1-q^{2j})\Big)^{\frac{n}{2}}\|\cdot\|_\rho^{(1)} \le \|\cdot\|_r \le \|\cdot\|_r^{(1)}. \tag{13}$$

Thus the families (8) and (11) of seminorms are equivalent, and so $\mathscr{O}_q^{\mathrm{hol}}(\mathbb{B}^n) \cong \mathscr{O}_q^{\mathrm{hol}}(\mathbb{D}^n)$, as required.

Note that, while the second inequality in (13) holds in the classical case $q = 1$ as well, the first inequality in (13) has no classical counterpart. Indeed, if we fix $r$ and take $\rho < r$ close enough to $r$, then the polydisk of radius $\rho$ will not be contained in the ball of radius $r$, and so the supremum over the polydisk will not be dominated by the supremum over the ball.

## 6. Relation to Fréchet algebra bundles

In this section we discuss in which sense $\mathscr{O}_q^{\mathrm{hol}}(\mathbb{B}^n)$ and $\mathscr{O}_q^{\mathrm{hol}}(\mathbb{D}^n)$ are deformations of the function algebras $\mathscr{O}^{\mathrm{hol}}(\mathbb{B}^n)$ and $\mathscr{O}^{\mathrm{hol}}(\mathbb{D}^n)$, respectively.

Recall some definitions from [14]. Let $X$ and $E$ be sets, and let $p\colon E \to X$ be a surjective map. Suppose that for each $x \in X$ the fiber $E_x = p^{-1}(x)$ is endowed with the structure of a vector space. A function $\|\cdot\|\colon E \to [0,+\infty)$ is a *seminorm* if its restriction to each $E_x$ is a seminorm in the usual sense. Let

$$E \times_X E = \{(x,y) \in E \times E : p(x) = p(y)\}.$$

Suppose now that $X$ is a locally compact, Hausdorff topological space, $A$ is a topological space, and $p\colon A \to X$ is a continuous, open surjection. Suppose also that each fiber $A_x = p^{-1}(x)$ is endowed with the structure of an algebra in such a way that the operations

$$\begin{aligned} A \times_X A &\to A, \quad (a,b) \mapsto a+b,\\ \mathbb{C} \times A &\to A, \quad (\lambda,a) \mapsto \lambda a,\\ A \times_X A &\to A, \quad (a,b) \mapsto ab, \end{aligned}$$

are continuous. Let $\mathscr{N} = \{\|\cdot\|_i : i \in I\}$ be a directed family of seminorms on $A$ having an at most countable cofinal subfamily. Assume that for each $x \in X$ the sets

$$\{a \in A : p(a) \in U,\ \|a\|_i < \varepsilon\} \qquad (i \in I,\ \varepsilon > 0,\ U \subseteq X \text{ is a neighbourhood of } x)$$

form a neighbourhood base of $0 \in A_x$ (this implies, in particular, that the topology on $A_x$ inherited from $A$ is determined by the seminorms $\|\cdot\|_i$, and that each seminorm $\|\cdot\|_i$ is an upper semicontinuous function on $A$). Finally, assume that each fiber $A_x$ is complete (so, according to the above remarks, $A_x$ is a Fréchet algebra). Under the above assumptions, the pair $(A, p)$ together with the family $\mathscr{N}$ is called a *Fréchet algebra bundle* over $X$.

**Theorem 6.1.**

(i) *For each $r \in (0, +\infty]$ there exists a Fréchet algebra bundle $(D, p)$ over $\mathbb{C}^\times$ such that for each $q \in \mathbb{C}^\times$ we have $D_q \cong \mathscr{O}_q^{\mathrm{hol}}(\mathbb{D}_r^n)$.*

(ii) *There exists a Fréchet algebra bundle $(B, p)$ over $\mathbb{C}^\times$ such that $B_1 \cong \mathscr{O}^{\mathrm{hol}}(\mathbb{B}^n)$ and such that for each $q \in (0, 1)$ we have $B_q \cong \mathscr{O}_q^{\mathrm{hol}}(\mathbb{B}^n)$.*

We will not give a detailed proof here for lack of space. Let us only explain how the bundles $(D, p)$ and $(B, p)$ are constructed. By a *Fréchet $\mathscr{O}(\mathbb{C}^\times)$-algebra* we mean a Fréchet algebra $R$ together with the structure of a left Fréchet $\mathscr{O}(\mathbb{C}^\times)$-module such that the multiplication $R \times R \to R$ is $\mathscr{O}(\mathbb{C}^\times)$-bilinear. For each $q \in \mathbb{C}^\times$, let $M_q = \{f \in \mathscr{O}(\mathbb{C}^\times) : f(q) = 0\}$. It is easy to see that, if $R$ is a Fréchet $\mathscr{O}(\mathbb{C}^\times)$-algebra, then $\overline{M_q \cdot R}$ is a closed two-sided ideal of $R$. The Fréchet algebra $R_q = R/\overline{M_q \cdot R}$ is called the *fiber* of $R$ over $q$. The following lemma explains the terminology.

**Lemma 6.2.** *Let $R$ be a Fréchet $\mathscr{O}(\mathbb{C}^\times)$-algebra, and let $A = \bigsqcup_{q \in \mathbb{C}^\times} R_q$. Define the projection $p\colon A \to \mathbb{C}^\times$ by $p(a) = q$ if $a \in R_q$. Let also $\{\|\cdot\|_i^R : i \in I\}$ be a directed defining family of seminorms on $R$. For each $q \in \mathbb{C}^\times$ and each $i \in I$, let $\|\cdot\|_{i,q}$ denote the corresponding quotient seminorm on $R_q$, and let $\|\cdot\|_i$ be the seminorm on $A$ whose restriction to $R_q$ is $\|\cdot\|_{i,q}$. Then there is a unique topology on $A$ making $(A, p)$ into a Fréchet algebra bundle over $\mathbb{C}^\times$ with respect to the family $\{\|\cdot\|_i : i \in I\}$ of seminorms.*

Now, in order to prove part (i) (respectively, (ii)) of Theorem 6.1, we need to construct a Fréchet $\mathscr{O}(\mathbb{C}^\times)$-algebra $R$ whose fiber over each $q \in \mathbb{C}^\times$ (respectively, over each $q \in (0, 1]$) is isomorphic to $\mathscr{O}_q^{\mathrm{hol}}(\mathbb{D}_r^n)$ (respectively, to $\mathscr{O}_q^{\mathrm{hol}}(\mathbb{B}^n)$). This can be done as follows. Let $z$ denote the complex coordinate on $\mathbb{C}^\times$, and let $I$ be the closed two-sided ideal of $\mathscr{O}(\mathbb{C}^\times, \mathscr{F}(\mathbb{D}_r^n))$ generated by the elements $\zeta_i\zeta_j - z\zeta_j\zeta_i$ $(i < j)$. The quotient $R = \mathscr{O}(\mathbb{C}^\times, \mathscr{F}(\mathbb{D}_r^n))/I$ is a Fréchet $\mathscr{O}(\mathbb{C}^\times)$-algebra in a natural way, and Proposition 5.2 easily implies that $R_q \cong \mathscr{O}_q^{\mathrm{hol}}(\mathbb{D}_r^n)$ for each $q \in \mathbb{C}^\times$. We can also use Taylor's $\mathscr{F}_n(r)$ instead of $\mathscr{F}(\mathbb{D}_r^n)$; the algebra $R$ will then be the same.

To prove part (ii), we have to replace $\mathscr{F}(\mathbb{D}_r^n)$ by the algebra $\mathscr{F}(\mathbb{B}^n)$ of "holomorphic functions on the free ball" introduced by G. Popescu [30][1]. By definition,

$$\mathscr{F}(\mathbb{B}^n) = \Big\{a = \sum_{\alpha \in W_n} c_\alpha \zeta_\alpha : \limsup_{k\to\infty} \Big(\sum_{|\alpha|=k} |c_\alpha|^2\Big)^{1/2k} \le 1\Big\}.$$

As was observed by Popescu [30], $\mathscr{F}(\mathbb{B}^n)$ is indeed an algebra with respect to the concatenation product.

Similarly to $\mathscr{F}(\mathbb{D}_r^n)$ and $\mathscr{F}_n(r)$, the algebra $\mathscr{F}(\mathbb{B}^n)$ can be used to construct a kind of "free holomorphic functional calculus" (cf. Proposition 4.5 and Remark 4.7). Let $H$ be a Hilbert space, and let $T = (T_1, \ldots, T_n)$ be an $n$-tuple in $\mathscr{B}(H)^n$. Following [30], we identify $T$ with the "row" operator acting from the Hilbert direct sum $H^n = H \oplus \cdots \oplus H$ to $H$. Thus we have $\|T\| = \|\sum_{i=1}^n T_i T_i^*\|^{1/2}$. If $\|T\| < 1$, then for each $f = \sum_\alpha c_\alpha \zeta_\alpha \in \mathscr{F}(\mathbb{B}^n)$ the series $\sum_\alpha c_\alpha T_\alpha$ converges in $\mathscr{B}(H)$ to an operator $f(T)$ [30], and we have an algebra homomorphism $\gamma_T^{\mathrm{free}} \colon \mathscr{F}(\mathbb{B}^n) \to \mathscr{B}(H)$, $f \mapsto f(T)$. In fact, $\gamma_T^{\mathrm{free}}$ exists under the weaker assumption that Bunce's "hilbertian" joint spectral radius [4] of $T$ is less than 1 (see [30, 4.1]), but we will not use this fact in the sequel.

There is a canonical topology on $\mathscr{F}(\mathbb{B}^n)$ defined as follows. Fix an infinite-dimensional Hilbert space $H$, and, for each $r \in (0,1)$ and each $f \in \mathscr{F}(\mathbb{B}^n)$, let

$$\|f\|_r^{(P)} = \sup\{\|f(T)\| : T \in \mathscr{B}(H)^n,\ \|T\| \le r\}.$$

By [30, 5.6], $\mathscr{F}(\mathbb{B}^n)$ is a Fréchet space with respect to the family $\{\|\cdot\|_r^{(P)} : 0 < r < 1\}$ of seminorms. Clearly, for each $T \in \mathscr{B}(H)^n$ with $\|T\| < 1$ the functional calculus $\gamma_T^{\mathrm{free}}$ (see above) is continuous.

In fact, it easily follows from the results of [30] that $\mathscr{F}(\mathbb{B}^n)$ admits the following simpler characterization.

**Proposition 6.3.** *We have*

$$\mathscr{F}(\mathbb{B}^n) = \Big\{a = \sum_{\alpha \in W_n} c_\alpha \zeta_\alpha : \|a\|_r = \sum_{k=0}^{\infty} \Big(\sum_{|\alpha|=k} |c_\alpha|^2\Big)^{1/2} r^k < \infty \ \forall 0 < r < 1\Big\}.$$

*The families* $\{\|\cdot\|_r : 0 < r < 1\}$ *and* $\{\|\cdot\|_r^{(P)} : 0 < r < 1\}$ *of seminorms are equivalent.*

Proposition 6.3 can also be interpreted as follows. Let $\ell_1^n$ (respectively, $\ell_2^n$) denote the $n$-dimensional version of $\ell_1$ (respectively, of $\ell_2$). Then Taylor's polydisk algebra $\mathscr{F}_n(1)$ is a weighted $\ell_1$-sum of the projective tensor powers $\ell_1^n \otimes_\pi \cdots \otimes_\pi \ell_1^n$, while $\mathscr{F}(\mathbb{B}^n)$ is a weighted $\ell_1$-sum of the Hilbert tensor powers $\ell_2^n \otimes_{\mathrm{hilb}} \cdots \otimes_{\mathrm{hilb}} \ell_2^n$.

Using Proposition 6.3, one can easily show that $\mathscr{F}(\mathbb{B}^n)$ is a Schwartz space, but is not a nuclear space (cf. Remark 4.7). Thus $\mathscr{F}(\mathbb{B}^n)$ is not an HFG algebra. For our purposes, $\mathscr{F}(\mathbb{B}^n)$ is useful because of the following fact.

---

[1] Note that our $\mathscr{F}(\mathbb{B}^n)$ is Popescu's $Hol(B(\mathcal{H})_1^n)$; we have changed the notation in order to emphasize the similarity between Popescu's algebras, our $\mathscr{F}(\mathbb{D}_r^n)$, and Taylor's $\mathscr{F}_n(r)$.

**Proposition 6.4.**

(i) *For each* $q \in (0,1)$, $\mathscr{O}_q^{\mathrm{hol}}(\mathbb{B}^n)$ *is topologically isomorphic to the quotient of* $\mathscr{F}(\mathbb{B}^n)$ *by the two-sided closed ideal generated by the elements* $\zeta_i\zeta_j - q\zeta_j\zeta_i$ *for all* $i < j$.

(ii) $\mathscr{O}^{\mathrm{hol}}(\mathbb{B}^n)$ *is topologically isomorphic to the quotient of* $\mathscr{F}(\mathbb{B}^n)$ *by the two-sided closed ideal generated by the elements* $\zeta_i\zeta_j - \zeta_j\zeta_i$ *for all* $i < j$.

Proposition 6.4 looks similar to Proposition 5.2, but its proof is less elementary. The proof of part (ii) involves, in particular, V. Müller's characterizations [26] of the joint spectral radius in commutative Banach algebras, as well as J.L. Taylor's holomorphic functional calculus on $\mathbb{B}^n$. The proof of part (i) is based on Theorem 5.4.

Repeating now the construction of the bundle $(D, p)$ with $\mathscr{F}(\mathbb{D}_r^n)$ replaced by $\mathscr{F}(\mathbb{B}^n)$, and using Proposition 6.4 instead of Proposition 5.2, we obtain the bundle $(B, p)$ whose existence was stated in part (ii) of Theorem 6.1.

**Acknowledgement**

This work was supported by the RFBR grant no. 12-01-00577. The author thanks A.Ya. Helemskii, D. Proskurin, and B. Solel for helpful discussions, and the referee for a careful reading and useful comments.

## References

[1] J. Alev and F. Dumas, *Sur le corps des fractions de certaines algèbres quantiques.* J. Algebra **170** (1994), no. 1, 229–265.

[2] K.A. Brown and K.R. Goodearl, *Lectures on algebraic quantum groups.* Advanced Courses in Mathematics. CRM Barcelona. Birkhäuser Verlag, Basel, 2002.

[3] J.L. Bueso, P. Jara, and A. Verschoren, *Compatibility, Stability, and Sheaves.* Monographs and Textbooks in Pure and Applied Mathematics, 185. Marcel Dekker, Inc., New York, 1995.

[4] J.W. Bunce, *Models for n-tuples of noncommuting operators.* J. Funct. Anal. **57** (1984), no. 1, 21–30.

[5] A. Connes, *Noncommutative Geometry.* Academic Press, Inc., San Diego, CA, 1994.

[6] A. Connes and M. Marcolli, *Noncommutative Geometry, Quantum Fields and Motives.* American Mathematical Society Colloquium Publications, 55. American Mathematical Society, Providence, RI, 2008.

[7] J. Cuntz, *Bivariante K-Theorie für lokalkonvexe Algebren und der Chern–Connes-Charakter.* Doc. Math. **2** (1997), 139–182.

[8] A.A. Dosiev, *Holomorphic functions of a basis of a nilpotent Lie algebra* (Russian). Funktsional. Anal. i Prilozhen. 34 (2000), no. 4, 82–84; translation in Funct. Anal. Appl. 34 (2000), no. 4, 302–304.

[9] A.A. Dosiev, *Algebras of power series of elements of a Lie algebra, and Słodkowski spectra* (Russian). Zap. Nauchn. Sem. St.-Peterburg. Otdel. Mat. Inst. Steklov.

(POMI) 290 (2002), Issled. po Linein. Oper. i Teor. Funkts. 30, 72–121, 179; translation in J. Math. Sci. (N. Y.) 124 (2004), no. 2, 4886–4908.

[10] A.A. Dosiev, *Homological dimensions of the algebra of entire functions of elements of a nilpotent Lie algebra* (Russian). Funktsional. Anal. i Prilozhen. 37 (2003), no. 1, 73–77; translation in Funct. Anal. Appl. 37 (2003), no. 1, 61–64.

[11] A. Dosiev, *Local left invertibility for operator tuples and noncommutative localizations.* J. K-Theory **4** (2009), no. 1, 163–191.

[12] A.A. Dosi (Dosiev), *Noncommutative holomorphic functions in elements of a Lie algebra and the absolute basis problem* (Russian). Izv. Ross. Akad. Nauk Ser. Mat. 73 (2009), no. 6, 77–100; translation in Izv. Math. 73 (2009), no. 6, 1149–1171.

[13] O. Forster, *Zur Theorie der Steinschen Algebren und Moduln*, Math. Z. **97** (1967), 376–405.

[14] G. Gierz, *Bundles of topological vector spaces and their duality.* Lecture Notes in Mathematics, 955. Springer-Verlag, Berlin-New York, 1982.

[15] J.M. Gracia-Bondía, J.C. Várilly, and H. Figueroa, *Elements of Noncommutative Geometry.* Birkhäuser Boston, Inc., Boston, MA, 2001.

[16] H. Grauert and R. Remmert, *Theory of Stein Spaces.* Springer-Verlag, Berlin-New York, 1979.

[17] R. Hartshorne, *Algebraic Geometry.* Graduate Texts in Mathematics, No. 52. Springer-Verlag, New York-Heidelberg, 1977.

[18] A.Ya. Helemskii, *Banach and Polynormed Algebras: General Theory, Representations, Homology*, Nauka, Moscow, 1989 (in Russian); English transl.: Oxford University Press, 1993.

[19] D.A. Jordan, *A simple localization of the quantized Weyl algebra.* J. Algebra **174** (1995), no. 1, 267–281.

[20] M. Khalkhali, *Basic Noncommutative Geometry.* EMS Series of Lectures in Mathematics. European Mathematical Society (EMS), Zurich, 2009.

[21] S. Klimek and A. Lesniewski, *A two-parameter quantum deformation of the unit disc.* J. Funct. Anal. **115** (1993), no. 1, 1–23.

[22] A. Klimyk and K. Schmüdgen, *Quantum groups and their representations.* Texts and Monographs in Physics. Springer-Verlag, Berlin, 1997.

[23] L. Le Bruyn, *Noncommutative geometry and Cayley-smooth orders.* Pure and Applied Mathematics (Boca Raton), 290. Chapman & Hall/CRC, Boca Raton, FL, 2008.

[24] D. Luminet, *A functional calculus for Banach PI-algebras.* Pacific J. Math. **125** (1986), no. 1, 127–160.

[25] E.A. Michael, *Locally multiplicatively-convex topological algebras.* Mem. Amer. Math. Soc., no. 11 (1952).

[26] V. Müller, *On the joint spectral radius.* Ann. Polon. Math. **66** (1997), 173–182.

[27] A.Yu. Pirkovskii, *Stably flat completions of universal enveloping algebras*, Dissertationes Math. (Rozprawy Math.) **441** (2006), 1–60.

[28] A.Yu. Pirkovskii, *Arens–Michael enveloping algebras and analytic smash products*, Proc. Amer. Math. Soc. **134** (2006), no. 9, 2621–2631.

[29] A.Yu. Pirkovskii, *Arens–Michael envelopes, homological epimorphisms, and relatively quasi-free algebras* (Russian), Tr. Mosk. Mat. Obs. **69** (2008), 34–125; English transl.: Trans. Moscow Math. Soc. 2008, 27–104.

[30] G. Popescu, *Free holomorphic functions on the unit ball of $B(\mathcal{H})^n$*. J. Funct. Anal. **241** (2006), no. 1, 268–333.

[31] D. Proskurin and Yu. Samoĭlenko, *Stability of the $C^*$-algebra associated with twisted CCR*. Algebr. Represent. Theory **5** (2002), no. 4, 433–444.

[32] W. Pusz and S.L. Woronowicz, *Twisted second quantization*. Rep. Math. Phys. **27** (1989), no. 2, 231–257.

[33] A.L. Rosenberg, *Noncommutative algebraic geometry and representations of quantized algebras*. Mathematics and its Applications, 330. Kluwer Academic Publishers Group, Dordrecht, 1995.

[34] J.L. Taylor, *Homology and cohomology for topological algebras*, Adv. Math. **9** (1972), 137–182.

[35] J.L. Taylor, *A general framework for a multi-operator functional calculus*, Adv. Math. **9** (1972), 183–252.

[36] J.L. Taylor, *Functions of several noncommuting variables*. Bull. Amer. Math. Soc. **79** (1973), 1–34.

[37] L.L. Vaksman, *Integral intertwining operators and quantum homogeneous spaces* (Russian). Teoret. Mat. Fiz. **105** (1995), no. 3, 355–363; translation in Theoret. and Math. Phys. **105** (1995), no. 3, 1476–1483.

[38] *Lectures on q-analogues of Cartan domains and associated Harish–Chandra modules*, L. Vaksman (ed.), Kharkov, Ukraine, 2001. Preprint arXiv:math.QA/0109198.

[39] L. Vaksman, *The maximum principle for "holomorphic functions" in the quantum ball* (Russian). Mat. Fiz. Anal. Geom. 10 (2003), no. 1, 12–28.

[40] L.L. Vaksman, *Quantum bounded symmetric domains*. Translations of Mathematical Monographs, 238. American Mathematical Society, Providence, RI, 2010.

[41] F.M.J. Van Oystaeyen, and A.H.M.J. Verschoren, *Noncommutative Algebraic Geometry. An introduction*. Lecture Notes in Mathematics, 887. Springer-Verlag, Berlin, 1981.

[42] F. Van Oystaeyen, *Algebraic geometry for associative algebras*. Monographs and Textbooks in Pure and Applied Mathematics, 232. Marcel Dekker, Inc., New York, 2000.

[43] J. Wess and B. Zumino, *Covariant differential calculus on the quantum hyperplane*. Recent advances in field theory (Annecy-le-Vieux, 1990). Nuclear Phys. B Proc. Suppl. **18B** (1990), 302–312.

[44] K.-W. Wiegmann, *Einbettungen komplexer Räume in Zahlenräume*. Invent. Math. **1** (1966), 229–242.

A.Yu. Pirkovskii
Faculty of Mathematics
National Research University Higher School of Economics
Vavilova 7, 117312 Moscow, Russia
e-mail: `aupirkovskii@hse.ru`
`pirkosha@gmail.com`

Operator Theory:
Advances and Applications, Vol. 233, 155–170


# Idempotent States on Locally Compact Groups and Quantum Groups

Pekka Salmi

*Dedicated to Professor Victor Shulman on the occasion of his 65th birthday*

**Abstract.** This is a short survey on idempotent states on locally compact groups and locally compact quantum groups. The central topic is the relationship between idempotent states, subgroups and invariant C*-subalgebras. We concentrate on recent results on locally compact quantum groups, but begin with the classical notion of idempotent probability measure. We also consider the 'intermediate' case of idempotent states in the Fourier–Stieltjes algebra: this is the dual case of idempotent probability measures and so an instance of idempotent states on a locally compact quantum group.

**Mathematics Subject Classification (2010).** Primary 46L89, Secondary 43A05, 43A35, 46L30, 60B15, 81R50.

**Keywords.** Idempotent state, idempotent probability measure, locally compact quantum group, compact quantum subgroup.

This is a short survey on idempotent states on locally compact groups and, more generally, on locally compact quantum groups. Idempotent states arise for example as limits of random walks (as we shall see in Section 1) and as limits in ergodic theorems for random walks [8]. Idempotent states are also connected to the construction of the Haar measure of a compact group: taking the Cesàro limit of convolution powers of a probability measure gives an idempotent probability measure, which is the Haar measure if the original measure is suitably chosen. The same process works in the case of compact quantum groups: this is the construction of Haar state due to Woronowicz [42, 45]. Very recently, idempotent states on locally compact quantum groups have also turned up in connection with Hopf images [3] and Poisson boundaries [16].

As we shall see, idempotent states are inherently related to subgroups. However, there is some evidence *against* the preceding claim, such as Pal's example of an idempotent state on the Kac–Paljutkin quantum group that does not arise from the Haar state of a subgroup [24]. In this paper we shall see that in fact also

Pal's example is associated with a subgroup (in a different way), and so perhaps there is still hope to connect all idempotent states to subgroups, quotient groups or combinations of these. There are not many new things in this survey article: only a cute new proof to a known result and the already-mentioned insight to the example of Pal. Many of the results, and much more, can be found in the recent papers due to (combinations of) Franz, Skalski, Tomatsu and the author [9, 10, 11, 31, 33].

## 1. Random walks and idempotent probability measures

Every probability measure $\mu$ on a discrete group $G$ determines a random walk: if we start from point $s \in G$, then the probability for taking a step to $t \in G$ is

$$P(s \mapsto t) = \mu_{ts^{-1}},$$

where $\mu_{ts^{-1}} := \mu(\{ts^{-1}\})$. Suppose that we start from the identity $e$ and $X_k$ is the random variable denoting the position after $k$ steps. We can use convolution to describe the random walk:

$$\begin{aligned} P(X_1 = t) &= \mu_t \\ P(X_2 = t) &= \sum_{s\in G} P(X_1 = s)P(s \mapsto t) = \sum_{s\in G} \mu_s \mu_{ts^{-1}} = (\mu \star \mu)_t \\ &\vdots \\ P(X_k = t) &= (\mu^{\star k})_t. \end{aligned}$$

In general the convolution of measures $\mu$ and $\nu$ on a locally compact group $G$ is defined by

$$\langle \mu \star \nu, f \rangle = \iint f(st)\, d\mu(s)\, d\nu(t) \qquad (f \in C_0(G));$$

here, and throughout the paper, we consider measures on $G$ as functionals on the C*-algebra $C_0(G)$ of continuous functions on $G$ vanishing at infinity. In the discrete case the convolution boils down to

$$\mu \star \nu = \sum_{s,t\in G} \mu_s \nu_t \delta_{st} = \sum_{s\in G} \Big( \sum_{t\in G} \mu_{st^{-1}} \nu_t \Big) \delta_s$$

where $\delta_s$ denotes the Dirac measure at $s$.

More generally, a probability measure on a locally compact group determines a random walk on that group. From this point of view, we shall see that random walks give rise to idempotent probability measures. An *idempotent probability measure* on a locally compact group $G$ is a probability measure $\mu$ on $G$ that is an idempotent under the convolution product:

$$\mu \star \mu = \mu.$$

Now suppose that $G$ is a compact group and $\nu$ is a probability measure on $G$. If the sequence of convolution powers of $\nu$ converges in the weak* topology, then the limit is an idempotent probability measure, which embodies the limit of the random walk. The convergence of such a sequence of convolution powers is widely studied in probability theory (see for example [12, 13]).

Taking a slightly different approach, consider the Cesàro averages

$$\frac{1}{n}\sum_{k=1}^{n} \nu^{\star k}.$$

The sequence of these averages always converges in the weak* topology. Moreover, the limit $\mu$ is an idempotent probability measure. This gives a useful way to generate idempotent probability measures: for example, the Haar measure of a compact group may be constructed this way. (In the case of non-compact groups the situation is more complicated and to make sure that the limit is non-zero, some form of tightness needs to be assumed for the sequence of convolution powers of $\nu$.)

## 2. Kawada–Itô theorem

Now that we have seen how idempotent probability measures may arise in practice, a natural question is how to characterise these measures. In the case of locally compact abelian groups, we may use the Fourier–Stieltjes transform to convert an idempotent measure to a characteristic function on the dual group. This trivialises the algebraic side of things, but now the positivity and the normalisation condition become non-trivial. Still, this is a useful approach and, as we shall see in the next section, leads to a simple characterisation. But let us first review the history of the general problem.

Already in 1940 Kawada and Itô [17] characterised idempotent probability measures on compact groups as the normalised Haar measures of compact subgroups. It seems that harmonic analysts were unaware of this paper, and Wendel [41] rediscovered the result in 1954 (truth be told, Wendel's main result is an interesting new proof for the existence of Haar measure on a compact group $G$, using idempotents in the compact semigroup formed by the probability measures on $G$). Trying to characterise all idempotent measures on a locally compact abelian group, Rudin [29, 30] showed that any idempotent measure is concentrated on a compact subgroup, thereby extending Wendel's result – or that of Kawada–Itô – to locally compact abelian groups. (The full description of idempotent measures on locally compact abelian groups is due to Cohen [4]; the non-abelian case is open.) Independently, Pym [28] and Loynes [23] characterised idempotent probability measures on locally compact groups (not necessarily abelian) as the normalised Haar measures of compact subgroups. However, since the problem took two separate paths, one starting from Kawada–Itô and another one from Wendel, it is perhaps not that surprising that the problem was solved already in 1954, by Kelley [18]. The three approaches, due to Kelley, Pym and Loynes, are all quite different: Kelley studies operators on $C_0(G)$, Pym idempotent measures on semigroups and Loynes operator-valued Fourier transform. There are also other generalisations, for example one due to Parthasarathy [25] to complete separable metric groups, which need not be locally compact.

**Theorem 1 (Kawada–Itô).** *Let $\mu$ be an idempotent probability measure on a locally compact group $G$. Then there is a compact subgroup $H$ of $G$ such that $\mu$ is the normalised Haar measure of $H$ (considered as a measure on $G$).*

The Kawada–Itô theorem gives a procedure to construct the Haar measure of a compact group. Start with a probability measure $\nu$ whose support generates the compact group $G$. As mentioned in the preceding section, the Cesàro averages of convolution powers of $\nu$ converge in the weak* topology to an idempotent probability measure. It follows from the Kawada–Itô theorem, that the limit is the normalised Haar measure of $G$ due to the choice of $\nu$.

As another example, consider the circle group $\mathbb{T}$ and the Dirac measure $\delta_z$ at some $z \in \mathbb{T}$. Then there is an idempotent probability measure $\mu$ such that $\frac{1}{n}\sum_{k=1}^{n} \delta_z^{*k} \to \mu$ weak*. If $z$ is a rational multiple of $\pi$, then $\mu$ is the counting measure of the finite subgroup of $\mathbb{T}$ generated by $z$. On the other hand, if $z$ is an irrational multiple of $\pi$, then $\mu$ is the normalised Lebesgue measure on $\mathbb{T}$. The latter statement amounts to the Weyl equidistribution theorem.

## 3. Idempotent states in the Fourier–Stieltjes algebra

Let $G$ be a locally compact group. The *Fourier–Stieltjes algebra* $\mathrm{B}(G)$ is the collection of all coefficient functions $(\,\pi(\cdot)\xi \mid \zeta\,)$ of strongly continuous unitary representations $\pi$ of $G$ (here $\xi$ and $\zeta$ are elements of the representation space $\mathcal{H}_\pi$). The Fourier–Stieltjes algebra is the dual space of the universal group C*-algebra $\mathrm{C}^*(G)$ (this determines the norm of $\mathrm{B}(G)$), and $\mathrm{B}(G)$ is a Banach algebra under the pointwise multiplication of functions. If $G$ is abelian, $\mathrm{B}(G)$ is isomorphic, via Fourier–Stieltjes transform, to the measure algebra $\mathrm{M}(\widehat{G})$ of Radon measures on the dual group $\widehat{G}$.

An *idempotent state* in $\mathrm{B}(G)$ is a state on $\mathrm{C}^*(G)$ that is an idempotent:

$$u^2 = u.$$

That $u$ is a state means that $u$ is a positive definite function with $u(e) = 1$, where $e$ denotes the identity of $G$. In the case of abelian $G$, the Fourier–Stieltjes transform takes an idempotent probability measure on $\widehat{G}$ to an idempotent state in $\mathrm{B}(G)$. This explains why the characterisation of idempotent states in $\mathrm{B}(G)$ may be viewed as the dual version of the Kawada–Itô theorem. In fact, many early studies on idempotents in $\mathrm{M}(G)$, with $G$ abelian, used the Fourier–Stieltjes transform to translate the problem to the dual setting.

Continuing from the work of Cohen [4], mentioned in the previous section, Host [14] characterised all idempotents in $\mathrm{B}(G)$ as characteristic functions of sets in the open coset ring of $G$. However, his characterisation does not immediately lead to the following characterisation of idempotent states, which is due to Ilie and

Spronk [15]. The short proof presented here is new (Ilie and Spronk obtained their result as a corollary of a more general characterisation of contractive idempotents).

**Theorem 2.** *Every idempotent state in* $\mathrm{B}(G)$ *is the characteristic function of an open subgroup of* $G$.

*Proof.* First of all, every idempotent in $\mathrm{B}(G)$ is a characteristic function of some open (and closed) set $H$, because $\mathrm{B}(G)$ consists of continuous functions. Denote the universal representation of $G$ by $\varpi$. If $s \in H$, then

$$u(s^{-1}) = \langle\, u, \varpi(s^{-1})\,\rangle = \langle\, u, \varpi(s)^*\,\rangle = \overline{\langle\, u, \varpi(s)\,\rangle} = 1,$$

so $s^{-1}$ is also in $H$. Moreover,

$$\langle\, u, \varpi(s)^*\varpi(s)\,\rangle = u(e) = 1,$$

because $u$ is a state, and

$$\langle\, u, \varpi(s)^*\,\rangle\langle\, u, \varpi(s)\,\rangle = 1\cdot 1 = 1.$$

It follows from Choi's theorem on multiplicative domains (see [26, Theorem 3.19]) that $u$ is multiplicative at $\varpi(s)$. So for every $t \in G$,

$$u(ts) = u(t)u(s)$$

which implies that $H$ is closed under multiplication. Hence $H$ is an open subgroup and $u = 1_H$. □

The fact that compact subgroups in the Kawada–Itô theorem have changed to open ones in the preceding result reflects subgroup duality. Suppose that $G$ is abelian and $H$ is a closed subgroup of $G$. Then the continuous characters on $G$ that are constant 1 on $H$ form a closed subgroup $H^\perp$ of $\widehat{G}$. Now $H$ is compact if and only if $H^\perp$ is open, and vice versa. Moreover, the Fourier–Stieltjes transform maps a measure on $G$ supported by $H$ to a function on $\widehat{G}$ that is constant on the cosets of $H^\perp$.

## 4. Locally compact quantum groups

Locally compact quantum groups provide a natural context to discuss the results in the previous sections in a unified manner. We shall walk through the definition due to Kustermans and Vaes [21]. Let $\mathbb{G}$ denote a locally compact quantum group. This means that we have a C*-algebra $\mathrm{C}_0(\mathbb{G})$, a nondegenerate $*$-homomorphism $\Delta\colon \mathrm{C}_0(\mathbb{G}) \to \mathrm{M}(\mathrm{C}_0(\mathbb{G}) \otimes \mathrm{C}_0(\mathbb{G}))$ (where the tensor product is the minimal C*-algebraic tensor product and $\mathrm{M}(\,\cdot\,)$ denotes the multiplier algebra) such that

$$(\mathrm{id}\otimes\Delta)\Delta = (\Delta\otimes\mathrm{id})\Delta \qquad \text{(coassociativity)};$$

and

$$\overline{\mathrm{span}}\,\Delta(\mathrm{C}_0(\mathbb{G}))(\mathrm{C}_0(\mathbb{G})\otimes 1) = \overline{\mathrm{span}}\,\Delta(\mathrm{C}_0(\mathbb{G}))(1\otimes \mathrm{C}_0(\mathbb{G})) = \mathrm{C}_0(\mathbb{G})\otimes \mathrm{C}_0(\mathbb{G}).$$

The map $\Delta$ is called the *comultiplication* of $\mathbb{G}$. We also need to assume that there exist left and right Haar weights on $\mathrm{C}_0(\mathbb{G})$, denoted by $\phi$ and $\psi$, respectively.

These are so-called KMS-weights, which are densely defined, faithful and lower semicontinuous. The important invariance properties, as with Haar measures, are that

$$\phi\big((\omega \otimes \mathrm{id})\Delta(a)\big) = \omega(1)\phi(a)$$

and

$$\psi\big((\mathrm{id} \otimes \omega)\Delta(b)\big) = \omega(1)\psi(b)$$

whenever $\omega \in C_0(\mathbb{G})^*_+$ and $a, b \in C_0(\mathbb{G})_+$ are such that $\phi(a) < \infty$ and $\psi(b) < \infty$. So a locally compact quantum group is given by a C*-algebra that has a suitable comultiplication and left and right Haar weights. It is convenient to use the suggestive notation $C_0(\mathbb{G})$ for the C*-algebra because in the commutative case the C*-algebra *is* $C_0(G)$ for some locally compact group $G$. In this case the comultiplication is given by dualised group multiplication:

$$\Delta(f)(s,t) = f(st) \qquad (f \in C_0(G), s, t \in G).$$

Note that $\Delta(f) \in C_b(G \times G) = M(C_0(G) \otimes C_0(G))$ but $\Delta(f) \notin C_0(G) \otimes C_0(G)$ unless $G$ is compact or $f = 0$. Of course the left and right Haar weights are given by integration against the left and right Haar measures, respectively. Whenever $\mathbb{G}$ is a locally compact quantum group such that $C_0(\mathbb{G})$ is commutative, it is of this form.

Next we consider the dual of the commutative case, which is known as the cocommutative case. Let $\lambda$ be the left regular representation of $G$. Then the reduced group C*-algebra $C^*_r(G)$ is generated by $\lambda(L^1(G))$ in $B(L^2(G))$. (Hopefully the reader is not confused by the two uses of 'B' as both the Fourier–Stieltjes algebra and the algebra of bounded operators; the distinction should be clear from the context.) We define a comultiplication on $C^*_r(G)$ by putting

$$\Delta(\lambda(s)) = \lambda(s) \otimes \lambda(s) \qquad (s \in G).$$

Note that actually $\lambda(s)$ is in $M(C^*_r(G))$ but the above does define a unique comultiplication on $C^*_r(G)$, because the linear span of $\lambda(G)$ is strictly dense in $M(C^*_r(G))$. In this case the left and right Haar weights coincide and are the so-called Plancherel weight. The construction of this weight uses Tomita–Takesaki theory [34, Section VII.3] (for discrete $G$, $\phi(a) = (\, a\delta_e \mid \delta_e \,)$ is the usual tracial state).

Both these examples may be considered as Kac algebras [7]. Every Kac algebra determines a locally compact quantum group, so the latter notion is more general. For an example of a locally compact quantum group that is not a Kac algebra, see Section 7, which includes the description of the quantum deformation of SU(2) defined by Woronowicz. For a more thorough introduction to locally compact quantum groups, see for example the survey by Kustermans and Tuset [19, 20] or the book by Timmermann [36].

From now on we shall concentrate on locally compact quantum groups $\mathbb{G}$ that are *coamenable*. That means that there is a state $\epsilon$ on $C_0(\mathbb{G})$, called the *counit* of $\mathbb{G}$, such that

$$(\mathrm{id} \otimes \epsilon)\Delta(a) = (\epsilon \otimes \mathrm{id})\Delta(a) = a$$

for every $a \in C_0(\mathbb{G})$. In the commutative case coamenability is a vacuous condition (every commutative quantum group is coamenable), but a co-commutative quantum group $\mathbb{G} = \widehat{G}$ is coamenable if and only if the locally compact group $G$ is amenable.

## 5. Classical cases as instances of idempotent states on locally compact quantum groups

The notion of idempotent state from the two classical cases – idempotent probability measures on groups and idempotent states in the Fourier–Stieltjes algebra – is easily generalised to the language of locally compact quantum groups. The dual space of the C*-algebra $C_0(\mathbb{G})$ carries a natural Banach algebra structure: the multiplication is defined by

$$\omega \star \sigma(a) = (\omega \otimes \sigma)\Delta(a) \qquad (\omega, \sigma \in C_0(\mathbb{G})^*, a \in C_0(\mathbb{G})).$$

An *idempotent state* on a locally compact quantum group $\mathbb{G}$ is a state $\omega$ on the C*-algebra $C_0(\mathbb{G})$ that is an idempotent under the product defined above: $\omega \star \omega = \omega$.

A much more difficult task than the definition is to unify the results from the classical cases to general results on locally compact quantum groups. Indeed, it is perhaps not even possible to do so. To even bring forth this discussion we need some further terminology.

A locally compact quantum group $\mathbb{H}$ is *compact* if $C_0(\mathbb{H})$ is unital, in which case we write $C(\mathbb{H})$ for $C_0(\mathbb{H})$. A *compact quantum subgroup* of a coamenable locally compact quantum group $\mathbb{G}$ is a compact quantum group $\mathbb{H}$ such that there exists a surjective $*$-homomorphism

$$\pi\colon C_0(\mathbb{G}) \to C(\mathbb{H}) \qquad (\pi \otimes \pi)\Delta_{\mathbb{G}} = \Delta_{\mathbb{H}}\pi.$$

(The reader is warned that there are other definitions of closed quantum subgroup [38, 39], and it is not clear whether they are all equivalent. In our situation, all the definitions coincide as they do in many other cases; see [5].) A compact quantum subgroup of $\mathbb{G}$ always gives rise to an idempotent state on $\mathbb{G}$. Indeed, when a locally compact quantum group $\mathbb{H}$ is compact, the left and right Haar weights are actually bounded functionals and coincide. By normalisation, there exists a unique state – the *Haar state* – $\phi_{\mathbb{H}}$ on $C(\mathbb{H})$ that is both left and right invariant. Using the subgroup morphism $\pi$, we may pull back $\phi_{\mathbb{H}}$ to obtain an idempotent state $\omega = \phi_{\mathbb{H}}\pi$ on $\mathbb{G}$. Obviously $\omega$ is a state and it is an idempotent due to invariance of $\phi_{\mathbb{H}}$:

$$\begin{aligned}\omega \star \omega &= \big((\phi_{\mathbb{H}}\pi) \otimes (\phi_{\mathbb{H}}\pi)\big)\Delta_{\mathbb{G}} = (\phi_{\mathbb{H}} \otimes \phi_{\mathbb{H}})(\pi \otimes \pi)\Delta_{\mathbb{G}} \\ &= (\phi_{\mathbb{H}} \otimes \phi_{\mathbb{H}})\Delta_{\mathbb{H}}\pi = \phi_{\mathbb{H}}(1)\phi_{\mathbb{H}}\pi = \omega.\end{aligned}$$

It should be noted that in the case of compact quantum groups, the existence of Haar state follows from the other axioms as shown by Woronowicz [42, 45]. Indeed, this may be done with a similar process of using Cesàro averages as mentioned after the Kawada–Itô theorem: the Cesàro averages of convolution powers of a faithful state converge to the Haar state (however, the resulting Haar state is not necessarily faithful). The assumption that there is a faithful state on the C*-algebra, which is true in the separable case, may be dropped, as shown by Van Daele [40].

Now it is possible to at least formulate the statement of the Kawada–Itô theorem: every idempotent state on a locally compact quantum group $\mathbb{G}$ is a *Haar idempotent*, that is, of the form $\omega = \phi_{\mathbb{H}}\pi$ where $\phi_{\mathbb{H}}$ is the Haar state of a compact quantum subgroup $\mathbb{H}$ of $\mathbb{G}$ and $\pi\colon \mathrm{C}_0(\mathbb{G}) \to \mathrm{C}(\mathbb{H})$ is the associated morphism. The problem is that this statement is false. It is, moreover, easily seen to be false. Let $G$ be an amenable locally compact group with a non-normal open subgroup $H$. Then $1_H$ is an idempotent state on $\mathrm{C}^*_r(G)$. If $1_H$ were a pullback of the Haar state of a compact quantum subgroup of $\mathrm{C}^*_r(G)$, then the compact quantum subgroup would necessarily be of the form $\mathrm{C}^*_r(G/H)$. But $H$ not being normal, this is not possible (see [10, Theorem 6.2] for the finite case and [31, Section 7] for a related discussion on left invariant C*-subalgebras in $\mathrm{C}^*_r(G)$).

Although this example, obtainable with finite groups, certainly seems to be the most straightforward counterexample of the Kawada–Itô theorem for quantum groups, it was not the first one. The first counterexample is due to Pal [24] and it comes from a genuine quantum group: the Kac–Paljutkin quantum group. We shall next describe this quantum group and Pal's example as well as provide a new insight to this example.

## 6. Pal's counterexample

The underlying C*-algebra of the Kac–Paljutkin quantum group $\mathbb{G}$ is

$$\mathbb{C} \oplus \mathbb{C} \oplus \mathbb{C} \oplus \mathbb{C} \oplus \mathrm{M}_2(\mathbb{C}),$$

the basis of which is given by the vectors

$$e_k = \delta_{1,k} \oplus \delta_{2,k} \oplus \delta_{3,k} \oplus \delta_{4,k} \oplus \begin{pmatrix} \delta_{5,k} & \delta_{8,k} \\ \delta_{7,k} & \delta_{6,k} \end{pmatrix}$$

$k = 1, 2, \ldots, 8$. The comultiplication of $\mathbb{G}$ is defined by

$$\begin{aligned}
\Delta(e_1) &= e_1 \otimes e_1 + e_2 \otimes e_2 + e_3 \otimes e_3 + e_4 \otimes e_4 \\
&\quad + \frac{1}{2}(e_5 \otimes e_5 + e_6 \otimes e_6 + e_7 \otimes e_7 + e_8 \otimes e_8) \\
\Delta(e_2) &= e_1 \otimes e_2 + e_2 \otimes e_1 + e_3 \otimes e_4 + e_4 \otimes e_3 \\
&\quad + \frac{1}{2}(e_5 \otimes e_6 + e_6 \otimes e_5 + ie_7 \otimes e_8 - ie_8 \otimes e_7) \\
\Delta(e_3) &= e_1 \otimes e_3 + e_3 \otimes e_1 + e_2 \otimes e_4 + e_4 \otimes e_2 \\
&\quad + \frac{1}{2}(e_5 \otimes e_6 + e_6 \otimes e_5 - ie_7 \otimes e_8 + ie_8 \otimes e_7)
\end{aligned}$$

$$\begin{aligned}
\Delta(e_4) &= e_1 \otimes e_4 + e_4 \otimes e_1 + e_2 \otimes e_3 + e_3 \otimes e_2 \\
&\quad + \frac{1}{2}(e_5 \otimes e_5 + e_6 \otimes e_6 - e_7 \otimes e_7 - e_8 \otimes e_8) \\
\Delta(e_5) &= e_1 \otimes e_5 + e_5 \otimes e_1 + e_2 \otimes e_6 + e_6 \otimes e_2 \\
&\quad + e_3 \otimes e_6 + e_6 \otimes e_3 + e_4 \otimes e_5 + e_5 \otimes e_4 \\
\Delta(e_6) &= e_1 \otimes e_6 + e_6 \otimes e_1 + e_2 \otimes e_5 + e_5 \otimes e_2 \\
&\quad + e_3 \otimes e_5 + e_5 \otimes e_3 + e_4 \otimes e_6 + e_6 \otimes e_4 \\
\Delta(e_7) &= e_1 \otimes e_7 + e_7 \otimes e_1 - ie_2 \otimes e_8 + ie_8 \otimes e_2 \\
&\quad + ie_3 \otimes e_8 - ie_8 \otimes e_3 - e_4 \otimes e_7 - e_7 \otimes e_4 \\
\Delta(e_8) &= e_1 \otimes e_8 + e_8 \otimes e_1 + ie_2 \otimes e_7 - ie_7 \otimes e_2 \\
&\quad - ie_3 \otimes e_7 + ie_7 \otimes e_3 - e_4 \otimes e_8 - e_8 \otimes e_4.
\end{aligned}$$

Pal's idempotent state is defined by

$$\omega\left(\sum_{k=1}^{8} \alpha_k e_k\right) = \frac{1}{4}\alpha_1 + \frac{1}{4}\alpha_4 + \frac{1}{2}\alpha_6.$$

As we shall see in Section 9, we can always associate a certain C*-subalgebra to an idempotent state. For Pal's idempotent state $\omega$, the associated C*-subalgebra $(\mathrm{id} \otimes \omega)\Delta(\mathrm{C}(\mathbb{G}))$ is spanned by the elements

$$a = e_1 + e_2 + e_3 + e_4 + e_5 + e_6 \quad \text{and} \quad b = e_1 - e_2 - e_3 + e_4 - e_5 + e_6.$$

Moreover, one can calculate that

$$\Delta(a) = a \otimes a \quad \text{and} \quad \Delta(b) = b \otimes b. \tag{1}$$

Now consider the quantum group $\widehat{\mathbb{Z}}_2$ given by the group C*-algebra $\mathrm{C}^*(\mathbb{Z}_2)$ (of course $\mathbb{Z}_2 \cong \widehat{\mathbb{Z}}_2$ but the chosen viewpoint suits us better). Then $\mathrm{C}^*(\mathbb{Z}_2)$ is spanned by $\lambda(0)$ and $\lambda(1)$, where $\lambda$ denotes the left regular representation of $\mathbb{Z}_2$. Define $\pi\colon \mathrm{C}^*(\mathbb{Z}_2) \to \mathrm{C}(\mathbb{G})$ by $\pi(\lambda(0)) = a$ and $\pi(\lambda(1)) = b$. By (1), we see that $\pi$ preserves the quantum group structure of $\widehat{\mathbb{Z}}_2$. There is also a conditional expectation $E$ onto $\pi(\mathrm{C}^*(\mathbb{Z}_2))$ defined by

$$\begin{gathered}
E(e_1) = E(e_4) = \frac{1}{8}(a+b) \qquad E(e_2) = E(e_3) = \frac{1}{8}(a-b) \\
E(e_5) = \frac{1}{4}(a-b) \qquad E(e_6) = \frac{1}{4}(a+b) \qquad E(e_7) = E(e_8) = 0.
\end{gathered}$$

The counit $\epsilon_{\widehat{\mathbb{Z}}_2}$ of $\widehat{\mathbb{Z}}_2$ is the constant function 1 (considered as an element of $\mathrm{B}(\mathbb{Z}_2)$). Finally, note that $\omega = \epsilon_{\widehat{\mathbb{Z}}_2} \circ \pi^{-1} \circ E$. What this shows is that Pal's idempotent state is of the similar form as the idempotent states $1_H$ on group C*-algebras: $1_H = \epsilon_{\widehat{H}} \circ \pi^{-1} \circ E$ where $\epsilon_{\widehat{H}}$ is the counit of $\widehat{H}$ (i.e., constant 1 on $H$), $\pi\colon \mathrm{C}^*_r(H) \to \mathrm{C}^*_r(G)$ is the natural embedding (i.e., zero extension), and $E\colon \mathrm{C}^*_r(G) \to \pi(\mathrm{C}^*_r(H))$ is the natural conditional expectation (i.e., restriction to $H$). So although Pal's idempotent is not like the idempotent states in the commutative case (i.e., not a

Haar idempotent), it is similar to the idempotent states in the co-commutative case. Thus it is associated with a subgroup but in a different way.

These examples of idempotent states that are not Haar idempotents show that a new approach is needed for general locally compact quantum groups. In Section 8 we consider another notion, that of left invariant C*-subalgebras, that is closely tied with idempotent states as we shall see. There is also the approach of Franz and Skalski, who show in [10] that every idempotent state on a finite quantum group arises from the Haar state of a so-called quantum sub*hyper*group. On a related note, Franz and Skalski [9] also show that idempotent states on a finite quantum group correspond to quantum pre-subgroups in the sense of [1]. In all these approaches one associates idempotent states with structures more general than subgroups, and that is what we shall do with left invariant C*-subalgebras in Section 9.

## 7. Positive examples from deformation quantum groups

In this section we shall consider idempotent states on some important examples of compact quantum groups, in particular on the quantum deformation of SU(2) introduced by Woronowicz [43, 42]. It turns out that on these deformations of classical groups, $SU_q(2)$, $U_q(2)$ and $SO_q(3)$, all idempotent states are Haar idempotents. The results in this section are due to Franz, Skalski and Tomatsu [11].

Define $C(SU_q(2))$ as the universal unital C*-algebra generated by elements $a$ and $c$ such that

$$\begin{pmatrix} a & -qc^* \\ c & a^* \end{pmatrix}$$

is formally a unitary matrix (a $2 \times 2$ matrix with entries in $C(SU_q(2))$). The comultiplication of $SU_q(2)$ is determined by the identity

$$\Delta \begin{pmatrix} a & -qc^* \\ c & a^* \end{pmatrix} = \begin{pmatrix} a & -qc^* \\ c & a^* \end{pmatrix} \otimes \begin{pmatrix} a & -qc^* \\ c & a^* \end{pmatrix}.$$

This identity is to be read as follows: on the left-hand side we apply $\Delta$ to each entry and on the right-hand side we take a formal matrix multiplication where we use the tensor product when 'multiplying' entries; then we just equate the entries of the two $2 \times 2$ matrices. Although the relations given above fully determine the structure of $SU_q(2)$, to prove that we actually get a compact quantum group takes some work. Recall however that the existence of the Haar state follows from the general theory of compact quantum groups.

Using the representation theory of $SU_q(2)$, Franz, Skalski and Tomatsu [11] calculated all the idempotent states on $SU_q(2)$ for $q \in (-1, 0) \cup (0, 1)$. It turns out that these are all Haar idempotents. Namely, the idempotent states on $SU_q(2)$ are the Haar state and the Haar idempotents coming from the subgroups $\mathbb{T}$ and $\mathbb{Z}_n$, $1 \leq n < \infty$. We see that $\mathbb{T}$ is a subgroup of $SU_q(2)$ by mapping the generator $a$ to the generator $z$ of $C(\mathbb{T})$ and $c$ to 0. Moreover, $\mathbb{Z}_n$'s are subgroups of $\mathbb{T}$. Already Podleś

[27] showed that these are all the closed quantum subgroups of $\mathrm{SU}_q(2)$. This result follows also from [11], but of course it takes more work to show that all idempotent states actually arise from these subgroups. Franz, Skalski and Tomatsu also give the complete list of idempotent states for the related deformation quantum groups $\mathrm{U}_q(2)$ and $\mathrm{SO}_q(3)$. Also in these cases all idempotent states are Haar idempotents.

## 8. Left invariant C*-subalgebras

In this section we shall consider another notion related to idempotent states besides subgroups. The notion is that of left invariant C*-subalgebra (here we could use alternative terminology and call these coideals or homogeneous spaces).

Let $\mathbb{G}$ be a coamenable locally compact quantum group. For $\omega \in \mathrm{C}_0(\mathbb{G})^*$, define the left and right convolution operators on $\mathrm{C}_0(\mathbb{G})$ by

$$\begin{aligned} L_\omega(a) &= (\omega \otimes \mathrm{id})\Delta(a) \\ R_\omega(a) &= (\mathrm{id} \otimes \omega)\Delta(a) \end{aligned} \qquad (a \in \mathrm{C}_0(\mathbb{G})).$$

A C*-subalgebra $X \subseteq \mathrm{C}_0(\mathbb{G})$ is said to be *left invariant* if $L_\omega(X) \subseteq X$ for all $\omega \in \mathrm{C}_0(\mathbb{G})^*$. A nondegenerate C*-subalgebra $X$ of $\mathrm{C}_0(\mathbb{G})$ is left invariant if and only if $\Delta \colon X \to \mathrm{M}(\mathrm{C}_0(\mathbb{G}) \otimes X)$. (A C*-subalgebra is nondegenerate if it contains a bounded approximate identity for the ambient C*-algebra.)

Consider the commutative case when $G$ is a locally compact group. Then a C*-subalgebra $X$ of $\mathrm{C}_0(G)$ is left invariant if and only if it is left translation invariant; that is, the function $L_s f(t) = f(st)$ is in $X$ whenever $f \in X$ and $s \in G$. Lau and Losert [22] have characterised left invariant C*-subalgebras of $\mathrm{C}_0(G)$: a C*-subalgebra $X \subseteq \mathrm{C}_0(G)$ is left invariant if and only if there is a compact subgroup $H$ of $G$ such that $X$ consists of all the functions in $\mathrm{C}_0(G)$ that are constant on left cosets of $H$. The latter statement means that $X$ is $*$-isomorphic to $\mathrm{C}_0(G/H)$. Earlier, de Leeuw and Mirkil [6] gave this characterisation in the case of locally compact *abelian* groups. Moreover, Takesaki and Tatsuuma [35] produced several related results, characterising closed (left) invariant self-adjoint subalgebras of $\mathrm{L}^\infty(G)$, the Fourier algebra $\mathrm{A}(G)$, the group von Neumann algebra $\mathrm{VN}(G)$ and the $\mathrm{L}^1$ group algebra (here the meaning of 'closed' depends on the context: with $\mathrm{A}(G)$ and $\mathrm{L}^1(G)$ it means norm-closed and with $\mathrm{L}^\infty(G)$ and $\mathrm{VN}(G)$ it means weak*-closed). The dual version of the Lau–Losert characterisation for a locally compact amenable group $G$ is given in [31]: a C*-subalgebra $X \subseteq \mathrm{C}_r^*(G)$ is invariant if and only if $X \cong \mathrm{C}_r^*(H)$ for some open subgroup $H$. One can also consider strictly closed left invariant C*-subalgebras of $\mathrm{C}_b(G)$ and $\mathrm{M}(\mathrm{C}_r^*(G))$ and obtain in both cases a one-to-one correspondence with closed subgroups of $G$ [32].

Finally, we also have the following result from [31], concerning left invariant C*-subalgebras of coamenable locally compact quantum groups. Recall that a conditional expectation on a C*-algebra $A$ is a norm 1 projection from $A$ onto a C*-subalgebra of $A$. The following result also employs a symmetry condition that is related to the problem brought out by Pal's counterexample. We postpone the formulation of this symmetry condition until after the theorem.

**Theorem 3.** *There is a one-to-one correspondence between compact quantum subgroups of* $\mathbb{G}$ *and symmetric, left invariant C*-subalgebras* $X$ *of* $C_0(\mathbb{G})$ *with a conditional expectation* $P$ *from* $C_0(\mathbb{G})$ *onto* $X$ *such that* $(\mathrm{id} \otimes P)\Delta = \Delta P$.

Let $G$ be an amenable locally compact group and $H$ an open subgroup of $G$. As noted above, $C_r^*(H)$ is an invariant C*-subalgebra of $C_r^*(G)$. However, $C_r^*(H)$ is not associated with a compact quantum subgroup unless $H$ is normal, in which case $C_r^*(H)$ is associated with $C_r^*(G/H)$. We shall need an analogue of this normality condition for more general quantum groups. This can be done through the so-called multiplicative unitary of a locally compact quantum group $\mathbb{G}$. There is a canonical way to define a unitary operator $W$ on $L^2(\mathbb{G}) \otimes L^2(\mathbb{G})$ such that $W$ determines the quantum group $\mathbb{G}$ [21, Proposition 3.17]. Here $L^2(\mathbb{G})$ denotes the Hilbert space obtained by applying the GNS-construction to the left Haar weight of $\mathbb{G}$. The C*-algebra $C_0(\mathbb{G})$ is faithfully represented on $L^2(\mathbb{G})$ and it is natural to identify $C_0(\mathbb{G})$ with its image in $B(L^2(\mathbb{G}))$. The multiplicative unitary $W$ determines the comultiplication via

$$\Delta(a) = W^*(1 \otimes a)W \qquad (a \in C_0(\mathbb{G})).$$

The notion of multiplicative unitary is central in the theory of locally compact quantum groups; seminal work in this area is due to Baaj and Skandalis [2] and Woronowicz [44].

We say that a C*-subalgebra $X$ of $C_0(\mathbb{G})$ is *symmetric* if

$$W(x \otimes 1)W^* \in M(X \otimes B_0(L^2(\mathbb{G})))$$

whenever $x \in X$ (here $B_0$ denotes the compact operators). Tomatsu [37] introduced this type of condition, calling it coaction symmetry (due to the fact that $X$ is symmetric if and only if it is closed under the natural action of the dual quantum group of $\mathbb{G}$). Returning to the co-commutative case, the left invariant C*-subalgebra $C_r^*(H)$ associated with an open subgroup $H$ of $G$ is symmetric if and only if $H$ is normal [31].

## 9. Idempotent states and left invariant C*-subalgebras

Although Pal's counterexample showed that we cannot associate all idempotent states to compact quantum subgroups, we may still have a chance of associating idempotent states to suitable left invariant C*-subalgebras. The results in this section are from [33], many of them generalisations from [9] or [11].

Let $\mathbb{G}$ be a coamenable locally compact quantum group. If $\omega$ is an idempotent state on $\mathbb{G}$, then $R_\omega(C_0(\mathbb{G}))$ is a left invariant C*-subalgebra of $C_0(\mathbb{G})$ and $R_\omega$ is a conditional expectation onto this C*-subalgebra. The following result generalises an earlier result due to Franz and Skalski [9] concerning compact quantum groups.

**Theorem 4.** *Suppose that $\mathbb{G}$ is unimodular (i.e., $\phi = \psi$). There is a one-to-one correspondence between idempotent states $\omega$ on $\mathbb{G}$ and left invariant $C^*$-subalgebras $X$ of $\mathrm{C}_0(\mathbb{G})$ with a conditional expectation $P$ from $\mathrm{C}_0(\mathbb{G})$ onto $X$ such that $\phi \circ P = \phi$. The correspondence is given by*

$$X_\omega = R_\omega(\mathrm{C}_0(\mathbb{G})), \qquad \omega_X = \epsilon P_X.$$

*where $\epsilon$ is the counit of $\mathbb{G}$.*

The preceding result leaves room for improvement: one would like to remove the unimodularity assumption in which case the conditional expectation should preserve both left and right Haar weights.

The following result characterises those idempotent states that arise from compact quantum subgroups. It also brings together the symmetry condition from the preceding section. The equivalence between the first and the third condition is proved for compact quantum groups in [11].

**Theorem 5.** *Let $\omega$ be an idempotent state on $\mathbb{G}$ and let $X_\omega = R_\omega(\mathrm{C}_0(\mathbb{G}))$. The following are equivalent:*

1. *$\omega$ is a Haar idempotent;*
2. *$X_\omega$ is symmetric;*
3. *$N_\omega := \{ a \in \mathrm{C}_0(\mathbb{G});\ \omega(a^*a) = 0 \}$ is an ideal.*

The set $N_\omega$ in the third condition of the preceding result is always a left ideal, so the condition is automatically satisfied if $\mathrm{C}_0(\mathbb{G})$ is commutative. Consequently, we get the Kawada–Itô theorem from Section 2 as a corollary.

**Corollary 6 (Kawada–Itô).** *If $\mathrm{C}_0(\mathbb{G})$ is commutative, then all idempotent states on $\mathbb{G}$ are Haar idempotents.*

Finally, we have the following correspondence result, which does not assume unimodularity, but works only for Haar idempotents.

**Theorem 7.** *There is a one-to-one correspondence between Haar idempotents $\omega$ on $\mathbb{G}$ and symmetric, left invariant $C^*$-subalgebras $X$ of $\mathrm{C}_0(\mathbb{G})$ with a conditional expectation $P$ from $\mathrm{C}_0(\mathbb{G})$ onto $X$ such that $\phi \circ P = \phi$ and $\psi \circ P = \psi$.*

Note that the preceding theorem improves Theorem 3 in the sense that the condition that $(\mathrm{id} \otimes P)\Delta = \Delta P$ may be replaced by the more natural condition that $P$ preserves both left and right Haar weights.

## Acknowledgement

This paper is based on a talk given in the conference 'Operator Theory and its Applications' held in honour of Victor Shulman in Gothenburg 2011; I thank the organisers Ivan Todorov and Lyudmila Turowska for a great conference. I thank Nico Spronk for generous support throughout my postdoctoral stay at University of Waterloo and in particular for enabling my conference visit. I thank the Emil Aaltonen Foundation for support during the preparation of this paper. I thank Adam Skalski and the referee for helpful comments improving the paper.

## References

[1] S. Baaj, E. Blanchard and G. Skandalis, *Unitaires multiplicatifs en dimension finie et leurs sous-objets*, Ann. Inst. Fourier (Grenoble) **49** (1999), 1305–1344.

[2] S. Baaj and G. Skandalis, *Unitaires multiplicatifs et dualité pour les produits croisés de $C^*$-algèbres*, Ann. Sci. École Norm. Sup. (4) **26** (1993), 425–488.

[3] T. Banica, U. Franz and A. Skalski, *Idempotent states and the inner linearity property*, Bull. Pol. Acad. Sci. Math. **60** (2012), 123–132.

[4] P.J. Cohen, *On a conjecture of Littlewood and idempotent measures*, Amer. J. Math. **82** (1960), 191–212.

[5] M. Daws, P. Kasprzak, A. Skalski and P.M. Sołtan, *Closed quantum subgroups of locally compact quantum groups*, Adv. Math. **231** (2012), 3473–3501.

[6] K. de Leeuw and H. Mirkil, *Translation-invariant function algebras on abelian groups*, Bull. Soc. Math. France **88** (1960), 345–370.

[7] M. Enock and J.-M. Schwartz, *Kac algebras and duality of locally compact groups*, Springer-Verlag, Berlin, 1992.

[8] U. Franz and A. Skalski, *On ergodic properties of convolution operators associated with compact quantum groups*, Colloq. Math. **113** (2008), 13–23.

[9] U. Franz and A. Skalski, *A new characterisation of idempotent states on finite and compact quantum groups*, C. R. Math. Acad. Sci. Paris **347** (2009), 991–996.

[10] U. Franz and A. Skalski, *On idempotent states on quantum groups*, J. Algebra **322** (2009), 1774–1802.

[11] U. Franz, A. Skalski and R. Tomatsu, *Idempotent states on compact quantum groups and their classification on $U_q(2)$, $SU_q(2)$, and $SO_q(3)$*, J. Noncommut. Geom. **7** (2013), 221–254.

[12] U. Grenander, *Probabilities on algebraic structures*, Almqvist & Wiksell, Stockholm, 1968.

[13] G. Högnäs and A. Mukherjea, *Probability measures on semigroups: convolution products, random walks, and random matrices*, second ed., Springer, New York, 2011.

[14] B. Host, *Le théorème des idempotents dans $B(G)$*, Bull. Soc. Math. France **114** (1986), 215–223.

[15] M. Ilie and N. Spronk, *Completely bounded homomorphisms of the Fourier algebras*, J. Funct. Anal. **225** (2005), 480–499.

[16] M. Kalantar, M. Neufang and Z.-J. Ruan, *Poisson boundaries over locally compact quantum groups*, Internat. J. Math. **24** (2013), 1350023, 21 pp.

[17] Y. Kawada and K. Itô, *On the probability distribution on a compact group. I*, Proc. Phys.-Math. Soc. Japan (3) **22** (1940), 977–998.

[18] J.L. Kelley, *Averaging operators on $C_\infty(X)$*, Illinois J. Math. **2** (1958), 214–223.

[19] J. Kustermans and L. Tuset, *A survey of $C^*$-algebraic quantum groups, part I*, Irish Math. Soc. Bull. (1999), no. 43, 8–63.

[20] J. Kustermans and L. Tuset, *A survey of $C^*$-algebraic quantum groups, part II*, Irish Math. Soc. Bull. (2000), no. 44, 6–54.

[21] J. Kustermans and S. Vaes, *Locally compact quantum groups*, Ann. Sci. École Norm. Sup. (4) **33** (2000), 837–934.

[22] A.T.-M. Lau and V. Losert, *Complementation of certain subspaces of $L_\infty(G)$ of a locally compact group*, Pacific J. Math. **141** (1990), 295–310.

[23] R.M. Loynes, *Fourier transforms and probability theory on a noncommutative locally compact topological group*, Ark. Mat. **5** (1963), 37–42.

[24] A. Pal, *A counterexample on idempotent states on a compact quantum group*, Lett. Math. Phys. **37** (1996), 75–77.

[25] K.R. Parthasarathy, *A note on idempotent measures in topological groups*, J. London Math. Soc. **42** (1967), 534–536.

[26] V. Paulsen, *Completely bounded maps and operator algebras*, Cambridge University Press, Cambridge, 2002.

[27] P. Podleś, *Symmetries of quantum spaces. Subgroups and quotient spaces of quantum* SU(2) *and* SO(3) *groups*, Comm. Math. Phys. **170** (1995), 1–20.

[28] J.S. Pym, *Idempotent measures on semigroups*, Pacific J. Math. **12** (1962), 685–698.

[29] W. Rudin, *Idempotent measures on Abelian groups*, Pacific J. Math. **9** (1959), 195–209.

[30] W. Rudin, *Measure algebras on abelian groups*, Bull. Amer. Math. Soc. **65** (1959), 227–247.

[31] P. Salmi, *Compact subgroups and left invariant $C^*$-subalgebras of locally compact quantum groups*, J. Funct. Anal. **261** (2011), 1–24.

[32] P. Salmi, *Subgroups and strictly closed invariant $C^*$-subalgebras*, preprint, arXiv: 1110.5459.

[33] P. Salmi and A. Skalski, *Idempotent states on locally compact quantum groups*, Q. J. Math. **63** (2012), 1009–1032.

[34] M. Takesaki, *Theory of operator algebras. II*, Springer-Verlag, Berlin, 2003.

[35] M. Takesaki and N. Tatsuuma, *Duality and subgroups*, Ann. of Math. (2) **93** (1971), 344–364.

[36] T. Timmermann, *An invitation to quantum groups and duality. From Hopf algebras to multiplicative unitaries and beyond*, European Mathematical Society (EMS), Zürich, 2008.

[37] R. Tomatsu, *A characterization of right coideals of quotient type and its application to classification of Poisson boundaries*, Comm. Math. Phys. **275** (2007), 271–296.

[38] S. Vaes, *A new approach to induction and imprimitivity results*, J. Funct. Anal. **229** (2005), 317–374.

[39] S. Vaes and L. Vainerman, *On low-dimensional locally compact quantum groups*, Locally compact quantum groups and groupoids (Strasbourg, 2002), IRMA Lect. Math. Theor. Phys., vol. 2, de Gruyter, Berlin, 2003, pp. 127–187.

[40] A. Van Daele, *The Haar measure on a compact quantum group*, Proc. Amer. Math. Soc. **123** (1995), 3125–3128.

[41] J.G. Wendel, *Haar measure and the semigroup of measures on a compact group*, Proc. Amer. Math. Soc. **5** (1954), 923–929.

[42] S.L. Woronowicz, *Compact matrix pseudogroups*, Comm. Math. Phys. **111** (1987), 613–665.

[43] S.L. Woronowicz, *Twisted* SU(2) *group. An example of a noncommutative differential calculus*, Publ. Res. Inst. Math. Sci. **23** (1987), 117–181.

[44] S.L. Woronowicz, *From multiplicative unitaries to quantum groups*, Internat. J. Math. **7** (1996), 127–149.

[45] S.L. Woronowicz, *Compact quantum groups*, Symétries quantiques (Les Houches, 1995), North-Holland, Amsterdam, 1998, pp. 845–884.

Pekka Salmi
Department of Mathematical Sciences
University of Oulu, PL 3000
FI-90014 Oulun yliopisto, Finland
e-mail: `pekka.salmi@iki.fi`

Operator Theory:
Advances and Applications, Vol. 233, 171–280


# Topological Radicals, V. From Algebra to Spectral Theory

Victor S. Shulman and Yurii V. Turovskii

*To the memory of Bill Arveson, a great mathematician and a great person*

**Abstract.** We introduce and study procedures and constructions of the theory of general topological radicals that are related to the spectral theory – the centralization, primitivity and socle procedures, the scattered radical, the radicals related to the continuity of the usual, joint and tensor radii. Among other applications we find some sufficient conditions of continuity of the spectrum and spectral radii of different types, and in particular prove that in a GCR C*-algebra the joint spectral radius is continuous on precompact sets and coincides with the Berger–Wang radius.

**Mathematics Subject Classification (2010).** Primary 46H20, 46H15, 47A10; Secondary 47L10, 22D25.

**Keywords.** Algebra, C*-algebra, $Q$-algebra, Engel algebra, radical, preradical, heredity, scattered radical, procedure, centralization, socle, primitivity procedure, operation, convolution procedure, superposition procedure, transfinite chain, primitive map, primitive ideal, Banach ideal, spectrum, spectral radius, joint spectral radius, tensor radius, Berger–Wang formula.

## 1. Introduction and preliminaries

### 1.1. Introduction

The general theory of radicals can be considered as a global structure theory of algebras which is aimed at the study of ideals and quotients simultaneously defined for a large class of algebras and related to some general properties of algebras.

Briefly speaking, a map $P$ defined on some class $\mathfrak{U}$ of algebras and sending each algebra $A \in \mathfrak{U}$ to its ideal $P(A)$ is called a radical if it satisfies some natural conditions of covariance and stability (see Section 2.1). The most popular example of a radical is the *Jacobson radical* rad defined on the class of all algebras; its restriction to the class of all Banach algebras is denoted by Rad. The importance of this map for the theory of Banach algebras cannot be overestimated.

The first significant achievements of the theory of radicals were related to the nilpotency and some close properties, and obtained by Baer, Levitzki, Köthe, Amitsur, Kurosh and others prominent algebraists. The functional analytic counterpart of this theory was initiated by Peter Dixon in his paper [D3], which contained basic definitions and presented first applications. Since then the theory was developed and applied to different problems of operator theory and Banach algebras in [ST2, ST5, ST6, ST7, KST]. The problems that turned out to be "solvable in radicals" had their origin in the theory of invariant subspaces, irreducible representations, semigroups of operators, tensor products, linear operator equations, joint spectral radius, Banach Lie algebras and other topics. They lead to introducing and study of corresponding radicals – the (*closed-*) *hypofinite radical* $\mathcal{R}_{\mathrm{hf}}$, the *hypocompact radical* $\mathcal{R}_{\mathrm{hc}}$, the *tensor radical* $\mathcal{R}_t$, the *compactly quasinilpotent radical* $\mathcal{R}_{\mathrm{cq}}$ and others (see Section 2 for definitions and discussion of these and other examples). It can be said that the essence of the radical approach to a problem (a property, a phenomenon) is to single out an ideal that accumulates elements related to this problem, and study the dependence of this ideal on the algebra. The construction of the corresponding radical is a typical result in the theory, and this is reasonable because all "well-behaved" radicals in their time find applications.

For functional analysts, the radicals on Banach algebras are most interesting. However, to have a more flexible technique, one has to consider topological radicals defined on non-complete normed algebras and also algebraic radicals defined on algebras without topology. For instance, the study of non-closed ideals of a Banach algebra ideals which are complete in a larger norm (we call them Banach ideals) turns out to be very useful. This is a reason to consider radicals in the wider context. So we outline three levels of consideration of radicals: algebraic, normed and Banach. The relations between these three theories are quite complicated. For example, the restriction of rad to the class of *all* normed algebras cannot be considered as a topological radical, because rad $(A)$ can be non-closed in $A$ if $A$ is not complete.

In this paper our main objects are spectral characteristics – spectrum, primitive ideals, spectral radii of different nature, socle, nilpotency, spectral continuity etc – and we construct special radicals and general procedures of the theory of radicals aiming at applications to the spectral theory in its algebraic and functional analytic aspects. The interplay between algebraic and functional analytic sides of the spectral theory forces us to devote a large part of the paper to understanding the links between algebraic and topological theories of radicals.

All main objects and constructions are related, to a greater or lesser extent, to the *scattered radical* $\mathcal{R}_s$.This radical associates with any Banach algebra the largest ideal whose elements have at most countable spectrum. From the viewpoint of the hierarchy of radicals the formulas

$$\mathcal{P}_\beta = \mathcal{R}_{\mathrm{hf}} \wedge \mathrm{Rad}, \tag{1.1}$$

$$\mathcal{R}_s = \mathcal{R}_{\mathrm{hf}} \vee \mathrm{Rad} \tag{1.2}$$

exactly determine the place of the signed radicals in Banach algebras: (1.1) and (1.2) mean that the *closed-Baer radical* $\mathcal{P}_\beta$ is the infimum of $\mathcal{R}_{\mathrm{hf}}$ and Rad, and $\mathcal{R}_s$ is the supremum of the same pair of radicals. All these radicals are topological, but $\mathcal{R}_s$ and Rad are the restrictions of some algebraic radicals to Banach algebras while $\mathcal{P}_\beta$ and $\mathcal{R}_{\mathrm{hf}}$ have algebraic analogs.

The paper is organized as follows. In the first section we gather the necessary preliminary information. The second section contains the basic definitions of the radical theory and the discussion of additional properties of radical-like maps. We present here also the basic examples of topological radicals. We introduce also radicals defined on the class of C*-algebras.

We call by a *procedure* any rule that transforms ideal maps. The simplest one which we use by default is the *restriction* of radicals to a subclass of algebras. Section 3 is devoted to the study of several important procedures. The first two of them (the *convolution procedure* $P \longmapsto P^*$ and *superposition procedure* $P \longmapsto P^\circ$) were introduced by Baer in the purely algebraic context, and by Dixon in the topological one. They produce a radical from an ideal map that lacks some stability property – an under radical and an over radical; we complement Dixon's results for the case of normed algebras. The third one (the *closure*) is very simple – from a map $A \longmapsto P(A)$ with non-necessarily closed ideals $P(A)$ it makes the map $A \longmapsto \overline{P(A)}$ which can be a topological radical or be transformed into a topological radical by means of the convolution procedure. Thus by means of the closure one can obtain a topological radical from an algebraic one. The fourth procedure is the *regularization* $P \longmapsto P^r$; it allows one to extend a radical from the class of all Banach algebras to a class of normed algebras by taking the completion.

In Section 4 we study some operations (multiplace procedures), that produce new radicals or radical-like maps from given families of radicals: supremum, infimum and two-place procedures – the convolution operation $*$ and superposition operation $\circ$. One of our aims here is to show that they all are closely related. For instance, the convolution and superposition procedures are the results of transfinite applications of the convolution and superposition operations. Supremum is reached by the convolution procedure; respectively, infimum is reached by the superposition procedure. We also study conditions which imply that a class of algebras is radical or semisimple.

We consider the problem of heredity of a radical that is obtained by closure and convolution procedures. A radical $P$ is called *hereditary* if $P(J) = J \cap P(A)$ for any appropriate ideal $J$ of an algebra. This property is very convenient for the study and use of a radical. Since many (almost all) important radicals are constructed by means of the closure and convolution, the heredity problem is one of the main inner problems of the theory. We give a criterion of heredity for the resulting radical which works for almost all known examples. In particular this approach allows us to answer in affirmative the question of Dixon [D3] about the heredity of the closed-Baer radical $\mathcal{P}_\beta$. This radical is the smallest one among topological radicals on Banach algebras for which all algebras with trivial multiplication are radical.

In Section 5 we deal with the property of commutativity modulo a radical. Namely, starting with a radical $P$ we define an under radical $P^a$, such that $P^a$-radical algebras are precisely those algebras that are commutative modulo $P$:

$$[A, A] \subset P(A).$$

We find the conditions on $P$ under which $P^a$ is a radical, and check that they are fulfilled for our main examples. For instance, this is true for all algebraic hereditary radicals and their topological analogs obtained by the closure and convolution procedures. In particular, $\mathrm{rad}^a$, $\mathcal{R}_s^a$, $\mathcal{P}_\beta^a$, $\mathcal{R}_{\mathrm{hf}}^a$, $\mathcal{R}_{\mathrm{cq}}^a$ and $\mathcal{R}_t^a$ are radicals. While the first four radicals satisfy our criterion, the last two need a separate consideration (the result for $\mathcal{R}_{\mathrm{cq}}^a$ was proved in [ST7]). This underlines the advantage of the joint consideration of algebraic and topological radicals. As the commutativity modulo rad, $\mathcal{R}_{\mathrm{cq}}$ and $\mathcal{R}_t$ is often used in the theory of spectral radii of different types, some important applications to spectral theory are indicated at the end of the section and also in Section 9. In particular, we consider the sufficient conditions for a Banach algebra to be Engel.

The subject of Section 6 is the very popular in the theory of algebras and Banach algebras notion of the socle of an algebra. We consider the ideal map $\mathrm{soc} : A \longmapsto \mathrm{socle}(A)$ and define a new procedure that transforms each radical $P$ into the convolution $\mathrm{soc} * P$ ("the socle modulo $P$"). By means of this procedure we establish some relations between, for example, hypofinite radical and Baer radical. Main applications of this construction will be given in Section 8 where we introduce the scattered radical.

Our object in Section 7 is more spectral: we study procedures related to the space $\mathrm{Prim}(A)$ of all primitive ideals of an algebra $A$. All of them begin with a choice of a subset $\Omega(A) \subset \mathrm{Prim}(A)$, but such a choice (a *primitive map*) must be done simultaneously for all algebras $A$ in the given class, and subjected to some natural restrictions. For example one can choose for $\Omega(A)$ all the space $\mathrm{Prim}(A)$, or $\varnothing$, or the set of primitive ideals of finite codimension. If a primitive map is fixed then the procedures we deal with are

- The $\Omega$-*hull-kernel closure* $P^{\mathrm{kh}_\Omega}$ of a preradical $P$ which sends $A$ to the intersection of those ideals $I \in \Omega(A)$ that contain $P(A)$;
- The $\Omega$-*primitivity extension* $P^{\mathrm{p}_\Omega}$ of $P$ that sends $A$ to the ideal of all elements $a \in A$ such that $a/I \in P(A/I)$, for all $I \in \Omega(A)$.

Our main aims here are to study the interrelations between these procedures and to find the conditions on $P$ under which the produced maps have sufficiently convenient properties. Note that the primitivity extension of the scattered radical take a central place in applications to the spectral continuity.

Section 8 which can be considered central in this paper, is devoted to the theory of the scattered radical $\mathcal{R}_s$. Almost all constructions of the previous sections find a reflection in or application to the theory of $\mathcal{R}_s$. We introduce and study $\mathcal{R}_s$ in its algebraic and Banach algebraic versions. The algebraic version of $\mathcal{R}_s$ is defined via the socle and convolution procedure: it equals $(\mathrm{soc} * \mathrm{rad})^*$. We show, using the connection between both versions, that $\mathcal{R}_s$ is a hereditary topological radical satisfying some additional flexibility conditions which allow one to use its centralization and primitivity extensions.

Another important fact is that the structure space of an $\mathcal{R}_s$-radical algebra is dispersed and that (the classes of the equivalence of) strictly irreducible representations of such algebras are uniquely determined by their kernels (= primitive ideals). We show that in hereditarily semisimple Banach algebras (in particular, in C*-algebras) $\mathcal{R}_s$ coincides with the hypocompact radical and that scattered C*-algebras can be characterized by many other equivalent conditions.

The subject of Section 9 is the continuity of spectral characteristics of an element or a family of elements of a Banach algebra. We show that the map $a \longmapsto \sigma(a)$ is continuous at points $a \in \mathcal{R}_s^p(A)$ where $\mathcal{R}_s^p$ is the primitivity extension of the scattered radical, and that this property is not fulfilled for elements of $\mathcal{R}_s^{p*}(A)$ and of $\mathcal{R}_s^a$. Among other results we obtain that if $A$ is a C*-algebra then $\mathcal{R}_s^{p*}(A)$ is the largest GCR-ideal of $A$ (so the radical $\mathcal{R}_s^{p*}$ extends the GCR-radical from C*-algebras to Banach algebras).

We construct topological radicals $\mathcal{R}_{\overrightarrow{\rho}}$, $\mathcal{R}_{\overrightarrow{\rho_j}}$ and $\mathcal{R}_{\overrightarrow{\rho_t}}$ that have the properties that elements, compact subsets and summable families in $\mathcal{R}_{\overrightarrow{\rho}}(A)$, $\mathcal{R}_{\overrightarrow{\rho_j}}(A)$ and $\mathcal{R}_{\overrightarrow{\rho_t}}(A)$, respectively, are the continuity points for spectral radius, joint spectral radius and tensor spectral radius, respectively. It is shown that $\mathcal{R}_{\overrightarrow{\rho}} \geq \mathcal{R}_s^{p*}$. Thus the spectral radius, unlike the spectrum, is continuous at all elements of $\mathcal{R}_s^{p*}(A)$. In particular, $\rho$ is continuous on any GCR C*-algebra.

In Section 10 we apply some of the listed results to the problem of recovering the spectral characteristics of an element or a family of elements of an algebra $A$ from the corresponding information about its images in some quotients $A/I$, where $I$ belong to a fixed family $\mathcal{F}$ of ideals in $A$ with trivial intersection. Our main interest is in the joint spectral radius $\rho(M)$ of a precompact set $M \subset A$ – a characteristic that attracts much interest not only in operator theory but in such branches of mathematics as the topological dynamics and fractal theory. We show that

$$\rho(M) = \sup_{I \in \mathcal{F}} \rho(M/I) \tag{1.3}$$

if $\mathcal{F}$ is finite, and extend this equality to the case of infinite family $\mathcal{F}$ for algebras that have non-zero compact elements. An especially interesting case is $\mathcal{F} = \mathrm{Prim}(A)$. We relate (1.3) to the equality $\rho(M) = r(M)$, where $r(M)$ is a spectral characteristic introduced by Berger and Wang [BW]; in the algebras where this equality holds for all precompact subsets (the *Berger–Wang algebras*) the analysis of many spectral problem becomes much easier.

We show that in the $\mathcal{R}_s^{p*}$-radical Berger–Wang algebras the joint spectral radius is continuous and find some criteria for (1.3) to hold. Then we study the case of C*-algebras and show that each GCR-algebra is a Berger–Wang algebra and therefore (if one takes into account that GCR-algebras are $\mathcal{R}_s^{p*}$-radical) the joint spectral radius is continuous on GCR-algebras.

It is natural that simultaneous consideration of various kinds of radicals – the algebraic ones, the radicals on Banach algebras, normed algebras, $Q$-algebras, C*-algebras and so on – can be difficult for the first acquaintance with the topic. However, apart from the reasons above, it gives the possibility not to repeat similar arguments many times. Anyway for the first reading it seems to be reasonable to restrict oneself to a concrete type of radicals and consider only Banach or C*-algebras. We can say that we did our best to simplify the text and, in particular, following the advice of Laurence Sterne, generously lighted up the dark places by stars.

### 1.2. Preliminaries

**1.2.1. Spaces.** Let $X, U$ be linear spaces, and let $Y$ and $Z$ be subspaces of $X$. If $Z \subset Y$ then $(X/Z)\,/\,(Y/Z) \cong X/Y$. If $(Y_\alpha)_{\alpha\in\Lambda}$ is a family of subspaces of $X$ and $Z \subset \cap_{\alpha\in\Lambda} Y_\alpha$ then

$$(\cap_{\alpha\in\Lambda} Y_\alpha)/Z = \cap_{\alpha\in\Lambda}(Y_\alpha/Z). \tag{1.4}$$

If $X$ is normed, let $\overline{Y}$, or more exactly $\overline{Y}^{(X)}$, denote the closure of $Y$ in $X$, and $\widehat{X}$ denote the completion of $X$. If $Z \subset Y$ is a closed subspace of $Y$ then $Y/Z \cong q\,(Y)$, where $q : X \longrightarrow X/\overline{Z}^{(X)}$ is the standard quotient map $x \longmapsto$ the coset of $x$.

Let $f : X \longrightarrow U$ be a linear map. If $M \subset U$ then

$$f^{-1}\,(M) := \{x \in X : f\,(x) \in M\}\,.$$

Clearly $N \subset f^{-1}\,(f\,(N))$ for every $N \subset X$; $N = f^{-1}\,(f\,(N)) \Leftrightarrow N = f^{-1}\,(M)$ for some $M \subset U$; in particular this holds if $f$ is invertible. If $X, U$ are normed, and $f$ is open and continuous, then

$$f^{-1}(\overline{M}) = \overline{f^{-1}(M)}. \tag{1.5}$$

If $X$ is a linear space then $L\,(X)$ is the algebra of all linear operators on $X$, $\mathcal{F}\,(X)$ the ideal of finite rank operators. If $X$ is normed then $\mathcal{B}\,(X)$ is the algebra of all bounded operators, $\mathcal{K}\,(X)$ is the ideal of compact operators.

**1.2.2. Algebras.** Let $A$ be an associative complex algebra; then $A^1$ equals $A$ if $A$ is unital, and is the algebra obtained from $A$ by adjoining the identity element otherwise. In what follows, an *ideal* of an algebra means a two-sided ideal. If $I$ is an ideal of $A$ then $q_I$ denotes the standard quotient map $A \longrightarrow A/I$ by default; we also write $a/I$ instead of $q_I\,(a)$ for any $a \in A$. Respectively, if $M = (a_\alpha)_\Lambda$ is a family in $A$ then $M/I$ denotes the family $(a_\alpha/I)_\Lambda$ in $A/I$. Define operators $\mathrm{L}_a$, $\mathrm{R}_a$ and $\mathrm{W}_a$ on $A$ by

$$\mathrm{L}_a x = ax,\ \mathrm{R}_a x = xa \text{ and } \mathrm{W}_a = \mathrm{L}_a\mathrm{R}_a;$$

again $\mathrm{L}_M$ is a family $(\mathrm{L}_{a_\alpha})_\Lambda$ of operators. Such rules act for sets by default.

**1.2.3. Representations.** A representation $\pi$ of an algebra $A$ on a linear space $X$ is called *strictly irreducible* if $\pi(A)\xi := \{\pi(a)\xi : a \in A\} = X$, for every $\xi \neq 0$. Representations $\pi$ and $\tau$, acting on the spaces $X$ and $Y$, respectively, are *equivalent* (write $\pi \sim \tau$) if there is a linear bijective operator $T : X \to Y$ satisfying the condition $T\pi(a) = \tau(a)T$ for all $a \in A$. The direct sum $\oplus_{i=1}^n \pi_i$ of representations $\pi_1, \dots, \pi_n$, acting on the spaces $X_1, \dots, X_n$, is the representation $\pi$ on $X = \oplus_{i=1}^n X_i$ defined by the formula

$$\pi(a)(\oplus_{i=1}^n \xi_i) = \oplus_{i=1}^n \pi_i(a)\xi_i.$$

It is well known (see for example [Ln, Chapter 17]) that if strictly irreducible representations $\pi_1, \dots, \pi_n$ are pairwise nonequivalent and $0 \neq \eta_i \in X_i$, $1 \leq i \leq n$, then the vector $\eta = \oplus_{i=1}^n \eta_i$ is cyclic for the representation $\pi$: $\pi(A)\eta = X$.

**Lemma 1.1.** *Let $\{\pi_i : 1 \leq i \leq n\}$ be pairwise non equivalent strictly irreducible representations. Then $\sum_{i=1}^n \operatorname{rank}(\pi_i(a)) \leq \dim aAa$ for any $a \in A$.*

*Proof.* Fix $a \in A$. We may suppose that $\dim aAa < \infty$ and that $\pi_i(a) \neq 0$ for all $i$.
Let $\pi = \oplus_{i=1}^n \pi_i$ and $X = \oplus_{i=1}^n X_i$. Clearly

$$\dim \pi(aAa)\xi \leq \dim aAa$$

for each $\xi \in X$. Let us choose vectors $\xi_i \in X_i$ such that $\eta_i := \pi_i(a)\xi_i \neq 0$, $1 \leq i \leq n$. Let $\xi = \oplus_{i=1}^n \xi_i$; then the vector $\eta = \pi(a)\xi = \oplus_{i=1}^n \pi_i(a)\xi_i$ is cyclic for $\pi(A)$. Therefore $\pi(aAa)\xi = \pi(a)\pi(A)\eta = \pi(a)X = \oplus_{i=1}^n \pi_i(a)X_i$ whence

$$\dim \pi(aAa)\xi = \sum \dim \pi_i(a_i)X_i = \sum_{i=1}^n \operatorname{rank}(\pi_i(a)). \qquad \square$$

Let $\operatorname{Irr}(A)$ be the set of all strictly irreducible representations of $A$, $\operatorname{Prim}(A)$ the set of all primitive ideals, i.e., the kernels of all $\pi \in \operatorname{Irr}(A)$. For any subset $E$ of $A$, let $\operatorname{h}(E; A)$ be the set of all $I \in \operatorname{Prim}(A)$ with $E \subset I$ and $\operatorname{kh}(E; A) = \cap_{I \in \operatorname{h}(E;A)} I$.

If $\pi \in \operatorname{Irr}(A)$ then let $X_\pi$ be the space on which $\pi$ acts (the representation space). It is well known that if $A$ is not unital then

$$\operatorname{Irr}(A^1) = \{\pi_{\text{triv}} : A^1 \longrightarrow A^1/A \cong \mathbb{C}\} \cup \{\pi^1 : \pi \in \operatorname{Irr}(A)\}, \tag{1.6}$$

$$\operatorname{Prim}(A^1) = \{A\} \cup \operatorname{Prim}(A) \text{ and } \operatorname{rad}(A^1) = \operatorname{rad}(A) \tag{1.7}$$

where every $\pi \in \operatorname{Irr}(A)$ extends to $\pi^1 \in \operatorname{Irr}(A^1)$ by setting $\pi^1(1_{A^1}) = 1_{X_\pi}$.

If $J$ is a proper ideal of $A$ then

$$\operatorname{Irr}(J) = \{\pi|_J : \pi \in \operatorname{Irr}(A), \pi(J) \neq 0\}, \tag{1.8}$$

$$\operatorname{Prim}(J) = \{I \cap J : I \in \operatorname{Prim}(A), I \cap J \neq J\}. \tag{1.9}$$

Every $\pi \in \operatorname{Irr}(J)$ can be uniquely extended to a representation $\widetilde{\pi} \in \operatorname{Irr}(A)$. To construct the extension it suffices to choose a non-zero vector $\zeta \in X_\pi$ and, for each $a \in A$, to set $\widetilde{\pi}(a)\xi = \pi(ab)\zeta$ for every $b \in J$ such that $\xi = \pi(b)\zeta$.

Every $\pi \in \mathrm{Irr}(A/J)$ induces $\pi_A \in \mathrm{Irr}(A)$ by setting $\pi_A(a) = \pi(a/J)$ for every $a \in A$. Every $\pi \in \mathrm{Irr}(A)$ with $\pi(J) = 0$ induces $\pi_{q_J} \in \mathrm{Irr}(A/J)$ by setting $\pi_{q_J}(q_J(a)) = \pi(a)$ for every $a \in A$. Hence

$$\mathrm{Prim}(A/J) = \{I/J : I \in \mathrm{Prim}(A), J \subset I\}. \tag{1.10}$$

Note that $A/J$ can be unital for an non-unital algebra $A$.

If $f : A \longrightarrow B$ is a homomorphism onto then $\pi \in \mathrm{Irr}(B)$ induces $\pi^f \in \mathrm{Irr}(A)$ by setting $\pi^f(a) = \pi(f(a))$ for every $a \in A$. So

$$\{f^{-1}(I) : I \in \mathrm{Prim}(B)\} \subset \mathrm{Prim}(A). \tag{1.11}$$

Let $A$ be a normed algebra, let $\mathrm{Irr}_{\mathrm{n}}(A)$ (respectively, $\mathrm{Irr}_{\mathrm{b}}(A)$) be the set of all continuous strictly irreducible representations of $A$ by bounded operators on a normed (respectively, Banach) space, and let $\mathrm{Prim}_{\mathrm{n}}(A) = \{\ker\pi : \pi \in \mathrm{Irr}_{\mathrm{n}}(A)\}$ and $\mathrm{Prim}_{\mathrm{b}}(A) = \{\ker\pi : \pi \in \mathrm{Irr}_{\mathrm{b}}(A)\}$.

It should be noted that for a C*-algebra, a primitive ideal is defined as the kernel of an irreducible *-representation on a Hilbert space. The fact that this definition is equivalent to the general one, is proved in [Dm, Corollary 2.9.6].

**1.2.4. $Q$-algebras.** A normed algebra $A$ is called a *$Q$-algebra* if the set of all invertible elements of $A^1$ is open. A normed algebra $A$ is a $Q$-algebra $\Leftrightarrow \sum_1^\infty a^n$ converges for any $a \in A$ with $\|a\| < 1 \Leftrightarrow$ every strictly irreducible representation of $A$ is equivalent to a continuous representation by bounded operators on a normed space [ST5, Theorem 2.1]. A normed algebra $A$ is called a *$Q_{\mathrm{b}}$-algebra* if every strictly irreducible representation of $A$ is equivalent to a continuous representation by bounded operators on a Banach space.

Let $A$ be an algebra and $a \in A$. The *spectrum* $\sigma(a)$, or more exactly $\sigma_A(a)$, is the set of all $\lambda \in \mathbb{C}$ such that $a - \lambda$ is not invertible in $A^1$; this is related to the definition in [BD]. So $\sigma_A(a)$ and $\sigma_{A^1}(a)$ determine the same set, but sometimes we prefer to write $\sigma_{A^1}(a)$ for exactness.

Let $B$ be a subalgebra of $A$; it is called a *spectral subalgebra* of $A$ if

$$\sigma_B(a) \setminus \{0\} = \sigma_A(a) \setminus \{0\}$$

for every $a \in B$. Every ideal of $A$ is a spectral subalgebra of $A$; a normed algebra $A$ is a $Q$-algebra $\Leftrightarrow$ $A$ is a spectral subalgebra of $\widehat{A}$, see [P, Theorem 4.2.10] and [KS, Lemma 20.9].

**1.2.5. The joint spectrum.** For the joint spectral theory in Banach algebras we refer to [H, M]. If $M = (a_\alpha)_\Lambda \subset A$ is a family in $A$ then the *left spectrum* $\sigma^l(M)$ is the set of all families $\lambda = (\lambda_\alpha)_\Lambda \subset \mathbb{C}$ such that the family $M - \lambda := (a_\alpha - \lambda_\alpha)_\Lambda$ generates the proper left ideal of $A^1$. The *right spectrum* $\sigma^r(M)$ is defined similarly by replacing 'left' by 'right'; $\sigma(M) = \sigma^l(M) \cup \sigma^r(M)$ is called the *Harte spectrum,* or simply the *spectrum.* We write $\sigma^l(a)$ and $\sigma^r(a)$ if $M = \{a\}$. Let $A$ be unital. A subalgebra $B$ of $A$ is called *unital* if it contains the identity element of $A$, and *inverse-closed* if $B$ contains $x^{-1}$ for every $x \in B$ which is invertible in $A$. If $B$ is inverse-closed then $\sigma_B(a) = \sigma_A(a)$ for every $a \in B$.

The following folklore lemma determines the operational possibilities of the joint spectra.

**Lemma 1.2.** *Let $A$ be an algebra and $M = (a_\alpha)_\Lambda$ be a family in $A$. Then*

(1) $\lambda \notin \sigma_A^l(M)$ *if and only if* $\lambda|_N \notin \sigma_A^l(N)$ *for some finite subfamily* $N \subset M$;
(2) *If $B$ is a unital subalgebra of $A^1$ and $M \subset B$ then* $\sigma_A^l(M) \subset \sigma_B^l(M)$;
(3) *If $f : A \longrightarrow C$ is a surjective homomorphism then* $\sigma_C^l(f(M)) \subset \sigma_A^l(M)$;
(4) $\sigma^l(M) = \sigma^l(M/\operatorname{rad}(A))$.

*Proof.* (1), (2) and (4) are standard and trivial.

(3) Consider only the case when $C$ is unital and $A$ is not. One can extend $f$ up to $f' : A^1 \longrightarrow C$ by setting $f'(1_A) = 1_C$, whence $\sigma_C^l(f'(M)) \subset \sigma_A^l(M)$, but $f'(M) = f(M)$. □

If $N = (T_\beta)$ is a family in $L(X)$ then $\sigma_X^p(N)$, or simply $\sigma^p(N)$, is a *point spectrum* of $N$, i.e., the set of all $\lambda = (\lambda_\beta) \subset \mathbb{C}$ such that there is a non-zero vector $\zeta \in X$ and $T_\beta\zeta = \lambda_\beta\zeta$; $\lambda$ is called the *eigenvalue* of $N$ corresponding to the *eigenvector* $\zeta$. If $X$ is normed then it is useful to use the *approximate point spectrum* $\sigma_X^a(N)$, or simply $\sigma^a(N)$; it is the set of all $\lambda = (\lambda_\beta) \subset \mathbb{C}$ such that there is a net $\zeta = (\zeta_\gamma) \subset X$ with $\lim_\gamma \|\zeta_\gamma\| > 0$ and $\lim_\gamma \|(T_\beta - \lambda_\beta)\zeta_\gamma\| = 0$ for every $\beta$; $\lambda$ is called the *approximate eigenvalue* of $N$ corresponding to the *approximate eigenvector* $\zeta$. If $A$ is a normed algebra and $M$ is a family in $A$ then set $\sigma_A^a(M) = \sigma_A^a(\mathrm{L}_M)$.

If $T \in L(X)$ then put $\sigma_X(T) := \sigma_{L(X)}(T)$. If $T$ is one-to-one and onto then $T$ is invertible in $L(X)$ whence $\sigma_X(T) = \sigma_X^p(T) \cup \sigma_{L(X)}^r(T)$. If $X$ is a Banach space, $T$ is bounded and the kernel and the image of $T$ satisfy the mentioned conditions then $T$ is invertible in $\mathcal{B}(X)$ by Banach's theorem, whence

$$\sigma_{\mathcal{B}(X)}(T) = \sigma_X^p(T) \cup \sigma_{\mathcal{B}(X)}^r(T) = \sigma_X^p(T) \cup \sigma_{L(X)}^r(T) = \sigma_X(T) \tag{1.12}$$

**Theorem 1.3.** *Let $A$ be an algebra and $M$ be a family in $A$. Then*

(1) $\sigma_A^l(M) = \cup_{\pi \in \operatorname{Irr}(A^1)} \sigma_{X_\pi}^p(\pi(M))$;
(2) *If $\pi$ is a representation of $A^1$ by bounded operators on a normed space $X$ then* $\sigma_X^a(\pi(M)) \subset \sigma_A^l(M)$;
(3) *If $A$ is a $Q$-algebra then* $\sigma_A^l(M) = \cup_{\pi \in \operatorname{Irr}_n(A^1)} \sigma_{X_\pi}^a(\pi(M))$.

*Proof.* (1) For $\lambda \in \sigma_A^l(M)$, there is a maximal left ideal $J$ of $A^1$ such that $M - \lambda = (a_\alpha - \lambda_\alpha) \subset J \neq A^1$. The representation $\pi$ of $A^1$ defined by $\pi(b) = \mathrm{L}_b|_X$ is strictly irreducible where $X = A^1/J$. For $\xi = 1/J \in X$ we obtain that $\pi(a_\alpha)\xi = \lambda_\alpha\xi$ for every $\alpha$. Therefore $\lambda \in \sigma_X^p(\pi(M))$. We proved the inclusion $\subset$.

Conversely, let $\lambda \in \sigma_{X_\pi}^p(\pi(M))$ for some $\pi \in \operatorname{Irr}(A^1)$. Then $\pi(a_\alpha - \lambda_\alpha)\xi = 0$ for some non-zero vector $\xi \in X_\pi$ and each $\alpha$. But $\{x \in A^1 : \pi(x)\xi = 0\}$ is a proper left ideal of $A^1$. Therefore $\lambda \in \sigma_A^l(M)$.

(2) Let $0 \in \sigma_{X_\pi}^a(\pi(M)) \backslash \sigma_A^l(M)$. By Lemma 1.2, there is a finite subfamily $N = (a_\alpha)_{\Lambda'} \subset M$ with $0 \notin \sigma_A^l(N)$. So there are $b_\alpha \in A^1$ for $\alpha \in \Lambda'$ such that

$\sum_{\Lambda'} b_\alpha a_\alpha = 1$. If $\zeta = (\zeta_\gamma) \subset X$ is an approximate eigenvector related to 0 then

$$0 < \lim_\gamma \|\zeta_\gamma\| \le \lim_\gamma \sum_{\alpha\in\Lambda'} \|\pi(b_\alpha a_\alpha)\zeta_\gamma\| \le \sum_{\alpha\in\Lambda'} \|\pi(b_\alpha)\| \lim_\gamma \|\pi(a_\alpha)\zeta_\gamma\| = 0,$$

a contradiction. Therefore $\sigma^a_{X_\pi}(\pi(M)) \subset \sigma^l_A(M)$.

(3) follows from (1) and (2). □

One can obtain the related statements for $\sigma^r_A(M)$ if we pass to the opposite algebra.

**1.2.6. Spectrum and primitive ideals.** If an algebra $A$ is normed then $\sigma(a)$ is not empty by the Gelfand–Mazur theorem; if $A$ is a $Q$-algebra then $\sigma(a)$ is a compact subset of $\mathbb{C}$.

**Theorem 1.4.** *Let $A$ be an algebra and $a \in A$. Then*

(1) $\sigma_A(a) \backslash \sigma^l_A(a) \subset \cup_{\pi\in\mathrm{Irr}(A^1)} \sigma^r_{L(X_\pi)}(\pi(a))$;

(2) *The following chain of equalities holds:*

$$\begin{aligned}\sigma_A(a) &= \cup_{I\in\mathrm{Prim}(A^1)} \sigma_{A^1/I}(a/I) = \cup_{\pi\in\mathrm{Irr}(A^1)} \sigma_{\pi(A^1)}(\pi(a))\\ &= \cup_{\pi\in\mathrm{Irr}(A^1)} \sigma_{X_\pi}(\pi(a)) = \cup_{\pi\in\mathrm{Irr}(A^1)} \left(\sigma^p_{X_\pi}(\pi(a)) \cup \sigma^r_{X_\pi}(\pi(a))\right);\end{aligned}$$

(3) *If $A$ is a $Q$-algebra then*

$$\sigma_A(a) = \cup_{\pi\in\mathrm{Irr}_{\mathrm{n}}(A^1)} \sigma_{\overline{\pi(A)}}(\pi(a)) = \cup_{\pi\in\mathrm{Irr}_{\mathrm{n}}(A^1)} \sigma_{\mathcal{B}(X_\pi)}(\pi(a));$$

(4) *If $A$ is a $Q_{\mathrm{b}}$-algebra then*

$$\sigma_A(a) = \cup_{\pi\in\mathrm{Irr}_{\mathrm{b}}(A^1)} \left(\sigma^p_{X_\pi}(\pi(a)) \cup \sigma^r_{\mathcal{B}(X_\pi)}(\pi(a))\right).$$

*Proof.* (1) Let $\lambda \in \sigma_A(a) \backslash \sigma^l_A(a)$; then there is $b \in A^1$ such that $b(a-\lambda) = 1$. Assume, to the contrary, that $\lambda \notin \sigma^r_{L(X_\pi)}(\pi(a))$ for every $\pi \in \mathrm{Irr}(A)$: there is an operator $T_\pi$ on $X_\pi$ such that $(\pi(a)-\lambda)T_\pi = 1$. As $\pi(b)(\pi(a)-\lambda) = 1$ then $T_\pi = \pi(b)$ for every $\pi \in \mathrm{Irr}(A^1)$. Hence $(a-\lambda)b - 1 \in \cap_{\pi\in\mathrm{Irr}(A^1)} \ker\pi = \mathrm{rad}(A^1)$ and $(a-\lambda)b = 1 + c$ for some $c \in \mathrm{rad}(A^1)$. So $(a-\lambda)b(1+c)^{-1} = 1$ and $\lambda \notin \sigma_A(a)$, a contradiction.

(2) By Lemma 1.2(2-3),

$$\begin{aligned}\sigma_A(a) &\supset \sigma_{A^1/I}(a/I) = \sigma_{\pi(A^1)}(\pi(a)) \supset \sigma_{L(X_\pi)}(\pi(a))\\ &= \sigma^p_{X_\pi}(\pi(a)) \cup \sigma^r_{L(X_\pi)}(\pi(a))\end{aligned}$$

where $\pi \in \mathrm{Irr}(A^1)$ and $I = \ker\pi$. The result follows from (1) and Theorem 1.3(1).

(3) If $A$ is a $Q$-algebra then every $\pi \in \mathrm{Irr}(A^1)$ is equivalent to a (bounded) strictly irreducible representation by bounded operators on a normed space $X_\pi$. As $\pi(A) \subset \overline{\pi(A)} \subset \mathcal{B}(X_\pi) \subset L(X_\pi)$ then

$$\sigma_{L(X_\pi)}(\pi(a)) = \sigma_{\mathcal{B}(X_\pi)}(\pi(a)) \subset \sigma_{\overline{\pi(A)}}(\pi(a)) \subset \sigma_{\pi(A)}(\pi(a))$$

for every $\pi \in \mathrm{Irr}(A^1)$, and the result follows from (2).

(4) If $A$ is a $Q_{\mathrm{b}}$-algebra then every $\pi \in \operatorname{Irr}(A^1)$ is equivalent to a (bounded) strictly irreducible representation by bounded operators on a Banach space $X_\pi$. Then the result follows from (3) and (1.12). □

The first equality in Theorem 1.4(2) was proved in [Z2, Proposition 1] (for Banach algebras).

**Corollary 1.5.** *Let $A$ be an algebra, and let $I$ be an ideal of $A$ and $a \in A$. Then*

(1) *If $(A/I)^1 \cong A^1/I$ then*

$$\sigma_{A/I}(a/I) = \sigma_{A^1/\mathrm{kh}(I;A^1)}\left(a/\mathrm{kh}\left(I;A^1\right)\right) = \cup_{J \in \mathrm{h}(I;A^1)} \sigma_{A^1/J}(a/J);$$

(2) *If $(A/I)^1 \ncong A^1/I$ then*

$$\sigma_{A/I}(a/I) = \sigma_{A/\mathrm{kh}(I;A)}(a/\mathrm{kh}(I;A)) = \cup_{J \in \mathrm{h}(I;A)} \sigma_{A/J}(a/J);$$

(3) *If $A$ is a $Q$-algebra then $\sigma_{A/I}(a/I) = \sigma_{A/\overline{I}}(a/\overline{I})$.*

*Proof.* (1) As $\mathrm{kh}(I;A^1) = q_I^{-1}\left(\mathrm{rad}\left(A^1/I\right)\right)$ then

$$A^1/\mathrm{kh}\left(I;A^1\right) \cong \left(A^1/I\right)/\mathrm{rad}\left(A^1/I\right) \cong \left((A/I)/\mathrm{rad}(A/I)\right)^1 \tag{1.13}$$

The first equality follows from Lemma 1.2(4). As

$$\mathrm{Prim}\left(A^1/I\right) = \left\{J/I : J \in \mathrm{h}\left(I;A^1\right)\right\}$$

by (1.10) and $\left(A^1/I\right)/(J/I) \cong A^1/J$ then, by Theorem 1.4(2),

$$\sigma_{A/I}(a/I) = \cup_{J \in \mathrm{h}(I;A^1)} \sigma_{(A^1/I)/(J/I)}(a/(J/I)) = \cup_{J \in \mathrm{h}(I;A^1)} \sigma_{A^1/J}(a/J).$$

(2) This is the case when $A/I$ is unital and $A$ is not unital. We see as in (1.13) that $A/\mathrm{kh}(I;A) = ((A/I)/\mathrm{rad}(A/I))^1$ is unital and the first equality follows. Furthermore, $\mathrm{Prim}(A/I) = \{J/I : J \in \mathrm{h}(I;A)\}$ by (1.10). If $J \in \mathrm{h}(I;A)$ then $A/J$ is also unital, whence as in (1) we have $\sigma_{A/I}(a/I) = \cup_{J \in \mathrm{h}(I;A)} \sigma_{A/J}(a/J)$.

(3) Consider only the case $(A/I)^1 \cong A^1/I$ because the other case is similar. As primitive ideals are closed, $\overline{I} \subset J$ for every $J \in \mathrm{Prim}\left(A^1\right)$ with $I \subset J$. Then $I \subset \overline{I} \subset \mathrm{kh}\left(I;A^1\right)$ implies $\sigma_{A^1/\mathrm{kh}(I;A^1)}\left(a/\mathrm{kh}\left(I;A^1\right)\right) \subset \sigma_{A/\overline{I}}(a/\overline{I}) \subset \sigma_{A/I}(a/I)$ and the result follows from (1). □

**Remark 1.6.** As a consequence of Corollary 1.5, the spectrum of $a/I$ in the quotient $A/I$ of a $Q$-algebra $A$ by a possibly unclosed ideal $I$ is a compact set in $\mathbb{C}$.

**1.2.7. Banach ideals.** Let $A$ be a normed algebra with norm $\|\cdot\|$; we write also $(A;\|\cdot\|)$. An ideal $I$ of $A$ is called *normed* if there are a norm $\|\cdot\|_I$ on $I$ and $t > 0$ such that

$$\|x\| \le t\,\|x\|_I$$

for every $x \in I$; the norm $\|\cdot\|_I$ is called *flexible* if $t = 1$. A normed ideal $I$ with a flexible norm is also called a flexible ideal; $I$ is called a Banach ideal if it is complete with respect to $\|\cdot\|_I$.

An ideal of $A$ with the norm inherited from $A$ is of course flexible. We will describe now a more interesting class of examples.

It is well known that the sum of two closed ideals $I$ and $J$ of a Banach algebra $A$ may be non-closed, see an excellent discussion in [D4]. One can consider this sum as a Banach ideal with the norm

$$\|a\|_{I+J} = \inf \{\|x\| + \|y\| : a = x + y,\, x \in J,\, y \in I\}$$

for every $a \in J + I$. If $J \cap I = 0$ then there is only one pair $x \in J$, $y \in I$ with $a = x + y$ and $\|a\|_{I+J} = \|x\| + \|y\|$. So we have in general for $x \in J$, $y \in I$ that

$$\|x + y\|_{I+J} = \inf \{\|x - z\| + \|y + z\| : z \in J \cap I\}.$$

It is clear that if $a$ lies in $I$ or $J$ then $\|a\| = \|a\|_{I+J}$. Therefore $J \cap I$ is a closed ideal of $K := (J + I; \|\cdot\|_{I+J})$. Then $K/(J \cap I)$ is a Banach algebra with the norm

$$\|a/(J \cap I)\|_{K/(J\cap I)} = \|x/(J \cap I)\| + \|y/(J \cap I)\|$$

for $a/(J \cap I) = x/(J \cap I) + y/(J \cap I)$ with $x \in J$, $y \in I$. Also, $I/(J \cap I)$ is a closed ideal of $K/(J \cap I)$, whence clearly

$$\begin{aligned} \|a/I\|_{K/I} &= \|(a/(J \cap I))/(I/(J \cap I))\|_{(K/(J\cap I))/(I/(J\cap I))} \\ &= \inf \{\|x/(J \cap I)\| + \|(y - z)/(J \cap I)\| : z \in I\} \\ &= \|x/(J \cap I)\|. \end{aligned} \tag{1.14}$$

On the other hand, as $I$ is a closed ideal of $K$ then

$$\begin{aligned} \|a/I\|_{K/I} &= \inf \left\{\|a - z\|_{I+J} : z \in I\right\} \\ &= \inf \{\|x\| + \|y\| : a = x + y + z,\, x \in J,\, y, z \in I\}. \end{aligned}$$

So we obtain the following

**Lemma 1.7.** *Let $(A, \|\cdot\|)$ be a normed algebra, and let $I, J$ be closed ideals of $A$. Then*

(1) *$J/(J \cap I)$ is isometrically isomorphic to $(J + I)/I$ with the norm*

$$\|a/I\|_{(J+I)/I} = \inf \{\|x\| + \|y\| : a = x + y + z,\, x \in J,\, y, z \in I\}.$$

(2) *$(J + I)/I$ with norm $\|\cdot\|_{(J+I)/I}$ is a flexible ideal of $A/I$.*

*Proof.* (1) There is a bounded isomorphism $\phi$ from $J/(J \cap I)$ onto $(J + I)/I$, so for every $x/(J \cap I)$ with $x \in J$ there is only one $a/I$ with $a \in J + I$. The above argument clearly works if $A$ is a normed algebra, and (1.14) shows that $\phi$ is an isometry.

(2) is obvious. □

**Remark 1.8.** It follows from the lemma that if $A$ is a Banach algebra then the ideal $(J + I)/I$ with norm $\|\cdot\|_{(J+I)/I}$ is a Banach ideal of $A/I$. Indeed, $J/(J \cap I)$ is complete whence $(J + I)/I$ with norm $\|\cdot\|_{(J+I)/I}$ is complete.

## 2. Radicals and other ideal maps

### 2.1. Definitions

Radicals are defined on classes of algebras satisfying some natural conditions (see [ST5]). For our aims it is sufficient to assume as a rule that each of these classes, $\mathfrak{U}$, is either the class $\mathfrak{U}_{\mathrm{a}}$ of all associative complex algebras for the algebraic case or one of the classes $\mathfrak{U}_{\mathrm{b}}$ and $\mathfrak{U}_{\mathrm{n}}$ of all Banach and all normed associative complex algebras, respectively, for the topological case. We consider also radicals defined on subclasses of these classes, for example on the subclass of $\mathfrak{U}_{\mathrm{n}}$ consisting of all $Q$-algebras and the subclass of $\mathfrak{U}_{\mathrm{b}}$ consisting of all C*-algebras.

We use the term *morphism* for a surjective homomorphism in the algebraic case and an open continuous surjective homomorphism in the topological case. In what follows, unless necessity for clarity we omit words 'algebraic' or 'topological' for notions of the radical theory in statements valid in both the algebraic and topological contexts. Sometimes we add necessary topological or algebraic specifications in square brackets. For instance, the expression "a [closed] ideal" means that the ideal in the proposition must be closed when we consider the proposition in the topological context.

A map $P$ on $\mathfrak{U}$ is called a *[closed] ideal map* if $P(A)$ is a [closed] ideal of $A$ for each [normed] $A \in \mathfrak{U}$. An *algebraic* (respectively, *topological*) *radical* is a (respectively, closed) ideal map $P$ that satisfies the following axioms:

**Axiom 1.** *$f(P(A)) \subset P(B)$ for every morphism $f : A \longrightarrow B$ of algebras from $\mathfrak{U}$;*

**Axiom 2.** *$P(A/P(A)) = 0$;*

**Axiom 3.** *$P(P(A)) = P(A)$;*

**Axiom 4.** *$P(I)$ is an ideal of $A$ contained in $P(A)$, for each ideal $I \in \mathfrak{U}$ of $A$.*

It is assumed in Axiom 4 that if $A$ is normed then the norm on $I$ is inherited from $A$. If a (closed) ideal map $P$ satisfies Axiom 1 on $\mathfrak{U}$ then $P$ is called a *preradical* on $\mathfrak{U}$; $P$ is called *pliant* if Axiom 1 on $\mathfrak{U}$ holds with algebraic morphisms. A (non-necessarily closed) ideal map $P$ is a *preradical with topological morphisms* if it defined on a class of normed algebras and satisfies Axiom 1 with topological morphisms (this class of maps contains the restrictions of arbitrary algebraic preradicals to classes of normed algebras). Algebraic preradicals are always pliant.

**Lemma 2.1.** *Let $P$ be a pliant preradical, and let $A$ be an algebra with two norms $\|\cdot\|_1$ and $\|\cdot\|_2$. Then $P(A; \|\cdot\|_1) = P(A; \|\cdot\|_2)$.*

*Proof.* Indeed, the identity map $(A; \|\cdot\|_1) \longrightarrow (A; \|\cdot\|_2)$ is an algebraic isomorphism. □

Sometimes the intermediate definitions are useful: a preradical $P$ on $\mathfrak{U} \subset \mathfrak{U}_n$ is called *strict* if

$$\overline{f(P(A))} = P(B) \tag{2.1}$$

for any continuous isomorphism $f : A \longrightarrow B$ of algebras from $\mathfrak{U}$, and called *strong* if $P$ satisfies Axiom 1 with respect to the continuous surjective homomorphisms of algebras from $\mathfrak{U}$. It is clear that every strict preradical is strong.

**Remark 2.2.** For radicals on $\mathfrak{U}_{\mathrm{b}}$ or $\mathfrak{U}_{\mathrm{a}}$ our definition coincides with Dixon's one, but for normed algebras the definitions differ:

(1) We do not assume that ideals in Axiom 4 are closed;
(2) In [D3] the morphisms are not necessarily open.

A preradical $P$ is called *hereditary* if

$$P(I) = I \cap P(A)$$

for every ideal $I$ of an algebra $A \in \mathfrak{U}$. The class of hereditary radicals is especially important; the heredity of a preradical $P$ implies the fulfillment of Axioms 3 and 4 for $P$. So one can define *hereditary radicals* as hereditary preradicals satisfying Axiom 2.

In several cases we will impose on a preradical $P$, defined on $\mathfrak{U}_{\mathrm{b}}$, a more strong condition – the *condition of Banach heredity*:

$$P(L, \|\cdot\|_L) = L \cap P(A) \text{ for any Banach ideal } (L, \|\cdot\|_L) \subset A, \tag{2.2}$$

where $A$ is a Banach algebra. One can introduce similarly the *condition of flexible heredity*, for normed algebras.

An algebra $A \in \mathfrak{U}$ is called *$P$-radical* if $A = P(A)$ and *$P$-semisimple* if $P(A) = 0$. Let $\mathbf{Rad}(P)$ denote the class of all $P$-radical algebras and $\mathbf{Sem}(P)$ the class of all $P$-semisimple algebras. A preradical $P$ is called *uniform* if every subalgebra $B \in \mathfrak{U}$ of a $P$-radical algebra is $P$-radical. It is clear that uniform radicals are hereditary.

By definition [D3], an *under radical* is a preradical which satisfies all axioms besides, possibly, Axiom 2. For instance, any hereditary preradical is an under radical; moreover, restrictions of algebraic radicals to appropriate classes of normed algebras are algebraic under radicals in general. The dual notion for under radicals is the notion of over radical. By definition [D3], an *over radical* is a preradical which satisfies all axioms besides, possibly, Axiom 3.

In the class of preradicals on $\mathfrak{U}$ there is a natural order: $P_1 \le P_2$ if $P_1(A) \subset P_2(A)$ for each $A \in \mathfrak{U}$. We write $P_1 < P_2$ if $P_1 \le P_2$ and $P_1 \ne P_2$. It is easy to see that

$$P_1 \le P_2 \implies \mathbf{Rad}(P_1) \subset \mathbf{Rad}(P_2) \text{ and } \mathbf{Sem}(P_2) \subset \mathbf{Sem}(P_1). \tag{2.3}$$

The following theorem is useful for comparing radicals.

**Theorem 2.3.** *Let $P_1$ and $P_2$ be over radicals on $\mathfrak{U}$, and let $R_1$ and $R_2$ be under radicals on $\mathfrak{U}$. Then*

(1) $\mathbf{Sem}(P_2) \subset \mathbf{Sem}(P_1)$ *if and only if* $P_1 \le P_2$*;*
(2) $\mathbf{Rad}(R_1) \subset \mathbf{Rad}(R_2)$ *if and only if* $R_1 \le R_2$.

*Proof.* Let $A$ be an algebra. Then $R_1(A) \in \mathbf{Rad}(R_1)$ by Axiom 3 and $A/P_2(A) \in \mathbf{Sem}(P_2)$ by Axiom 2.

(1) If $\mathbf{Sem}(P_2) \subset \mathbf{Sem}(P_1)$ then $A/P_2(A) \in \mathbf{Sem}(P_1)$. As $P_1$ is a preradical,

$$q(P_1(A)) \subset P_1(A/P_2(A)) = 0$$

for the standard quotient map $q : A \longrightarrow A/P_2(A)$. Therefore $P_1(A) \subset P_2(A)$, i.e., $P_1 \leq P_2$, and (2.3) completes the proof.

(2) If $\mathbf{Rad}(R_1) \subset \mathbf{Rad}(R_2)$ then $R_1(A) \in \mathbf{Rad}(R_2)$. As $R_1(A)$ is an ideal of $A$,

$$R_1(A) = R_2(R_1(A)) \subset R_2(A)$$

by Axiom 4, i.e., $R_1 \leq R_2$, and the converse follows by (2.3). □

**Corollary 2.4.** *Let $P_1$ and $P_2$ be radicals on $\mathfrak{U}$. If* $\mathbf{Sem}(P_1) = \mathbf{Sem}(P_2)$ *or* $\mathbf{Rad}(P_1) = \mathbf{Rad}(P_2)$ *then* $P_1 = P_2$.

### 2.2. Classes of algebras

**2.2.1. Base classes of algebras.** Let us say more about classes of algebras on which the radicals are defined. It is convenient to assume by default that a class of algebras contains all appropriate [topologically] isomorphic images of its elements.

A class $\mathfrak{U}$ of algebras is called *algebraically universal* if it contains quotients and ideals of algebras from $\mathfrak{U}$. A class $\mathfrak{U}$ of normed algebras is called *universal* if it contains quotients by closed ideals and ideals of algebras from $\mathfrak{U}$, and *ground* if it contains quotients by closed ideals and closed ideals of algebras from $\mathfrak{U}$.

For instance, $\mathfrak{U}_{\mathrm{a}}$ is algebraically universal, $\mathfrak{U}_{\mathrm{n}}$ is universal, and $\mathfrak{U}_{\mathrm{b}}$ is ground. These classes are main in this paper. We also use the following classes of algebras:

(1) $\mathfrak{U}_{\mathrm{q}}$ and $\mathfrak{U}_{\mathrm{q}_b}$ are the class of all $Q$-algebras and $Q_{\mathrm{b}}$-algebras, respectively. These classes are universal [ST5].
(2) $\mathfrak{U}_{\mathrm{b}}^{u}$ is the smallest universal class containing all Banach algebras. An algebra $A$ is called a *subideal* of an algebra $B$ if there is a finite series of algebras $A = I_0 \subset \cdots \subset I_n = B$ such that $I_{i-1}$ is an ideal of $I_i$ for $i = 1, \ldots, n$; in such a case $A$ is called an *$n$-subideal* of $B$. By [ST5, Theorem 2.24], $\mathfrak{U}_{\mathrm{b}}^{u}$ is the class of all subideals of Banach algebras.
(3) $\mathfrak{U}_{\mathrm{c}^*}$ is the class of all C*-algebras. Recall that every closed ideal of a C*-algebra is automatically self-adjoint [Dm, Proposition 1.8.2]. So this class is ground.

**2.2.2. About the definition of radicals in C*-algebras.** We will also consider the restrictions of radicals of Banach algebras to $\mathfrak{U}_{\mathrm{c}^*}$. The work with C*-algebra has several advantages:

(1) The sum of two closed ideals of a C*-algebra is closed [Dv, Corollary 1.5.6];
(2) Each irreducible *-representation of a C*-algebra is strictly irreducible [Dm, Theorem 2.8.3];
(3) A closed ideal of a closed ideal of a C*-algebra is an ideal of the algebra [Dm, Proposition 1.8.5].

So Axiom 4 for radicals on C*-algebras is equivalent to the following:

**Axiom 4 for $\mathfrak{U}_{c^*}$.** $P(I) \subset P(A)$ for any closed ideal $I$ of $A$.

As C*-algebras are semisimple, every morphism in $\mathfrak{U}_{c^*}$ is automatically continuous by Johnson's theorem [J]. It seems to be natural to consider all *-epimorphisms (i.e., *-morphisms) as morphisms in $\mathfrak{U}_{c^*}$. The following result shows that this does not change the class of preradicals.

**Theorem 2.5.** *Let $P$ be a closed ideal map on $\mathfrak{U}_{c^*}$. If $f(P(A)) \subset P(B)$ for each *-morphism $f : A \longrightarrow B$ then the same is true for all morphisms.*

*Proof.* Let $f : A \longrightarrow B$ be an epimorphism, $J = \ker f$ and $C = A/J$. Then $f = g \circ q_J$, where $g$ is an isomorphism of $C$ onto $B$. So it suffices to show that $q_J(P(A)) \subset P(C)$ and $g(P(C)) \subset P(B)$. The first inclusion is evident because $q_J$ is a *-epimorphism.

One can consider $B$ as an operator C*-algebra, that is, a closed *-subalgebra of $\mathcal{B}(H)$ where $H$ is a Hilbert space. It was proved by T. Gardner [Gr] that $g = g_1 \circ g_2$ where $g_2$ is a *-isomorphism of $C$ onto $B$, and $g_1 : B \longrightarrow B$ acts by the rule

$$g_1(x) = vxv^{-1}$$

where $v$ is an invertible positive operator on $H$. So, to prove that $g(P(C)) \subset P(B)$, it suffices to show that $g_1$ preserves closed ideals of $B$.

Let $S$ be the operator on $\mathcal{B}(H)$ defined by $Sx = vxv^{-1}$ for all $x \in \mathcal{B}(H)$. By the assumption, $S$ preserves $B$ and $g_1 = S|_B$. Clearly $S$ is a product of two commuting operators: $S = \mathrm{L}_v \mathrm{R}_{v^{-1}}$. Since $\mathrm{L}_v$ and $\mathrm{R}_{v^{-1}}$ have positive spectra on $\mathcal{B}(H)$, the same is true for $S$ on $\mathcal{B}(H)$.

Let us denote by log the holomorphic extension of the function

$$\phi(z) = \sum_{k=1}^{\infty} \frac{(-1)^{k-1}}{k}(z-1)^k$$

from the disk $D = \{z : |z-1| < 1$ to $\mathbb{C} \setminus (-\infty, 0]$. By the functional calculus, this function can be applied to every element of a Banach algebra whose spectrum does not intersect $(-\infty, 0] \subset \mathbb{R}$. As $E := \{K \in \mathcal{B}(\mathcal{B}(H))) : KB \subset B\}$ is a Banach algebra and $S \in E$ has spectrum outside of $(-\infty, 0] \subset \mathbb{R}$ then the operator $T = \log S$ preserves $B$.

Moreover, $T = \log \mathrm{L}_v + \log \mathrm{R}_{v^{-1}}$. Indeed, to prove that

$$\begin{aligned} &\log\left(1 + \lambda\left(\mathrm{L}_v - 1\right)\right)\left(1 + \lambda\left(\mathrm{R}_{v^{-1}} - 1\right)\right) \\ &\quad = \log\left(1 + \lambda\left(\mathrm{L}_v - 1\right)\right) + \log\left(1 + \lambda\left(\mathrm{R}_{v^{-1}} - 1\right)\right) \end{aligned}$$

for $\lambda = 1$, it suffices to show that the equality holds for sufficiently small $|\lambda|$. Clearly the equality $\phi(KL) = \phi(K) + \phi(L)$ is checked for commuting operators $K, L$ having spectra in $D$ by the calculation as in the case of numerical series.

Let $u = \log v$. As $0 = \log 1 = \log v + \log v^{-1}$ and $\log \mathrm{L}_v = \mathrm{L}_{\log v}$ then $\log \mathrm{L}_v = \mathrm{L}_u$ and $\log \mathrm{R}_{v^{-1}} = -\mathrm{R}_u$. We proved that the operator $T$ acts by the formula

$$Tx = ux - xu$$

for $x \in \mathcal{B}(H)$ and preserves $B$.

The restriction of $T$ to $B$ is a derivation of $B$ and therefore preserves closed ideals of $B$. Indeed, if $I$ is a closed ideal of $B$ then it is a C*-algebra, so it has a bounded approximate identity whence, by [BD, Theorem 11.10], each $a \in J$ can be written in the form $a = bc$ for $a, b \in J$, and

$$T(a) = bT(c) + T(b)c \in I.$$

Therefore $S = \exp(T)$ also preserves all closed ideals of $B$. □

As a result, Axiom 1 for radicals on C*-algebras is equivalent to the following:

**Axiom 1 for $\mathfrak{U}_{\mathrm{c}^*}$.** $f(P(A)) \subset P(B)$ for any *-morphism $f : A \longrightarrow B$.

### 2.3. Some important examples

**2.3.1. The Jacobson radical.** The most famous radical on $\mathfrak{U}_{\mathrm{a}}$ is the *Jacobson radical* rad which is defined by

$$\mathrm{rad}(A) = \ker \mathrm{Prim}(A^1) := \cap\{I : I \in \mathrm{Prim}(A^1)\}$$

for every algebra $A$; rad is an algebraic hereditary radical. As usual, $A$ is called *radical* if $A = \mathrm{rad}(A)$, and *semisimple* if $\mathrm{rad}(A) = 0$. By Johnson's theorem [J], the topology of a complete norm in a semisimple Banach algebra is unique.

**Theorem 2.6.** *Let $A, B$ be Banach algebras, and let $P$ be a topological over radical such that* rad $\leq P$ *on Banach algebras. Then*

(1) *If $f : A \longrightarrow B$ is an algebraic morphism then $f(P(A)) \subset P(B)$;*
(2) *If $f : A \longrightarrow B$ is an algebraic isomorphism then $f(P(A)) = P(B)$.*

*Proof.* (1) Let $B' = B/\mathrm{rad}(B)$ and $q : B \longrightarrow B'$ be the standard quotient map. Then $q \circ f : A \longrightarrow B'$ is a topological morphism by Johnson's result [A, Corollary 5.5.3]. Therefore $(q \circ f)(P(A)) \subset P(B')$, whence $f(P(A)) \subset q^{-1}(P(B'))$.

Let $g : B/\mathrm{rad}(B) \to B/P(B)$ be the natural morphism, then

$$g(P(B/\mathrm{rad}(B))) \subset P(B/P(B)) = 0$$

whence $P(B') = P(B/\mathrm{rad}(B)) \subset \ker g = P(B)/\mathrm{rad}(B) = q(P(B))$. Thus

$$f(P(A)) \subset q^{-1}(q(P(B))) = P(B).$$

(2) follows from (1) since $f^{-1} : B \longrightarrow A$ is also an algebraic morphism. □

As a consequence, topological radicals larger than (or equal to) rad on Banach algebras do not depend on the complete norm topology.

Let Rad be the restriction of rad to Banach algebras: $\mathrm{Rad} = \mathrm{rad}|_{\mathfrak{U}_{\mathrm{b}}}$. Then Rad is a pliant hereditary topological radical, while the restriction $\mathrm{rad}|_{\mathfrak{U}_{\mathrm{n}}}$ is not even a topological radical [D3]. The reason was already discussed: a normed algebra can have nonclosed primitive ideals. It is clear that Rad has a unique hereditary extension to $\mathfrak{U}_{\mathrm{b}}^{u}$, namely rad $|_{\mathfrak{U}_{\mathrm{b}}^{u}}$. The hereditary extension of Rad to normed algebras is given by the *regular Jacobson radical* $\mathrm{Rad}^{r}$ (see Section 3.3):

$$\mathrm{Rad}^{r}(A) = \left\{a \in A : \rho(ab) = 0 \ \forall b \in \widehat{A}\right\} \tag{2.4}$$

where $\rho(a)$ is a (geometric) spectral radius $\inf_n \|a^n\|^{1/n}$. Below we consider the other topological hereditary extensions of Rad to $\mathfrak{U}_{\mathrm{n}}$.

**2.3.2. Primitive maps and related radicals.** Let $\mathfrak{U}$ be a base class of algebras. A rule that indicates, for each algebra $A \in \mathfrak{U}$, a subset $\Omega(A)$ of $\mathrm{Prim}(A)$ is called a *primitive map* on $\mathfrak{U}$ if

($1_{pm}$) $\Omega(B) = \{f(I) : I \in \Omega(A)\}$ for an injective morphism $f : A \longrightarrow B$ of algebras from $\mathfrak{U}$;

($2_{pm}$) $\Omega(J) = \{J \cap I : I \in \Omega(A), J \cap I \neq J\}$ for every ideal $J \in \mathfrak{U}$ of $A$;

($3_{pm}$) $\Omega(A/J) = \{J/I : I \in \Omega(A), J \subset I\}$ for every [closed] ideal $J \in \mathfrak{U}$ of $A$.

If ($1_{pm}$) holds for arbitrary isomorphisms of algebras $A, B \in \mathfrak{U}$, then we say that $\Omega$ is *pliant*; in other words, a primitive map $\Omega$ on $\mathfrak{U}$ is pliant if the following condition holds:

($4_{pm}$) $\Omega(A)$ does not depend on the choice of a norm in $A$, for every $A$ from $\mathfrak{U}$ (in short: on the choice of $\mathfrak{U}$-norm).

Clearly all primitive maps on $\mathfrak{U}_{\mathrm{a}}$ and on $\mathfrak{U}_{\mathrm{c}^*}$ are pliant.

**Proposition 2.7.** *All primitive maps defined on Banach algebras are pliant.*

*Proof.* Let $A$ be a Banach algebra. It follows from ($3_{pm}$) that $\Omega(A)$ is uniquely determined by $\Omega(A/\mathrm{Rad}(A))$, but $A/\mathrm{Rad}(A)$ has only one complete norm topology by Johnson's theorem. □

The following proposition is straightforward.

**Proposition 2.8.** *A primitive map defined on some class $\mathfrak{U}$ of normed algebras is pliant if and only if it does not depend on the choice of a $\mathfrak{U}$-norm on semisimple algebras from $\mathfrak{U}$.*

It is easy to see from the results of Section 1.2.3 that setting $\Omega(A) = \mathrm{Prim}(A)$ for each $A$, we obtain a primitive map on $\mathfrak{U}_{\mathrm{a}}$; it will be denoted by Prim.

**Theorem 2.9.** *Let $\mathcal{F} = (\Omega_\alpha)$ be a family of (pliant) primitive maps. Then*

(1) *$\Omega_{\cap\mathcal{F}} : A \longrightarrow \cap_\alpha \Omega_\alpha(A)$ and $\Omega_{\cup\mathcal{F}} : A \longrightarrow \cup_\alpha \Omega_\alpha(A)$ are (pliant) primitive maps;*

(2) *If $\Omega_i$ is a (pliant) primitive map for $i = 1, 2$ then so is $\Omega_1 \backslash \Omega_2 : A \longmapsto \Omega_1(A) \backslash \Omega_2(A)$.*

*Proof.* (1) Since

$$\begin{aligned}\cup_{\alpha}\Omega_{\alpha}(J) &= \cup_{\alpha}\{J\cap I : I\in\Omega_{\alpha}(A), J\cap I\neq J\}\\ &= \{J\cap I : I\in\cup_{\alpha}\Omega_{\alpha}(A), J\cap I\neq J\}\end{aligned}$$

we obtain $(2_{pm})$ for $\Omega_{\cup\mathcal{F}}$. The remaining assertions can be proved similarly.

(2) Since

$$\begin{aligned}(\Omega_1\backslash\Omega_2)(A/J) &= \{J/I : I\in\Omega_1(A), J\subset I\}\backslash\{J/I : I\in\Omega_2(A), J\subset I\}\\ &= \{J/I : I\in\Omega_1(A)\backslash\Omega_2(A), J\subset I\}\end{aligned}$$

we get $(3_{pm})$ for $\Omega_1\backslash\Omega_2$. The other conditions can be checked in a similar way. □

As usual for subsets of $\mathrm{Prim}(A)$, $\ker\Omega(A)$ is simply $\cap\{I : I\in\Omega(A)\}$. We define an ideal map $\Pi_{\Omega}$ by

$$\Pi_{\Omega}(A) = \ker\Omega(A)$$

for every $A\in\mathfrak{U}$, meaning that $\Pi_{\Omega}(A) = A$ if $\Omega(A) = \varnothing$, or, more formally, by $\Pi_{\Omega}(A) = A\cap\ker\Omega(A)$. Not all primitive maps determine radicals (see Example 2.14). The obstacle is that the ideals $\Pi_{\Omega}(A)$ can be non-closed.

**Theorem 2.10.** *Let $\Omega$ be a primitive map on $\mathfrak{U}$. Then*

(1) *If ($\Omega$ is pliant and) $\mathfrak{U}\subset\mathfrak{U}_{\mathrm{q}}$ then $\Pi_{\Omega}$ is a (pliant) topological hereditary radical;*
(2) *If $\Omega$ is pliant and $\mathfrak{U}_{\mathrm{q}}\subset\mathfrak{U}\subset\mathfrak{U}_{\mathrm{n}}$ then $\Pi_{\Omega}$ is a pliant hereditary preradical;*
(3) *If $\mathfrak{U}=\mathfrak{U}_{\mathrm{a}}$ then $\Pi_{\Omega}$ is an algebraic hereditary radical.*

*Proof.* (3) Let $A\in\mathfrak{U}$ be an algebra, and let $J\in\mathfrak{U}$ be an ideal of $A$. It follows from $(2_{pm})$ that $\Pi_{\Omega}(J) = J\cap\Pi_{\Omega}(A)$. So $\Pi_{\Omega}$ is hereditary.

Let $J = \Pi_{\Omega}(A)$. Then $J\subset I$ for every $I\in\Omega(A)$. By (1.4),

$$\Pi_{\Omega}(A/J) = \ker\Omega(A)/J = \Pi_{\Omega}(A)/J = J/J = 0.$$

As every morphism is represented as the superposition of a standard quotient map and an injective morphism, $(1_{pm})$ and $(3_{pm})$ imply that $\Pi_{\Omega}$ is a pliant radical on $\mathfrak{U}$.

(1) Indeed, every strictly irreducible representation of a normed $Q$-algebra $A$ is equivalent to a bounded representation by bounded operators on a normed space [ST5, Theorem 2.1]. So $\mathrm{Prim}(A)$ consists of closed ideals and $\Pi_{\Omega}$ is a closed ideal map on normed $Q$-algebras. As in (3), $\Pi_{\Omega}$ is a hereditary radical.

(2) is evident. □

**Remark 2.11.** An algebra $A\in\mathfrak{U}$ is $\Pi_{\Omega}$-radical if and only if $\Omega(A) = \varnothing$, and $A$ is $\Pi_{\Omega}$-semisimple if and only if $\cap\{I : I\in\Omega(A)\} = 0$; it is clear that every $\Pi_{\Omega}$-semisimple algebra is rad-semisimple, i.e., $\mathrm{rad}\leq\Pi_{\Omega}$ on $\mathfrak{U}$. In particular, Theorem 2.6 holds for $P = \Pi_{\Omega}$.

The following statement is obvious.

**Corollary 2.12.** *Let $\Omega$ be a pliant primitive map on a universal class $\mathfrak{U}$ of normed algebras. Then $\Pi_\Omega$ satisfies the condition of flexible heredity (in particular, the condition of Banach heredity) and $f(P(A)) = P(B)$ for an algebraic injective morphism $f : A \longrightarrow B$ of any algebras $A, B \in \mathfrak{U}$.*

See the definitions of $\mathrm{Prim}_{\mathrm{n}}(A)$ and $\mathrm{Prim}_{\mathrm{b}}(A)$ for a normed algebra $A$ in Section 1.2.3.

**Example 2.13.** $\mathrm{Prim}_{\mathrm{n}}$ and $\mathrm{Prim}_{\mathrm{b}}$ are primitive maps on $\mathfrak{U}_{\mathrm{n}}$.

*Proof.* In virtue of Section 1.2.3, it is sufficient to check that the declared properties of representations from $\mathrm{Irr}_{\mathrm{n}}(A)$ or $\mathrm{Irr}_{\mathrm{b}}(A)$ are preserved under passing to an ideals, quotients, actions of isomorphisms and taking the converse manipulations. But this is straightforward in virtue of [ST5, Lemmas 2.3 and 2.4]. □

The topological radicals $\Pi_{\mathrm{Prim}_{\mathrm{n}}}$ and $\Pi_{\mathrm{Prim}_{\mathrm{b}}}$ coincide with the introduced in [D3, ST5] extensions $\mathrm{rad}_{\mathrm{b}}$ and $\mathrm{rad}_{\mathrm{b}}$ of Rad to normed algebras. On the base of $\mathrm{Prim}_{\mathrm{n}}$ and $\mathrm{Prim}_{\mathrm{b}}$ one can construct other useful primitive maps by imposing additional requirements on representations $\pi$ from $\mathrm{Irr}_{\mathrm{n}}(A)$ or $\mathrm{Irr}_{\mathrm{b}}(A)$.

**Example 2.14.** By Theorem 2.9, $\Omega_{\mathrm{n}} := \mathrm{Prim} \setminus \mathrm{Prim}_{\mathrm{n}}$ and $\Omega_{\mathrm{b}} := \mathrm{Prim} \setminus \mathrm{Prim}_{\mathrm{b}}$ are primitive maps on $\mathfrak{U}_{\mathrm{n}}$, but $\Pi_{\Omega_{\mathrm{n}}}$ and $\Pi_{\Omega_{\mathrm{b}}}$ are not closed ideal maps on $\mathfrak{U}_{\mathrm{n}}$; $Q$-algebras are $\Pi_{\Omega_{\mathrm{n}}}$-radical, and $\Pi_{\Omega_{\mathrm{n}}}(A)$ is a $Q$-algebra for every normed algebra $A$.

The simplest characteristic of a representation is its dimension. Recall that the *dimension* of $\pi \in \mathrm{Irr}(A)$ is the dimension of $X_\pi$. Let $N$ be a subset of $\mathbb{N} \cup \{\infty\}$, and let $\mathrm{Irr}^{\dim\in N}(A)$ be the set of all representations $\pi \in \mathrm{Irr}(A)$ such that $\dim X_\pi \in N$. Respectively, let $\mathrm{Prim}^{\dim\in N}(A) = \left\{\ker\pi : \pi \in \mathrm{Irr}^{\dim\in N}(A)\right\}$. The simplest special cases of the relation $\dim \in N$ are $\dim = 1$, $\dim > 1$, $\dim < \infty$.

**Theorem 2.15.** *Let $N \subset \mathbb{N} \cup \{\infty\}$. Then $\mathrm{Prim}^{\dim\in N}$ is a pliant hereditary primitive map.*

*Proof.* It suffices to note that the representation spaces are not changed when we extend a representation from an ideal to the algebra, restrict to an ideal or induce the strictly irreducible representations from a quotient. □

**Corollary 2.16.** *Let $N \subset \mathbb{N} \cup \{\infty\}$. Then the map*

$$\mathrm{rad}^{\dim\in N} : A \longmapsto \Pi_{\mathrm{Prim}^{\dim\in N}}(A)$$

*is a hereditary radical on $\mathfrak{A}_a$.*

Note that $\mathrm{rad}^{\dim\in N}$ is a pliant hereditary topological radical on $Q$-algebras.

**Example 2.17.** An algebra $A$ is $\mathrm{rad}^{\dim>1}$-radical $\Leftrightarrow$ $A$ admits only one-dimensional strictly irreducible representations, and $A$ is $\mathrm{rad}^{\dim=1}$-semisimple $\Leftrightarrow$ the intersection of kernels of one-dimensional strictly irreducible representations is trivial.

Respectively, $A$ is $\mathrm{rad}^{\dim=\infty}$-radical $\Leftrightarrow$ $A$ admits only finite-dimensional strictly irreducible representations, and $A$ is $\mathrm{rad}^{\dim<\infty}$-semisimple $\Leftrightarrow$ the intersection of kernels of finite-dimensional strictly irreducible representations is trivial.

Let us finish by showing that in terms of primitive maps one can define one of the most important objects of the theory of C*-algebras – the largest GCR-ideal.

Recall that a C*-algebra $A$ is called *elementary* if it is isomorphic to the algebra $\mathcal{K}(H)$ of all compact operators on some Hilbert space $H$. Furthermore, $A$ is a *CCR-algebra* (a *GCR-algebra*) if for every irreducible representation $\pi$ of $A$, the image, $\pi(A)$, consists of compact operators (respectively, contains a non-zero compact operator, or, equivalently, all compact operators). A simplest example of a CCR-algebra is $\mathcal{K}(H)$.

For $A \in \mathfrak{U}_{\mathrm{c}^*}$, let $\Omega_{\mathfrak{gcr}}(A)$ denote the set of all $I \in \mathrm{Prim}(A)$ such that $A/I$ has no elementary ideals.

**Theorem 2.18.** *$\Pi_{\Omega_{\mathfrak{gcr}}}$ is a topological radical on $\mathfrak{U}_{\mathrm{c}^*}$; $\Pi_{\Omega_{\mathfrak{gcr}}}(A)$ is the largest GCR-ideal for every $A \in \mathfrak{U}_{\mathrm{c}^*}$.*

*Proof.* It is easy to check that $\Omega_{\mathfrak{gcr}}$ is a primitive map whence $\Pi_{\Omega_{\mathfrak{gcr}}}$ is a topological radical on $\mathfrak{U}_{\mathrm{c}^*}$.

Let $A$ be a C*-algebra and $J_0 = \Pi_{\Omega_{\mathfrak{gcr}}}(A) = \cap\{I \in \Omega_{\mathfrak{gcr}}(A)\}$. To show that $J_0$ is a GCR-algebra, take an irreducible representation $\pi$ of $J_0$ on a Hilbert space $H$ and extend it to an irreducible representation (again $\pi$) of $A$ on $H$. Then $\ker \pi \notin \Omega_{\mathfrak{gcr}}(A)$, so $\pi(A)$ contains an elementary ideal $K$. Since $K$ is irreducible and $\mathcal{K}(H)$ is a CCR-algebra, $K = \mathcal{K}(H)$. Then

$$0 \neq \pi(J_0)\mathcal{K}(H) \subset \pi(J_0)\pi(A) \subset \pi(J_0),$$

so $\pi(J_0)$ contains a non-zero compact operator. This means that $J_0$ is a GCR-algebra.

Let now $J$ be a GCR-ideal of $A$. If $I \in \Omega_{\mathfrak{gcr}}(A)$ then $I = \ker \pi$ for an irreducible representation $\pi$ of $A$ such that $\pi(A)$ has no elementary ideals. But the restriction of $\pi$ to $J$ is also irreducible (if non-zero), so $\pi(J)$ must contain $\mathcal{K}(H)$. Therefore $\mathcal{K}(H) \subset \pi(A)$, a contradiction. Thus $\pi(J) = 0$ and then $J \subset I$. So

$$J \subset \cap\{I : I \in \Omega_{\mathfrak{gcr}}(A)\} = J_0. \qquad \square$$

Another way to establish that the map sending each C*-algebra $A$ to its largest GCR-ideal is a hereditary topological radical, will be considered in Section 2.3.7.

**2.3.3. The tensor Jacobson radical $\mathcal{R}_t$.** It is defined via the tensor spectral radius. Let $A$ be a normed algebra, $M = (a_i)_1^\infty$ and $N = (b_j)_1^\infty$ be *summable* families in $A$, i.e., $\|M\|_+ := \sum_i \|a_i\| < \infty$ and $\|N\|_+ := \sum_i \|b_i\| < \infty$. The norm of a family does not depend on the order, so one can consider the corresponding equivalence relation $\simeq$ between families (see details in [ST6]). Define the product $MN$ within up to $\simeq$ as a family $H = (c_k)_1^\infty$ where $c_k = a_i b_j$ for $k = \phi(i,j)$ and $\phi$ is an arbitrary bijection of $\mathbb{N}\times\mathbb{N}$ onto $\mathbb{N}$. Then the power $M^{n+1}$ is defined by $M^{n+1} \simeq MM^n$ for every $n > 0$ and the *tensor (spectral) radius* $\rho_t(M)$ is defined by

$$\rho_t(M) = \inf_n \|M^n\|_+^{1/n} = \lim_{n\to\infty} \|M^n\|_+^{1/n}.$$

Let us define by $N \sqcup M$ the disjunct union of families $M = (a_i)$ and $N = (b_j)$ (that is a family equivalent to $(b_1, a_1, b_2, a_2, \ldots)$); then $\mathcal{R}_t(A)$ is defined as the set of all $a \in A$ such that $\rho_t(\{a\} \sqcup M) = \rho_t(M)$ for every summable family $M$ in $A$; $\mathcal{R}_t$ is a uniform topological radical [ST5, ST6].

The term *tensor* is justified by the fact that $a \in \mathcal{R}_t(A)$ if and only if $a \otimes b \in \mathrm{Rad}\left(A \widehat{\otimes} B\right)$ for every normed algebra $B$ and $b \in B$ [ST6, Theorem 3.36]. By [ST6, Theorem 3.29],

$$\rho_t(M) = \rho_t(M/\mathcal{R}_t(A)). \tag{2.5}$$

There is a problem whether $\mathcal{R}_t = \mathrm{Rad}$ on Banach algebras. In the algebraic case it is known that the tensor product of a radical algebra and the other algebra can be not radical.

**Problem 2.19.** *Is there an algebraic radical $P$ on $\mathfrak{U}_a$ such that $P = \mathcal{R}_t$ on $\mathfrak{U}_n$?*

**2.3.4. The compactly quasinilpotent radical $\mathcal{R}_{cq}$.** Let $A$ be a normed algebra, and let $M \subset A$ be bounded. Define the *norm* $\|M\| = \sup_{a\in M} \|a\|$ and the *joint spectral radius*

$$\rho(M) = \inf_n \|M^n\|^{1/n} = \lim_{n\to\infty} \|M^n\|^{1/n}$$

where $M^n = \{a_1 \cdots a_n : a_i \in M\}$ [RS]. Let $\mathfrak{k}(A)$ be the set of all precompact subsets of $A$. Then $\mathcal{R}_{cq}(A)$ is the set of all $a \in A$ such that $\rho(\{a\} \cup M) = \rho(M)$ for every $M \in \mathfrak{k}(A)$; $\mathcal{R}_{cq}$ is a uniform topological radical [ST5].

A normed algebra $A$ is called *compactly quasinilpotent* if $\rho(M) = 0$ for every $M \in \mathfrak{k}(A)$; the completion of compactly quasinilpotent algebra is again compactly quasinilpotent [ST5, Lemma 4.11]. Note that $\mathcal{R}_{cq}(A)$ is the largest compactly quasinilpotent ideal [ST5, Corollary 4.21].

An important fact related to this radical is that every compact operator $a$ in the Jacobson radical of a closed operator algebra $A$ generates a compactly quasinilpotent ideal of $A$; a close argument was a key in the proof that $A$ has a hyperinvariant subspace if $a \neq 0$ [S]. The other invariant subspace results [T2, ST1] are connected with the calculation of the joint spectral radius. It was established in [ST1] that for a precompact set $M$ of compact operators on a Banach space $X$

$$\rho(M) = r(M) \tag{2.6}$$

where the *Berger–Wang radius* $r(M)$ is defined by [BW]:

$$r(M) = \limsup_{n\to\infty} \sup\{\rho(a) : a \in M^n\}^{1/n}.$$

We call (2.6) the *Berger–Wang formula* because the matrix version of (2.6), for $\dim X < \infty$, was established in [BW]. A useful contribution in the calculation of $\rho$ is the equality

$$\rho(M) = \rho(M/\mathcal{R}_{cq}(A)) \tag{2.7}$$

for every $M \in \mathfrak{k}(A)$ [ST5, Theorem 4.18].

There is a version of the joint spectral radius for bounded countable families $M = (a_n)_1^\infty$ in $A$, where $M^n$ is calculated by rules of families [ST6].

**2.3.5. The hypocompact radical $\mathcal{R}_{\mathrm{hc}}$.** Let $A$ be a normed algebra. An element $a \in A$ is called *compact* if $\mathrm{W}_a := \mathrm{L}_a\mathrm{R}_a$ (see Section 1.2.3) is a compact operator on $A$; $A$ is *compact* if it consists of compact elements, *bicompact* if $\mathrm{L}_a\mathrm{R}_b$ is a compact operator on $A$ for every $a, b \in A$, and *hypocompact* if every non-zero quotient of $A$ by a closed ideal has a non-zero compact element. Then $\mathcal{R}_{\mathrm{hc}}(A)$ is defined as the largest hypocompact ideal of $A$ (equivalently, as the smallest closed ideal $J$ of $A$ such that $A/J$ has no non-zero compact elements); $\mathcal{R}_{\mathrm{hc}}$ is a hereditary topological radical [ST7].

This radical plays an important role in the theory of the joint spectral radius. It was proved in [ST7] that the following equality (an *algebra version of the joint spectral radius formula*):

$$\rho(M) = \max\{\rho(M/\mathcal{R}_{\mathrm{hc}}(A)), r(M)\} \tag{2.8}$$

holds for every normed algebra $A$ and $M \in \mathfrak{k}(A)$. In particular, it follows from (2.8) that

$$\mathcal{R}_{\mathrm{hc}}(A) \cap \mathrm{Rad}^r(A) \subset \mathcal{R}_{\mathrm{cq}}(A) \tag{2.9}$$

for every normed algebra $A$. It should be noted that $A \longmapsto \mathcal{R}_{\mathrm{hc}}(A) \cap \mathrm{Rad}^r(A)$ is also a hereditary topological radical on normed algebras. This radical is called the *Jacobson hypocompact radical* and is denoted by $\mathcal{R}_{\mathrm{jhc}}$.

If $A = \mathcal{B}(X)$ for a Banach space $X$, then $\mathcal{R}_{\mathrm{hc}}(A) \supset \mathcal{K}(X)$, so that (2.8) implies (2.6) and a stronger result, the *operator version of the joint spectral radius formula*:

$$\rho(M) = \max\{\rho_e(M), r(M)\} \tag{2.10}$$

that holds for every $M \in \mathfrak{k}(\mathcal{B}(X))$, where the *essential spectral radius* $\rho_e(M)$ is defined as the joint spectral radius $\rho(M/\mathcal{K}(X))$ in the Calkin algebra $\mathcal{B}(X)/\mathcal{K}(X)$ (see details in [ST7]).

There is the largest of topological radicals $P$ which can change $\mathcal{R}_{\mathrm{hc}}$ in (2.8) [ST7]; it is denoted by $\mathcal{R}_{\mathrm{bw}}$ and called the *Berger–Wang radical.*

A normed algebra $A$ is called a *Berger–Wang algebra* if (2.6) holds for every $M \in \mathfrak{k}(A)$.

**2.3.6. The hypofinite radicals $\mathcal{R}_{\mathrm{hf}}$ and $\mathfrak{R}_{\mathrm{hf}}$.** We begin with the algebraic hypofinite radical $\mathfrak{R}_{\mathrm{hf}}$ and describe its construction more transparently because of the lack of references. An element $a$ of an algebra $A$ is called a *finite rank element* if $\dim aAa < \infty$. An algebra $A$ is called *finite* if it consists of finite rank elements, *bifinite* if $\dim aAb < \infty$ for every $a, b \in A$, and *hypofinite* if $A/I$ contains a non-zero finite rank element for every ideal $I \neq A$. We transfer these notions to ideals. Let $\digamma(A)$ be the set of all finite rank elements of $A$.

**Theorem 2.20.** *Let $A$ be an algebra, and let $J$ be an ideal of $A$. Then*

1. $f(\digamma(A)) \subset \digamma(B)$ *for every morphism* $f : A \longrightarrow B$;
2. *If* $\digamma(J) \neq 0$ *then* $J \cap \digamma(A) \neq 0$;

(3) *If $x \in F(A)$ and $I$ is an ideal generated by $x$ then* $\mathrm{L}_a\mathrm{R}_b$ *is a finite rank operator on $A$ for every $a, b \in I$;*
(4) *If $A$ is hypofinite and $I$ a non-zero (one-sided) ideal then $I \cap F(A) \neq 0$;*
(5) *The following conditions are equivalent:*
 (a) *$J$ is a hypofinite ideal of $A$;*
 (b) *for every morphism $f : A \longrightarrow B$, either $f(J) = 0$ or $f(J) \cap F(B) \neq 0$;*
 (c) *there is an increasing transfinite sequence $(J_a)_{\alpha \leq \gamma}$ of ideals of $A$ such that $J_0 = 0$, $J_\gamma = J$, $J_\alpha = \cup_{\alpha' < \alpha} J_{\alpha'}$ for every limit ordinal $\alpha$, and all $J_{\alpha+1}/J_\alpha$ are bifinite.*
(6) *The following conditions are equivalent:*
 (a) *$A$ is hypofinite;*
 (b) *all ideals and quotients of $A$ are hypofinite;*
 (c) *$A/I$ and $I$ are hypofinite for some ideal $I$ of $A$.*
(7) *There is a largest hypofinite ideal of $A$ (which is denoted by $\mathfrak{R}_{\mathrm{hf}}(A)$);*
(8) *$\mathfrak{R}_{\mathrm{hf}}(J) = J \cap \mathfrak{R}_{\mathrm{hf}}(A)$;*
(9) *$A/\mathfrak{R}_{\mathrm{hf}}(A)$ has no non-zero hypofinite ideals and finite rank elements.*

*Proof.* (1) $\mathrm{W}_{f(a)}B = f(\mathrm{W}_aA)$ is finite dimensional if $a \in F(A)$.

(2) It is clear that

$$\mathrm{W}_{ba} = \mathrm{L}_b\mathrm{W}_a\mathrm{R}_b = \mathrm{R}_a\mathrm{W}_b\mathrm{L}_a \tag{2.11}$$

for every $a, b \in A$. If $a \in F(J)$ then $\mathrm{W}_{ba} \in F(A)$, whence $JF(J) \subset J \cap F(A)$. If $JF(J) = 0$ then clearly $F(J) \subset J \cap F(A)$.

(3) follows from (2.11).

(4) Assume that $I$ is a left ideal. Let $K = \{a \in A : aI = 0\}$; $K$ is an ideal of $A$. If $K = A$ then $I \subset F(A)$, otherwise $A/K$ has a non-zero finite rank element $b$. Let $q_K : A \longrightarrow A/K$ be the standard quotient map and $q_K(a) = b$. Clearly there is $x \in J$ such that $ax \neq 0$, it is easy to check that $ax \in F(A)$.

(5a) $\Longrightarrow$ (5b) Let $I = \ker f$ and $K = I \cap J$. Assume that $f(J) \neq 0$; then there is a non-zero $a \in J/K \cap F(A/K)$ by (2). Let $b \in A$ be such that $a = q_K(b)$; then $0 \neq f(b) \in F(B)$.

(5b) $\Longrightarrow$ (5c) Assume that we already have built $J_\beta$ by transfinite induction. If $J \neq J_\beta$ then $J/J_\beta$ has a non-zero finite rank element by definition, and contains a non-zero finite rank element $a$ of $A$ by (2). Then $J/J_\beta$ contains a non-zero bifinite ideal $I$ of $A$ by (3). Take the preimage of $I$ in $A$ as $J_{\beta+1}$. This proves the implication.

(5c) $\Longrightarrow$ (5a) Let $I$ be an ideal of $J$ and $I \neq J$. Then there is the first ordinal $\alpha$ such that $J_\alpha$ does not lie in $I$. Then $q_I(J_\alpha) \subset F(J/I)$.

(6a) $\Longrightarrow$ (6b) It is clear that the quotients of a hypofinite algebra are hypofinite. Let $K$ be an ideal of $A$, $f : A \longrightarrow B$ a morphism and $I = \ker f$. Assume that $f(K) \neq 0$; then $q_I(K) \neq 0$. As $A$ is hypofinite then there is $a \in K$ such that $0 \neq q_I(a) \in F(A/I)$ by (4). Then $0 \neq f(a) \in F(B)$ by (1). Therefore $K$ is hypofinite by (5).

(6b) $\Longrightarrow$ (6c) is obvious, and (6c) $\Longrightarrow$ (6a) follows by (5): one can build a sequence of ideals for $A$ as in (5c) having such sequences for $I$ and $A/I$.

(7) Let $K$ be the sum of all hypofinite ideals of $A$. Then an arbitrary morphism $f : A \longrightarrow B$ vanishes on $K$ if and only if it vanishes on each hypofinite ideal of $A$. Therefore $K$ is hypofinite by (5).

(8) As $J \cap \mathfrak{R}_{\mathrm{hf}}(A)$ is a hypofinite ideal of $J$, we obtain that $J \cap \mathfrak{R}_{\mathrm{hf}}(A) \subset \mathfrak{R}_{\mathrm{hf}}(J)$. Let $I$ be the ideal of $A$ generated by $\mathfrak{R}_{\mathrm{hf}}(J)$. Then clearly $I^3 \subset \mathfrak{R}_{\mathrm{hf}}(J)$ is hypofinite and $I/I^3$ is bifinite, whence $I$ is hypofinite by (6). Therefore $\mathfrak{R}_{\mathrm{hf}}(J) \subset I \subset \mathfrak{R}_{\mathrm{hf}}(A)$.

(9) If $I$ is a non-zero hypofinite ideal of $A/\mathfrak{R}_{\mathrm{hf}}(A)$ then its preimage $K$ in $A$ is hypofinite by (6), whence $\mathfrak{R}_{\mathrm{hf}}(A) \subsetneqq I$, a contradiction. Therefore $A/\mathfrak{R}_{\mathrm{hf}}(A)$ has no non-zero finite rank elements. □

**Corollary 2.21.** $\mathfrak{R}_{\mathrm{hf}}$ *is a hereditary radical.*

*Proof.* By Theorem 2.20(8-9), it suffices to show that $\mathfrak{R}_{\mathrm{hf}}$ is an algebraic preradical. Let $f : A \longrightarrow B$ be a morphism, and let $q : B \longrightarrow B/\mathfrak{R}_{\mathrm{hf}}(B)$ be the standard quotient map. Then $q \circ f$ is a morphism $A \longrightarrow B/\mathfrak{R}_{\mathrm{hf}}(B)$ and it follows from Theorem 2.20(5b) that either $(q \circ f)(\mathfrak{R}_{\mathrm{hf}}(A)) = 0$ or $B/\mathfrak{R}_{\mathrm{hf}}(B)$ has a non-zero finite rank element which is impossible by Theorem 2.20(9). Therefore $(q \circ f)(\mathfrak{R}_{\mathrm{hf}}(A)) = 0$ and $f(\mathfrak{R}_{\mathrm{hf}}(A)) \subset \mathfrak{R}_{\mathrm{hf}}(B)$. □

The radical $\mathfrak{R}_{\mathrm{hf}}$ is called an (*algebraic*) *hypofinite radical.*

The radical $\mathcal{R}_{\mathrm{hf}}$ is a topological hereditary radical and can be defined as the largest closed-hypofinite ideal: a normed algebra (or ideal) $A$ is called (*closed-*) *hypofinite* if every non-zero quotient of $A$ by a *closed* ideal has a non-zero finite rank element of $A$, and *approximable* [ST6] if $A = \overline{\mathcal{F}(A)}$. Elements of $\overline{\mathcal{F}(A)}$ are called *approximable elements* of $A$. Theorem 2.20 and Corollary 2.21 hold for $\mathcal{R}_{\mathrm{hf}}$ if one replaces the terms for $\mathfrak{R}_{\mathrm{hf}}$ by appropriate terms for $\mathcal{R}_{\mathrm{hf}}$.

**2.3.7. Radicals on C*-algebras.** Let $P$ be a radical. It is known ([ST5, Theorems 2.9 and 2.10]) that the class $\mathbf{Rad}(P)$ is stable under extensions and quotients, and that the [closure of] the union of an up-directed net of ideals $I_\lambda \in \mathbf{Rad}(P)$ of some algebra also belongs to $\mathbf{Rad}(P)$. Moreover, if $P$ is hereditary then each ideal of $A \in \mathbf{Rad}(P)$ clearly belongs to $\mathbf{Rad}(P)$. The following result shows that the converse holds for C*-algebras: a class of C*-algebras with the listed above properties is $\mathbf{Rad}(P)$ for some (hereditary) radical $P$.

**Theorem 2.22.** *Let* $\mathfrak{C}$ *be a class of* $C^*$*-algebras containing *-isomorphic images of its elements and having the following properties:*

(a) *If* $A \in \mathfrak{C}$ *then* $A/J \in \mathfrak{C}$ *for every closed ideal* $J$ *of* $A$;
(b) *If* $J \in \mathfrak{C}$ *is an ideal of a* $C^*$*-algebra* $A$ *and* $A/J \in \mathfrak{C}$ *then* $A \in \mathfrak{C}$;
(c) *If* $(J_\alpha)_\alpha \subset \mathfrak{C}$ *is an up-directed net of ideals of a* $C^*$*-algebra* $A$ *then* $\overline{\cup_\alpha J_\alpha} \in \mathfrak{C}$.

*Then*

(1) *Each* $C^*$*-algebra* $A$ *has a largest ideal* $R_{\mathfrak{C}}(A) \in \mathfrak{C}$;
(2) $R_{\mathfrak{C}}$ *is a radical on* $\mathfrak{U}_{c^*}$ *such that* $\mathbf{Rad}(R_{\mathfrak{C}}) = \mathfrak{C}$;

(3) *The radical $R_{\mathfrak{C}}$ is hereditary if and only if $\mathfrak{C}$ satisfies the condition*

(d) *Any closed ideal of an algebra $A \in \mathfrak{C}$ belongs to $\mathfrak{C}$;*

(4) *$R_{\mathfrak{C}}$ is uniform if and only if $B \in \mathfrak{C}$ for any $C^*$-subalgebra $B$ of every $A \in \mathfrak{C}$.*

*Proof.* (1) If ideals $I, J$ of $A$ belong to $\mathfrak{C}$ then

$$(I + J)/I = J/(I \cap J) \in \mathfrak{C}$$

whence $I + J \in \mathfrak{C}$ by (a).

Let $\Lambda$ be the set of all finite families of ideals of $A$ that belong to $\mathfrak{C}$; it is ordered by the inclusion. For $\omega \in \Lambda$, let $J_\omega$ be the sum of all ideals in $\omega$. Then $(I_\omega)_{\omega\in\Lambda} \subset \mathfrak{C}$ is an up-directed net of ideals of $A$, whence the closure of their sum is also in $\mathfrak{C}$ by (c). Clearly it is a largest ideal in $A$ that belongs to $\mathfrak{C}$.

(2) If $f : A \to B$ is a *-morphism and $J \in \mathfrak{C}$ is a ideal of $A$ then $f(J)$ is an ideal of $B$ *-isomorphic to a quotient of $J$. Hence $f(J) \in \mathfrak{C}$ is an ideal of $B$ and therefore $f(J) \subset R_{\mathfrak{C}}(B)$.

In particular, $f(R_{\mathfrak{C}}(A)) \subset R_{\mathfrak{C}}(B)$. We proved that the map $A \to R_{\mathfrak{C}}(A)$ is a preradical.

As $R_{\mathfrak{C}}(A) \in \mathfrak{C}$ then $R_{\mathfrak{C}}(R_{\mathfrak{C}}(A)) = R_{\mathfrak{C}}(A)$.

If $I$ is a closed ideal of $A$ then $R_{\mathfrak{C}}(I)$ belongs to $\mathfrak{C}$, whence $R_{\mathfrak{C}}(A) \subset I \cap R_{\mathfrak{C}}(A)$.

Let $N = R_{\mathfrak{C}}(A/R_{\mathfrak{C}}(A))$ and $q = q_{R_{\mathfrak{C}}(A)}$. Setting $J = q^{-1}(N)$, we have that $J$ is a closed ideal of $A$ and $J/R_{\mathfrak{C}}(A) = q(J) = N$ whence $J \in \mathfrak{C}$ by (b). Therefore $J \subset R_{\mathfrak{C}}(A)$ and $N = 0$.

We proved that the map $R_{\mathfrak{C}}$ is a radical on $\mathfrak{U}_{c^*}$; the equality $\mathbf{Rad}(R_{\mathfrak{C}}) = \mathfrak{C}$ is obvious.

(3) If $R_{\mathfrak{C}}$ is hereditary then for each closed ideal $I$ of an algebra $A \in \mathfrak{C}$, we have that $R_{\mathfrak{C}}(I) = I \cap R_{\mathfrak{C}}(A) = I \cap A = I$, whence $I \in \mathfrak{C}$.

Conversely, let $\mathfrak{C}$ contain all closed ideals of all algebras in $\mathfrak{C}$. Since $R_{\mathfrak{C}}(A) \in \mathfrak{C}$ then the ideal $I \cap R_{\mathfrak{C}}(A)$ of $R_{\mathfrak{C}}(A)$ belongs to $\mathfrak{C}$ whence $I \cap R_{\mathfrak{C}}(A) \subset R_{\mathfrak{C}}(I)$. We proved that $R_{\mathfrak{C}}$ is hereditary.

(4) obviously follows from (2). □

The above theorem permits to construct important examples of C*-radicals. Recall that a C*-algebra $A$ is called *a GCR-algebra* (or C*-algebra of type I) if $\pi(A)$ contains a non-zero compact operator for any irreducible representation $\pi$ of $A$. Furthermore $A$ is *nuclear* if, for each C*-algebra $B$, there is only one C*-norm on the algebraic tensor product $A \otimes B$. If $A$ contains an up-directed net of finite-dimensional *-subalgebras with dense union then $A$ is called an *AF-algebra*.

**Corollary 2.23.** *Each C*-algebra $A$ has the largest GCR-ideal $\mathcal{R}_{\mathfrak{gcr}}(A)$, the largest nuclear ideal $\mathcal{R}_{\mathfrak{nc}}(A)$ and the largest AF-ideal $\mathcal{R}_{\mathfrak{af}}(A)$. The maps $\mathcal{R}_{\mathfrak{gcr}}$, $\mathcal{R}_{\mathfrak{nc}}$ and $\mathcal{R}_{\mathfrak{af}}$ are hereditary topological radicals on $\mathfrak{U}_{c^*}$.*

*Proof.* We should check the properties (a)–(d) for the corresponding classes. For GCR-algebras they are well known, as well as the existence of the largest GCR-ideal (see [Dm, Section 4.3]).

Clearly a quotient of an AF-algebra is an AF-algebra. The proof of the fact that a closed ideal of an AF-algebra is an AF-algebra see in [Dv, Lemma III.4.1]. The property (b) for AF-algebras was established by Brown [Br]. The property (c) for AF-algebras is obvious.

For nuclear algebras, the proof of (a), (b) and (d) can be found in [Tk2, Corollary 15.3.4]. The property (c) follows immediately from the fact that an inductive limit of nuclear C*-algebras is nuclear; it was established in the first paper on the subject [Tk1]. □

The radical $\mathcal{R}_{\mathfrak{gcr}}$ is uniform, this follows from [Dm, Proposition 4.3.5]. The radicals $\mathcal{R}_{\mathfrak{nc}}$ and $\mathcal{R}_{\mathfrak{af}}$ are not uniform; the construction of a non-nuclear C*-subalgebra of a nuclear C*-algebra can be found in [Ws, Example 2.4]; for an example of a non-AF C*-subalgebra of AF-algebra one can take an irrational rotation algebra (see [Dv, Theorem 6.5.2]).

## 3. Procedures

We understand procedures in the theory of radicals as transforms of ideal maps. In other words, a *procedure* is a mapping $P \longmapsto P'$ from a class of ideal maps that act on a class $\mathfrak{U}$ of algebras, into another class of ideal maps acting on a possibly different class $\mathfrak{U}'$ of algebras.

### 3.1. The convolution and superposition procedures

Let $A$ be a [normed] algebra. Increasing and decreasing transfinite sequences $(I_\alpha)$ and $(J_\alpha)$, respectively, of [closed] ideals of $A$ are called *transfinite chains* if

$$I_\alpha = \cup_{\alpha'<\alpha} I_{\alpha'} \qquad [I_\alpha = \overline{\cup_{\alpha'<\alpha} I_\alpha}], \tag{3.1}$$
$$J_\alpha = \cap_{\alpha'<\alpha} J_{\alpha'}$$

for every limit ordinal $\alpha$. The quotients $I_{\alpha+1}/I_\alpha$ and $J_\alpha/J_{\alpha+1}$ are called the *gap-quotients* of $(I_\alpha)$ and $(J_\alpha)$, respectively.

Let $P$ and $R$ be preradicals satisfying Axiom 4. For each $A \in \mathfrak{U}$, we define transfinite chains of ideals $(I_\alpha)$ and $(J_\alpha)$ by the conditions

$$I_0 = 0, \quad I_{\alpha+1} = q_{I_\alpha}^{-1}(R(A/I_\alpha)),$$
$$J_0 = A, \quad J_{\alpha+1} = P(J_\alpha),$$

respectively, for every ordinal $\alpha$, where $q_{I_\alpha} : A \longrightarrow A/I_\alpha$ is the standard quotient map.

Let $R_\alpha(A) = I_\alpha$ and $P_\alpha(A) = J_\alpha$ for every $\alpha$. Transfinite sequences $(R_\alpha)$ and $(P_\alpha)$ of ideal maps are called the *convolution* and *superposition chains* for $R$ and $P$, respectively. The transfinite chains of ideals stabilize at some steps $\delta = \delta(A)$ and $\gamma = \gamma(A)$:

$$I_\delta = I_{\delta+1} \text{ and } J_\gamma = J_{\gamma+1}.$$

It was proved in [D3, Theorem 6.6 and Theorem 6.10] that the map $R^*$ defined by

$$R^*(A) := I_\delta$$

satisfies Axiom 2, and the map $P^{\circ}$ defined by

$$P^{\circ}(A) := J_{\gamma}$$

satisfies Axiom 3.

The maps $R \longmapsto R^{*}$ and $P \longmapsto P^{\circ}$ are called the *convolution* and *superposition procedures*, respectively. We underline that the convolution procedure for algebraic preradicals on $\mathfrak{U}_{\mathrm{a}}$ differs from the one for topological preradicals on $\mathfrak{U}_{\mathrm{n}}$ because the latter includes the operation of the closure. So we have two different – *algebraic* and *topological* – convolution procedures.

The following theorem was proved by Dixon [D3, Corollary 6.12] for algebraic (over/under) radicals and for topological (over/under) radicals on normed algebras in assumption that ideals in Axiom 4 are closed, in particular for (over/under) radicals on Banach algebras. We extend the Dixon results to preradicals of normed algebras satisfying Axiom 4 without any restrictions.

**Theorem 3.1.** *Let $R$, $P$ be preradicals satisfying Axiom 4. Then*

(1) *$R^{*}$ is a smallest over radical larger than or equal to $R$.*
 *If $R$ is an under radical then $R^{*}$ is a radical;*

(2) *$P^{\circ}$ is a largest under radical smaller than or equal to $P$.*
 *If $P$ is an over radical then $P^{\circ}$ is a radical.*

*Proof.* Using the transfinite induction, Dixon proved [D3] that if $R$ is an under radical and $P$ is an over radical then all $R_{\alpha}$ and $P_{\alpha}$ are under and over radicals, respectively, under his assumption on ideals. It is easy to see from Dixon's proof that if $R$ and $P$ are preradicals satisfying Axiom 4 restricted to closed ideals then so are $R_{\alpha}$ and $P_{\alpha}$ for every $\alpha$.

Now let $R$ and $P$ satisfy Axiom 4. We have to check Axiom 4 for all maps from $(R_{\alpha})$ and $(P_{\alpha})$; then $R^{*}$ and $P^{\circ}$ will also satisfy Axiom 4. For other assertions of this theorem we refer to [D3].

Let $A$ be a normed algebra, and let $K$ be an ideal of $A$. Using the transfinite induction, we fix an ordinal $\alpha$ and assume that $(R_{\alpha'})_{\alpha'<\alpha}$ and $(P_{\alpha'})_{\alpha'<\alpha}$ consist of preradicals satisfying Axiom 4.

If $\alpha$ is a limit ordinal then all $R_{\alpha'}(K) \subset R_{\alpha'}(A)$ and $P_{\alpha'}(K) \subset P_{\alpha'}(A)$ are ideals of $A$ for $\alpha' < \alpha$, whence

$$\begin{aligned} R_{\alpha}(K) &= \overline{\cup_{\alpha'<\alpha} R_{\alpha'}(K)}^{(K)} \subset \overline{\cup_{\alpha'<\alpha} R_{\alpha'}(A)}^{(A)} = R_{\alpha}(A), \\ P_{\alpha}(K) &= \cap_{\alpha'<\alpha} P_{\alpha'}(K) \subset \cap_{\alpha'<\alpha} P_{\alpha'}(A) = P_{\alpha}(A) \end{aligned} \tag{3.2}$$

are ideals of $A$. Indeed, to check this for $R_{\alpha}(K)$ note that $I := \cup_{\alpha'<\alpha} R_{\alpha'}(K)$ is an ideal of $A$, and for every $x \in R_{\alpha}(K)$ there is a sequence $(y_n) \subset I$ such that $y_n \to x$ as $n \to \infty$. For every $a \in A$, one has $ay_n, y_n a \in I$ for each $n$, whence

$$ax, xa \in K \cap \overline{I}^{(A)} = \overline{I}^{(K)} = R_{\alpha}(K).$$

Therefore $R_{\alpha}(K)$ is an ideal of $A$; the same assertion for $P_{\alpha}$ is trivial.

Let now $\alpha = \alpha' + 1$, for some ordinal $\alpha'$. Since $P_{\alpha'}(K) \subset P_{\alpha'}(A)$ is an ideal of $A$ then $P_\alpha(K) = P(P_{\alpha'}(K))$ is an ideal of $A$. As $P_{\alpha'}(K)$ is an ideal of $P_{\alpha'}(A)$, we obtain that

$$P_\alpha(K) = P(P_{\alpha'}(K)) \subset P(P_{\alpha'}(A)) = P_\alpha(A).$$

By the hypothesis of induction (valid for any $A$), $J := R_{\alpha'}(K) \subset R_{\alpha'}(A)$ is an ideal of $A$. Then, as $K$ is an ideal of $\overline{K}$,

$$J \subset R_{\alpha'}\left(\overline{K}\right) \tag{3.3}$$

is an ideal of $\overline{K}$. Since $J = K \cap \overline{J}$, the map $x/J \longmapsto q_{\overline{J}}(x)$ is an injective map from $K/J$ onto the ideal $q_{\overline{J}}(K)$ of $A/\overline{J}$. So

$$R_\alpha(K) = \{x \in K : q_{\overline{J}}(x) \in R(q_{\overline{J}}(K))\} \tag{3.4}$$

whence $R_\alpha(K)$ is an ideal of $A$. As $q_{\overline{J}}(K)$ is an ideal of $\overline{K}/\overline{J}$, we obtain that

$$R(q_{\overline{J}}(K)) \subset R\left(\overline{K}/\overline{J}\right). \tag{3.5}$$

Since $\overline{J} \subset R_{\alpha'}\left(\overline{K}\right)$ is an ideal of $\overline{K}$ by (3.3), it follows that

$$q\left(R\left(\overline{K}/\overline{J}\right)\right) \subset R\left(\overline{K}/R_{\alpha'}\left(\overline{K}\right)\right) = R_\alpha\left(\overline{K}\right)/R_{\alpha'}\left(\overline{K}\right). \tag{3.6}$$

for the natural map $q : \overline{K}/\overline{J} \longrightarrow \overline{K}/R_{\alpha'}\left(\overline{K}\right)$. It follows from (3.4), (3.5) and (3.6) that $R_\alpha(K) \subset R_\alpha\left(\overline{K}\right)$. As $\overline{K}$ is a closed ideal of $A$ then $R_\alpha\left(\overline{K}\right) \subset R_\alpha(A)$ by [D3, Theorem 6.6]. Therefore $R_\alpha(K) \subset R_\alpha(A)$. This completes the proof. □

We outline a part of the proof of Theorem 3.1 in the form of a separate statement.

**Proposition 3.2.** *Let $R, R_1, P, P_1$ be preradicals satisfying Axiom 4. Then*

(1) *If $R_2$ is defined on $\mathfrak{U}$ by $R_2(A) = q_{R_1(A)}^{-1}(R(A/R_1(A)))$ then $R_2$ is a preradical satisfying Axiom 4;*
   (a) *If $R, R_1$ are under radicals then $R_2$ is an under radical;*

(2) *If $P_2$ is defined on $\mathfrak{U}$ by $P_2(A) = P(P_1(A))$ then $P_2$ is a preradical satisfying Axiom 4;*
   (a) *If $P, P_1$ are over radicals then $P_2$ is an over radical.*

*Proof.* This is the step $\alpha \mapsto \alpha + 1$ of the transfinite induction applied in Theorem 3.1. □

**Theorem 3.3.** *Let $R$ be an under radical, and let $P$ be an over radical. Then*

(1) $\mathbf{Sem}(R^*) = \mathbf{Sem}(R)$;
(2) $\mathbf{Rad}(P^\circ) = \mathbf{Rad}(P)$.

*Proof.* Let $A$ be an algebra, and let $(R_\alpha)$ and $(P_\alpha)$ be the convolution and superposition chains of $R$ and $P$, respectively.

(1) If $R(A) = 0$ then $R_1(A) = R(A) = 0$, whence $R^*(A) = 0$.

(2) If $A = P(A)$ then $P_1(A) = P(A) = A$, whence $P^\circ(A) = A$. □

**Theorem 3.4.** *Let $R, R_1$ be under radicals, and let $P, P_1$ be over radicals. Then*

(1) *If $R_1 \leq R$ then $R_1^* \leq R^*$;*
(2) *If $P_1 \leq P$ then $P_1^\circ \leq P^\circ$;*
(3) *If $R \leq P$ then $R^* \leq P^\circ$;*
(4) *If $P \leq R$ then $P^\circ \leq R^*$.*

*Proof.* (1) $R^*$ is a radical and $R_1 \leq R \leq R^*$. As $R_1^*$ is the smallest radical among radicals $R'$ such that $R_1 \leq R'$ then $R_1^* \leq R^*$.

(2) is similar to (1): $P_1^\circ$ is a radical and $P_1^\circ \leq P_1 \leq P$. As $P^\circ$ is the largest radical among radicals $P'$ such that $P' \leq P$ then $P_1^\circ \leq P^\circ$.

(3) This is [D3, Theorem 6.11], but the following simple observation somewhat shortens the proof: as $R \leq P$, we have that

$$\mathbf{Sem}\,(P) \subset \mathbf{Sem}\,(R) = \mathbf{Sem}\,(R^*)\,.$$

Since $R^*$ is an over radical then $R^* \leq P$ by Theorem 2.3. As $R^*$ is a radical and $P^\circ$ is a largest radical smaller than or equal to $P$ then $R^* \leq P^\circ$.

(4) is trivial. □

In other words, (1) and (2) of Proposition 3.4 state that the convolution and superposition procedures are isotone.

### 3.2. The closure procedure

The *closure procedure* $P \longmapsto \overline{P}$ makes a preradical $P$ on a class $\mathfrak{U}$ (containing $\mathfrak{U}_\mathrm{n}$) into a topological preradical $\overline{P}$ on $\mathfrak{U}_\mathrm{n}$ by the rule

$$\overline{P}\,(A) = \overline{P\,(A)}$$

for every $A \in \mathfrak{U}_\mathrm{n}$; $\overline{P}$ is called the *closure* of $P$. Dixon pointed out in [D3, Example 6.4] that if $P$ is a radical then the closure $\overline{P}$ may be not a topological radical, but always is a topological under radical. Recall that the restrictions of algebraic radicals from $\mathfrak{U}_\mathrm{a}$ to a class of normed algebras are algebraic under radicals. One can state Dixon's result [D3, Theorem 6.3] in the following stronger form.

**Theorem 3.5.** *The closure of the restriction of an algebraic under radical to a class of normed algebras is a strict topological under radical.*

*Proof.* We only prove the strictness for $\overline{P}$. Let $f : A \longrightarrow B$ be a continuous isomorphism of normed algebras. Then $f\,(P\,(A)) = P\,(B)$ by definition whence $f\left(\overline{P\,(A)}\right)$ is dense in $\overline{P\,(B)}$. Therefore $\overline{f\,(\overline{P}\,(A))} = \overline{P}\,(B)$. □

Starting with a (restricted) algebraic radical $P$, one can take its closure and then apply the topological convolution procedure. The following theorem shows that the result will not change if we do the same with the algebraic convolution of $P$.

**Theorem 3.6.** *Let $P$ be an algebraic under radical. Then $\overline{P}^* = \overline{P^*}^*$ and $\overline{P}^*$ is a strong topological radical.*

*Proof.* By Theorem 3.3, $\mathbf{Sem}(P) = \mathbf{Sem}(P^*)$. Then

$$\mathbf{Sem}\left(\overline{P}^*\right) = \mathbf{Sem}\left(\overline{P}\right) = \mathfrak{U}_{\mathrm{n}} \cap \mathbf{Sem}(P) = \mathfrak{U}_{\mathrm{n}} \cap \mathbf{Sem}(P^*)$$
$$= \mathbf{Sem}\left(\overline{P^*}\right) = \mathbf{Sem}\left(\overline{P^*}^*\right).$$

As $\overline{P}^*$ and $\overline{P^*}^*$ are radicals, $\overline{P}^* = \overline{P^*}^*$ by Corollary 2.4.

Let $f : A \longrightarrow B$ be a continuous surjective homomorphism of normed algebras, and let $(P_\alpha)$ be the convolution series of $\overline{P}$. As $P_1 = \overline{P}$ is strict by Theorem 3.5, it is strong. Let us prove by transfinite induction that every $P_\alpha$ is strong.

Let $\alpha$ be a limit ordinal, and let all $P_{\alpha'}$ be strong for $\alpha' < \alpha$. Then

$$f(P_\alpha(A)) = f\left(\overline{\cup_{\alpha'<\alpha} P_{\alpha'}(A)}\right) \subset \overline{f(\cup_{\alpha'<\alpha} P_{\alpha'}(A))}$$
$$= \overline{\cup_{\alpha'<\alpha} f(P_{\alpha'}(A))} \subset \overline{\cup_{\alpha'<\alpha} P_{\alpha'}(B)} = P_\alpha(B).$$

The step $\alpha \mapsto \alpha + 1$ of the induction is reduced to the consideration of $\overline{P}$: it suffices to check that

$$f\left(q_I^{-1}\left(\overline{P}(A/I)\right)\right) \subset q_J^{-1}\left(\overline{P}(B/J)\right) \tag{3.7}$$

whenever $f(I) \subset J$ where $I = P_\alpha(A)$ and $J = P_\alpha(B)$. As $\overline{P}$ is strong, (3.7) is easily checked. Indeed, $f$ induces a continuous homomorphism $f'$ from $A/I$ onto $B/J$ such that $f' \circ q_I = q_J \circ f$; then

$$f\left(q_I^{-1}\left(\overline{P}(A/I)\right)\right) \subset q_J^{-1}\left(q_J\left(f\left(q_I^{-1}\left(\overline{P}(A/I)\right)\right)\right)\right)$$
$$= q_J^{-1}\left(f'\left(\overline{P}(A/I)\right)\right) \subset q_J^{-1}\left(\overline{P}(B/J)\right).$$

As a consequence, $\overline{P}^*$ is strong. □

We apply the above theorem to classical algebraic radicals related to the property of nilpotency. An algebra $A$ is called *nilpotent* if there is an integer $n > 0$ such that $a_1 \cdots a_n = 0$ for every $a_1, \ldots, a_n \in A$, *locally nilpotent* if every finite subset of $A$ generates a nilpotent subalgebra, *nil* if every element of $A$ is nilpotent. All these notions transfer to ideals.

An algebra $A$ is called *semiprime* if it has no non-zero ideals with zero square. Each semiprime algebra has no non-zero one-sided ideals with zero square [BD, Lemma 30.4].

One can define the *Baer* (*prime* or *lower nil-*) *radical* $\mathfrak{P}_\beta(A)$ of $A$ as the smallest ideal $I$ of $A$ with semiprime quotient $A/I$; the *Levitzki* (or *locally nilpotent*) and *Köte* (or *upper nil-*) *radicals* $\mathfrak{P}_\lambda(A)$ and $\mathfrak{P}_\kappa(A)$ are defined as the largest locally nilpotent and nil ideals of $A$, respectively. They all are hereditary radicals and

$$\mathfrak{P}_\beta < \mathfrak{P}_\lambda < \mathfrak{P}_\kappa. \tag{3.8}$$

Let $\mathcal{P}_\beta = \overline{\mathfrak{P}_\beta}^*$, $\mathcal{P}_\lambda = \overline{\mathfrak{P}_\lambda}^*$ and $\mathcal{P}_\kappa = \overline{\mathfrak{P}_\kappa}^*$. They are strong topological radicals by Theorems 3.5 and 3.1, and are called the *closed-Baer*, *closed-Levitzki* and *closed-Köte radical*, respectively.

Let $A$ be an algebra in $\mathfrak{U}_a$ and let $\Sigma_\beta(A)$ be the sum of all nilpotent ideals of $A$. It is clear that $\Sigma_\beta$ is an algebraic under radical on $\mathfrak{U}_\mathrm{a}$; by construction, the (algebraic) Baer radical $\mathfrak{P}_\beta$ is equal to $\Sigma_\beta^*$.

Theorem 3.6 yields

**Corollary 3.7.** $\mathcal{P}_\beta = \overline{\Sigma_\beta}^*$.

**Theorem 3.8.** *$\mathcal{P}_\beta < \mathcal{P}_\lambda < \mathcal{P}_\kappa$ on $\mathfrak{U}_\mathrm{n}$ and $\mathcal{P}_\beta = \mathcal{P}_\lambda = \mathcal{P}_\kappa$ on $\mathfrak{U}_\mathrm{b}$.*

*Proof.* As all procedures are isotone, the non-strict inequalities follows from (3.8). Also, $\mathcal{P}_\beta < \mathcal{P}_\lambda$ follows from [D3, Proposition 3.3 and Corollary 9.4], and for the proof of $\mathcal{P}_\lambda < \mathcal{P}_\kappa$ it is sufficient to point out a normed nil algebra with zero Levitzki radical; it was done in [T1] basing on Golod's solution of the Bernside problem.

By [D1],

$$\mathfrak{P}_\beta = \mathfrak{P}_\lambda = \mathfrak{P}_\kappa = \Sigma_\beta$$

on $\mathfrak{U}_\mathrm{b}$. Therefore $\mathcal{P}_\beta = \mathcal{P}_\lambda = \mathcal{P}_\kappa$ on $\mathfrak{U}_\mathrm{b}$. □

The common radical on $\mathfrak{U}_\mathrm{b}$ is denoted by $\mathcal{P}_{\mathrm{nil}}$ and called the *closed-nil radical*. One can consider $\mathcal{P}_\beta$, $\mathcal{P}_\lambda$, $\mathcal{P}_\kappa$ as topological extensions of $\mathcal{P}_{\mathrm{nil}}$ to normed algebras.

Let $A$ be an algebra, and let $\Sigma_{\mathrm{hf}}(A)$ be the sum of all bifinite ideals of $A$. If $A$ is normed, let $\Sigma_{\mathrm{hc}}(A)$ be the closure of sum of all bicompact ideals of $A$. It is easy to see that $\Sigma_{\mathrm{hf}}$ and $\Sigma_{\mathrm{hc}}$ are under radicals.

In the following theorem we apply the typical method of comparing radicals.

**Theorem 3.9.** *$\mathfrak{R}_{\mathrm{hf}} = \Sigma_{\mathrm{hf}}^*$ and $\mathcal{R}_{\mathrm{hf}} = \overline{\mathfrak{R}_{\mathrm{hf}}}^* = \overline{\Sigma_{\mathrm{hf}}}^* \le \mathcal{R}_{\mathrm{hc}} = \Sigma_{\mathrm{hc}}^*$.*

*Proof.* Let $B$ be an algebra. As every bifinite ideal is hypofinite, $\Sigma_{\mathrm{hf}}^*(B) \subset \mathfrak{R}_{\mathrm{hf}}$. Since $\mathfrak{R}_{\mathrm{hf}}$ is a radical, $\Sigma_{\mathrm{hf}}^* \le \mathfrak{R}_{\mathrm{hf}}$. As every $\Sigma_{\mathrm{hf}}^*$-semisimple algebra is $\mathfrak{R}_{\mathrm{hf}}$-semisimple, $\mathfrak{R}_{\mathrm{hf}} \le \Sigma_{\mathrm{hf}}^*$. One can show similarly that $\mathcal{R}_{\mathrm{hc}} = \Sigma_{\mathrm{hc}}^*$.

Let $A$ be a normed algebra and $J = \mathfrak{R}_{\mathrm{hf}}(A)$. Let $(J_\alpha)_{\alpha\le\gamma}$ be the increasing transfinite chain of ideals of $A$ constructed in Theorem 2.20(5) with $J_0 = 0$ and $J_\gamma = J$. As each $J_{\alpha+1}/J_\alpha$ is bifinite, $\overline{J_{\alpha+1}}/\overline{J_\alpha}$ is approximable, so $\overline{J}$ is closed-hypofinite (see [ST6, Proposition 3.48]). As $\mathcal{R}_{\mathrm{hf}}(A)$ is the largest closed-hypofinite ideal of $A$ then $\overline{J} \subset \mathcal{R}_{\mathrm{hf}}(A)$. Therefore $\overline{\mathfrak{R}_{\mathrm{hf}}} \le \mathcal{R}_{\mathrm{hf}}$, whence $\overline{\mathfrak{R}_{\mathrm{hf}}}^* \le \mathcal{R}_{\mathrm{hf}}$. But every $\overline{\mathfrak{R}_{\mathrm{hf}}}^*$-semisimple algebra has no non-zero finite rank elements and is therefore $\mathcal{R}_{\mathrm{hf}}$-semisimple, whence $\mathcal{R}_{\mathrm{hf}} \le \overline{\mathfrak{R}_{\mathrm{hf}}}^*$ by Theorem 2.3.

By Theorem 3.6, $\mathcal{R}_{\mathrm{hf}} = \overline{\Sigma_{\mathrm{hf}}}^*$. As $\overline{\Sigma_{\mathrm{hf}}} \le \Sigma_{\mathrm{hc}}$ then $\mathcal{R}_{\mathrm{hf}} = \overline{\Sigma_{\mathrm{hf}}}^* \le \Sigma_{\mathrm{hc}}^* = \mathcal{R}_{\mathrm{hc}}$ by Theorem 3.4. □

### 3.3. The regularization procedure

Let $P$ be a hereditary topological radical defined on some class $\mathfrak{U}$ of normed algebras such that $\mathfrak{U}_{\mathrm{b}} \subset \mathfrak{U}$. By [ST5, Theorem 2.21], the map $P^r$ defined by

$$P^r(A) = A \cap P\left(\widehat{A}\right) \tag{3.9}$$

for each $A \in \mathfrak{U}_{\mathrm{n}}$, where $\widehat{A}$ is a completion of $A$, is a hereditary topological radical on $\mathfrak{U}_{\mathrm{n}}$. The map $P \longmapsto P^r$ is called the *regularization procedure*; it is clear that

$$P^{rr} = P^r.$$

A hereditary topological radical $P$ on $\mathfrak{U}_{\mathrm{n}}$ is called *regular* if $P = P^r$. One of the important reasons for the consideration of over radicals is the fact [ST5] that preradicals obtained from topological radicals by the regular procedure are over radicals.

**Theorem 3.10.** *The radicals $\mathcal{R}_{\mathrm{cq}}$ and $\mathcal{R}_t$ are regular.*

*Proof.* Let $A$ be a normed algebra. By definition, $A \cap \mathcal{R}_{\mathrm{cq}}\left(\widehat{A}\right) \subset \mathcal{R}_{\mathrm{cq}}(A)$. On the other hand, the completion of a compactly quasinilpotent algebra is again compactly quasinilpotent [ST5, Lemma 4.11]. So $\overline{\mathcal{R}_{\mathrm{cq}}(A)}^{(\widehat{A})}$ is compactly quasinilpotent, and as it is an ideal of $\widehat{A}$ then $\overline{\mathcal{R}_{\mathrm{cq}}(A)}^{(\widehat{A})} \subset \mathcal{R}_{\mathrm{cq}}\left(\widehat{A}\right)$. But $A \cap \overline{\mathcal{R}_{\mathrm{cq}}(A)}^{(\widehat{A})} = \mathcal{R}_{\mathrm{cq}}(A)$ because $\mathcal{R}_{\mathrm{cq}}(A)$ is a closed ideal of $A$. Thus $\mathcal{R}_{\mathrm{cq}}$ is regular.

The proof for $\mathcal{R}_t$ is similar. For the other proof we refer to Proposition 5.17. ☐

**Theorem 3.11.** *The radicals $\mathcal{R}_\beta$ and $\mathcal{R}_{\mathrm{hf}}$ are not regular.*

*Proof.* Let $B$ be the Banach space $L^1[0,1]$ considered as a Banach algebra with multiplication $(f*g)(t) = \int_0^t f(s)g(t-s)ds$. It can be shown that nilpotent elements are dense in $B$ (and therefore all elements of $B$ are quasinilpotent since $B$ is commutative): indeed, it follows from the Titchmarsh convolution theorem that a function $f \in B$ is nilpotent if and only if it vanishes a.e. on some interval $(0, a)$. Therefore $\mathcal{R}_\beta(B) = B$. Moreover, $f$ is an element of finite rank if it is nilpotent of index 2 (equivalently if $f(t) = 0$ for $t \in (0, 1/2)$), so $\mathcal{R}_{\mathrm{hf}}(B) \neq 0$. In fact, $\mathcal{R}_{\mathrm{hf}}(B) = B$.

Let $A$ be the subalgebra of $B$ that consists of polynomials. It is dense in $B$ so $B$ can be considered as the completion of $A$. Clearly $A$ has no nilpotent elements (a non-zero polynomial cannot vanish on an interval), so $\mathcal{R}_\beta(A) = 0$. As a quasinilpotent finite rank element is nilpotent, $A$ has no finite rank elements and $\mathcal{R}_{\mathrm{hf}}(A) = 0$. ☐

**Problem 3.12.** *Is the hypocompact radical $\mathcal{R}_{\mathrm{hc}}$ regular?*

It is clear that the regular Jacobson radical $\mathrm{Rad}^r$ defined by (2.4) is a regular radical. Note that

$$\mathcal{R}_{\mathrm{cq}} \le \mathcal{R}_t \le \mathrm{Rad}^r < \mathrm{rad}_{\mathrm{n}} < \mathrm{rad}_{\mathrm{b}}; \tag{3.10}$$

also $\mathrm{rad} = \mathrm{rad}_{\mathrm{n}}$ on $\mathfrak{U}_{\mathrm{q}}$ and $\mathrm{rad} = \mathrm{rad}_{\mathrm{b}}$ on $\mathfrak{U}_{\mathrm{q_b}}$ [ST5, Theorem 2.18].

Consider the *closed-Jacobson radical* $\overline{\mathrm{rad}}^*$; then $\overline{\mathrm{rad}}^* = \mathrm{Rad}$ on $\mathfrak{U}_{\mathrm{b}}$. Define also the map $\Pi_{\mathrm{pc}}$ on normed algebras by

$$\Pi_{\mathrm{pc}}(A) = \cap \left\{ I \in \mathrm{Prim}\left(A^1\right) : I \text{ is closed in } A \right\}. \tag{3.11}$$

**Remark 3.13.** The map $A \longmapsto \left\{ I \in \mathrm{Prim}\left(A^1\right) : I \text{ is closed in } A \right\}$ is not a primitive map because, for an ideal $J$ of a normed algebra $A$, there can exist a non-closed primitive ideal $I$ of $A$ such that $I \cap J$ is a closed primitive ideal of $J$.

**Lemma 3.14.** *$\Pi_{\mathrm{pc}}$ is a topological over radical on normed algebras, and* $\overline{\mathrm{rad}} \leq \Pi_{\mathrm{pc}}$.

*Proof.* Let $A$ be a normed algebra, and let $I$ be an ideal of $A$. If $J \in \mathrm{Prim}\left(A^1\right)$ is closed then $I \cap J \in \mathrm{Prim}\left(I^1\right)$ is closed in $I$, whence $\Pi_{\mathrm{pc}}(I) \subset \Pi_{\mathrm{pc}}(A)$ is an ideal of $A$ by (1.9).

Let $B$ be a normed algebra, and let $f : A \longrightarrow B$ be a morphism. Let $\pi \in \mathrm{Irr}\left(B^1\right)$ and $K := \ker \pi$ is closed. Then $\pi \circ f \in \mathrm{Irr}\left(A^1\right)$ and $\ker(\pi \circ f) = f^{-1}(K)$ is closed.

Let $\mathcal{F}_A$ be the set of all closed primitive ideals of $A^1$. Assume that $I \subset \Pi_{\mathrm{pc}}(A)$ is closed. Then $I \subset J$ for every $J \in \mathcal{F}_A$ and

$$\begin{aligned} \Pi_{\mathrm{pc}}(A)/I &= \cap \{J : J \in \mathcal{F}_A\}/I = \cap \{J/I : J \in \mathcal{F}_A\} \\ &= \cap \left\{J/I : J/I \in \mathcal{F}_{A/I}\right\} = \Pi_{\mathrm{pc}}(A/I) \end{aligned}$$

by (1.9) and (1.10). Hence $\Pi_{\mathrm{pc}}$ is an over radical.

Furthermore, $\overline{\mathrm{rad}}(A) = \overline{\cap_{I \in \mathrm{Prim}(A^1)} I} \subset \Pi_{\mathrm{pc}}(A)$ in virtue of (3.11). □

As a consequence, $\Pi_{\mathrm{pc}}^{\circ}$ is a topological radical which also extends Rad to normed algebras; it is called the *primitively closed Jacobson radical.*

**Theorem 3.15.** $\overline{\mathrm{rad}}^* < \Pi_{\mathrm{pc}}^{\circ} \leq \mathrm{rad}_{\mathrm{n}}$*;* $\overline{\mathrm{rad}}^*$ *and* $\mathrm{Rad}^r$ *are not comparable.*

*Proof.* Indeed, it is clear that $\overline{\mathrm{rad}}(A) \subset \Pi_{\mathrm{pc}}(A) \subset \mathrm{rad}_{\mathrm{n}}(A)$ for every normed algebra $A$. Then $\overline{\mathrm{rad}}^* \leq \Pi_{\mathrm{pc}}^{\circ} \leq \mathrm{rad}_{\mathrm{n}}$ by Theorems 2.3 and 3.4.

Let $a$ be a quasinilpotent element of a Banach algebra and $a^n \neq 0$ for every $n > 0$. Let $A$ be the algebra generated by $a$, and $B = \overline{A}$. Then $\mathrm{rad}(A) = 0$, but $\mathrm{rad}(B) = B$. Hence $\overline{\mathrm{rad}}^*(A) = \overline{\mathrm{rad}}(A) = 0$, but $\mathrm{Rad}^r(A) = A \cap \mathrm{rad}(B) = A$.

Also, $A$ has no closed primitive ideals in $\mathrm{Prim}\left(A^1\right)$ besides of $A$. Indeed, a primitive ideal $I$ is the kernel of multiplicative functional $f$ on $A^1$. If $I \neq A$ is closed then $f(a) \neq 0$, $f$ is continuous and $f(a)$ in the spectrum $\sigma_B(a)$, a contradiction. Therefore $\Pi_{\mathrm{pc}}^{\circ}(A) = \Pi_{\mathrm{pc}}(A) = A$ and $\overline{\mathrm{rad}}^* < \Pi_{\mathrm{pc}}^{\circ}$.

By [D3, Example 9.3], there are Banach algebras $C$, $D$ and an injective embedding $\phi : C \longrightarrow D$ with the dense image such that $\mathrm{rad}(C) = C$ and $\mathrm{rad}(D) = 0$. Let $E = \phi(C)$. As $\phi$ and $\phi^{-1}$ are algebraic morphisms for appropriate algebras then $\mathrm{rad}(E) = E$ and $\overline{\mathrm{rad}}^*(E) = E$, but $\mathrm{Rad}^r(E) = E \cap \mathrm{rad}(D) = 0$. So the observations with algebras $A$ and $E$ show that $\overline{\mathrm{rad}}^*$and $\mathrm{Rad}^r$ are not comparable. □

As we will see in Theorem 4.31, the closed-Jacobson radical $\overline{\mathrm{rad}}^*$ is hereditary.

## 4. Operations with radicals

Operations are multiplace procedures that act on a class of ideal maps.

### 4.1. Supremum and infimum

Let $\mathcal{F}$ be a non-empty family (set, class) of radicals. Then the maps $\mathrm{H}_{\mathcal{F}}$ and $\mathrm{B}_{\mathcal{F}}$ defined by

$$\mathrm{H}_{\mathcal{F}}(A) = \sum_{P \in \mathcal{F}} P(A) \qquad \left[\mathrm{H}_{\mathcal{F}}(A) = \overline{\sum_{P \in \mathcal{F}} P(A)}\right]$$

$$\mathrm{B}_{\mathcal{F}}(A) = \cap_{P \in \mathcal{F}} P(A)$$

are the under and over radicals, respectively (see [ST7, Lemma 3.2]). We extend the statement to under/over radicals as follows.

**Theorem 4.1.** (1) *If $\mathcal{F}_u$ is a family of under radicals then $\mathrm{H}_{\mathcal{F}_u}$ is an under radical, and*

$$\mathbf{Sem}(\mathrm{H}_{\mathcal{F}_u}) = \cap_{P \in \mathcal{F}_u} \mathbf{Sem}(P).$$

(2) *If $\mathcal{F}_o$ is a family of over radicals then $\mathrm{B}_{\mathcal{F}_o}$ is an over radical, and*

$$\mathbf{Rad}(\mathrm{B}_{\mathcal{F}_o}) = \cap_{P \in \mathcal{F}_o} \mathbf{Rad}(P).$$

*Proof.* The equalities for radical and semisimple classes follow from the definitions of $\mathrm{H}_{\mathcal{F}_u}$ and $\mathrm{B}_{\mathcal{F}_o}$. It is clear that $\mathrm{H}_{\mathcal{F}_u}$ and $\mathrm{B}_{\mathcal{F}_o}$ satisfy Axioms 1 and 4.

Let $A$ be an algebra. If $P \in \mathcal{F}_u$ then $P(A)$ is an ideal in $\mathrm{H}_{\mathcal{F}_u}(A)$ and therefore

$$P(A) = P(P(A)) \subset P(\mathrm{H}_{\mathcal{F}_u}(A)) \subset \mathrm{H}_{\mathcal{F}_u}(\mathrm{H}_{\mathcal{F}_u}(A))$$

whence $\mathrm{H}_{\mathcal{F}_u}(A) \subset \mathrm{H}_{\mathcal{F}_u}(\mathrm{H}_{\mathcal{F}_u}(A))$. This proves that $\mathrm{H}_{\mathcal{F}_u}$ is an under radical.

Let $I = \mathrm{B}_{\mathcal{F}_o}(A)$ and $J = q_I^{-1}(\mathrm{B}_{\mathcal{F}_o}(A/I))$ where $q_I : A \longrightarrow A/I$ is the standard quotient map. Let $R \in \mathcal{F}_o$; then $R(A) \subset I$ and there is an [open continuous] epimorphism $p : A/I \longrightarrow A/R(A)$ such that $q = p \circ q_I$ where $q : A \longrightarrow A/R(A)$ is the standard quotient map. Then

$$q(J) = (p \circ q_I)(J) = p(\mathrm{B}_{\mathcal{F}_o}(A/I)) \subset p(R(A/I)) \subset R(A/R(A)) = 0,$$

whence $J \subset R(A)$ for every $R \in \mathcal{F}_o$. Then $J \subset I$ and $\mathrm{B}_{\mathcal{F}_o}(A/I) = q_I(J) = 0$. This means that $\mathrm{B}_{\mathcal{F}_o}$ is an over radical. □

From the viewpoint of the order, $\mathrm{H}_{\mathcal{F}_u}$ is the supremum of $\mathcal{F}_u$ in the class of under radicals and $\mathrm{B}_{\mathcal{F}_o}$ is the infimum of $\mathcal{F}_o$ in the class of over radicals.

In general, for a family $\mathcal{F}$ of preradicals satisfying Axiom 4, let us define the ideal maps $\vee\mathcal{F}$ and $\wedge\mathcal{F}$ by

$$\vee\mathcal{F} = \mathrm{H}_{\mathcal{F}}^{*} \quad \text{and} \quad \wedge\mathcal{F} = \mathrm{B}_{\mathcal{F}}^{\circ},$$

respectively; if $\mathcal{F}$ is $\{P_1, P_2\}$ then instead we write $P_1 \vee P_2$ or $P_1 \wedge P_2$, respectively.

**Remark 4.2.** It is easy to check (arguing as in Theorems 4.1 and 3.1) that, for a family $\mathcal{F}$ of preradicals satisfying Axiom 4, $\vee\mathcal{F}$ determines the *smallest over radical* larger than or equal to each $R \in \mathcal{F}$ and $\wedge\mathcal{F}$ determines the largest *under radical* smaller than or equal to each $P \in \mathcal{F}$.

**Corollary 4.3.** *The operations* $\vee$ *and* $\wedge$ *produce supremum and infimum of a family of radicals in the class of radicals.*

*Proof.* Let $\mathcal{F}$ be a family of radicals. Let $P$ be a radical such that $R \leq P$ for every $R \in \mathcal{F}$. Then $\mathrm{H}_{\mathcal{F}} \leq P$, whence $\vee\mathcal{F} \leq P$ by Theorem 4.1 and Proposition 3.4. As $R \leq \vee\mathcal{F}$ for every $R \in \mathcal{F}$, we conclude that $\vee\mathcal{F}$ is the supremum of $\mathcal{F}$.

Similarly, $\wedge\, \mathcal{F}$ is the infimum of $\mathcal{F}$. □

Taking into account Theorems 2.3 and 3.3, one can reformulate Theorem 3.1 as follows.

**Theorem 4.4.** *Let* $R$ *be an under radical, and let* $P$ *be an over radical. Then*

(1) $R^*$ *is the only radical* $S$ *such that* $\mathbf{Sem}\,(S) = \mathbf{Sem}\,(R)$;
(2) $P^\circ$ *is the only radical* $T$ *such that* $\mathbf{Rad}\,(T) = \mathbf{Rad}\,(P)$.

In assumptions of Theorem 4.4, we also recall that

$$\mathbf{Rad}\,(R) \subset \mathbf{Rad}\,(R^*) \text{ and } \mathbf{Sem}\,(P) \subset \mathbf{Sem}\,(P^\circ)\,.$$

**Corollary 4.5.** *Let* $V$ *be a class of algebras. Then there exist the largest radical* $S$ *such that* $\mathbf{Sem}\,(S) \supset V$, *and the smallest radical* $T$ *such that* $\mathbf{Rad}\,(T) \supset V$.

(1) $V = \mathbf{Sem}\,(S)$ *if and only if there is a family* $\mathcal{F}_1$ *of under radicals such that* $\mathbf{Sem}\,(\mathrm{H}_{\mathcal{F}_1}) = V$;
(2) $V = \mathbf{Rad}\,(T)$ *if and only if there is a family* $\mathcal{F}_2$ *of over radicals such that* $\mathbf{Rad}\,(\mathrm{B}_{\mathcal{F}_2}) = V$.

*Proof.* Let $\mathcal{F}_1$ be the family of all under radicals $P$ such that $\mathbf{Sem}\,(P) \supset V$, and let $\mathcal{F}_2$ be the family of all over radicals $P$ such that $\mathbf{Rad}\,(P) \supset V$. By Theorems 4.4 and 4.1, there exist radicals $S$ and $T$ such that $\mathbf{Sem}\,(S) = \cap_{P\in\mathcal{F}_1}\mathbf{Sem}\,(P) \supset V$ and $\mathbf{Rad}\,(T) = \cap_{P\in\mathcal{F}_2}\mathbf{Rad}\,(P) \supset V$. □

For instance, let $W_\rho$ be the class of all normed algebras on which the usual spectral radius $a \longmapsto \rho(a)$ is continuous; this class contains all commutative normed algebras and Banach algebras whose elements have at most countable spectrum. Then there exists a largest topological radical $S_\rho$ such that $W_\rho \subset \mathbf{Sem}\,(S_\rho)$, but it is a problem whether there is a largest topological radical $T$ for which every $T$-radical normed algebra has the property of spectral radius continuity, i.e., lies in $W_\rho$.

Let $V$ be a class of algebras. For being a semisimple or radical class it is necessary that $V$ is stable under extensions.

**Theorem 4.6.** *Let* $A$ *be an algebra, and let* $I$ *be a* [*closed*] *ideal of* $A$. *Then*

(1) *If* $P$ *is an under radical and* $I$, $A/I$ *are* $P$*-semisimple then* $A$ *is* $P$*-semisimple;*
(2) *If* $P$ *is an over radical and* $I$, $A/I$ *are* $P$*-radical then* $A$ *is* $P$*-radical.*

*Proof.* (1) As $q_I(P(A)) \subset P(A/I) = 0$ then $P(A) \subset I$, whence $P(A)$ is an ideal of $I$ and $P(A) = P(P(A)) \subset P(I) = 0$.

(2) As $I \subset P(A)$ then there is a morphism $p : A/I \longrightarrow A/P(A)$ such that $q_{P(A)} = p \circ q_I$ where $q_{P(A)}$ and $q_I$ are standard quotient maps. Then

$$A/P(A) = p(A/I) = p(P(A/I)) \subset P(A/P(A)) = 0,$$

whence $A = P(A)$. □

In other words, the class **Rad**$(P)$ (respectively, **Sem**$(P)$) is stable under extensions if $P$ is an over radical (respectively, under radical). We will need also partially converse statements.

**Lemma 4.7.** *Let $R$ be an under radical, and let $P$ be an over radical. Then*

(1) *If* **Rad**$(R)$ *is stable under extensions then $R$ is a radical;*
(2) *If* **Sem**$(P)$ *is stable under extensions then $P$ is a radical.*

*Proof.* (1) Let $J = q_{R(A)}^{-1}(R(A/R(A)))$. Then the ideal $J$ is an extension of $R(A)$ by $R(A/R(A))$: both algebras are $R$-radical because $R$ is an under radical. By our assumption, $J$ is a $R$-radical ideal of $A$. Therefore $J \subset R(A)$ which means that $R(A/R(A)) = 0$, i.e., $R$ is a radical.

(2) Let $J = P(P(A))$. Then the algebra $A/J$ is an extension of $P(A)/J$ by $(A/J)/(P(A)/J) \cong A/P(A)$. Both algebras are $P$-semisimple because $P$ is an over radical. By our assumption, $A/J$ is $P$-semisimple. Then

$$P(A)/J = q_J(P(A)) \subset P(A/J) = 0,$$

whence $J = P(A)$, i.e., $P$ is a radical. □

We apply this result to preradicals $\overline{\Pi_{\Omega_{\mathrm{n}}}}$ and $\overline{\Pi_{\Omega_{\mathrm{b}}}}$, generated by the primitive maps $\Omega_{\mathrm{n}} := \mathrm{Prim} \setminus \mathrm{Prim}_{\mathrm{n}}$ and $\Omega_{\mathrm{b}} := \mathrm{Prim} \setminus \mathrm{Prim}_{\mathrm{b}}$ on $\mathfrak{U}_{\mathrm{n}}$ (see Example 2.14).

**Theorem 4.8.** *$\overline{\Pi_{\Omega_{\mathrm{n}}}}$ and $\overline{\Pi_{\Omega_{\mathrm{b}}}}$ are radicals on $\mathfrak{U}_{\mathrm{n}}$; for every normed algebra $A$, $\overline{\Pi_{\Omega_{\mathrm{n}}}}(A)$ is the largest $Q$-ideal of $A$ and $\overline{\Pi_{\Omega_{\mathrm{b}}}}(A)$ is the largest $Q_{\mathrm{b}}$-ideal of $A$.*

*Proof.* As $\Pi_{\Omega_{\mathrm{n}}}$ is a hereditary preradical with topological morphisms then $\overline{\Pi_{\Omega_{\mathrm{n}}}}$ is an under radical by Theorem 3.5. Since $\Pi_{\Omega_{\mathrm{n}}}(A)$ is a $Q$-algebra, $\overline{\Pi_{\Omega_{\mathrm{n}}}}(A) := \overline{\Pi_{\Omega_{\mathrm{n}}}(A)}$ is a $Q$-algebra. Taking into account that every $Q$-algebra is a $\Pi_{\Omega_{\mathrm{n}}}$-radical algebra by definition, and the class of all $Q$-algebras is stable under extensions (see for instance [ST5, Theorem 2.5]), we conclude that $\overline{\Pi_{\Omega_{\mathrm{n}}}}$ is a radical, by Lemma 4.7. It is clear that $\overline{\Pi_{\Omega_{\mathrm{n}}}}(A)$ is the largest $Q$-ideal of $A$.

A similar argument gives the result for $\overline{\Pi_{\Omega_{\mathrm{b}}}}$. □

### 4.2. Superposition and convolution operations

The *superposition* of preradicals $P$ and $R$ satisfying Axiom 4 is defined by the usual rule:

$$(P \circ R)(A) = P(R(A))$$

for each algebra $A$. This map is a preradical satisfying Axiom 4 by Proposition 3.2. The superposition operation is clearly associative.

Let $A$ be an algebra, and let $P$ and $R$ be ideal maps. For a [closed] ideal $I$ of $A$, define the ideal $P * I$ of $A$ by

$$P * I := q_I^{-1}(P(A/I)),$$

where $q_I : A \longrightarrow A/I$ is the standard quotient map; in particular $P * 0 = P(A)$, $P*P(A) = q_{P(A)}^{-1}(P(A/P(A)))$ and $P*A = A$. Sometimes we will write $(P * I; A)$ instead of $P * I$ to avoid a misunderstanding.

Define the *convolution* $P*R$ of ideal maps $P, R$ by

$$(P*R)(A) := P * R(A) = q_{R(A)}^{-1} P(A/R(A)). \tag{4.1}$$

Proposition 3.2 shows that if $P$ and $R$ are preradicals satisfying Axiom 4 then $P*R$ is a preradical satisfying Axiom 4.

The operations of superposition and convolution are related by the dual procedure $P \longmapsto P^\dagger$, where $P^\dagger$ is a *quotient-valued map* corresponding to $P$:

$$P^\dagger(A) = A/P(A)$$

for every algebra $A$. This relation reflected in Proposition 4.9 below reminds the Fourier transform which relates convolution and product of functions. Rewrite (4.1) as

$$(P*R)(A) = q_{R(A)}^{-1}\left(P\left(R^\dagger(A)\right)\right)$$

for every $A$.

Let $S$ be a quotient-valued map, i.e., let $S$ send every algebra $A$ to $A/I$ for some ideal $I$. The dependence of $I$ on $A$ can be written as $I = S^\dagger(A)$, so we have an ideal-valued map $S^\dagger$ such that

$$S(A) = A/S^\dagger(A) = q_{S^\dagger(A)}(A)$$

for every algebra $A$. If $T, S$ are quotient-valued maps then their *superposition* $T \circ S$ is defined by

$$(T \circ S)(A) := A/q_{S^\dagger(A)}^{-1}(T^\dagger(S(A))) \cong S(A)/T^\dagger(S(A)) = T(S(A))$$

for every algebra $A$.

**Proposition 4.9.** *Let $P, R$ be ideal-valued maps. Then*

$$(R*P)^\dagger = R^\dagger \circ P^\dagger.$$

*Proof.* If $J$ is a [closed] ideal of $A$ and $I$ is a [closed] ideal of $A/J$ then one identifies in a standard way the algebras $(A/J)/I$ and $A/q_J^{-1}(I)$. Therefore

$$\begin{aligned} R^\dagger \circ P^\dagger(A) &= (A/P(A))/R(A/P(A)) = A/q_{P(A)}^{-1}(R(A/P(A))) \\ &= A/(R*P)(A) = (R*P)^\dagger(A). \end{aligned}$$

□

**Lemma 4.10.** *Let $P$ and $R$ be preradicals. Then*

(1) *$P*R$ is a preradical;*
(2) *The convolution operation is associative.*

*Proof.* (1) Let $f : A \longrightarrow B$ be a morphism. Then $f(P(A)) \subset P(B)$, so there is a morphism $g : q_{P(A)}(A) \longrightarrow q_{P(B)}(B)$ such that $g \circ q_{P(A)} = q_{P(B)} \circ f$; then $g\left(R\left(q_{P(A)}(A)\right)\right) \subset R\left(q_{P(B)}(B)\right)$. Let $a \in (P{*}R)(A)$; then

$$\left(g \circ q_{P(A)}\right)(a) \in R\left(q_{P(B)}(B)\right),$$

whence $\left(q_{P(B)} \circ f\right)(a) \in R\left(q_{P(B)}(B)\right)$ and $f(a) \in (P{*}R)(B)$.

(2) Let $K$ be a preradical. We have that

$$(R * K) * P(A) = q_{I_1}^{-1}(R * K(A/I_1)) = q_{I_1}^{-1}\left(q_{I_2}^{-1}(R((A/I_1)/I_2))\right)$$

where $I_1 = P(A)$, $I_2 = K(A/I_1)$. Also, $R * (K * P)(A) = q_{I_3}^{-1}(R(A/I_3))$ where

$$I_3 = K * P(A) = q_{I_1}^{-1}(K(A/I_1)) = q_{I_1}^{-1}(I_2).$$

It remains to note that $q_{I_3} = q_{I_2} \circ q_{I_1}$ and $(A/I_1)/I_2 \cong A/q_{I_1}^{-1}(I_2) = A/I_3$. □

The following result is a direct consequence of Proposition 3.2.

**Corollary 4.11.**

(1) *If $P$ and $R$ are under radicals then $P{*}R$ is an under radical.*
(2) *If $P$ and $R$ are over radicals then $P{\circ}R$ is an over radical.*

**Theorem 4.12.** *If $P$ and $R$ are hereditary under radicals defined on a universal class $\mathfrak{U}$, and $P$ is pliant then $P{*}R$ is a hereditary under radical.*

*Proof.* Let $A$ be an algebra, $J$ an ideal of $A$. As $P$ and $R$ are under radicals then $P{*}R$ is also an under radical by Corollary 4.11. Therefore

$$(P{*}R)(J) \subset J \cap (P{*}R)(A).$$

We have to prove the converse inclusion.

Since $R(J) = J \cap R(A)$, setting $f(x/R(J)) = q_{R(A)}(x)$ we obtain an isomorphism $f$ of the algebra $J/R(J)$ onto the ideal $q_{R(A)}(J)$ of $A/R(A)$. Since $P$ is pliant, $f(P(J/R(J))) = P(q_{R(A)}(J))$.

If $x \in J \cap (P{*}R)(A)$ then

$$q_{R(A)}(x) \in P(A/R(A)) \cap q_{R(A)}(J) = P(q_{R(A)}(J)) = f(P(J/R(J))).$$

Thus $f(x/R(J)) \in f(P(J/R(J)))$, whence $x/R(J) \in P(J/R(J))$ and $x \in (P{*}R)(J)$. Therefore $J \cap (P{*}R)(A) \subset (P{*}R)(J)$ and the proof is complete. □

**Remark 4.13.** The proof of Theorem 4.12 shows that it remains true if $P, Q$ are topological radicals on Banach algebras and $P$ satisfies the condition of Banach heredity (2.2).

For brevity, let

$$P + R := \mathrm{H}_{\{P,R\}} \text{ and } P \cdot R := \mathrm{B}_{\{P,R\}}$$

for preradicals $P$ and $R$.

**Lemma 4.14.**

(1) *Let* $P$ *and* $R$ *be under radicals. Then*

$$P+R\leq P{*}R\leq(P+R)*(P+R).$$

(2) *Let* $P$ *and* $R$ *be over radicals. Then*

$$(P\cdot R)\circ(P\cdot R)\leq P{\circ}R\leq P\cdot R.$$

*Proof.* (1) Let $A$ be an arbitrary algebra, $I=P(A)$, $J=(P+R)(A)$, and let $q_I:A\longrightarrow A/I$ and $q_J:A\longrightarrow A/J$ be standard quotient maps. As

$$q_I(R(A))\subset R(A/I)=q_I((P{*}R)(A))$$

then $P+R\leq P{*}R$. As $I\subset J$, there is a morphism $q:A/I\longrightarrow A/J$ such that $q\circ q_I=q_{J}$.. Therefore

$$q_J((P{*}R)(A))=q(R(A/I))\subset R(A/J)\subset(P+R)(A/J)$$

whence $(P{*}R)(A)\subset((P+R)*(P+R))(A)$.

(2) Clearly

$$(P{\circ}R)(A)=P(R(A))\subset P(A)\cap R(A)=(P\cdot R)(A).$$

Let $I=P(A)\cap R(A)$. As $I$ is an ideal of $R(A)$ then

$$(P\cdot R)(I)\subset P(I)\subset P(R(A)).$$

□

**Corollary 4.15.**

(1) *Let* $P$ *and* $R$ *be under radicals. Then*

$$(P{*}R)^*=(R*P)^*=P\vee R=P^*\vee R^*.$$

(2) *Let* $P$ *and* $R$ *be over radicals. Then*

$$(P\circ R)^\circ=(R\circ P)^\circ=P\wedge R=P^\circ\wedge R^\circ.$$

*Proof.* (1) follows from Lemma 4.14 if one takes into account that

$$((P+R)*(P+R))^*=P\vee R$$

by associativity of the convolution.

(2) follows from Lemma 4.14 if one takes into account that

$$((P\cdot R)\circ(P\cdot R))^\circ=P\wedge R$$

by associativity of the superposition. □

This corollary can be easily extended to finite and even transfinite "products" of corresponding maps.

It is straightforward that if $P$ and $R$ are hereditary radicals then $P\cdot R$ is a hereditary radical. As a consequence of this, Lemma 4.14(2) and Corollary 4.15, we obtain the following

**Corollary 4.16.** *Let* $P$ *and* $R$ *be the hereditary radicals. Then* $P\circ R$ *is a hereditary radical and*

$$P\circ R=R\circ P=P\wedge R=R\cdot P.$$

### 4.3. Transfinite chains of ideals and radicals

We start with a simple observation on the quotient algebras related to preradicals.

**Proposition 4.17.** *Let $A$ be an algebra, and let $I$ be an ideal of $A$. Then*

(1) *If $P$ is an over radical and*
   (a) *$P(A) \subset I$ then $I/P(A)$ is $P$-semisimple;*
   (b) *$I \subset P(A)$ [for closed $I$] then $P(A/I) = P(A)/I$.*

(2) *If $P$ is a hereditary preradical and $I \subset P(A)$ [for closed $I$] then $P(A)/I$ is $P$-radical.*

*Proof.* (1a) $P(I/P(A)) \subset P(A/P(A)) = 0$.

(1b) There is a morphism $p : A/I \longrightarrow A/P(A)$ such that $q_{P(A)} = p \circ q_I$ where $q_{P(A)}$, $q_I$ are the corresponding standard quotient maps. Then

$$p(P(A/I)) \subset P(A/P(A)) = 0$$

whence $P(A/I) \subset \ker p = P(A)/I = q_I(P(A)) \subset P(A/I)$.

(2) Indeed, $P(A)/I = q_I(P(A)) \subset P(A/I)$. Hence, by heredity of $P$,

$$P(P(A)/I) = (P(A)/I) \cap P(A/I) = P(A)/I. \qquad \square$$

Let $P$ be a preradical, and let $A$ be an algebra. A [closed] ideal $I$ of $A$ is called *$P$-absorbing* if $A/I$ is $P$-semisimple.

**Theorem 4.18.** *Let $A$ be an algebra, $P$ an over radical, $R$ an under radical. Then*

(1) *[The closure of] the sum of any family of $R$-radical ideals is $R$-radical;*
(2) *If $(J_\alpha)_{\alpha \le \gamma}$ is an increasing transfinite chain of ideals of $A$ with $J_0 = 0$ and all gap-quotients are $R$-radical then $J_\gamma$ is $R$-radical;*
(3) *The intersection of any number of $P$-absorbing ideals is $P$-absorbing;*
(4) *If $(I_\alpha)_{\alpha \le \delta}$ is a decreasing transfinite chain $(I_\alpha)_{\alpha \le \gamma}$ of ideals of $A$ with $I_0 = A$ and all $I_\alpha/I_{\alpha+1}$ are $P$-semisimple then $I_\delta$ is $P$-absorbing.*

*Proof.* (1) Let $I$ be the [closure of the] sum of a family $(J_\alpha)_{\alpha\in\Lambda}$ of $P$-radical ideals of $A$. Then $J_\alpha = R(J_\alpha) \subset R(I) \subset I$ for every $\alpha$, whence $I = R(I)$.

(2) follows by transfinite induction. The step $\alpha \mapsto \alpha+1$ follows from Theorem 4.6, and the case of limit ordinals follows from (1).

(3) Let $I = \cap_{\alpha\in\Lambda} J_\alpha$ be the intersection of $P$-absorbing ideals of $A$. As $I \subset J_\alpha$ for every $\alpha$, then, for the standard morphism $q_\alpha : A/I \longrightarrow A/J_\alpha$, we have $q_\alpha(P(A/I)) \subset P(A/J_\alpha) = 0$. Therefore $P(A/I) \subset \cap_\alpha \ker q_\alpha = \cap_{\alpha\in\Lambda} J_\alpha/I = (\cap_{\alpha\in\Lambda} J_\alpha)/I = 0$.

(4) follows by transfinite induction. The step $\alpha \mapsto \alpha+1$ follows from Theorem 4.6, and the case of limit ordinals follows from (3). $\square$

**Theorem 4.19.** *Let $\mathcal{F}_u$ be a family of under radicals, let $\mathcal{F}_o$ be a family of over radicals, and let $A$ be a [normed] algebra. Then*

(1) *There is an increasing transfinite chain $(I_\alpha)_{\alpha \le \gamma}$ of [closed] ideals of $A$ such that $I_0 = 0$, $I_\gamma = (\vee\mathcal{F}_u)(A)$ and every gap-quotient $I_{\alpha+1}/I_\alpha$ of the chain is $P$-radical and is equal to $P(A/I_\alpha)$ for some $P \in \mathcal{F}_u$;*

(2) *There is an increasing transfinite chain* $(J_\alpha)_{\alpha\le\delta}$ *of [closed] ideals of* $A$ *such that* $J_0 = A$, $J_\delta = (\wedge\mathcal{F}_o)(A)$ *and every gap-quotient of the chain is* $P$*-semisimple for some* $P \in \mathcal{F}_o$.

*Proof.* (1) Let $I = (\vee\mathcal{F}_u)(A)$. Arguing by transfinite induction, assume that we already have the required chain $(I_{\alpha'})_{\alpha'\le\alpha}$ but with the only distinction $I_\alpha \subset I$. If $I_\alpha \ne I$ then $I/I_\alpha$ is a non-zero $\vee\mathcal{F}_u$-radical algebra by Proposition 4.17(1b), because $\vee\mathcal{F}_u$ is a radical. Therefore $\mathrm{H}_{\mathcal{F}_u}(I/I_\alpha) \ne 0$. Then $\mathrm{H}_{\mathcal{F}_u}(A/I_\alpha) \ne 0$, and there is some $P \in \mathcal{F}_u$ such that $P(A/I_\alpha) \ne 0$. Let $I_{\alpha+1} = q_{I_\alpha}^{-1}(P(A/I_\alpha))$. Then $I_{\alpha+1}/I_\alpha$ is $P$-radical and

$$I_{\alpha+1}/I_\alpha = P(A/I_\alpha) \subset (\vee\mathcal{F}_u)(A/I_\alpha) = (\vee\mathcal{F}_u)(A)/I_\alpha = I/I_\alpha$$

by Proposition 4.17(1b). So $I_{\alpha+1} \subset I$.

(2) Let $J = (\wedge\mathcal{F}_o)(A)$. Assume, by induction, that we already have the required chain $(J_{\alpha'})_{\alpha'\le\alpha}$ but with $J \subset J_\alpha$. If $J_\alpha \ne J$ then $J_\alpha/J$ is a non-zero $\wedge\mathcal{F}_o$-semisimple algebra by Proposition 4.17(1a). Then $\mathrm{H}_{\mathcal{F}_o}(J_\alpha/J) = 0$, and there is some $P \in \mathcal{F}$ such that $P((J_\alpha/J)) \ne J_\alpha/J$. Let $J_{\alpha+1} = q_J^{-1}(P((J_\alpha/J)))$. Then $J_\alpha/J_{\alpha+1} \cong (J_\alpha/J)/P(J_\alpha/J)$ is $P$-semisimple. □

Let $\mathfrak{U}$ be a base class, and let $V$ be a class of [normed] algebras from $\mathfrak{U}$. We say that $V$ is *closed under increasing transfinite chains* if $V$ contains any [normed] algebra $A$ for which there is an increasing transfinite chain $(I_\alpha)_{\alpha\le\gamma}$ of [closed] ideals $A$ such that $I_0 = 0$, $I_\gamma = A$ and all gap-quotients of the chain are algebras from $V$; we say that $V$ is *closed under decreasing transfinite chains* if $V$ contains any [normed] algebra $A$ for which there is a decreasing transfinite chain $(J_\alpha)_{\alpha\le\delta}$ of [closed] ideals of $A$ such that $J_0 = A$, $J_\delta = 0$ and all gap-quotients of the chain are algebras from $V$.

**Corollary 4.20.** *Let* $V$ *be a class of [normed] algebras from* $\mathfrak{U}$. *Then*

(1) *If* $V$ *is closed under increasing transfinite chains then there is the largest radical on* $\mathfrak{U}$ *whose radical algebras lie in* $V$;
(2) *If* $V$ *is closed under decreasing transfinite chains then there is the smallest radical on* $\mathfrak{U}$ *whose semisimple algebras lie in* $V$.

*Proof.* (1) Let $\mathcal{F}$ be the family of all radicals $P$ on $\mathfrak{U}$ with $\mathbf{Rad}(P) \subset V$. Let $R = \vee\mathcal{F}$, and let $A$ be an $R$-radical algebra. By Theorem 4.19, there is an increasing transfinite chain $(I_\alpha)_{\alpha\le\gamma}$ of [closed] ideals of $A$ such that $I_0 = 0$, $I_\gamma = A$ and every gap-quotient of the chain is $P$-radical for some $P \in \mathcal{F}$. Then all gap-quotients of the chain lie in $V$, and $R \in \mathcal{F}$.

(2) The proof is similar to the proof of (1). □

A transfinite sequence $(S_\alpha)_{\alpha\le\delta}$ of over radicals is called a *decreasing transfinite chain* of over radicals if $(S_\alpha(A))_{\alpha\le\delta}$ is a decreasing transfinite chain of ideals of $A$ for every algebra $A$; similarly, a transfinite sequence $(T_\alpha)_{\alpha\le\gamma}$ of under radicals is called an *increasing transfinite chain* of under radicals if $(T_\alpha(A))_{\alpha\le\gamma}$ is an increasing transfinite chain of ideals of $A$ for every algebra $A$.

Examples:

(1) A *convolution chain* $(T_\alpha)_{\alpha\le\gamma}$ obtained from family $\mathcal{F}_u$ of under radicals by the rule $T_0 := \mathcal{P}_0 : A \longmapsto 0$, *the zero radical,* and $T_{\alpha+1} = P * T_\alpha$ for some $P \in \mathcal{F}_u$;

(2) A *superposition chain* $(S_\alpha)_{\alpha\le\delta}$ obtained from a family $\mathcal{F}_o$ of over radicals by the rule $S_0 := \mathcal{P}_1 : A \longmapsto A$, *the identity radical,* and $S_{\alpha+1} = R \circ S_\alpha$ for some $R \in \mathcal{F}_o$.

Theorem 4.1 and Corollary 4.11 guarantee that $(T_\alpha)_{\alpha\le\gamma}$ consists of over radicals and $(S_\alpha)_{\alpha\le\delta}$ consists of under radicals.

**Theorem 4.21.** *Let $\mathcal{F}_u^*$ be the family of all convolution chains of a family $\mathcal{F}_u$ of under radicals, and let $\mathcal{F}_o^\circ$ be the family of all superposition chains of a family $\mathcal{F}_o$ of over radicals. Then*

(1) *$T_\gamma \le \vee\mathcal{F}_u$ for every $(T_\alpha)_{\alpha\le\gamma} \in \mathcal{F}_u^*$ and $S_\delta \ge \wedge\mathcal{F}_o$ for every $(S_\alpha)_{\alpha\le\delta} \in \mathcal{F}_o^\circ$.*

(2) *For any algebras $A_1, \ldots, A_n$, there exist*

  (a) *$(T_\alpha)_{\alpha\le\gamma} \in \mathcal{F}_u^*$ such that $T_\gamma(A_i) = (\vee\mathcal{F}_u)(A_i)$, for $i \le n$;*

  (b) *$(S_\alpha)_{\alpha\le\delta} \in \mathcal{F}_o^\circ$ such that $S_\delta(A_i) = (\wedge\mathcal{F}_o)(A_i)$, for $i \le n$.*

*Proof.* (1) follows by transfinite induction: the step $\alpha \mapsto \alpha + 1$ follows by Lemma 4.14 and the case of limit ordinals is obvious.

(2a) Construct $(T_\alpha) \in \mathcal{F}_u^*$ taking $T_{\alpha+1} = P * T_\alpha$ for some $P \in \mathcal{F}_u$ with $P(A_1/T_\alpha(A_1)) \ne 0$. The chain $(T_\alpha(A_1))$ of ideals of $A_1$ must be stabilized: there is $T_{\gamma_1}$ such that $P(A_1/T_{\gamma_1}(A_1)) = 0$ for every $P \in \mathcal{F}_u$. Continuing the construction of $(T_\alpha)_{\alpha\le\gamma_1}$ first for $A_2, \ldots, A_n$ and then for every other algebra $B$ in such a way if necessary, one can find $(T_\alpha)_{\alpha\le\gamma_i}$ with $\gamma_1 \le \gamma_2 \le \cdots \le \gamma_n = \gamma$ such that $P(A_i/T_{\gamma_i}(A_i)) = 0$ for every $P \in \mathcal{F}_u$, and define a preradical $T$ such that $T(A_i) = T_\gamma(A_i)$ for $i = 1, \ldots, n$. By Theorem 4.1 and Corollary 4.11, $T$ is an under radical and $T \le \vee\mathcal{F}_u$ by (1). It is clear that $T(B/T(B)) = 0$, i.e., $T$ is a radical. As

$$q_{T(B)}(P(B)) \subset P(B/T(B)) = 0$$

for every $P \in \mathcal{F}_u$, then $P(B) \subset T(B)$, for every algebra $B$. Hence $\vee\mathcal{F}_u \le T$.

(2b) Construct $(S_\alpha) \in \mathcal{F}_o^\circ$ taking $S_{\alpha+1}(A_1) = R(S_\alpha(A_1)) \ne S_\alpha(A_1)$ for some $R \in \mathcal{F}_o$. The chain $(S_\alpha(A_1))$ of ideals must be stabilized: there is $S_{\delta_1}$ such that $R(S_{\delta_1}(A_1)) = S_{\delta_1}(A_1)$ for every $R \in \mathcal{F}_o$. Continuing the construction of $(S_\alpha)$ first for $A_2, \ldots, A_n$ and then for every other algebra $B$ in such a way if necessary, one can find $(S_\alpha)_{\alpha\le\delta_i}$ with $\delta_1 \le \delta_2 \le \cdots \le \delta_n = \delta$ such that $R(A_i/S_{\delta_i}(A_i)) = 0$ for every $R \in \mathcal{F}_o$, and define a preradical $S$ such that $S(A_i) = S_\delta(A_i)$ for $i = 1, \ldots, n$. By Theorem 4.1 and Corollary 4.11, $S$ is an over radical and $S \ge \wedge\mathcal{F}_o$ by (1). It is clear that $S(S(B)) = S(B)$, i.e., $S$ is a radical. As $S(B) = R(S(B)) \subset R(B)$ for every $R \in \mathcal{F}_o$ and every algebra $B$, then $S \le \wedge\mathcal{F}_o$. □

### 4.4. The heredity problem

The closure procedure does not preserve the class of hereditary radicals: $\overline{P}$ can be non-hereditary when $P$ is a hereditary preradical (see [D3, Example 7.1]). The operation of the closure (of an ideal) is involved also in the construction of $P^*$; this leads to a possible loss of heredity. The following lemma shows that it is not the case for the convolution chains in $\mathfrak{U}_a$ (in particular, for the algebraic convolution procedure) and for the superposition chains in general.

**Lemma 4.22.** *Let $\delta$ be a limit ordinal. Then*

(1) *If $(P_\alpha)_{\alpha\leq\delta}$ is a decreasing transfinite chain of preradicals and $P_\alpha$ is hereditary for $\alpha<\delta$, then $P_\delta$ is hereditary.*
(2) *If $(P_\alpha)_{\alpha\leq\delta}$ is an increasing transfinite chain of algebraic preradicals and $P_\alpha$ is hereditary for $\alpha<\delta$, then $P_\delta$ is hereditary.*

*Proof.* (1) Let $A$ be an algebra, and let $J$ be an ideal of $A$. By definition, the chain $(P_\alpha(A))_{\alpha\leq\delta}$ is a decreasing transfinite chain of [closed] ideals. Then

$$P_\delta(A)=\cap_{\alpha<\delta}P_\alpha(A)$$

by (3.1), and $P_\delta(J)=J\cap P_\delta(A)$.

(2) By definition, $(P_\alpha(A))_{\alpha\leq\delta}$ is an increasing transfinite chain of ideals. So

$$\begin{aligned}P_\delta(J)&=\cup_{\alpha<\delta}P_\alpha(J)=\cup_{\alpha<\delta}(J\cap P_\alpha(A))=J\cap(\cup_{\alpha<\delta}P_\alpha(A))\\&=J\cap P_\delta(A).\end{aligned}$$

□

**Corollary 4.23.** *If $P$ is an algebraic hereditary preradical then $P^*$ is a hereditary radical.*

*Proof.* It follows from Theorem 4.11 and Lemma 4.22 that the convolution chain generated by $P$ consists of hereditary preradicals. Therefore $P^*$ is hereditary. □

Let $\mathcal{F}$ be a family of hereditary radicals. Then $\wedge\mathcal{F}$ is a hereditary radical and therefore is the infimum of $\mathcal{F}$ in the class of hereditary radicals. Besides of this $\wedge\mathcal{F}$ satisfies the following equality [ST7, Lemma 3.2]

$$\wedge\mathcal{F}=\mathrm{B}_{\mathcal{F}}. \tag{4.2}$$

**Theorem 4.24.** *The operations $\vee$ and $\wedge$ restricted to the class of algebraic hereditary radicals produce supremum and infimum in this class; moreover $\wedge$ produces infimum in the class of topological hereditary radicals.*

*Proof.* Let $\mathcal{F}$ be a family of algebraic hereditary radicals, $A$ an algebra, and let $J$ be an ideal of $A$. By Theorem 4.21, there is a convolution chain $(T_\alpha)_{\alpha\leq\gamma}$ obtained from $\mathcal{F}$ such that $(\vee\mathcal{F})(A)=T_\gamma(A)$ and $(\vee\mathcal{F})(J)=T_\gamma(J)$. By Theorem 4.11 and Lemma 4.22, $(T_\alpha)_{\alpha\leq\gamma}$ consists of hereditary preradicals. Therefore $T_\gamma(J)=J\cap T_\gamma(A)$. This shows that $\vee\mathcal{F}$ is a hereditary radical. Therefore $\vee\mathcal{F}$ is the supremum of $\mathcal{F}$ in the class of algebraic hereditary radicals.

The assertion about the infimum is evident. □

**Theorem 4.25.** *Let $\mathcal{F}$ be a family of hereditary radicals and $V=\cup_{P\in\mathcal{F}}\mathbf{Rad}(P)$. Let $\mathcal{F}'$ be the family of all hereditary radicals $R$ such that $\mathbf{Rad}(R)\supset V$. Then the supremum of $\mathcal{F}$ in the class of hereditary radicals is equal to*

$$\wedge\mathcal{F}'=\mathrm{B}_{\mathcal{F}'}.$$

*Proof.* Let $T = \wedge\mathcal{F}'$. Then $T$ is a hereditary radical by [ST7, Lemma 3.2], and $T \leq R$, for every $R \in \mathcal{F}'$, by definition. As $\wedge\mathcal{F}' = \mathrm{B}_{\mathcal{F}'}$ then, for every $P \in \mathcal{F}$, we obtain that

$$\mathbf{Rad}\,(T) = \cap_{R\in\mathcal{F}'}\mathbf{Rad}\,(R) \supset V \supset \mathbf{Rad}\,(P)$$

and therefore $T \geq P$. So $T$ is the supremum of $\mathcal{F}$ in the class of hereditary radicals. □

Now we consider a way to show heredity of some radicals that are constructed via the closure procedure and the topological convolution procedure.

Let $A$ be an algebra, and let $I$ be an ideal of $A$. Let $I^{\perp} = \{x \in A : xI = 0\}$, the left annihilator of $I$ in $A$. A preradical $P$ is called *nil-exact* if

$$I^{\perp}\,(P * I) \subset P\,(A)$$

for every [closed] ideal $I$ of any algebra $A$.

A preradical $P$ is called *weakly hereditary* if, for any $P$-radical algebra $A$, $P\,(I) \neq 0$ for every non-zero ideal $I$ of $A$.

Recall that $\Sigma_{\beta}\,(A)$ means the sum of all nilpotent ideals of $A$; if $P$ is a topological preradical then $\Sigma_{\beta} \leq P$ means that the inequality holds for normed algebras on which $P$ is defined.

**Lemma 4.26.** *Let $P$ be an algebraic under radical. Then*

1. *If $P$ is nil-exact then $\overline{P}$ is nil-exact;*
2. *If $P$ is weakly hereditary and $\Sigma_{\beta} \leq P$ then $\overline{P}$ is weakly hereditary.*

*Proof.* (1) Let $A$ be a normed algebra, and let $I$ be a closed ideal of $A$. We have that $I^{\perp}\,(P * I) \subset P\,(A)$, i.e., $I^{\perp}q_I^{-1}\,(P\,(A/I)) \subset P\,(A)$. By (1.5),

$$q_I^{-1}\,(\overline{P}\,(A/I)) = q_I^{-1}\left(\overline{P\,(A/I)}\right) = \overline{q_I^{-1}\,(P\,(A/I))},$$

so

$$I^{\perp}q_I^{-1}\,(\overline{P}\,(A/I)) = I^{\perp}\overline{q_I^{-1}\,(P\,(A/I))} \subset \overline{I^{\perp}q_I^{-1}\,(P\,(A/I))} \subset \overline{P(A)} = \overline{P}(A),$$

i.e., $I^{\perp}\left(\overline{P} * I\right) \subset \overline{P}(A)$.

(2) Let $A$ be a $\overline{P}$-radical algebra, and let $I$ be a non-zero ideal of $A$. Set $J = IA$.

If $J \neq 0$ then $P(J) \neq 0$ by the weak heredity of $P$. As $J$ is an ideal of $I$, it follows that $P(I) \neq 0$, whence $\overline{P}(I) \neq 0$.

If $IA = 0$ then $I^2 = 0$, whence $I = \Sigma_{\beta}(I) \subset P\,(I)$ and $\overline{P}(I) = I \neq 0$. □

**Lemma 4.27.** *Let $P$ be a weakly hereditary under radical, and let $A$ be an algebra and $I$ an ideal of $A$. Then*

1. *If $I \cap P\,(A) \neq 0$ then $P\,(I) \neq 0$;*
2. *If $I \subset P\,(A)$ then $I$ is $P^*$-radical.*

*Proof.* (1) Let $J = I \cap P(A) \neq 0$. Then $J$ is a non-zero ideal of the $P$-radical algebra $P(A)$, whence $P(J) \neq 0$ by weak heredity. As $J$ is an ideal of $I$ then $P(J) \subset P(I) \neq 0$.

(2) Let $A$ be an algebra, and let $I$ be a non-zero ideal of $A$ and $I \subset P(A)$. Then $I$ is an ideal of the $P$-radical algebra $B_1 := P(A)$. Let $I_1 = P * 0 = P(I)$ with respect to the algebra $I$; then $I_1 \neq 0$ is an ideal of $B_1$. Let $I_1'$ be $I_1$ for the algebraic case and the closure of $I_1$ in $B_1$ for the topological case. Then $B_1/I_1' = q_{I_1'}(P(A)) \subset B_2 := P(A/I_1')$ and $I/I_1 \cong q_{I_1'}(I)$.

If $I/I_1 \neq 0$ then $q_{I_1'}(I)$ is a non-zero ideal of the $P$-radical algebra $B_2$. Let $I_2 = P * I_1$ with respect to the algebra $I$: $I_2 = q^{-1}(P(I/I_1))$ where $q : I \longrightarrow I/I_1$ is the standard quotient map. If $q_{I_1'}(I) \neq 0$ then $P\left(q_{I_1'}(I)\right) \neq 0$ by the assumptions and therefore $I_2 \neq I_1$.

Using transfinite induction, we can build the increasing transfinite chain of ideals $(I_\alpha)$ of ideals of $A$ such that all $I_\alpha \subset I$ and $I_{\alpha+1} = P * I_\alpha \neq I_\alpha$ if $I \neq I_\alpha$, and also the corresponding transfinite chain $(I_\alpha')$ (consisting of closures of $I_\alpha$'s for the topological case) such that $q_{I_\alpha'}(I)$ is an ideal of the $P$-radical algebra $P(A/I_\alpha')$. The chain $(I_\alpha)$ must be stabilized: there is an ordinal $\gamma$ such that $I = I_\gamma$. By construction, $I$ is $P^*$-radical. □

**Lemma 4.28.** *Let $P$ be a nil-exact and weakly hereditary under radical, and let $I$ be an ideal of an algebra $A$. If $IP^*(A) \neq 0$ then $P(I) \neq 0$.*

*Proof.* If $IP(A) \neq 0$ then $P(I) \neq 0$ by Lemma 4.27(1). So one may assume that $IP(A) = 0$. For the convolution chain $(P_\alpha)$ (with $P_0 = \mathcal{P}_0$ and $P_{\alpha+1} = P * P_\alpha$ for all $\alpha$), one can find the first ordinal $\gamma$ such that $I(P * P_\gamma(A)) \neq 0$ (this is possible because it is clear that if $IP_{\alpha'}(A) = 0$ for $\alpha' < \alpha$ and $\alpha$ is a limit ordinal then $IP_\alpha(A) = 0$). Let $J = P_\gamma(A)$. Then $Iq_J^{-1}(P(A/J)) \neq 0$ and $IJ = 0$ by the choice of $\gamma$. By nil-exactness of $P$, $Iq_J^{-1}(P(A/J)) \subset P(A)$. As $Iq_J^{-1}(P(A/J)) \subset I$ then $I \cap P(A) \neq 0$, whence $P(I) \neq 0$ by Lemma 4.27(1). □

**Lemma 4.29.** *Let $P$ be a weakly hereditary and nil-exact under radical, and $\Sigma_\beta \leq P$. Then $P^*$ is weakly hereditary.*

*Proof.* Let $A$ be a $P^*$-radical algebra, and let $I$ be a non-zero ideal of $A$. If $I^2 = 0$ then $I = \Sigma_\beta(I) \subset P(I) \subset P^*(I)$ and $P^*(I) = I \neq 0$. If $I^2 \neq 0$ then $IP^*(A) \neq 0$ and $P(I) \neq 0$ by Lemma 4.28. Therefore $P(I) \subset P^*(I) \neq 0$. □

**Theorem 4.30.** *Let $P$ be a weakly hereditary and nil-exact under radical, and $\Sigma_\beta \leq P$. Then $P^*$ is hereditary.*

*Proof.* Let $A$ be an algebra, and let $I$ be an ideal of $A$. By Lemma 4.29, $P^*$ is weakly hereditary. We apply Lemma 4.27(2) to $P^*$ instead of $P$. As $P^*$ is a radical (in particular, $P^{**} = P^*$), every ideal $J$ of $A$ with $J \subset P^*(A)$ is $P^*$-radical. Then $J := I \cap P^*(A)$ is $P^*$-radical; as $J$ is an ideal of $I$, it follows that

$$J = P^*(J) \subset P^*(I) \subset I \cap P^*(A) = J,$$

i.e., $P^*(I) = I \cap P^*(A)$. Therefore $P^*$ is hereditary. □

As is well known (see for instance [AR]), the radicals $\mathfrak{P}_\beta$, $\mathfrak{P}_\lambda$, $\mathfrak{P}_\kappa$ and rad are hereditary.

**Theorem 4.31.** *$\mathcal{P}_\beta$, $\mathcal{P}_\lambda$, $\mathcal{P}_\kappa$ and $\overline{\mathrm{rad}}^*$ are hereditary.*

*Proof.* To apply Lemma 4.26 and Theorem 4.30, one has to show that the algebraic under radicals $\Sigma_\beta$ (see Corollary 3.7), $\mathfrak{P}_\lambda$, $\mathfrak{P}_\kappa$ and rad are weakly hereditary and nil exact. Since any ideal of a $\Sigma_\beta$-radical, locally nilpotent, nil or Jacobson-radical algebra have the same structure, all the under radicals are weakly hereditary. It remains to establish nil exactness. Let $A$ be an algebra, and let $I$ be an ideal of $A$, $B = A/I$ and $a \in I^\perp$.

Let $x \in q_I^{-1}(\Sigma_\beta(B))$; then $q_I(x) = b_1 + \ldots + b_n$, where $b_i$ is in a nilpotent ideal of $B$, for each $i$. We take $x_i \in A$ such that $b_i = q_I(x_i)$. Then $q(ax_i)$ lies in a nilpotent ideal of $B$ for every $i$: there is $n_i > 0$ such that $q_I(ax_iA)^{n_i} = 0$, i.e., $(ax_iA)^{n_i} \subset I$. As $ax_iA \subset I^\perp = 0$ then $(ax_iA)^{n_i+1} = 0$. Hence $ax_i \in \Sigma_\beta(A)$ for every $i$ and therefore $ax = ax_1 + \cdots + ax_n \in \Sigma_\beta(A)$.

Let $N \subset q_I^{-1}(\mathfrak{P}_\lambda(B))$ be finite, and let $A_{aN}$ be the subalgebra generated by $aN$. Then $q_I(aN) \subset \mathfrak{P}_\lambda(B)$ generates the nilpotent subalgebra: there is $m > 0$ such that $q_I(A_{aN})^m = 0$, i.e., $(A_{aN})^m \subset I$. But $A_{aN} \subset I^\perp$, whence $(A_{aN})^{m+1} = 0$. As $N$ is arbitrary, we conclude that $aq_I^{-1}(\mathfrak{P}_\lambda(B)) \subset \mathfrak{P}_\lambda(A)$.

Let $x \in q_I^{-1}(\mathfrak{P}_\kappa(B))$, and let $J$ be the ideal of $A$ generated by $ax$. Then $q_I(J)$ consists of nilpotents. In particular, if $b \in J$ is arbitrary then there is $k > 0$ such that $b^k \in I$. As $b \in I^\perp$ then $b^{k+1} = 0$. This means that $ax \in \mathfrak{P}_\kappa(A)$.

Let $x \in q_I^{-1}(\mathrm{rad}(B))$, and let $\pi \in \mathrm{Irr}(A)$ be arbitrary. If $\pi(I) \neq 0$ then $\pi(I)$ is a strictly irreducible algebra of operators and $\pi(a)\pi(I) = \pi(aI) = 0$ implies that $\pi(a) = 0$, whence $\pi(ax) = 0$. If $\pi(I) = 0$ then there exists $\tau \in \mathrm{Irr}(B)$ such that $\pi = \tau \circ q_I$; as $\tau(q_I(ax)) = 0$ then $\pi(ax) = 0$. Therefore $ax \in \mathrm{rad}(A)$.

All under radicals in consideration are nil exact. By Lemma 4.26 and Theorem 4.30, $\mathcal{P}_\beta$, $\mathcal{P}_\lambda$, $\mathcal{P}_\kappa$ and $\overline{\mathrm{rad}}^*$ are hereditary. □

Establishing heredity of the radical $\mathcal{P}_\beta$, Theorem 4.31 answers a question posed by Dixon [D3, Page 188].

Recall that, for any algebra $A$, $\mathcal{F}(A)$ is the set of all finite rank elements of $A$.

**Lemma 4.32.** *Let $A$ be a semiprime algebra. Then*

(1) $\mathcal{F}(A) \cap \mathrm{rad}(A) = 0$.
(2) *If $A$ is normed then* $\overline{\mathcal{F}(A)} \cap \overline{\mathrm{rad}(A)} = 0$.

*Proof.* (1) Assume that $x \in \mathcal{F}(A) \cap \mathrm{rad}(A)$ is nonzero, and let $n$ be the rank of $\mathrm{W}_x$. Then $xAx$ is $n$-dimensional and the subalgebra generated by $x$ in $A^1$ is at most $(n+2)$-dimensional. So $x$ is an algebraic element in $\mathrm{rad}(A)$, whence there is a minimal polynomial $p(x)$ on $x$ with degree at most $n+2$. The roots of $p$ lie in the spectrum of $x$ with respect to $A^1$ by [Bo, Section 1.1.2]. As $x$ is in $\mathrm{rad}(A)$, these roots are equal to 0, whence $x^{n+2} = 0$.

Let $a \in A^1$ be arbitrary. As $\mathrm{W}_{ax} = \mathrm{L}_a\mathrm{W}_x\mathrm{R}_a$, the rank of $\mathrm{W}_{ax}$ is at most $n$. As $ax \in \mathrm{rad}(A)$, it follows that $(ax)^{n+2} = 0$ by above. Since $A^1x$ consists of nilpotents of fixed degree, $A^1x$ is a nilpotent algebra by the Nagata–Higman theorem (in fact, Dubnov–Nagata–Higman theorem, see [Fr]). As $A^1x$ is a non-zero left nilpotent ideal of $A$, we obtain that $A$ is not semiprime, a contradiction. Therefore $\mathcal{F}(A) \cap \mathrm{rad}(A) = 0$.

(2) Let $x, y \in I := \overline{\mathcal{F}(A)} \cap \overline{\mathrm{rad}(A)}$ be arbitrary. Let $(a_n) \subset \mathcal{F}(A)$ and $(b_m) \subset \mathrm{rad}(A)$ be such that $a_n \to x$ and $b_n \to y$ as $n \to \infty$. Then $a_n b_n = 0$ for all $n$ by (1), but $a_n b_n \to xy$ as $n \to \infty$, whence $xy = 0$ and $I^2 = 0$. As $I$ is an ideal of $A$ and $A$ is semiprime, $I = 0$. □

Let us, for brevity, call a subset $M$ of an algebra $A$ *square zero* if $ab = 0$ for all $a, b \in M$.

**Theorem 4.33.** $\mathfrak{P}_\beta = \mathfrak{R}_{\mathrm{hf}} \wedge \mathrm{rad}$ *and* $\mathcal{R}_\beta = \mathcal{R}_{\mathrm{hf}} \wedge \overline{\mathrm{rad}}^* \le \mathcal{R}_{\mathrm{jhc}} = \mathcal{R}_{\mathrm{hc}} \wedge \mathcal{R}_{\mathrm{cq}}$.

*Proof.* Let $A$ be an algebra, and let $\Sigma_0(A)$ denote the sum of all square zero left ideals of $A$. If $L$ is a square zero left ideal of $A$ then $La$ is also a square zero left ideal of $A$: $L(aL)a \subset (LL)a = 0$. So $\Sigma_0(A)$ is an ideal of $A$. It is easy to check that $\Sigma_0$ is an under radical. If $I$ is a nilpotent ideal of $A$ then it is obvious that $I \subset \Sigma_0^*(A)$. Therefore $\Sigma_\beta(A) \subset \Sigma_0^*(A)$. As $\Sigma_0^*$ is a radical then $\mathfrak{P}_\beta = \Sigma_\beta^* \subset \Sigma_0^*$. On the other hand, every semiprime algebra is $\Sigma_0^*$-semisimple, whence

$$\mathfrak{P}_\beta = \Sigma_0^*.$$

Note that every square zero left ideal consists of zero rank elements of $A$. So $\Sigma_0(A)$ is an ideal of $A$ consisting of finite rank elements. Therefore $\Sigma_0 \le \mathfrak{R}_{\mathrm{hf}}$, whence

$$\mathfrak{P}_\beta = \Sigma_0^* \le \mathfrak{R}_{\mathrm{hf}}.$$

It is well known that $\mathfrak{P}_\beta \le \mathrm{rad}$. Thus

$$\mathfrak{P}_\beta \le \mathfrak{R}_{\mathrm{hf}} \wedge \mathrm{rad}\,.$$

Let now $A$ be semiprime. We must to prove that $A$ is $(\mathfrak{R}_{\mathrm{hf}} \wedge \mathrm{rad})$-semisimple. Assume, to the contrary, that

$$I := (\mathfrak{R}_{\mathrm{hf}} \wedge \mathrm{rad})(A) = \mathfrak{R}_{\mathrm{hf}}(A) \cap \mathrm{rad}(A) \ne 0$$

by (4.2). By heredity of $\mathfrak{R}_{\mathrm{hf}}$, $I$ is an $\mathfrak{R}_{\mathrm{hf}}$-radical ideal of $A$. By Theorem 2.20(5), there is a non-zero finite rank element $x$ of $A$ such that $x \in \mathrm{rad}(A)$, a contradiction to Lemma 4.32. So $A$ is $(\mathfrak{R}_{\mathrm{hf}} \wedge \mathrm{rad})$-semisimple, whence

$$\mathfrak{R}_{\mathrm{hf}} \wedge \mathrm{rad} \le \mathfrak{P}_\beta.$$

It is clear that $\overline{\mathfrak{P}_\beta}^* \le \overline{\mathfrak{R}_{\mathrm{hf}}}^* \wedge \overline{\mathrm{rad}}^*$, whence $\mathcal{P}_\beta \le \mathcal{R}_{\mathrm{hf}} \wedge \overline{\mathrm{rad}}^*$.

Let $A$ be a $\mathcal{P}_\beta$-semisimple normed algebra; then $A$ is semiprime. Assume, to the contrary, that $I := \left(\mathcal{R}_{\mathrm{hf}} \wedge \overline{\mathrm{rad}}^*\right)(A) \ne 0$. By Corollary 4.16, $I = \overline{\mathrm{rad}}^*(\mathcal{R}_{\mathrm{hf}}(A))$. Therefore $J := \overline{\mathrm{rad}}(\mathcal{R}_{\mathrm{hf}}(A)) \ne 0$. Arguing as above, one has that there is a non-zero element $x \in \overline{\mathcal{F}(A)}$ such that $x \in J \subset \overline{\mathrm{rad}}(A)$, whence $x = 0$ by Lemma 4.32, a contradiction. Hence $A$ is $\left(\mathcal{R}_{\mathrm{hf}} \wedge \overline{\mathrm{rad}}^*\right)$-semisimple. Therefore

$$\mathcal{P}_\beta \le \mathcal{R}_{\mathrm{hf}} \wedge \overline{\mathrm{rad}}^* \le \mathcal{P}_\beta.$$

By definition, $\mathcal{R}_{\mathrm{jhc}} = \mathcal{R}_{\mathrm{hc}} \wedge \mathrm{Rad}^r$. By (2.9), $\mathcal{R}_{\mathrm{hc}} \wedge \mathrm{Rad}^r \dot{\leq} \mathcal{R}_{\mathrm{cq}}$, whence

$$\mathcal{R}_{\mathrm{jhc}} = (\mathcal{R}_{\mathrm{hc}} \wedge \mathrm{Rad}^r) \wedge \mathcal{R}_{\mathrm{cq}} = \mathcal{R}_{\mathrm{hc}} \wedge \mathcal{R}_{\mathrm{cq}}.$$

As every nilpotent ideal of $A$ lies in $\mathcal{R}_{\mathrm{cq}}(A)$ then $\mathfrak{P}_\beta \leq \mathcal{R}_{\mathrm{cq}}$, whence also $\mathcal{P}_\beta \leq \mathcal{R}_{\mathrm{cq}}$. Taking into consideration that $\mathcal{P}_\beta = \mathcal{R}_{\mathrm{hf}} \wedge \overline{\mathrm{rad}}^* < \mathcal{R}_{\mathrm{hc}}$, we obtain that $\mathcal{P}_\beta \leq \mathcal{R}_{\mathrm{jhc}}$. □

## 5. Centralization

### 5.1. Commutative ideals and centralization of radicals

Let $A \in \mathfrak{U}_{\mathrm{a}}$ be an algebra, and let $Z(A)$ be the center of $A$. An ideal $J$ of $A$ is called *central* if $J \subset Z(A)$. Let $\Sigma_a(A)$ be the sum of all commutative ideals of $A$.

**Lemma 5.1.**

(1) *$\Sigma_a$ is a preradical.*
(2) *If $A$ is a semiprime algebra then $\Sigma_a(A)$ is the largest central ideal of $A$ and*
   (a) *$\Sigma_a(A) = \{x \in A : x[a,b] = 0 \ \forall a,b \in A\}$;*
   (b) *$\Sigma_a(J) = J \cap \Sigma_a(A)$ for every ideal $J$ of $A$.*

*Proof.* (1) is straightforward.

(2) Let $J$ be a commutative ideal of $A$, and let $x, y \in J$, $a, b \in A$ be arbitrary. Then

$$[x,a]y = x(ay) - axy = (ay)x - axy = a[y,x] = 0.$$

As $b[x,a] \in J$, it follows that $[x,a]b[x,a] = 0$, i.e., $[x,a]A[x,a] = 0$. As $A$ is semiprime, $[x,a] = 0$, i.e., $J \subset Z(A)$, whence $\Sigma_a(A) \subset Z(A)$. So $\Sigma_a(A)$ is a commutative ideal of $A$. Then, by definition, $\Sigma_a(A)$ is the largest ideal that lies in $Z(A)$.

(a) Let $K = \{x \in A : x[a,b] = 0 \ \forall a,b \in A\}$. Let $x \in K$, $a,b,c \in A$ be arbitrary. It is clear that $K$ is an ideal of $A$: indeed, one has

$$xa[b,c] = xa[b,c] + x[a,c]b = x[ab,c] = 0.$$

Then $[x,a]b \in K$, whence $[x,a]b[x,a] = 0$. Since $A$ is semiprime, $[x,a] = 0$, i.e., $x \in Z(A)$. Therefore $K \subset \Sigma_a(A)$.

Let $y \in \Sigma_a(A)$ be arbitrary. As $y, ya \in Z(A)$, we obtain that

$$y[a,b] = [ya,b] - [y,b]a = 0.$$

Thus $\Sigma_a(A) \subset K$.

(b) Obviously $J \cap \Sigma_a(A) \subset \Sigma_a(J)$. As $J$ is a semiprime algebra, $\Sigma_a(J)$ is described as in (a) for $A = J$. Then $\Sigma_a(J)$ is a left ideal of $A$ that lies in the center of $J$. By symmetry, $\Sigma_a(J)$ is a commutative ideal of $A$ and so $\Sigma_a(J) \subset \Sigma_a(A)$ by definition. □

**Remark 5.2.** It follows from Lemma 5.1 (2) that if $A$ is a semiprime normed algebra then $\Sigma_a(A)$ is closed.

Let $P$ be a preradical and let

$$P^a = \Sigma_a * P;$$

$P^a$ is called the *centralization* of $P$. The map $P \longmapsto P^a$ is called the *centralization procedure.*

Let $P$ be a radical such that $P \geq \mathfrak{P}_\beta$. If $P$ is topological on $\mathfrak{U}_{\mathrm{n}}$ or $\mathfrak{U}_{\mathrm{b}}$, the inequality $P \geq \mathfrak{P}_\beta$ is assumed to keep on $\mathfrak{U}_{\mathrm{n}}$ or $\mathfrak{U}_{\mathrm{b}}$, respectively; in such a case the inequality $P \geq \mathcal{P}_\beta$ also holds.

**Theorem 5.3.** *Let $P$ be a radical such that $P \geq \mathfrak{P}_\beta$. Then*

(1) *$P^a$ is an under radical;*
(2) *If $P$ is hereditary then $P^a$ is hereditary.*

*Proof.* Note that $P^a$ is a preradical by Lemma 4.10. Remark 5.2 guarantees that $P^a$ is a topological preradical in the topological case.

Let $A$ be an algebra, and let $J$ be an ideal of $A$. We claim that $A/P(A)$ is semiprime: indeed, $\mathfrak{P}_\beta(A/P(A)) \subset P(A/P(A)) = 0$. Similarly, $J/P(J)$ is a semiprime algebra. It follows from Lemma 5.1(2a) that

$$\begin{aligned} P^a(A) &= \{x \in A : x[a,b] \in P(A) \ \forall a,b \in A\}, \\ P^a(J) &= \{x \in J : x[y,z] \in P(J) \ \forall y,z \in J\}. \end{aligned} \tag{5.1}$$

Let $x \in P^a(J)$ be arbitrary, and let $q : A/P(J) \longrightarrow A/P(A)$ be the standard quotient map. Then

$$\left(q \circ q_{P(J)}\right)(x) \in \Sigma_a\left(q\left(J/P(J)\right)\right) \subset \Sigma_a\left(A/P(A)\right)$$

by Lemma 5.1(2b), whence $x \in P^a(A)$. So

$$P^a(J) \subset P^a(A).$$

(2) Let $x \in J \cap P(A)$. Then $x[a,b] \in J \cap P(A)$ for every $a,b \in A$, in particular for every $a,b \in J$. As $J \cap P(A) = P(J)$, it follows that $x \in P^a(J)$. Therefore, $P^a(J) = J \cap P^a(A)$. Thus $P^a$ is a hereditary preradical, in particular, it is an under radical. This completes the proof of (2).

(1) It follows from (5.1) applied to the opposite algebra of $A$ that $P^a(J)$ is an ideal of $A$. It is clear that $P(P^a(A)) \subset P(A) \subset P^a(A)$, and then we obtain that $P(A) = P(P(A)) \subset P(P^a(A))$, whence

$$P(A) = P(P^a(A)). \tag{5.2}$$

It follows from Lemma 5.1(2) that $[a,b] \in P(A)$ for every $a,b \in P^a(A)$, and from (5.1) for $J = P^a(A)$ and (5.2) that

$$P^a(P^a(A)) = \{x \in P^a(A) : x[a,b] \in P(A) \ \forall a,b \in P^a(A)\} = P^a(A).$$

This completes the proof of (1). □

We see from Theorem 5.3 that in general $P^a$ is an under radical (if $P \geq \mathfrak{P}_\beta$), and one can apply the convolution procedure to obtain the corresponding radical $P^{a*}$ (*the centralized $P$ radical*). But $P^{a*}$ has a much more complicated description than $P^a$ which can be transparently defined by the condition: $P^a(A)$ is the largest ideal of $A$ commutative modulo $P(A)$.

Below we study the question: *when $P^a$ itself is a radical?* Earlier the answer was obtained in the case of the compactly quasinilpotent radical $R_{\mathrm{cq}}$ – it was proved in [ST7] that $\mathcal{R}_{\mathrm{cq}}^a$ is a topological radical on $\mathfrak{U}_{\mathrm{n}}$.

Theorem 5.3 and Lemma 4.7 immediately imply the following result.

**Lemma 5.4.** *Let $P$ be a radical such that $P \geq \mathfrak{P}_\beta$. Then $P^a$ is a radical if and only if, for every [closed] ideal $J$ of an algebra $A$ such that the algebras $A/J$ and $J$ are $P^a$-radical, the algebra $A$ is $P^a$-radical.*

In the following proposition we obtain a tool for the proof that $P^a$ is a radical, which turns out to be convenient for many important examples of radicals.

**Proposition 5.5.** *Let $P$ be a radical such that $P \geq \mathfrak{P}_\beta$. Then $P^a$ is a radical if and only if*

(∗) *any $P$-semisimple algebra $B$ with $P^a$-radical $B/\Sigma_a(B)$, is commutative.*

*Proof.* Suppose that (∗) is true. By Lemma 5.4, it suffices to show that if a [closed] ideal $J$ of an algebra $A$ and the quotient $A/J$ are $P^a$-radical then $A$ is $P^a$-radical.

Let $B = A/P(A)$ and $I$ be the [closure of the] image of $J$ under the standard map $A \longrightarrow A/P(A)$. Then $B$ is $P$-semisimple, in particular semiprime. Since $J$ is commutative modulo $P(J) \subset P(A)$ then $I$ is a commutative ideal of $B$. Therefore $I \subset \Sigma_a(B)$ and, using the standard quotient map $q : A/J \longrightarrow B/I \longrightarrow B/\Sigma_a(B)$, we get

$$B/\Sigma_a(B) = q(A/J) = q(P^a(A/J)) \subset P^a(B/\Sigma_a(B))$$

i.e., $B/\Sigma_a(B)$ is $P^a$-radical.

By (∗), $B$ is commutative whence $B = \Sigma_a(B)$ and

$$A = q_{P(A)}^{-1}(\Sigma_a(A/P(A))) = P^a(A),$$

i.e., $A$ is $P^a$-radical. By Proposition 5.4, $P^a$ is a radical.

Conversely, let $P^a$ be a radical, and let $A$ be a $P$-semisimple algebra such that $A/\Sigma_a(A)$ is $P^a$-radical. Let now $q : A/\Sigma_a(A) \to A/P^a(A)$ be the natural morphism. Since the first algebra is $P^a$-radical then its image is contained in $P^a(A/P^a(A)) = 0$. Therefore $A/P^a(A) = 0$, whence $A = P^a(A)$. Since $P(A) = 0$, we get that $P^a(A) = \Sigma_a(A)$. Thus $A = \Sigma_a(A)$, i.e., $A$ is commutative. We proved (∗). □

**Lemma 5.6.** *Let $A$ be an algebra, and let $J$ be an ideal of $A$ such that $\Sigma_a(A) \subset J$. If $A$ is semiprime and $J/\Sigma_a(A)$ is commutative or nilpotent then $J = \Sigma_a(A)$.*

*Proof.* Let $\Sigma_a(A) \neq 0$. Suppose, to the contrary, that $J \neq \Sigma_a(A)$. If $J/\Sigma_a(A)$ is nilpotent then there exists an ideal $K$ of $A$ such that $\Sigma_a(A) \subsetneqq K \subset J$ and $K^2 \subset \Sigma_a(A)$. Therefore $[K,K] \subset \Sigma_a(A)$. This is also true in the case when $J/\Sigma_a(A)$ is commutative: in this case we can take $K = J$.

Let $x, y \in K$ be arbitrary. Then $z := [x,y] \in \Sigma_a(A)$. By Lemma 5.1,

$$z^2 = z[x,y] = 0.$$

If $z \neq 0$ then $\mathfrak{P}_\beta(\Sigma_a(A)) \neq 0$, whence $\mathfrak{P}_\beta(A) \neq 0$, a contradiction to the assumption that $A$ is semiprime. Therefore $z = 0$, whence $K$ is a commutative ideal of $A$. By definition, $K \subset \Sigma_a(A)$, a contradiction. Therefore $J = \Sigma_a(A)$. □

**Theorem 5.7.** *Let $P \geq \mathfrak{P}_\beta$ be a weakly hereditary under radical. If one of the following conditions holds:*

(1) *$\mathfrak{U}$ is [algebraically] universal and $P$ is pliant or strict;*
(2) *$P$ is defined on Banach algebras and satisfies the condition of Banach heredity;*

*then $P^{*a}$ is a radical.*

*Proof.* First we prove the following statement:

(∗) *Let $A$ be a $P^*$-semisimple algebra and $B = A/\Sigma_a(A)$. If $B$ is commutative modulo $P(B)$ then $A = \Sigma_a(A)$.*

Indeed, $A$ is $P$-semisimple and semiprime. Let $J := \Sigma_a(A)$, $I = J^\perp$ (the left annihilator of $J$). Then $[A, A] \subset I$ by Lemma 5.1, hence either $A$ is commutative or $I \neq 0$.

So we assume, to the contrary, that $I \neq 0$. If $I \cap J \neq 0$ then it is a nilpotent ideal of $A$, a contradiction to semiprimeness of $A$.

Therefore $I \cap J = 0$, and the restriction $f$ of $q_J : A \longrightarrow B$ to $I$ is one-to-one onto $K := q_J(I) \cong (I + J)/J$. If $K$ is commutative then $I + J$ is commutative modulo $J$, whence $I + J$ is commutative by Lemma 5.6, because $A$ is semiprime. This contradicts to the maximality of $\Sigma_a(A)$. So $K$ is not commutative. Since $[K, K] \subset P(B)$, we see that

$$K \cap P(B) \neq 0.$$

In the case when $A$ is a Banach algebra, $K$ is a Banach ideal of $B$ (see Section 1.2.7). If $P$ satisfies the condition of Banach heredity then

$$P(K; \|\cdot\|_K) = K \cap P(B) \neq 0.$$

Taking the norm $\|\cdot\|_K$ as in Section 1.2.7, we have that $f$ is an isometric isomorphism between $I$ and $(K; \|\cdot\|_K)$. Then $P(I) \neq 0$ whence $P(A) \neq 0$, a contradiction.

Let $P$ be pliant or strict, and let $L = K \cap P(B)$. As $L$ is a non-zero ideal of the $P$-radical algebra $P(B)$ then

$$P(L) \neq 0$$

by the weak heredity of $P$.

Since $f$ is a [bounded] isomorphism $I \longrightarrow K$, then $f\left(P\left(f^{-1}(L)\right)\right) = P(L)$ whenever $P$ is pliant, or $\overline{f\left(P(f^{-1}(L))\right)} = P(L)$ whenever $P$ is strict. In any case

$$P\left(f^{-1}(L)\right) \neq 0.$$

As $f^{-1}(L)$ is an ideal of $I$, this implies that

$$0 \neq P\left(f^{-1}(L)\right) \subset P(I) \subset P(A),$$

a contradiction. The proof of the statement (∗) is completed.

Let now $B$ be commutative modulo $P^*(B)$, and let $(P_\alpha)$ be the convolution chain of $P$; then $P_1 = P$. Let $A_1 = \{x \in A : x/J \in P(B)\}$. Then $A_1$ is an ideal of $A$; therefore $A_1$ is $P^*$-semisimple, and $J \subset \Sigma_a(A_1) = J \cap A_1$ by Lemma 5.1. So

$$\Sigma_a(A_1) = J.$$

Let $B_1 = A_1/\Sigma_a(A_1)$. As $B_1 = P(B)$ then $B_1 = P(B_1)$, in particular $B_1$ is commutative modulo $P(B_1)$. Then $A_1$ is commutative by $(*)$. Therefore $A_1 = J$ and $P(B) = 0$. Then $P^*(B) = 0$, $B$ is commutative and $A$ is commutative by Lemma 5.6. As $P^*$ is a radical, $P^{*a}$ is a radical by Proposition 5.5. □

**Remark 5.8.** In the assumptions of Theorem 5.7, if $P^*$ is hereditary then $P^{*a}$ is hereditary by Theorem 5.3. If $P$ is an algebraic hereditary radical and $P \geq \mathfrak{P}_\beta$, then $\overline{P}$ is a strict and weakly hereditary under radical by Theorem 3.5 and Lemma 4.26, whence $\overline{P}^{*a}$ is a radical.

### 5.2. Centralization of classical radicals

Now we apply Proposition 5.7 and Remark 5.8 to the classical radicals.

**Corollary 5.9.** $\mathrm{rad}^a$, $\overline{\mathrm{rad}}^{*a}$ *and* $\mathrm{Rad}^a$ *are hereditary radicals.*

*Proof.* The radical Rad satisfies the condition of Banach heredity, rad is pliant and $\overline{\mathrm{rad}}$ is weakly hereditary. By Theorem 5.7, the centralizations of them are radicals. As $\overline{\mathrm{rad}}^*$ is hereditary, $\overline{\mathrm{rad}}^{*a}$ is a hereditary radical. □

**Theorem 5.10.** $\mathrm{Rad}^a(A) = \mathrm{rad}^{\dim>1}(A) = \{x \in A : \rho(x[a,b]) = 0\ \forall a, b \in A\}$ *for every Banach algebra* $A$.

*Proof.* Set $J = \mathrm{Rad}^a(A)$. As $\mathrm{rad}^{\dim>1}(A)$ admits only one-dimensional strictly irreducible representations, $\mathrm{rad}^{\dim>1}(A) \subset J$. If $\pi(J) \neq 0$ for some $\pi \in \mathrm{Irr}^{\dim>1}(A)$ then $\dim \pi(A) = \dim \pi(J) = 1$ by Jacobson's density theorem and commutativity of $J$ modulo $\mathrm{Rad}(A)$, a contradiction. So $J \subset \mathrm{rad}^{\dim>1}(A)$.

Set $I = \{x \in A : \rho(x[a,b]) = 0\ \forall a, b \in A\}$. Then $J \subset I$ by (5.1) for $P = \mathrm{Rad}$.

Let $x \in I$ be arbitrary. Assume that $\pi(x)\xi = \zeta \neq 0$ for some $\pi \in \mathrm{Irr}^{\dim>1}(A)$. Taking a vector $\eta$ such that $\zeta$ and $\eta$ are linearly independent, one can find $a, b \in A$ by Jacobson's density theorem such that $\pi(a)\eta = \xi$, $\pi(a)\zeta = 0$ and $\pi(b)\zeta = \eta$. Then $\pi([a,b])\zeta = \xi$, whence

$$\pi(x[a,b])\zeta = \zeta,$$

a contradiction with $\rho(x[a,b]) = 0$. Therefore $\pi(x) = 0$ for every $\pi \in \mathrm{Irr}^{\dim>1}(A)$. This means that $x \in \mathrm{rad}^{\dim>1}(A)$ and $I \subset J$. □

**Remark 5.11.** $\mathrm{Rad}^{ar}(A) = \mathrm{Rad}^{ra}(A) = \left\{x \in A : \rho(x[a,b]) = 0\ \forall a, b \in \widehat{A}\right\}$ for every normed algebra $A$.

Now we turn to the radicals of Baer, Levitzki, Köthe, to the hypofinite radical, and to their topological counterparts.

**Corollary 5.12.** $\mathfrak{P}_\beta^a$, $\mathfrak{P}_\lambda^a$, $\mathfrak{P}_\kappa^a$, $\mathfrak{R}_{\mathrm{hf}}^a$, $\mathcal{P}_\beta^a$, $\mathcal{P}_\lambda^a$, $\mathcal{P}_\kappa^a$ *and* $\mathcal{R}_{\mathrm{hf}}^a$ *are hereditary radicals.*

*Proof.* Follows from Theorem 5.7 and Remark 5.8 because all radicals are hereditary, and the closures of $\mathfrak{P}_\beta$, $\mathfrak{P}_\lambda$, $\mathfrak{P}_\kappa$, $\mathfrak{R}_{\mathrm{hf}}$ are strict and weakly hereditary. □

Recall that the closed-nil radical $\mathcal{P}_{\mathrm{nil}}$ is the restriction of $\mathcal{P}_\beta$ to Banach algebras.

**Corollary 5.13.** *$\mathcal{P}^a_{\mathrm{nil}}$ is a hereditary radical.*

**Lemma 5.14.** *Let $P_1$ and $P_2$ be radicals such that $\mathfrak{P}_\beta \le P_1 < P_2$. If $P_1 = P_2$ on commutative algebras then $P^a_1 < P^a_2$.*

*Proof.* Let $A$ be an algebra such that $P_1(A) \neq P_2(A)$. Assume, to the contrary, that $P^a_1(A) = P^a_2(A)$. Let $B = A/P_1(A)$. Then $B$ is semiprime and therefore commutative by Lemma 5.1. On the other hand, $B$ is $P_1$-semisimple and therefore $P_2$-semisimple by our assumption. But $B$ is not $P_2$-semisimple because $P_2(A)/P_1(A)$ is $P_2$-radical, non-zero and lies in $P_2(B)$, a contradiction. □

**Corollary 5.15.** $\mathrm{rad}^a_{\mathrm{n}} < \mathrm{rad}^a_{\mathrm{b}} < \mathrm{Rad}^{ra}$, $\mathfrak{P}^a_\beta < \mathfrak{P}^a_\lambda < \mathfrak{P}^a_\kappa$ *and* $\mathcal{P}^a_\beta < \mathcal{P}^a_\lambda < \mathcal{P}^a_\kappa$.

*Proof.* The result follows from Lemma 5.14, (3.10), (3.8), Theorem 3.8, the coincidence of $\mathrm{Rad}^{ra}$ with $\mathrm{rad}^a_{\mathrm{n}}$ on commutative algebras (by Theorem 5.10), and also the coincidence of $\mathfrak{P}^a_\kappa$ with $\mathfrak{P}^a_\beta$ on commutative algebras. □

**Problem 5.16.** *Is $\mathcal{R}^a_{\mathrm{hc}}$ a radical?*

### 5.3. The centralization of the tensor Jacobson radical

It is not clear if $\mathcal{R}_{\mathrm{cq}}$ and $\mathcal{R}_t$ are strict, so we cannot apply Theorem 5.7 to them. However, $\mathcal{R}^a_{\mathrm{cq}}$ is a radical by [ST7], and it will be proved that $\mathcal{R}^a_t$ is a radical.

First we note that the maps $\mathcal{R}_t$ and $\mathcal{R}^a_t$ are regular.

**Theorem 5.17.** *Let $A$ be a normed algebra. Then $\mathcal{R}_t(B) = B \cap \mathcal{R}_t(A)$ and $\mathcal{R}^a_t(B) = B \cap \mathcal{R}^a_t(A)$ for every dense subalgebra $B$ of $A$.*

*Proof.* As $\mathcal{R}_t(B)$ is the set of all $a \in B$ such that $a \otimes c \in \mathrm{Rad}\left(B \widehat{\otimes} C\right) = \mathrm{Rad}\left(A \widehat{\otimes} C\right)$ for every normed algebra $C$ and every $c \in C$ [ST6], then clearly

$$\mathcal{R}_t(B) = B \cap \mathcal{R}_t(A).$$

By (5.1), $\mathcal{R}^a_t(B) = \{x \in B : x[a,b] \in \mathcal{R}_t(B)\ \forall a, b \in B\}$, whence we obtain that

$$\mathcal{R}^a_t(B) = B \cap \mathcal{R}^a_t(A). \qquad \square$$

We need here and will use also in the further sections a result on tensor radius of operator families, analogous to the result about the joint spectral radius established in [ST1, Lemma 4.2]. If $T$ is an operator on a space $X$ and $Y$ is an invariant subspace for $T$ then $T|_Y$ denotes the restriction of $T$ to $Y$, and $T|_{X/Y}$ – the operator induced by $T$ on $X/Y$. If $Y$ is invariant for a set $M$ of operators then $M|_Y = \{T|_Y : T \in M\}$ and $M|_{X/Y} = \left\{T|_{X/Y} : T \in M\right\}$; this is transferred similarly to families of operators.

Recall that the tensor spectral radius $\rho_t$ is defined in Section 2.3.3 and

$$\rho_t(M^m) = \rho_t(M)^m$$

for every $m > 0$ and every summable family $M$ in a normed algebra.

**Theorem 5.18.** *Let $M = (a_i)_1^\infty$ be a summable family of operators in $\mathcal{B}(X)$, and let $Y$ be an invariant closed subspace for $M$. Then*

$$\rho_t(M) = \max\left\{\rho_t\left(M|_{X/Y}\right), \rho_t(M|_Y)\right\}.$$

*Proof.* It is clear that $\max\left\{\rho_t\left(M|_{X/Y}\right), \rho_t(M|_Y)\right\} \le \rho_t(M)$. Let us prove the converse inequality. Let $q : X \longrightarrow V := X/Y$ be the standard quotient map. Note that

$$\left\|M^{2n}\right\|_+ = \sum_{i=1}^{\infty}\sum_{j=1}^{\infty}\|b_ib_j\|$$

where $M^n = (b_i)_1^\infty$. For every $\varepsilon > 0$, there is $x_{ij} \in X$ with $\|x_{ij}\| = 1$ such that $\|b_ib_j\| \le \|b_ib_jx_{ij}\| + \varepsilon$. Clearly $\|b_ib_jx_{ij}\| \le \|b_iy\| + \|b_i\|\,\|b_jx_{ij} - y\|$ for any $y \in Y$. Take $y = y_{ij}$ such that $\|b_jx_{ij} - y_{ij}\| \le \|q(b_jx_{ij})\| + \varepsilon$. Then

$$\|y_{ij}\| \le \|b_jx_{ij}\| + \|b_jx_{ij} - y_{ij}\| \le \|b_jx_{ij}\| + \|q(b_jx_{ij})\| + \varepsilon.$$

Hence

$$\begin{aligned}\|b_ib_j\| &\le \|b_ib_jx_{ij}\| + \varepsilon \le \|b_iy_{ij}\| + \|b_i\|\,\|b_jx_{ij} - y_{ij}\| + \varepsilon\\ &\le \|b_i|_Y\|\left(\|b_jx_{ij}\| + \|q(b_jx_{ij})\| + \varepsilon\right) + \|b_i\|\left(\|q(b_jx_{ij})\| + \varepsilon\right) + \varepsilon\end{aligned}$$

Taking the limit for $\varepsilon \to 0$, we get

$$\|b_ib_j\| \le \|b_i|_Y\|\left(\|b_j\| + \|b_j|_V\|\right) + \|b_i\|\,\|b_j|_V\|.$$

Therefore

$$\begin{aligned}\left\|M^{2n}\right\|_+ &\le \sum_i\sum_j\left(\|b_i|_Y\|\,\|b_j\| + \|b_i|_Y\|\,\|b_j|_V\| + \|b_i\|\,\|b_j|_V\|\right)\\ &\le \|M^n|_Y\|_+\|M^n\|_+ + \|M^n|_Y\|_+\|M^n|_V\|_+ + \|M^n\|_+\|M^n|_V\|_+.\end{aligned}$$

Let $n = 2k$ in this inequality; as $\|M^n\|_+ \le \left\|M^k\right\|_+^2$ and

$$\begin{aligned}\|M^n|_Y\|_+ &\le \left\|M^k|_Y\right\|_+\left\|M^k|_Y\right\|_+ \le \left\|M^k|_Y\right\|_+\left\|M^k\right\|_+,\\ \|M^n|_V\|_+ &\le \left\|M^k|_V\right\|_+\left\|M^k|_V\right\|_+ \le \left\|M^k|_V\right\|_+\left\|M^k\right\|_+\end{aligned}$$

then

$$\begin{aligned}\left\|M^{2n}\right\|_+ &\le \left\|M^k\right\|_+^2\left(\left\|M^k|_Y\right\|_+^2 + \left\|M^k|_Y\right\|\left\|M^k|_V\right\| + \left\|M^k|_V\right\|_+^2\right)\\ &\le \left\|M^k\right\|_+^2\left(\left\|M^k|_Y\right\|_+ + \left\|M^k|_V\right\|_+\right)^2\\ &\le 4\left\|M^k\right\|_+^2\max\left\{\left\|M^k|_Y\right\|_+, \left\|M^k|_V\right\|_+\right\}^2\end{aligned}$$

Taking roots and setting $k \to \infty$, we obtain

$$\rho_t(M) = \rho_t(M)^{1/2} \max\{\rho_t(M|_Y), \rho_t(M|_V)\}^{1/2},$$

i.e., $\rho_t(M) \le \max\{\rho_t(M|_Y), \rho_t(M|_V)\}$. □

Let $A$ be a normed algebra. Recall that, for every element $a \in A$, $\mathrm{L}_a$ and $\mathrm{R}_a$ are defined as operators $x \longmapsto ax$ and $x \longmapsto xa$ on $A$, respectively. If $M = (a_n)_1^\infty$ is a family in $A$, let $\mathrm{L}_M$ and $\mathrm{R}_M$ denote the operator families $(\mathrm{L}_{a_n})_1^\infty$ and $(\mathrm{R}_{a_n})_1^\infty$, respectively.

**Theorem 5.19.** *Let $A$ be a normed algebra, and let $M = (a_n)_1^\infty$ be a summable family in $A$. Then $\rho_t(M) = \rho_t(\mathrm{L}_M) = \rho_t(\mathrm{R}_M)$ and $\rho_t(\mathrm{L}_M\mathrm{R}_M) = \rho_t(M)^2$.*

*Proof.* It is clear that $\rho_t(\mathrm{L}_M) \le \rho_t(M)$, and $\rho_t(\mathrm{L}_M\mathrm{R}_M) \le \rho_t(\mathrm{L}_M)\rho_t(\mathrm{R}_M) \le \rho_t(M)^2$ by [ST6, Proposition 3.4]. Further,

$$\begin{aligned}\left\|M^{m+1}\right\|_+ = \|\mathrm{L}_M^m(M)\|_+ &= \sum_{i_1=1}^{\infty}\cdots\sum_{i_n=1}^{\infty}\sum_{i_{n+1}=1}^{\infty}\left\|\mathrm{L}_{a_{i_1}}\cdots\mathrm{L}_{a_{i_n}}a_{i_{n+1}}\right\| \\ &\le \sum_{i_1=1}^{\infty}\cdots\sum_{i_n=1}^{\infty}\sum_{i_{n+1}=1}^{\infty}\left\|\mathrm{L}_{a_{i_1}}\cdots\mathrm{L}_{a_{i_n}}\right\|\left\|a_{i_{n+1}}\right\| = \|\mathrm{L}_M^m\|_+\|M\|_+\end{aligned}$$

for every $m > 0$; this implies $\rho_t(M) \le \rho_t(\mathrm{L}_M)$. Similarly,

$$\left\|M^{2m+1}\right\|_+ \le \|(\mathrm{L}_M\mathrm{R}_M)^m\|_+\|M\|_+,$$

for every $m > 0$, implies $\rho_t(M)^2 = \rho_t(M^2) \le \rho_t(\mathrm{L}_M\mathrm{R}_M)$. □

Recall that $M/I$ is the image of $M$ in $A/I$.

**Corollary 5.20.** *Let $A$ be a normed algebra, and let $I$ be a closed ideal of $A$. Then $\rho_t(M) = \max\{\rho_t(M/I), \rho_t(\mathrm{L}_M|_I)\}$ for every summable family $M = (a_n)_1^\infty$ in $A$.*

*Proof.* Indeed, it follows from Theorems 5.19 and 5.18 that

$$\begin{aligned}\rho_t(M) = \rho_t(\mathrm{L}_M) &= \max\left\{\rho_t\left(\mathrm{L}_M|_{A/I}\right), \rho_t(\mathrm{L}_M|_I)\right\} \\ &= \max\left\{\rho_t\left(\mathrm{L}_{M/I}\right), \rho_t(\mathrm{L}_M|_I)\right\} = \max\{\rho_t(M/I), \rho_t(\mathrm{L}_M|_I)\}.\end{aligned}$$ □

For a summable family $M = (a_n)_1^\infty$ in $A$, let

$$M|^k = (a_n)_1^k, \quad M|_{k+1} = (a_n)_{k+1}^\infty \quad \text{and} \quad \rho_+(M) = \sum_1^\infty \rho(a_n).$$

As $\rho(a_n) \le \|a_n\|$ for every $n > 0$, then $\rho_+(M) < \infty$ for every summable family $M$.

**Remark 5.21.** The inequality $\rho_+(M) \le \rho_t(M)$ does not hold even for commutative Banach algebra. Indeed, if $M = (p, 1-p, 0, \ldots)$ for a non-trivial idempotent $p$ then $\rho_t(M) = 1$ while $\rho_+(M) = 2$. See the calculation of $\rho_t(M)$ for commutative finite families $M$ via joint spectra in [M].

**Lemma 5.22.** *Let $M = (a_n)_1^\infty$ be a commutative summable family in $A$. Then $\rho_t(M) \le \rho_+(M)$.*

*Proof.* As $M|^k$ and $M|_{k+1}$ commute for every $k > 0$, and $M = M|^k \sqcup M|_{k+1}$, we obtain that

$$\rho_t(M) \le \rho_t\left(M|^k\right) + \rho_t(M|_{k+1}) \le \rho_+\left(M|^k\right) + \rho_t(M|_{k+1})$$

by [ST6, Proposition 3.4]. For every $\varepsilon > 0$ there is $n > 0$ such that $\rho_t(M|_{n+1}) \le \|M|_{n+1}\|_+ < \varepsilon$, whence

$$\rho_t(M) \le \sup_k \rho_t\left(M|^k\right) = \sup_k \rho_+\left(M|^k\right) = \rho_+(M). \qquad \square$$

**Corollary 5.23.** *Let $A$ be a normed algebra, $I$ a closed ideal of $A$, and let $M = (a_n)_1^\infty$ be a summable family in $A$. Then*

(1) *If $A/I$ is commutative then $\rho_t(M) \le \max\{\rho_+(M), \rho_t(\mathrm{L}_M|_I)\}$;*
(2) *If $I$ is central then, for every $k > 0$,*

$$\rho_t(M) \le \max\left\{\rho_t(M/I), \rho_t\left(M|^k\right) + \rho_t(M|_{k+1})\right\}.$$

*Proof.* (1) It follows from Lemma 5.22 and Corollary 5.20 that

$$\rho_t(M) \le \max\{\rho_+(M/I), \rho_t(\mathrm{L}_M|_I)\} \le \max\{\rho_+(M), \rho_t(\mathrm{L}_M|_I)\}$$

since $\rho(a_n/I) \le \rho(a_n)$ for every $n$.

(2) It is easy to check that $\mathrm{L}_M|_I$ is commutative. Then

$$\begin{aligned}\rho_t(\mathrm{L}_M|_I) &\le \rho_t\left(\mathrm{L}_{M|^k}|_I\right) + \rho_t\left(\mathrm{L}_{M|_{k+1}}|_I\right) \le \rho_t\left(M|^k/I\right) + \rho_t(M|_{k+1}/I)\\ &\le \rho_t\left(M|^k\right) + \rho_t(M|_{k+1})\end{aligned}$$

by Lemma 5.22, and

$$\begin{aligned}\rho_t(M) &= \max\{\rho_t(M/I), \rho_t(\mathrm{L}_M|_I)\}\\ &\le \max\left\{\rho_t(M/I), \rho_t\left(M|^k\right) + \rho_t(M|_{k+1})\right\}. \qquad \square\end{aligned}$$

**Corollary 5.24.** *Let $A$ be a normed algebra. If $A/\mathcal{R}_t(A)$ is commutative then $\rho_t(M) \le \rho_t\left(M|^k\right) + \rho_t(M|_{k+1})$ for every summable family $M = (a_n)_1^\infty$ in $A$ and every $k > 0$.*

*Proof.* By (2.5) and Lemma 5.22,

$$\rho_t(M) = \rho_t(M/\mathcal{R}_t(A)) \le \rho_t\left(M|^k\right) + \rho_t(M|_{k+1}). \qquad \square$$

**Lemma 5.25.** *Let $A$ be an $\mathcal{R}_t$-semisimple algebra, and let $I$ be a closed commutative ideal of $A$. If $A/I$ is commutative modulo $\mathcal{R}_t(A/I)$ then $\rho_t(\{a\} \sqcup M) \le \rho(a) + \rho_t(M)$ for every $a \in A$ and every summable family $M = (a_n)_1^\infty$ in $A$.*

*Proof.* As $A$ is a semiprime algebra, $I$ is a central ideal of $A$ by Lemma 5.1. By Corollary 5.23,

$$\rho_t(\{a\} \sqcup M) \le \max\{\rho_t((\{a\} \sqcup M)/I), \rho(a) + \rho_t(M)\}.$$

By Corollary 5.24,

$$\rho_t\left(\left(\{a\} \sqcup M\right)/I\right) \leq \rho\left(a/I\right) + \rho_t\left(M/I\right) \leq \rho\left(a\right) + \rho_t\left(M\right).$$

Therefore $\rho_t\left(\{a\} \sqcup M\right) \leq \rho\left(a\right) + \rho_t\left(M\right)$. □

In the following theorem we will use the inequality

$$\rho_t\left(\{a+b\} \sqcup M\right) \leq \rho_t\left(\{a\} \sqcup \{b\} \sqcup M\right) \tag{5.3}$$

for every $a, b \in A$ and every summable family $M$ in $A$; this is a special case of [ST6, Proposition 3.3].

**Theorem 5.26.** *$\mathcal{R}_t^a$ is a uniform regular topological radical on $\mathfrak{U}_{\mathrm{n}}$.*

*Proof.* Let $A$ be an $\mathcal{R}_t$-semisimple algebra, and let $I$ be a closed commutative ideal of $A$ such that $A/I$ is commutative modulo $\mathcal{R}_t\left(A/I\right)$. By Proposition 5.5, to prove that $\mathcal{R}_t^a$ is a topological radical, it suffices to show that $A$ is commutative.

Assume that $A$ is a Banach algebra. Let $M = \left(a_n\right)_1^\infty$ be a summable family in $A$, and $a, b \in A$ be arbitrary. Put $c_\lambda = \exp\left(\lambda \operatorname{ad}\left(b\right)\right)\left(a\right)$ for every $\lambda \in \mathbb{C}$, and put $f\left(\lambda\right) = \left(c_\lambda - a\right)/\lambda$; then $\lambda \longmapsto f\left(\lambda\right)$ is an analytic function in $A$ and $f\left(0\right) = [b, a]$. As $\rho\left(c_\lambda\right) = \rho\left(a\right)$ then, by (5.3) and Lemma 5.25,

$$\rho_t\left(\{c_\lambda - a\} \sqcup M\right) \leq \rho_t\left(\{c_\lambda\} \sqcup \{a\} \sqcup M\right) \leq 2\rho\left(a\right) + \rho_t\left(M\right)$$

for every $\lambda \in \mathbb{C}$. Replace $a$ by $a/\lambda$ for $\lambda \neq 0$; we obtain that

$$\rho_t\left(\{f\left(\lambda\right)\} \sqcup M\right) \leq 2\rho\left(a\right)/\lambda + \rho_t\left(M\right) \tag{5.4}$$

The function $\lambda \longmapsto \rho_t\left(\{f\left(\lambda\right)\} \sqcup M\right)$ is subharmonic by [ST6, Theorem 3.16] and bounded on $\mathbb{C}$ by (5.4); therefore it is constant and

$$\rho_t\left(\{[b,a]\} \sqcup M\right) = \rho_t\left(\{f\left(0\right)\} \sqcup M\right) = \lim_{|\lambda| \to \infty} \rho_t\left(\{f\left(\lambda\right)\} \sqcup M\right) \leq \rho_t\left(M\right).$$

As $M$ is an arbitrary summable family in $A$ then $[b, a] \in \mathcal{R}_t\left(A\right) = 0$, therefore $A$ is commutative.

As $\mathcal{R}_t$ is hereditary, it follows from Theorem 5.3 and Proposition 5.5 that $\mathcal{R}_t^a$ is a hereditary topological radical on $\mathfrak{U}_{\mathrm{b}}$. As $\mathcal{R}_t^a$ is regular by Theorem 5.17, $\mathcal{R}_t^a$ is a hereditary topological radical on $\mathfrak{U}_{\mathrm{n}}$.

Let $A$ be an $\mathcal{R}_t^a$-radical algebra, and let $B$ be a subalgebra of $A$. If $a, b \in B$ then $\rho_t\left(\{[b,a]\} \sqcup M\right) = \rho_t\left(M\right)$ for every summable family in $B$. Therefore $B$ is $\mathcal{R}_t^a$-radical and $\mathcal{R}_t^a$ is uniform. □

### 5.4. Spectral applications

Let $\mathfrak{U}$ be a class of algebras. A property of algebras from $\mathfrak{U}$ is called *radical* (respectively, *semisimple*) in $\mathfrak{U}$ if algebras with this property form the radical (respectively, semisimple) class for some radical.

We start with $\mathrm{Rad}^a$-radical algebras. It is well known that Banach algebras commutative modulo the Jacobson radical share with the commutative ones the advantages of easy calculation of spectra and the continuity of the functions $a \to \sigma(a)$ and $a \to \rho(a)$. Moreover, it was shown in [PZ, Z1] that a Banach algebra is commutative modulo radical if and only if one (or all) of the following conditions holds:

($1_a$) The function $a \to \rho(a)$ is submultiplicative;
($2_a$) The function $a \to \rho(a)$ is subadditive;
($3_a$) The function $a \to \rho(a)$ is uniformly continuous.

Using the fact that $\mathrm{Rad}^a$ is a radical (Corollary 5.9), we obtain the following result:

**Corollary 5.27.** *Let $A$ be a Banach algebra, let $I$ be a closed ideal of $A$ and let $\mathcal{F}$ be a family of closed ideals of $A$ with dense sum in $A$. If either $I$ and $A/I$, or all ideals from $\mathcal{F}$ have one of the properties ($1_a$), ($2_a$), ($3_a$) then $A$ has the same property.*

In other words, the uniform continuity of $a \to \rho(a)$ is a radical property of Banach algebras.

**Problem 5.28.** *Will the statement of Corollary 5.27 stay true if one replaces the uniform continuity of $a \to \rho(a)$ by the continuity of this map?*

Some results related to this problem can be found in Section 9.

In the theory of joint spectra and spectral radii the commutativity modulo the radical $\mathcal{R}_{\mathrm{cq}}$ or $\mathcal{R}_t$ plays a role which is at least partially similar to the role of commutativity modulo Rad in the "individual" spectral theory.

**Theorem 5.29.** *Let $A$ be a normed algebra. Then*

(1) *If $M = \{a_\alpha : \alpha \in \Lambda\}$ is a family in $A$ which is precompact and commutative modulo $\mathcal{R}_{\mathrm{cq}}(A)$ then $\rho(M) = \sup\left\{|\lambda| : \lambda \in \sigma^l_{\widehat{A}}(M)\right\}$; (Here $|\lambda| = \sup_\alpha |\lambda_\alpha|$.)*

(2) *If $M = (a_n)_1^\infty$ is a summable family which is commutative modulo $\mathcal{R}_t(A)$ then $\rho_t(M) = \sup\left\{|\lambda|_+ : \lambda \in \sigma^l_{\widehat{A}}(M)\right\} = \sup\left\{|\lambda|_+ : \lambda \in \sigma^r_{\widehat{A}}(M)\right\}$. (Here $|\lambda|_+ = \sum_1^\infty |\lambda_n|$.)*

*Proof.* As $\mathcal{R}_{\mathrm{cq}}(A) = A \cap \mathcal{R}_{\mathrm{cq}}\left(\widehat{A}\right)$, $\mathcal{R}_t(A) = A \cap \mathcal{R}_t\left(\widehat{A}\right)$ and the values of the joint spectral and tensor radii do not change under taking the completion of $A$, one may assume that $A$ is a Banach algebra.

(1) As $\rho(M) = \rho(M/\mathcal{R}_{\mathrm{cq}}(A))$ by [ST6, Theorem 3.29], one may assume that $M$ is commutative. By [ST7, Theorem 6.3], the joint spectral radius is continuous at $M$ and

$$\rho(M) = \sup\{\rho(N) : N \text{ is a finite subfamily of } M\}. \tag{5.5}$$

The first equality below holds by [M, Theorem 35.5] while the other relations are obvious:

$$\begin{aligned}\rho(N) &= \sup\left\{|\lambda| : \lambda \in \sigma^l_A(N)\right\} \\ &\le \sup\left\{|\lambda| : \lambda \in \sigma^l_A(M)\right\} = \sup_{a_\alpha \in M} \rho(a_\alpha) \le \rho(M).\end{aligned} \tag{5.6}$$

The result follows from (5.5) and (5.6).

(2) As $\rho(M) = \rho(M/\mathcal{R}_{\mathrm{t}}(A))$ by [ST5, Theorem 4.18], one may assume that $M$ is commutative. As $M$ is summable then for every $\varepsilon > 0$ there is $n > 0$ such that $\sum_{n+1}^{\infty} \|a_k\| < \varepsilon$. By Corollary 5.24,

$$\rho(M|^n) \leq \rho(M) \leq \rho(M|^n) + \rho(M|_{n+1}) \leq \rho(M|^n) + \sum_{n+1}^{\infty} \|a_k\| < \rho(M|^n) + \varepsilon$$

whence $\rho(M) = \lim_{n\to\infty} \rho(M|^n)$. By [M, Theorem 35.6],

$$\rho(M|^n) = \sup\left\{|\lambda|_+ : \lambda \in \sigma^l_{\widehat{A}}(M|^n)\right\},$$

whence the result follows. □

It should be noted that the results of Müller, used in the above proof, were formulated in [M] for finite families in Banach algebras and for the Harte spectrum, but they hold also for the left and right spectra due to [M, Proposition 35.2].

It follows from Theorem 5.29(1) that $\rho(M) = \sup_{a_\alpha \in M} \rho(a_\alpha)$ under the posed conditions (this was proved also in [ST7]). As a consequence, we obtain that each $\mathcal{R}^a_{\mathrm{cq}}$-radical algebra is a Berger–Wang algebra (i.e., $\rho(M) = r(M)$ for all precompact sets $M$ in $A$). By [ST7, Corollary 5.15], $\mathcal{R}_{\mathrm{hc}} \vee \mathcal{R}^a_{\mathrm{cq}} \leq \mathcal{R}_{\mathrm{bw}}$ and in particular for every normed algebra $A$ the following *algebra version of the joint spectral radius formula*

$$\rho(M) = \max\left\{\rho\left(M/\left(\mathcal{R}_{\mathrm{hc}} \vee \mathcal{R}^a_{\mathrm{cq}}\right)(A)\right), r(M)\right\}$$

holds for every precompact set $M$ in $A$. We do not know however whether every Berger–Wang algebra is $\mathcal{R}_{\mathrm{bw}}$-radical.

**Proposition 5.30.** *Let $A$ be a normed algebra. If $A/\mathcal{R}_{\mathrm{bw}}(A)$ is a Berger–Wang algebra then $A$ is a Berger–Wang algebra.*

*Proof.* Indeed,

$$\rho(M) = \max\{\rho(M/\mathcal{R}_{\mathrm{bw}}(A)), r(M)\} = \max\{r(M/\mathcal{R}_{\mathrm{bw}}(A)), r(M)\} = r(M)$$

for every precompact set $M$ in $A$. Therefore $A$ is a Berger–Wang algebra. □

The following theorem supplies us with a class of examples of Banach algebras which are $\mathcal{R}^a_{\mathrm{cq}}$-radical (and therefore $\mathcal{R}^a_t$-radical); these algebras hold an important place in the theory of joint spectra of Lie representations (see for instance [F, BS, Ds]).

Recall that algebras can be considered as Lie algebras with respect to the Lie bracket $[a, b] = ab - ba$. A Lie subalgebra of an algebra $A$ is a subspace $L \subset A$ with the property that $[a, b] \in L$ for all $a, b \in L$.

Operators $\mathrm{ad}_L(a) : x \longmapsto [a, x]$ on a Lie algebra $L$, for $a \in L$, are called *adjoint operators* of $L$.

For a Lie algebra $L$, one defines the upper (lower) central series $L^{[n]}$ (respectively $L_{[n]}$) by setting $L_{[1]} = L^{[1]} = L$, $L_{[n+1]} = [L, L_{[n]}]$ and $L^{[n+1]} = [L^{[n]}, L^{[n]}]$; $L$ is called *nilpotent* (*solvable*) if $L_{[n]} = 0$ (respectively $L^{[n]} = 0$), for some $n$.

**Theorem 5.31.** *Let $A$ be a normed algebra and let $L$ be a Lie subalgebra of $A$. If one of the following conditions holds:*

(1) *$L$ is a nilpotent Lie algebra, and the inverse-closed subalgebra generated by $L$ is dense in $A^1$;*

(2) *$L$ is a finite-dimensional solvable Lie algebra, and the subalgebra generated by $L$ is dense in $A$;*

*then $A$ is commutative modulo $\mathcal{R}_{\mathrm{cq}}(A)$, i.e., $A = \mathcal{R}_{\mathrm{cq}}^{a}(A)$.*

*Proof.* Let $A' = A/\mathcal{R}_{\mathrm{cq}}(A)$, and let $L'$ be the image of $L$ in $A'$.

(1) If $L'$ is not commutative then, by the Kleinecke–Shirokov theorem, there is a non-zero quasinilpotent element $x \in [L', L']$ in the center of $L'$. Let $B_0$ be the subalgebra of $A'$ generated by $L'$. Then clearly $x$ is in the center of $B_0$. Let $B_1$ be the subalgebra generated by $B_0$ and all inverses of elements from $B_0$. Then $x$ is again in the center of $B_1$. If we have already built $B_n$ and proved that $x$ is in the center of $B_n$, then $B_{n+1}$ is defined as a subalgebra generated by $B_n$ and all inverses of elements from $B_n$, and it is clear that $x$ is in the center of $B_{n+1}$.

Let $C = \cup_{n=0}^{\infty} B_n$. Then $C$ is an inverse-closed subalgebra of $A'$; indeed, if $a \in B_n$ has an inverse $a^{-1}$ in $A'$ then $a^{-1} \in B_{n+1}$. It is clear that $C$ is dense in $A'$. Thus $x$ is in the center of $A'$. As $x$ commutes with elements of each $M \in \mathfrak{k}(A')$ then

$$\rho(\{x\} \cup M) \le \max\{\rho(x), \rho(M)\} = \rho(M)$$

since $x$ is a quasinilpotent element. Therefore $x \in \mathcal{R}_{\mathrm{cq}}(A') = 0$ (see Section 2.3.4), a contradiction.

So $L'$ is commutative, whence $A$ is commutative modulo $\mathcal{R}_{\mathrm{cq}}(A)$.

(2) Let $I$ be the set of nilpotents in $L'$. By [BS, Proposition 24.1], $I$ is a Lie ideal of $L'$. As $I$ is finite dimensional, the nilpotency indexes of elements of $I$ are uniformly bounded, so $I$ is a nilpotent Lie algebra (by the algebraic Engel condition) and, moreover, generates nilpotent subalgebra in $A$, whence $I^n = 0$ for some $n > 0$. As $L'I \subset IL' + I$, it follows that $I$ generates a nilpotent ideal in the subalgebra generated by $L'$ and generates therefore a closed nilpotent ideal in $A'$. But every nilpotent ideal lies in $\mathcal{R}_{\mathrm{cq}}(A') = \mathcal{R}_{\mathrm{cq}}(A/\mathcal{R}_{\mathrm{cq}}(A)) = 0$, whence $I = 0$. This implies that $L'$ is a nilpotent Lie algebra; indeed, every eigenvector $x$ of an adjoint operator $\mathrm{ad}_{L'}(a)$ corresponding to a non-zero eigenvalue is nilpotent which is impossible. By (1), $A' = \mathcal{R}_{\mathrm{cq}}^{a}(A')$, whence $A'$ is commutative and then $A = \mathcal{R}_{\mathrm{cq}}^{a}(A)$. □

Now we will discuss conditions under which the algebras generated (in the above sense) by nilpotent Lie subalgebras belong to a special subclass of the class of $\mathrm{Rad}^a$-radical algebras; this subclass occupies an important place in the theory of linear operator equations and, more generally, in the study of multiplication operators on Banach algebras (see [ST6]).

A normed algebra $A$ is called an *Engel algebra* if all operators $\mathrm{ad}_A(a)$, $a \in A$, are quasinilpotent. It was proved in [ST4] (and can be deduced from earlier results of [AM]) that each Engel Banach algebra $A$ is commutative modulo $\mathrm{Rad}(A)$. The converse is not true even for finite-dimensional algebras, for instance for the algebra of all upper-triangular matrices. Clearly the class of all Engel algebras is stable under taking closed ideals and quotients; it evidently contains all commutative and all radical Banach algebras; as a consequence, this class is not stable under extensions.

An operator $T$ on an algebra $A$ is call *elementary* if $T = \sum_{i=1}^{n} \mathrm{L}_{a_i} \mathrm{R}_{b_i}$ for some $a_1, b_1, \ldots, a_n, b_n \in A^1$. For instance, the identity operator on $A$ is also elementary. Let $\mathcal{E\ell}(A)$ be the algebra of all elementary operators on $A$.

**Lemma 5.32.** *Let $A$ be a Banach algebra, $L$ be a nilpotent Lie subalgebra of $A$. Then the closed, inverse-closed subalgebra $B$ of $\mathcal{B}(A^1)$ generated by $\mathrm{L}_L + \mathrm{R}_L$ is commutative modulo $\mathcal{R}_{\mathrm{cq}}(B)$.*

*Proof.* Since the closure of an inverse-closed subalgebra of a Banach algebra is inverse-closed and $\mathrm{L}_L + \mathrm{R}_L$ is a nilpotent Lie algebra, we obtain that $B$ is commutative modulo $\mathcal{R}_{\mathrm{cq}}(B)$ by Theorem 5.31. □

**Theorem 5.33.** *Let $A$ be a Banach algebra, and let $L$ be a nilpotent Lie subalgebra of $A$ such that the inverse-closed subalgebra generated by $L$ is dense in $A^1$. Then the following statements are equivalent:*

(1) *$A$ is an Engel algebra;*

(2) *For every elementary operator $\sum \mathrm{L}_{a_i} \mathrm{R}_{b_i}$,*

$$\sigma_{\mathcal{B}(A^1)}\left(\sum_1^n \mathrm{L}_{a_i} \mathrm{R}_{b_i}\right) = \sigma_{A^1}\left(\sum_1^n a_i b_i\right);$$

(3) *The closed subalgebra generated by $\mathrm{ad}_A(A)$ is compactly quasinilpotent;*

(4) *$\mathrm{ad}_A(a)$ is quasinilpotent for every $a \in L \setminus [L, L]$.*

*Proof.* (2) ⇒ (1) ⇒ (4) and (3) ⇒ (1) are obvious; (1) ⇒ (3) follows from Theorem 5.31 and Lemma 5.32.

(1) ⇒ (2) Let $B$ be the closed, inverse-closed subalgebra of $\mathcal{B}(A^1)$ generated by $\mathrm{L}_L + \mathrm{R}_L$, and let $T = \sum_1^n \mathrm{L}_{a_i}(\mathrm{R}_{b_i} - \mathrm{L}_{b_i})$ and $c = \sum_1^n a_i b_i$. As $B$ is inverse-closed,

$$\sigma_{\mathcal{B}(A^1)}\left(\sum_1^n \mathrm{L}_{a_i} \mathrm{R}_{b_i}\right) = \sigma_B\left(\sum_1^n \mathrm{L}_{a_i} \mathrm{R}_{b_i}\right) = \sigma_B(T + \mathrm{L}_c).$$

By Lemma 5.32, $B$ is commutative modulo $\mathcal{R}_{\mathrm{cq}}(B)$. As $\mathcal{R}_{\mathrm{cq}}(B) \subset \mathrm{Rad}(B)$ then $B$ is commutative modulo $\mathrm{Rad}(B)$, whence the set of all quasinilpotent elements of $B$ coincides with $\mathrm{Rad}(B)$. Then, as $A$ is Engel, it follows that $T \in \mathrm{Rad}(B)$ and $\sigma_B(T + \mathrm{L}_c) = \sigma_B(\mathrm{L}_c)$. But $\sigma_B(\mathrm{L}_c) = \sigma_{\mathcal{B}(A^1)}(\mathrm{L}_c) = \sigma_{A^1}(c)$ by [BD, Proposition 3.19], and the result follows.

(4) ⇒ (1) As $[L, L]$ consists of quasinilpotents, it follows that $\mathrm{ad}_A(L)$ consists of quasinilpotents. Let $E = \{a \in A^1 : \rho(\mathrm{ad}_{A^1}(a)) = 0\}$. It is easy to see that $E$ is an algebra by using Lemma 5.32. If $x \in E$ is invertible in $A$ then

$$\rho(\mathrm{ad}_{A^1}(x^{-1})) = \rho(\mathrm{L}_{x^{-1}} \mathrm{ad}_{A^1}(x) \mathrm{R}_{x^{-1}}) \leq \rho(\mathrm{L}_{x^{-1}}) \rho(\mathrm{ad}_{A^1}(x)) \rho(\mathrm{R}_{x^{-1}}) = 0$$

by Lemma 5.32, whence $E$ is an inverse-closed subalgebra of $A$. As $L \subset E$ then $E$ is dense in $A$.

Let $B$ be defined as in Lemma 5.32; as $B$ is commutative modulo $\mathrm{Rad}(B)$ then the spectral radius is continuous on $B$. Then $E$ is closed and $A = E$. □

**Corollary 5.34.** *Let $A$ be a Banach algebra, and let $L$ be a nilpotent Lie subalgebra of $A$ such that the inverse-closed subalgebra generated by $L$ is dense in $A^1$. If $\sigma_{A^1}(a)$ is at most countable for every $a \in L\backslash[L, L]$, then $A$ is an Engel algebra.*

*Proof.* Let $A_0$ be the subalgebra of $A^1$ generated by $L$ and the identity element 1 of $A^1$. It follows that $\sigma_{\mathcal{B}(A^1)}(\mathrm{ad}_{A^1}(a))$ is at most countable for every $a \in L\backslash[L, L]$. As $\mathrm{ad}_{A^1}(a)$ is nilpotent on $L$, it is locally nilpotent on $A_0$ and is quasinilpotent on the closure $B$ of $A_0$ by [ST4, Corollary 3.7]. To apply Theorem 5.33, it is sufficient to show $B$ is an inverse-closed subalgebra of $A$.

Let $x \in B$ and $x^{-1} \in A$. Then there is a sequence $x_n$ of elements of $A_0$ such that $x_n \to x$ as $n \to \infty$. As the map $a \longmapsto a^{-1}$ is continuous on the set of all invertible elements of $A$ and this set is open, one can assume that for every $n$ there is $x_n^{-1} \in A$ and the sequence $x_n^{-1} \to x^{-1}$ as $n \to \infty$.

Note that, for every $n$, there is a polynomial $p_n$ such that $x_n = p_n(a_1, \ldots, a_k)$ for some $a_1, \ldots, a_k \in L\backslash[L, L]$. By the spectral mapping theorem [T1], we obtain that

$$\sigma_{A^1}(p_n(a_1, \ldots, a_k)) = p_n(\sigma_{A^1}(a_1, \ldots, a_k)) \subset p_n(\sigma_{A^1}(a_1) \times \cdots \times \sigma_{A^1}(a_k)).$$

As $\sigma_{A^1}(a_1) \times \cdots \times \sigma_{A^1}(a_k)$ is at most countable, it follows that the spectrum $\sigma_{A^1}(x_n)$ is also at most countable. By [BD, Theorem 5.11], $x_n^{-1}$ lies in the closed subalgebra of $A^1$ generated by $x_n$ and 1. As $x_n, 1 \in B$ then $x_n^{-1} \in B$ for every $n$. Therefore $x^{-1} \in B$.

We showed that $B$ is inverse-closed. Therefore $B = A^1$. As $\mathrm{ad}_A(a)$ is quasinilpotent for every $a \in L\backslash[L, L]$, it follows that $A$ is an Engel algebra by Theorem 5.33. □

Some related results are obtained in [C].

**Problem 5.35.** *Does every Banach algebra have the largest Engel ideal?*

## 6. Socle procedure and radicals

### 6.1. Socle

Let $A \in \mathfrak{U}_\mathrm{a}$ be an algebra. The (left) *socle* $\mathrm{soc}(A)$ of $A$ is the sum of all minimal left ideals of $A$. If $A$ has no minimal left ideals then $\mathrm{soc}(A) = 0$; $\mathrm{soc}(A)$ is an ideal of $A$ [BD, Lemma 30.9]. If $I$ is a minimal left ideal of $A$ and $I^2 \neq 0$ then there is a projection $p \in I$ such that $I = Ap$, and every such projection $p$ is minimal [BD, Lemma 30.2]. Recall that a non-zero projection $p$ of $A$ is *minimal* if $pAp$ is a division algebra.

Let $\mathrm{Min}(A)$ be the set of minimal projections of $A$. If $A$ is semiprime then $\mathrm{soc}(A)$ equals the sum of all minimal right ideals of $A$ (i.e., the *right socle*) [BD, Proposition 30.10] and $L$ is a left (right) minimal ideal of $A$ if and only if $L = Ap$ (respectively, $L = pA$) for some $p \in \mathrm{Min}(A)$.

**Remark 6.1.** If $B \in \mathfrak{U}_{\mathrm{n}}$ is a division algebra then $B$ is isomorphic to $\mathbb{C}$ by the Gelfand–Mazur theorem. If $A$ is a semiprime Banach algebra then $\mathrm{soc}(A)$ is closed if and only if $A$ is finite dimensional [T], $\overline{\mathrm{soc}(A)} \cap \mathrm{Rad}\,(A) = 0$ [GR, Lemma 4] and $\mathrm{soc}(A)$ is the largest ideal of algebraic elements of $A$ [GR, Theorem 5].

**Proposition 6.2.**

(1) *The map* soc *is a preradical on* $\mathfrak{U}_{\mathrm{a}}$.
(2) *If* $A \in \mathfrak{U}_{\mathrm{a}}$ *is semiprime then* $\mathrm{soc}\,(J) = J \cap \mathrm{soc}\,(A)$ *for each ideal* $J$ *of* $A$*; in particular,* $\mathrm{rad}\,(A) \cap \mathrm{soc}\,(A) = 0$.

*Proof.* (1) is straightforward.

(2) Let $E = \{p \in \mathrm{Min}(A) : Jp \neq 0\}$. Clearly $J$ is a semiprime algebra, so the minimal left ideals of $J$ are determined by $\mathrm{Min}(J)$. Let $p \in \mathrm{Min}(J)$ and $L = Jp$. Then $L = Jp \subset Ap \subset L$ because $p = pp \in Jp = L$. Hence $Jp = Ap$ is a left ideal of $A$ and $p \in \mathrm{Min}(A)$ because $pJp = pAp$ is a division algebra. So $Jp$ is a minimal left ideal of $A$. This proves that $\mathrm{soc}\,(J) \subset J \cap \mathrm{soc}\,(A)$ and $\mathrm{Min}(J) \subset E$. Let us prove the converse inclusions.

If $p \in E$ then $Ap$ is a minimal left ideal of $A$. As $Jp \subset Ap$ is a non-zero left ideal of $A$, it follows that $Jp = Ap$, whence $pJp = pAp$ is a division algebra. Therefore $E \subset \mathrm{Min}(J)$. We proved that

$$\mathrm{Min}(J) = \{p \in \mathrm{Min}(A) : Jp \neq 0\}. \tag{6.1}$$

Let $a \in J \cap \mathrm{soc}\,(A)$ be arbitrary. Then $a = \sum b_i p_i + \sum c_j p_j'$ (both sums are finite) for all $b_i, c_j \in A$, where all $p_i \in \mathrm{Min}(J)$ and all $p_j' \in \mathrm{Min}(A)$ with $Jp_j' = 0$. In other words, $a = b + c$, where

$$b := \sum b_i p_i = \sum (b_i p_i)\, p_i \in \mathrm{soc}\,(J)$$

and $c := \sum c_j p_j' = a - b \in J$. By condition, $Jp_j' = 0$ for all $j$, so one has $Jc = 0$. But the set $K = \{d \in J : Jd = 0\}$ is an ideal of $A$ with $K^2 = 0$, whence $K = 0$ and therefore $c = 0$ and $a = b \in \mathrm{soc}\,(J)$. Thus $J \cap \mathrm{soc}\,(A) \subset \mathrm{soc}\,(J)$.

If $\mathrm{rad}\,(A) \cap \mathrm{soc}\,(A) \neq 0$ then $\mathrm{rad}\,(A)$ has a non-zero socle and there is a non-zero projection $p$ in $\mathrm{rad}\,(A)$; this is impossible. □

Let $P$ be a preradical. Define the map $P^{\mathrm{soc}}$ by

$$P^{\mathrm{soc}} = \mathrm{soc} * P,$$

so $P^{\mathrm{soc}}$ is the preimage of $\mathrm{soc}\,(A/P\,(A))$ in $A$, for each algebra $A$. Let

$$\mathrm{Min}_P(A) = \left\{x \in A : q_{P(A)}\,(x) \in \mathrm{Min}(A/P\,(A))\right\}$$

where $q_{P(A)} : A \longrightarrow A/P\,(A)$ is the standard quotient map.

**Theorem 6.3.** *Let* $P$ *be a radical such that* $P \geq \mathfrak{P}_\beta$. *Then*

(1) $\overline{P^{\mathrm{soc}}}$ *is a topological under radical;*
(2) *If* $P$ *is topological then* $\overline{P^{\mathrm{soc}}} = P^{\overline{\mathrm{soc}}}$;
(3) *If* $P$ *is algebraic then* $P^{\mathrm{soc}}$ *is an algebraic under radical, and if* $P$ *is hereditary then* $P^{\mathrm{soc}\,*}$ *is hereditary.*

*Proof.* In (2) the equality $\overline{P^{\mathrm{soc}}} = P^{\overline{\mathrm{soc}}}$ is evident; (1) is a consequence of (3) and Theorem 3.5 in the algebraic case and is proved similarly to (3) otherwise. So we will prove only (3).

(3) $P^{\mathrm{soc}}$ is a preradical by Lemma 4.10. Let $A$ be an algebra, and let $J$ be an ideal of $A$. Since $A/P(A)$ and $J/P(J)$ are semiprime then, to prove the inclusion $P^{\mathrm{soc}}(J) \subset P^{\mathrm{soc}}(A)$, it suffices to show that

$$\mathrm{Min}_P(J) \subset \mathrm{Min}_P(A) \cup \{0\}. \tag{6.2}$$

Let $q' : J \longrightarrow J/P(J)$ be the quotient map, $q = q_{P(A)}$ (as above), and let $g : J/P(J) \longrightarrow A/P(A)$ be the homomorphism $q'(a) \longmapsto q(a)$ for $a \in J$. Let $e \in \mathrm{Min}_P(J)$ be arbitrary. Then $q'(eJe)$ is a division algebra. If $q(e) \neq 0$ then $q(eJe) = g(q'(eJe))$ is also a division algebra and $q(e) \in \mathrm{Min}(q(J)) \subset \mathrm{Min}(A/P(A))$ by (6.1). In all cases $e \in \mathrm{Min}_P(A) \cup \{0\}$. This completes the proof of (6.2), so the inclusion $P^{\mathrm{soc}}(J) \subset P^{\mathrm{soc}}(A)$ holds.

If $a \in A$ is arbitrary, then $q(a)q(Je) \subset q(Je)$ and $q(Je)q(a) \subset q(Jea) = q(Jeea)$. As $q(ea) \in q(J)$ then, by [BD, Lemma 30.7], $q(Je(ea))$ is a minimal left ideal of $q(J)$ or zero. So $P^{\mathrm{soc}}(J)$ is an ideal of $A$.

Let $B = P^{\mathrm{soc}}(A)$. Then $P(B) \subset P(A)$. But $P(A)$ is a $P$-radical ideal of $B$; so $P(B) = P(A)$. Hence $\mathrm{soc}(B/P(B)) = \mathrm{soc}(A/P(A))$. This shows that $P^{\mathrm{soc}}(P^{\mathrm{soc}}(A)) = P^{\mathrm{soc}}(A)$. We proved that $P^{\mathrm{soc}}$ is an under radical.

Let now $P$ be hereditary. Since $P(J) = J \cap P(A)$ then $g : q'(x) \longmapsto q(x)$ is an isomorphism of $J/P(J)$ onto the ideal $q(J)$ of $A/P(A)$. Therefore

$$g(\mathrm{Min}(J/P(J))) = \mathrm{Min}(q(J)) = \{q(y) \in \mathrm{Min}(A/P(A)) : q(Jy) \neq 0\}$$

by (6.1). So if $y \in J$ and $q(y) \in \mathrm{Min}(A/P(A))$ then $q'(y) \in \mathrm{Min}(J/P(J))$. It follows that $J \cap P^{\mathrm{soc}}(A) \subset P^{\mathrm{soc}}(J)$. Then $P^{\mathrm{soc}}$ is hereditary. By Corollary 4.23, $P^{\mathrm{soc}*}$ is hereditary. □

As a consequence of Theorem 6.3, $P^{\mathrm{soc}*}$ and $\overline{P^{\mathrm{soc}}}^{*}$ are (algebraic and, respectively, topological) radicals.

### 6.2. Some applications

**Lemma 6.4.** *Let $A$ be an algebra, and let $I$ be a one-sided ideal of $A$. Then*

(1) $\mathrm{rad}(I) = I \cap \mathrm{rad}(A)$.
(2) *If $J$ is a one-sided ideal of $I$ then* $\mathrm{rad}(J) = J \cap \mathrm{rad}(A)$.

*Proof.* (1) Let $I$ be a left ideal of $A$. As $\mathrm{rad}(A)$ is the largest left ideal consisting of left quasi-invertible elements of $A$ [BD, Proposition 24.16], $\mathrm{rad}(I) \subset \mathrm{rad}(A)$ and also $I \cap \mathrm{rad}(A) \subset \mathrm{rad}(I)$.

(2) By (1), we have that $\mathrm{rad}(J) = J \cap \mathrm{rad}(I)$. Then

$$\mathrm{rad}(J) = J \cap \mathrm{rad}(I) = J \cap I \cap \mathrm{rad}(A) = J \cap \mathrm{rad}(A).$$ □

**Proposition 6.5.** *Let $A$ be an algebra, and let $J$ be a one-sided ideal of $A$. Then* $\mathrm{rad}^{\mathrm{soc}}(J) \subset \mathrm{rad}^{\mathrm{soc}}(A)$.

*Proof.* Let $J$ be a left ideal. As $\operatorname{rad}(J) = J \cap \operatorname{rad}(A)$ by Lemma 6.4, the standard map $g : J/\operatorname{rad}(J) \longrightarrow B := A/\operatorname{rad}(A)$ defined by $g : x/\operatorname{rad}(J) \longmapsto x/\operatorname{rad}(A)$ is one-to-one. Let $I = g(J/\operatorname{rad}(J))$. Then $I$ is a left ideal of $B$ and one may identify the socle of $J/\operatorname{rad}(J)$ with the socle of $I$ (because $g$ and $g^{-1}$ are morphisms).

If $L$ is a minimal left ideal of $I$ there is $p \in \operatorname{Min}(I)$ such that $L = Ip$. Then $p \in L$, $BL$ is a left ideal of $B$ and $BL \subset L$. So either $BL = 0$ or $BL = L$. But the equality $BL = 0$ is impossible since $B$ is semisimple. So $L$ is a minimal left ideal of $B$. This proves $\operatorname{soc}(I) \subset \operatorname{soc}(B)$ and the result follows. □

Recall that $\mathcal{F}(A)$ is the set of all finite rank elements of $A$.

**Lemma 6.6.** *Let $A$ be a semiprime normed algebra. Then*

(1) $\overline{\operatorname{soc}(A)} = \overline{\mathcal{F}(A)}$;
(2) $\overline{\operatorname{soc}(A)} \cap \operatorname{rad}(A) = 0$;
(3) *If $a, b \in \mathcal{F}(A)$ then $\mathrm{L}_a \mathrm{R}_b$ is a finite rank operator on $A$.*

*Proof.* (1) Let $p \in \operatorname{Min}(A)$ be arbitrary. As $pAp$ is a division normed algebra, $pAp$ is one-dimensional by the Gelfand–Masur theorem, so $p \in \mathcal{F}(A)$, whence $Ap \subset \mathcal{F}(A)$ and $\operatorname{soc}(A) \subset \mathcal{F}(A)$.

Let $x \in \mathcal{F}(A)$. Then the algebra $xA^1x$ is finite dimensional. As $A^1x$ is a left ideal of $A^1$ then $\operatorname{rad}(A^1x) = A^1x \cap \operatorname{rad}(A)$ by Lemma 6.4. Let $B = xA^1x + \mathbb{C}x$. Then $B$ is a finite-dimensional algebra and $B \subset \mathcal{F}(A)$.

As $B$ is a right ideal of $A^1x$, it follows that

$$\operatorname{rad}(B) = B \cap \operatorname{rad}(A) \subset \mathcal{F}(A) \cap \operatorname{rad}(A) = 0$$

by Lemmas 6.4 and 4.32. As $B$ is a finite-dimensional semisimple algebra then $B = \operatorname{soc}(B)$. Therefore $x \in L_1 + \cdots + L_n$ where $L_i$ is a minimal left ideal of $B$ for every $i$. Then there is $p_i \in \operatorname{Min}(B)$, for every $i$, such that $L_i = Bp_i$. As

$$p_iA^1p_i = p_i\left(p_iA^1p_i\right)p_i \subset p_iBp_i \subset p_iA^1p_i$$

and $p_iBp_i = \mathbb{C}p_i$ then $p_i \in \operatorname{Min}\left(A^1\right)$ and $A^1p_i$ is a minimal left ideal of $A^1$, whence $x \in \operatorname{soc}\left(A^1\right)$. It remains to note that if $A$ is infinite-dimensional then clearly $\operatorname{soc}\left(A^1\right) = \operatorname{soc}(A)$, and if $A$ is finite dimensional then $A$ is semisimple and $A = A^1$ by the classical Wedderburn results.

(2) Follows from (1) and Lemma 4.32.

(3) One may assume that $A$ is unital. Let $p_1, p_2 \in \operatorname{Min}(A)$ be arbitrary. As $p_1 + p_2 \in \operatorname{soc}(A)$ then $\mathrm{L}_{p_1+p_2}\mathrm{R}_{p_1+p_2}$ is a finite rank operator, whence $\mathrm{L}_{p_1}\mathrm{R}_{p_2} + \mathrm{L}_{p_2}\mathrm{R}_{p_1}$ is a finite rank operator on $A$. Then

$$\mathrm{L}_{1-p_2}\left(\mathrm{L}_{p_1}\mathrm{R}_{p_2} + \mathrm{L}_{p_2}\mathrm{R}_{p_1}\right) = \mathrm{L}_{p_1}\mathrm{R}_{p_2} - \mathrm{L}_{p_2}\mathrm{L}_{p_1}\mathrm{R}_{p_2}$$

is a finite rank operator. As $\mathrm{L}_{p_2}\mathrm{L}_{p_1}\mathrm{R}_{p_2} = \left(\mathrm{L}_{p_2}\mathrm{R}_{p_2}\right)\mathrm{L}_{p_1}$ is a finite rank operator, $\mathrm{L}_{p_1}\mathrm{R}_{p_2}$ is a finite rank operator.

Let $a, b \in \operatorname{soc}(A)$. Then there are $p_1, \ldots, p_n \in \operatorname{Min}(A)$, $x_1, y_1, \ldots, x_n, y_n \in A$ such that $a = \sum_i x_i p_i$ and $b = \sum_j y_j p_j$. Then

$$\mathrm{L}_a \mathrm{R}_b = \sum_{i=1}^{n} \sum_{j=1}^{n} \mathrm{L}_{x_i p_i} \mathrm{R}_{y_j p_j} = \sum_{i=1}^{n} \sum_{j=1}^{n} \mathrm{L}_{x_i} \left( \mathrm{L}_{p_i} \mathrm{R}_{p_j} \right) \mathrm{R}_{y_j}$$

is a finite rank operator. □

**Remark 6.7.** (1) and (3) of Lemma 6.6 were proved for not necessary normed, semiprime algebras with lower socle instead of socle in [BE, Lemma 3.1 and Theorem 3.3]; the lower socle of $A$ is defined as the ideal generated by all minimal projections $p$ with $pAp < \infty$. It should be noted that our proofs are completely different from ones in [BE].

Lemma 6.6(3) yields

**Corollary 6.8.** *Every finite semiprime normed algebra is bifinite.*

It follows that if $A$ is a semiprime normed algebra and $\mathcal{R}_{\mathrm{hf}}(A) \neq 0$ then $\mathcal{E\ell}(A)$ has nonzero finite rank operators. The converse is also true.

**Theorem 6.9.** *Let $A$ be an algebra. If there is a finite rank elementary operator $T$ on $A$ then the image of $T$ is contained in $\mathfrak{R}_{\mathrm{hf}}(A)$.*

*Proof.* Let $T = \sum_{i=1}^{n} \mathrm{L}_{a_i} \mathrm{R}_{b_i}$, $I = \mathfrak{R}_{\mathrm{hf}}(A)$ and $B = A/I$. Then the operator $S = \sum_{i=1}^{n} \mathrm{L}_{a_i/I} \mathrm{R}_{b_i/I}$ is a finite rank elementary operator on $B$. By [BT, Lemma 7.1], if $S \neq 0$ then there is a non-zero finite rank element of $B$ which is impossible. Therefore $S = 0$ and the image of $T$ is contained in $\mathfrak{R}_{\mathrm{hf}}(A)$. □

In particular, if $A$ is normed then $\mathfrak{R}_{\mathrm{hf}}(A) \subset \mathcal{R}_{\mathrm{hf}}(A)$ and the image of $T$ lies in $\mathcal{R}_{\mathrm{hf}}(A)$.

**Remark 6.10.** By [BT, Theorem 8.4], if $A$ is a Banach algebra and $T$ is a compact elementary operator on $A$ then the image of $T$ is contained in $\mathcal{R}_{\mathrm{hc}}(A)$. It should be noted that this result holds also for normed algebras, with the same proof.

**Problem 6.11.** *Let $A$ be a Banach algebra, and let $T = \sum_{i=1}^{\infty} \mathrm{L}_{a_i} \mathrm{R}_{b_i}$ be a compact operator on $A$ with $\sum_i^{\infty} \|a_i\| \|b_i\| < \infty$ for some $a_i$, $b_i \in A$. Is the image of $T$ contained in $\mathcal{R}_{\mathrm{hc}}(A)$?*

**Theorem 6.12.** $\mathcal{R}_{\mathrm{hf}} = \mathcal{P}_{\beta}^{\overline{\mathrm{soc}}*}$.

*Proof.* Let $A$ be a normed algebra. If $A$ is $\mathcal{P}_{\beta}^{\overline{\mathrm{soc}}*}$-semisimple then $A$ is semiprime and $\operatorname{soc}(A) = 0$. By Lemma 6.6, $A$ has no non-zero finite rank elements, whence $A$ is $\mathcal{R}_{\mathrm{hf}}$-semisimple. Therefore $\mathcal{R}_{\mathrm{hf}} \leq \mathcal{P}_{\beta}^{\overline{\mathrm{soc}}*}$ by Theorem 2.3.

Conversely, if $A$ has no non-zero finite rank elements then $A$ is semisimple and $\operatorname{soc}(A) = 0$. Therefore $\mathcal{P}_{\beta}^{\overline{\mathrm{soc}}*} \leq \mathcal{R}_{\mathrm{hf}}$ by Theorem 2.3. □

**Theorem 6.13.** $\mathcal{R}_{\mathrm{hc}} = \mathcal{R}_{\mathrm{hf}} \vee \mathcal{R}_{\mathrm{jhc}} = \mathcal{R}_{\mathrm{jhc}}^{\overline{\mathrm{soc}}*} \leq \mathcal{R}_{\mathrm{cq}}^{\overline{\mathrm{soc}}*}$ *on Banach algebras.*

*Proof.* Let $A$ be a Banach algebra. If $A$ is $\mathcal{R}_{\mathrm{jhc}}^{\overline{\mathrm{soc}}*}$-semisimple then $A$ has no non-zero compact elements which lie in $\mathrm{Rad}\,(A)$, and $\mathrm{soc}\,(A) = 0$. Hence $A$ is semiprime, so $A$ has no non-zero finite rank elements by Lemma 6.6. Thus $A$ is $(\mathcal{R}_{\mathrm{hf}} \vee \mathcal{R}_{\mathrm{jhc}})$-semisimple.

It follows that $A$ has no non-zero compact elements. Indeed, if $a \in A$ is a compact element then every spectral projection of $a$ corresponding to a non-zero eigenvalue is clearly a finite rank element. So $a$ is quasinilpotent. But $A^1 a$ consists of compact elements, whence $A^1 a \subset \mathrm{Rad}\,(A)$. By the assumption, $a = 0$. Thus $A$ is $\mathcal{R}_{\mathrm{hc}}$-semisimple.

By Theorem 2.3,

$$\mathcal{R}_{\mathrm{jhc}}^{\overline{\mathrm{soc}}*} \leq \mathcal{R}_{\mathrm{hf}} \vee \mathcal{R}_{\mathrm{jhc}} \leq \mathcal{R}_{\mathrm{hc}}$$

on Banach algebras. Assume now that $A$ has no non-zero compact elements. Then it is clear that $A$ is semiprime, has no non-zero finite rank elements and $\mathrm{soc}\,(A) = 0$. This proves the converse.

Let now $A$ be $\mathcal{R}_{\mathrm{cq}}^{\overline{\mathrm{soc}}*}$-semisimple. As $\mathcal{R}_{\mathrm{cq}}\,(A) = 0$ then $A$ is semiprime, and as $\mathrm{soc}\,(A) = 0$ then $A$ has no non-zero finite rank elements by Lemma 6.6. As above, if $a \in A$ is a compact element then $a \in \mathrm{Rad}\,(A)$. But

$$\mathcal{R}_{\mathrm{hc}}\,(A) \cap \mathrm{Rad}\,(A) \subset \mathcal{R}_{\mathrm{cq}}\,(A) = 0$$

by (2.9) (the inclusion follows by the algebraic version of the joint spectral radius formula (2.8)), whence $a = 0$. Therefore $A$ has no non-zero compact elements, i.e., $A$ is $\mathcal{R}_{\mathrm{hc}}$-semisimple. By Theorem 2.3, $\mathcal{R}_{\mathrm{hc}} \leq \mathcal{R}_{\mathrm{cq}}^{\overline{\mathrm{soc}}*}$ on Banach algebras. □

## 7. The kernel-hull closures of radicals and the primitivity procedure

### 7.1. The kernel-hull closures of radicals

Let $A$ be an algebra; $\mathrm{Prim}\,(A)$ is called the *structure space* of $A$. For any subset $E$ of $A$, let $\mathrm{hull}\,(E;A)$ or simply $\mathrm{h}(E;A)$ be the set of all $I \in \mathrm{Prim}\,(A)$ with $E \subset I$; this set is called a *hull* of $E$. For any set $W \subset \mathrm{Prim}\,(A)$, let $\mathrm{ker}(W;A)$ or $\mathrm{k}(W;A)$ be defined by $\mathrm{k}(W;A) = \cap_{I \in W} I$; this ideal is called a *kernel* of $W$. Note that the Jacobson radical rad coincides with the *kernel-hull closure* (briefly, kh-*closure*) of zero:

$$\mathrm{rad}\,(A) := \cap\,\mathrm{Prim}\,(A) = \mathrm{kh}\left(\{0\}\,;A^1\right) := \mathrm{k}\left(\mathrm{h}\left(\{0\}\,;A^1\right);A^1\right)$$

for each $A \in \mathfrak{U}_{\mathrm{a}}$. The operation of the *hull-kernel closure*

$$M \longmapsto \mathrm{hk}\,(M;A) := \mathrm{h}\,(\mathrm{k}\,(M;A)\,;A)\,,$$

for $M \subset \mathrm{Prim}\,(A)$, determines the *Jacobson topology* on $\mathrm{Prim}\,(A)$ [Bo, Section 1.1]: closed sets are the sets of form $\mathrm{hk}\left(M;A^1\right)$. As is known, this topology is not Hausdorff in general.

Let $\Omega$ be a primitive map, and let $\Pi_\Omega$ be the related ideal map on $\mathfrak{U}$ (Section 2.3.2). Let $A$ be an algebra, and let

$$\mathrm{Irr}_\Omega\,(A) = \{\pi \in \mathrm{Irr}\,(A) : \ker \pi \in \Omega\,(A)\}\,.$$

For every ideal $J$ of $A$, put

$$\mathrm{h}_{\Omega}(J;A) = \mathrm{h}(J;A) \cap \Omega(A).$$

**Lemma 7.1.** *Let $A$ be an algebra and $I, J$ be ideals of $A$. Then*

(1) *If $I \subset J$ then* $\mathrm{kh}_{\Omega}(I;A) \subset \mathrm{kh}_{\Omega}(J;A)$;
(2) $\mathrm{kh}_{\Omega}(I \cap J;J) = J \cap \mathrm{kh}_{\Omega}(I;A)$;
(3) $\mathrm{kh}_{\Omega}(J;A) = q_J^{-1}(\Pi_{\Omega}(A/J))$;
(4) *If $A$ is a $Q$-algebra then* $I \subset \overline{I} \subset \mathrm{kh}_{\Omega}(I;A) = \mathrm{kh}_{\Omega}(\overline{I};A)$.

*Proof.* (1) It is clear that $\mathrm{kh}(I;A) \subset \mathrm{kh}(J;A)$, whence $\mathrm{kh}_{\Omega}(I;A) \subset \mathrm{kh}_{\Omega}(J;A)$.

(2) Let $K = I \cap J$, and let $\tau \in \mathrm{Irr}_{\Omega}(J)$ be arbitrary with $\tau(K) = 0$. Then, for any representation $\pi \in \mathrm{Irr}_{\Omega}(A)$ extending $\tau$ (on the same representation space), we have that

$$\pi(I)\tau(J) = \pi(IJ) = \tau(IJ) = 0.$$

It follows that $\pi(I) = 0$. Conversely, the restriction $\tau$ of each $\pi \in \mathrm{Irr}_{\Omega}(A)$ with $\pi(I) = 0$ to $J$ vanishes on $K$ and if $\tau \neq 0$ then $\tau \in \mathrm{Irr}_{\Omega}(J)$.

If $a \in J \cap \mathrm{kh}_{\Omega}(I;A)$ then $\pi(a) = 0$ for every $\pi \in \mathrm{Irr}_{\Omega}(A)$ with $\pi(I) = 0$. For $\tau \in \mathrm{Irr}_{\Omega}(J)$ corresponding to $\pi$, we have that $\tau(K) = 0$ and $\tau(a) = \pi(a) = 0$, whence $a \in \mathrm{kh}_{\Omega}(K;J)$.

If $a \in \mathrm{kh}_{\Omega}(K;J)$ then $\tau(a) = 0$ for every $\tau \in \mathrm{Irr}_{\Omega}(J)$ with $\tau(K) = 0$. For $\pi \in \mathrm{Irr}_{\Omega}(A)$ corresponding to $\tau$, we have that $\pi(I) = 0$ and either $\pi(J) = 0$ (whence $\pi(a) = 0$) or $\pi|_J \in \mathrm{Irr}_{\Omega}(J)$ with $\pi(a) = \tau(a) = 0$, whence $a \in J \cap \mathrm{kh}_{\Omega}(I;A)$.

(3) is obvious.

(4) As each $J \in \mathrm{h}_{\Omega}(I;A)$ is closed then $\overline{I} \subset J$, whence $\overline{I} \subset \mathrm{kh}_{\Omega}(I;A)$. □

Let $\Omega$ be a primitive map, and let $P$ be a preradical. Define the map $P^{\mathrm{kh}_{\Omega}}$ by

$$P^{\mathrm{kh}_{\Omega}}(A) = \mathrm{kh}_{\Omega}(P(A);A)$$

for every algebra $A$. It follows from Lemma 7.1(3) that

$$P^{\mathrm{kh}_{\Omega}}(A) = (\Pi_{\Omega} * P)(A) = q_{P(A)}^{-1}(\Pi_{\Omega}(A/P(A))) \tag{7.1}$$

for every algebra $A$.

In the following theorem and corollary we assume that $\Omega$ is a primitive map defined on a base class $\mathfrak{U}$, and one of the following conditions, listed in Theorem 2.10, holds:

($1_{kh}$) $\Omega$ is pliant and either $\mathfrak{U} = \mathfrak{U}_{\mathrm{a}}$ or $\mathfrak{U}_{\mathrm{q}} \subset \mathfrak{U} \subset \mathfrak{U}_{\mathrm{n}}$ (correspondingly, $\Pi_{\Omega}$ is either a hereditary radical or a hereditary preradical);
($2_{kh}$) $\mathfrak{U} \subset \mathfrak{U}_{\mathrm{q}}$ ($\Pi_{\Omega}$ is a hereditary topological radical).

**Theorem 7.2.** *Let $\Omega$ be a primitive map satisfying ($1_{\mathrm{kh}}$) or ($2_{\mathrm{kh}}$), and let $P$ be a preradical. Then*

(1) *$P^{\mathrm{kh}_{\Omega}}$ is a preradical, and $P^{\mathrm{kh}_{\Omega}} = \overline{P}^{\mathrm{kh}_{\Omega}}$ on $Q$-algebras;*
(2) *If $P$ is an under radical then $P^{\mathrm{kh}_{\Omega}}$ is an under radical;*
(3) *If $P$ is hereditary then $P^{\mathrm{kh}_{\Omega}}$ is hereditary;*
(4) *If $\Omega$ and $P$ are pliant then $P^{\mathrm{kh}_{\Omega}}$ is pliant.*

*Proof.* (1) If $f : A \longrightarrow B$ is a morphism then $f(P(A)) \subset P(B)$ and $f(P(A))$ is an ideal of $B$. We will show that

$$f(\mathrm{kh}_\Omega(P(A);A)) \subset \mathrm{kh}_\Omega(f(P(A));B).$$

Indeed, let $\pi \in \mathrm{Irr}_\Omega(B)$ be arbitrary and $\pi(P(B)) = 0$. Then $\tau = \pi \circ f \in \mathrm{Irr}_\Omega(A)$ and $\tau(P(A)) = 0$. Then $\tau(a) = 0$ for any $a \in \mathrm{kh}_\Omega(P(A);A)$, whence $\pi(f(a)) = 0$. Therefore $f(a) \in \mathrm{kh}_\Omega(P(B), B)$.

As $\mathrm{kh}_\Omega(f(P(A));B) \subset \mathrm{kh}_\Omega(P(B);B)$ by Lemma 7.1 then

$$f\left(P^{\mathrm{kh}_\Omega}(A)\right) = f(\mathrm{kh}_\Omega(P(A);A)) \subset \mathrm{kh}_\Omega(P(B);B) = P^{\mathrm{kh}_\Omega}(B)$$

and $P^{\mathrm{kh}_\Omega}$ is a preradical.

If $A$ is a $Q$-algebra then

$$P^{\mathrm{kh}_\Omega}(A) = \mathrm{kh}_\Omega(P(A);A) = \mathrm{kh}_\Omega\left(\overline{P(A)};A\right) = \mathrm{kh}_\Omega\left(\overline{P}(A);A\right) = \overline{P}^{\mathrm{kh}_\Omega}(A)$$

by Lemma 7.1(4).

It remains to note that primitive ideals of a $Q$-algebra are closed. Hence $P^{\mathrm{kh}_\Omega}$ is topological on $Q$-algebras.

(2) follows from (7.1) and Theorem 4.11.

(3) Let $A$ be an algebra, and let $J$ be an ideal of $A$. If $P(J) = J \cap P(A)$ then

$$P^{\mathrm{kh}_\Omega}(J) = \mathrm{kh}_\Omega(P(J);J) = J \cap \mathrm{kh}_\Omega(P(A);A) = J \cap P^{\mathrm{kh}_\Omega}(A)$$

by Lemma 7.1, whence $P^{\mathrm{hk}_\Omega}$ is hereditary.

(4) is obvious. □

**Corollary 7.3.** *Let $\Omega$ be a primitive map satisfying $(1_{\mathrm{kh}})$ or $(2_{\mathrm{kh}})$, and let $P$ be an under radical. Then*

(1) $P^{\mathrm{kh}_\Omega *} = \Pi_\Omega \vee P = \Pi_\Omega^* \vee P^*$;
(2) $P^{\mathrm{kh}_\Omega *}$ *is a radical, and* $P^{\mathrm{kh}_\Omega *} = \overline{P}^{\mathrm{kh}_\Omega *}$ *on $Q$-algebras;*
(3) *If $P$ is a hereditary preradical then* $\overline{P}^{\mathrm{kh}_\Omega *}$ *is hereditary on $Q$-algebras.*
(4) *If $\Omega$ and $P$ are pliant then* $P^{\mathrm{kh}_\Omega *}$ *is pliant.*

*Proof.* (1) is clear.

(2) Assume that $P$ is an (algebraic/topological) under radical. It is clear that

$$P^{\mathrm{kh}_\Omega *} = \mathrm{rad} \vee P = (\mathrm{rad} \vee P) \vee \mathrm{rad} = \left(P^{\mathrm{kh}_\Omega} * \mathrm{rad}\right)^*.$$

Let $(R_\alpha)$ and $(S_\alpha)$ be the algebraic and topological convolution chains of under radicals generated by $P^{\mathrm{kh}_\Omega} * \Pi_\Omega$ and by $\overline{P}^{\mathrm{kh}_\Omega} * \Pi_\Omega$, respectively.

Let $A$ be a $Q$-algebra. By Theorem 7.2(1),

$$P^{\mathrm{kh}_\Omega} * \Pi_\Omega(A) = \overline{P}^{\mathrm{kh}_\Omega} * \Pi_\Omega(A).$$

We prove by transfinite induction that $R_{\alpha+1}(A) = S_{\alpha+1}(A)$ for every ordinal $\alpha$, and $R_\alpha^{\mathrm{kh}_\Omega}(A) = S_\alpha^{\mathrm{kh}_\Omega}(A)$ for a limit ordinal $\alpha$. Indeed, the step $\alpha \mapsto \alpha+1$ is easy if $R_\alpha(A) = S_\alpha(A)$. Assume that $R_\alpha^{\mathrm{kh}_\Omega}(A) = S_\alpha^{\mathrm{kh}_\Omega}(A)$ for some limit ordinal $\alpha$. Then

$$\begin{aligned} R_{\alpha+1}(A) &= P^{\mathrm{kh}_\Omega} * \Pi_\Omega(R_\alpha(A)) = P^{\mathrm{kh}} * \Pi_\Omega\left(R_\alpha^{\mathrm{kh}}(A)\right) \\ &= \overline{P}^{\mathrm{kh}} * \Pi_\Omega\left(S_\alpha^{\mathrm{kh}}(A)\right) = \overline{P}^{\mathrm{kh}} * \Pi_\Omega(S_\alpha(A)) \\ &= S_{\alpha+1}(A). \end{aligned}$$

So we proved the step $\alpha \mapsto \alpha + 1$.

So it remains to show that $R_\alpha^{\mathrm{kh}_\Omega}(A) = S_\alpha^{\mathrm{kh}_\Omega}(A)$ for each limit ordinal $\alpha$. Assume by induction, that $R_{\alpha'+1}(A) = S_{\alpha'+1}(A)$ for every ordinal $\alpha' < \alpha$. Then

$$\begin{aligned} R_\alpha^{\mathrm{kh}_\Omega}(A) &= \mathrm{kh}_\Omega(\cup_{\alpha'<\alpha} R_{\alpha'+1}(A); A) = \mathrm{kh}_\Omega(\cup_{\alpha'<\alpha} S_{\alpha'+1}(A); A) \\ &= \mathrm{kh}_\Omega\left(\overline{\cup_{\alpha'<\alpha} S_{\alpha'+1}(A)}; A\right) = S_\alpha^{\mathrm{kh}}(A) \end{aligned}$$

by Lemma 7.1(4). Therefore there is an ordinal $\gamma$ such that

$$P^{\mathrm{kh}_\Omega *}(A) = R_{\gamma+1}(A) = R_{\gamma+2}(A) = S_{\gamma+2}(A) = S_{\gamma+1}(A) = \overline{P}^{\mathrm{kh}_\Omega *}(A).$$

(3) follows from (2) and Corollary 4.23.

(4) follows from Theorem 7.2(4). □

**Remark 7.4.** In (3) of Corollary 7.3 we state the equality of two radicals on $Q$-algebras if one of them is obtained by application of the algebraic convolution procedure while the other one is obtained by application of the topological convolution procedure. So the resulting radical can be considered as an algebraic radical as well as a topological radical, simultaneously. This gathers all advantages of both classes of radicals.

### 7.2. Primitivity procedure

Let $\Omega$ be a primitive map, and let $R$ be a preradical. Define $R^{p_\Omega}$ by

$$R^{p_\Omega}(A) = \cap\{(R * I; A) : I \in \Omega(A) \cup \{A\}\}$$

for every algebra $A$; we recall that

$$(R * I; A) = q_I^{-1}(P(A/I)).$$

In other words, $R^{p_\Omega}(A)$ consists of all elements $a \in A$ such that $a/I \in P(A/I)$ for all $I \in \Omega(A)$.

If $\Omega(A) = \mathrm{Prim}(A)$ then we write $R^p$ instead of $R^{p_\Omega}$. The procedure $R \longmapsto R^{p_\Omega}$ is called the *$\Omega$-primitivity procedure*.

**Remark 7.5.** If $R \leq \Pi_\Omega$ then $A/I$ is $\Pi_\Omega$-semisimple and therefore $R$-semisimple, whence $(R * I; A) = I$ for every $I \in \Omega(A) \cup \{A\}$; thus $R^p(A) = \Pi_\Omega(A)$. In particular this is the case for $R \leq \mathrm{rad}$.

For $R^{p_\Omega}$ to have sufficiently convenient properties, one has to impose additional requirements on $R$. One of possible approaches is to require that $R$ well behaves on Banach ideals. We mean the condition of Banach heredity (see (2.2)).

**Lemma 7.6.** *Let $R$ be a preradical on $\mathfrak{U}_{\mathrm{b}}$ satisfying the condition of Banach heredity. If $A$ is a Banach algebra, $I$ and $J$ are closed ideals of $A$, then*

$$(R * (J \cap I) ; J) = J \cap (R * I; A) .$$

*Proof.* Let $K = (J + I) / I$ with the norm $\|\cdot\|_K$ defined in Lemma 1.7. Then the natural map $\phi : J/ (J \cap I) \longrightarrow (K; \|\cdot\|_K)$ is an isometry. Let $q : J \longrightarrow J/ (J \cap I)$ and $q_I : A \longrightarrow A/I$ be standard quotient maps. Considering $\phi$ as an algebraic morphism, one may write that $q_I|_J = \phi \circ q$. By the assumption,

$$\phi (R ((J/ (J \cap I)))) = R (K; \|\cdot\|_K) = K \cap R (A/I)$$

As $q_I|_J = \phi \circ q$ then

$$q^{-1} (R ((J/ (J \cap I)))) = (q_I|_J)^{-1} (K \cap R (A/I)) .$$

It is clear that $(q_I|_J)^{-1} (K) = (J + I) \cap J = J$. Then

$$J \cap (q_I|_J)^{-1} (R (A/I)) = J \cap q_I^{-1} (R (A/I))$$

and the result follows. □

Another way to investigate the primitivity procedure is connected with modifications of Axiom 1. Namely, we may require that $R$ be either pliant or strict (i.e., satisfies (2.1)); clearly if $\mathfrak{U} = \mathfrak{U}_{\mathrm{b}}$ then this holds for all preradicals.

In the following theorem and corollary we join both approaches.

**Theorem 7.7.** *Let $\Omega$ be a primitive map on $\mathfrak{U}$ satisfying $(1_{kh})$ or $(2_{kh})$, and let $R$ be a preradical on $\mathfrak{U}$. Then*

(1) *If $\Omega$ and $R$ are pliant then*
   (a) *$R^{p_\Omega}$ is a pliant preradical;*
   (b) *If $R$ is a radical then $R^{p_\Omega}$ satisfies Axiom 4;*
   (c) *If $R$ is a hereditary preradical then $R^{p_\Omega}$ is a hereditary preradical.*
(2) *If $R$ is strict, $\mathfrak{U}$ is universal and contained in $\mathfrak{U}_{\mathrm{q}}$ then*
   (a) *$R^{p_\Omega}$ is a strict preradical;*
   (b) *If $R$ is a topological radical then $R^{p_\Omega}$ satisfies Axiom 4;*
   (c) *If $R$ is a hereditary topological preradical then $R^{p_\Omega}$ is a hereditary topological preradical.*
(3) *If $R$ is defined on $\mathfrak{U}_{\mathrm{b}}$ and satisfies the condition of Banach heredity then $R^{p_\Omega}$ is a hereditary preradical on $\mathfrak{U}_{\mathrm{b}}$.*

*Proof.* (1a) and (2a) Let $f : A \longrightarrow B$ be an algebraic morphism [bounded epimorphism]. As $(\pi \circ f) \in \mathrm{Irr}_\Omega (A)$ for every $\pi \in \mathrm{Irr}_\Omega (B)$, then $J := f^{-1} (I) \in \Omega (A)$ for every $I \in \Omega (B)$. Then $f$ induces an algebraic morphism [bounded epimorphism] $f_0 : A/J \longrightarrow B/I$ such that $f_0 \circ q_J = q_I \circ f$. Hence

$$\begin{aligned} f (R * J) &= f \left(q_J^{-1} (R (A/J))\right) \subset q_I^{-1} (f_0 (R (A/J))) \\ &\subset q_I^{-1} (R (B/I)) = R * I \end{aligned} \tag{7.2}$$

for every $I \in \Omega (B)$. So $f (R^{p_\Omega} (A)) \subset R^{p_\Omega} (B)$.

Assume that $\mathfrak{U}$ is universal and $R$ satisfies (2.1) for continuous isomorphisms. Let $f : A \longrightarrow B$ be a continuous isomorphism. Then $f^{-1} : B \longrightarrow A$ is an algebraic morphism and $f_0 : A/J \longrightarrow B/I$ is a continuous isomorphism, where $J = f^{-1}(I)$. Then

$$f(R * J) = f\left((f_0 \circ q_J)^{-1}(f_0(R(A/J)))\right) = q_I^{-1}((R(A/J))) = R * I.$$

As $\Omega(A) = \left\{f^{-1}(I) : I \in \Omega(B)\right\}$ then

$$\begin{aligned} f\left(R^{p\Omega}(A)\right) &= \cap_{I \in \Omega(B) \cup \{B\}} f\left(R * f^{-1}(I); A\right) \\ &= \cap_{I \in \Omega(B) \cup \{B\}} (R * I; B) = R^{p\Omega}(B). \end{aligned}$$

Therefore $R^{p\Omega}$ satisfies (2.1) for continuous isomorphisms.

(1b) and (2b) If $K$ is an ideal of $A$ then

$$\Omega(K) \cup \{K\} = \{K \cap I : I \in \Omega(A) \cup \{A\}\}.$$

As $K \cap I$ is an ideal of $A$ then $(R * (K \cap I); K)$ is an ideal of $A$. Therefore $R^{p\Omega}(K)$ is an ideal of $A$.

Let $q_I' : K \longrightarrow K/(K \cap I)$ and $q_I : A \longrightarrow A/I$ be standard quotient maps. There is a [bounded] injective imbedding $g_I$ of $K/(K \cap I)$ onto an ideal $q_I(K)$ of $A/I$, whence

$$g_I(R(K/(K \cap I))) \subset R(q_I(K)) \subset R(A/I). \tag{7.3}$$

As $q_I|_K = g_I \circ q_I'$ then

$$\begin{aligned} R^{p\Omega}(K) &= \cap_{I \in \Omega(A) \cup \{A\}} \left(q_I'\right)^{-1}(R(K/(K \cap I))) \\ &= \cap_{I \in \Omega(A) \cup \{A\}} \left(g_I \circ q_I'\right)^{-1}(g_I(R(K/(K \cap I)))) \\ &\subset \cap_{I \in \Omega(A) \cup \{A\}} g_I^{-1}(R(A/I)) = R^{p\Omega}(A). \end{aligned} \tag{7.4}$$

Therefore $R^{p\Omega}$ satisfies Axiom 4.

(1c) and (2c) As $R$ is hereditary, and also is algebraic or satisfies (2.1) for continuous isomorphisms, then we rewrite (7.3) as follows:

$$g_I(R(K/(K \cap I))) = R(q_I(K)) = q_I(K) \cap R(A/I)$$

for every $I \in \Omega(A) \cup \{A\}$. It follows from (7.4) that

$$\begin{aligned} R^{p\Omega}(K) &= \cap_{I \in \Omega(A) \cup \{A\}} \left(q_I|_K\right)^{-1}(q_I(K) \cap R(A/I)) \\ &= \cap_{I \in \Omega(A) \cup \{A\}} \left(K \cap q_I^{-1}(R(A/I))\right) \\ &= K \cap \left(\cap_{I \in \Omega(A) \cup \{A\}} q_I^{-1}(R(A/I))\right) = K \cap R^{p\Omega}(A). \end{aligned}$$

Therefore $R^{p\Omega}$ is a hereditary preradical.

(3) Let $f : A \longrightarrow B$ be a topological morphism of Banach algebras. Then $(\pi \circ f) \in \mathrm{Irr}_\Omega(A)$ for every $\pi \in \mathrm{Irr}_\Omega(B)$, whence $J := f^{-1}(I) \in \Omega(A) \cup \{A\}$ for every $I \in \Omega(B) \cup \{B\}$. Then $f$ induces a topological morphism $f_0 : A/J \longrightarrow B/I$ such that $f_0 \circ q_J = q_I \circ f$. Hence

$$f(R * J) \subset R * I$$

as in (7.2), for every $I \in \Omega(B) \cup \{B\}$. So $f\left(R^{p\Omega}(A)\right) \subset R^{p\Omega}(B)$.

By our assumption, for a closed ideal $K$ of $A$,

$$\begin{aligned} R^{p\Omega}(K) &= \cap_{I\in\Omega(A)\cup\{A\}}(R*(K\cap I);K) \\ &= \cap_{I\in\Omega(A)\cup\{A\}} K\cap(R*I;A) \\ &= K\cap\left(\cap_{I\in\Omega(A)\cup\{A\}}(R*I;A)\right) = K\cap R^{p\Omega}(A) \end{aligned}$$

by Lemma 7.6. □

Now we apply the convolution procedure to obtain a radical.

**Corollary 7.8.** *Let $\Omega$ be a primitive map on $\mathfrak{U}$ satisfying $(1_{\mathrm{kh}})$ or $(2_{\mathrm{kh}})$, and let $R$ be a preradical. Then*

(1) *If $\Omega$ and $R$ are pliant then $R^{p\Omega*}$ is a pliant hereditary radical;*
(2) *If $R$ is a closed ideal map satisfying (2.1) (i.e., $R$ is strict), $\mathfrak{U}$ is universal and contained in $\mathfrak{U}_{\mathrm{q}}$ then $R^{p\Omega*}$ is a topological radical on $\mathfrak{U}$;*
(3) *If $R$ is defined on $\mathfrak{U}_{\mathrm{b}}$ and satisfies the condition of Banach heredity then $R^{p\Omega*}$ is a radical on Banach algebras;*
(4) $\Pi_\Omega\vee R=\Pi_\Omega^*\vee R^*\leq R^{p\Omega*}$.

*Proof.* (1)–(3) follow from the properties of convolution procedure.

(4) It is clear that

$$\Pi_\Omega(A)=\cap_{I\in\Omega(A)\cup\{A\}}q_I^{-1}(0)\subset\cap_{I\in\Omega(A)\cup\{A\}}q_I^{-1}(R(A/I))=R^{p\Omega}(A)$$

for an algebra $A$, and every $R$-radical algebra is $R^{p\Omega}$-radical. This implies the result. □

## 8. Scattered algebras

Our aim in this section is to give a consistent exposition of the theory of scattered radical. This radical was announced in [ST2] and already applied in [BT, ST6].

### 8.1. Thin ideals

Let $K\subset\mathbb{C}$ be a bounded set. Recall that the *polynomially convex hull* $\mathrm{pc}(K)$ of $K$ is the set of all $z\in\mathbb{C}$ such that

$$|p(z)|\leq\sup_{w\in K}|p(w)|$$

for every polynomial $p$, and $K$ is *polynomially convex* if $K=\mathrm{pc}(K)$. By [G, Lemma 3.1.3], a compact subset $K\subset\mathbb{C}$ is polynomially convex if and only if the complement of $K$ is connected. We say that $H$ is a *polynomially convex neighborhood* of $K$ if $K$ is contained in the interior of $H$ and $H=\mathrm{pc}(H)$ (hence $H$ is a compact subset of $\mathbb{C}$).

An algebra $A$ is called *inessential* (or *Riesz* [BMS]) if $\sigma(a)$ is finite or countable with the only one limit point at 0 for every $a\in A$; $A$ is *scattered* if $\sigma(a)$ is at most countable for every $a\in A$. These notions are transferred to ideals. As ideals are spectral subalgebras, it is not important with respect to what – the algebra or the ideal – the spectrum of an element of the ideal is considered. The algebras

$\mathcal{F}(X)$ of finite rank operators and $\mathcal{K}(X)$ of compact operators on a Banach space $X$ are inessential, and they are inessential ideals of $\mathcal{B}(X)$.

By [A, Corollary 5.7.6], if $I$ is an inessential ideal of a Banach algebra $A$ then $\overline{I}$ and $\mathrm{kh}(I;A)$ are also inessential ideals of $A$ and have the same set of projections as $I$.

**Lemma 8.1.** *Let $A$ be a Banach algebra, and let $I$ be an inessential ideal of $A$. If $a \in A$ and $V$ is a polynomially convex neighborhood of $\sigma(a/I)$ then there is a finite set $Z$ such that $\sigma(a) \subset V \cup Z$.*

*Proof.* Let $\Delta$ be the set of all isolated points in $\sigma(a)$ with Riesz (spectral) projections in $I$, and $D = \sigma(a) \setminus \Delta$. By [A, Theorem 5.7.4] and Corollary 1.5,

$$D \subset \mathrm{pc}\left(\sigma\left(a/\overline{I}\right)\right) = \mathrm{pc}(\sigma\left(a/I\right)).$$

Then $D \subset V$, and only a finite number of points in $\sigma(a)$ can be outside of $V$ because otherwise $\sigma(a)$ would have a limit point outside of $D$ which contradicts the definition of $D$. Therefore

$$\sigma(a) \subset V \cup Z$$

for some finite set $Z \subset \mathbb{C}$. □

Let us say that an ideal $I$ of an algebra $A$ is *thin* if for each $a \in A$ there is a countable set $Z \subset \mathbb{C}$ such that

$$\sigma(a) \subset \mathrm{pc}(\sigma(a/I)) \cup Z. \tag{8.1}$$

**Lemma 8.2.** *Let $A$ be an algebra, and let $J \subset I$ be ideals of $A$. If $I$ is a thin ideal of $A$ then $J$ is a thin ideal of $A$.*

*Proof.* It is sufficient to note that $\sigma(a/I) \subset \sigma(a/J)$ for every $a \in A$. □

If $I$ is a thin ideal of a $Q$-algebra then $\overline{I}$ is also thin by Corollary 1.5. Recall that the spectrum of each element of a $Q$-algebra is a compact set in $\mathbb{C}$.

**Lemma 8.3.** *Let $A$ be a $Q$-algebra. An ideal $I$ of $A$ is thin if and only if for each $a \in A$ and for every polynomially convex neighborhood $V$ of $\sigma(a/I)$, there is a countable set $Z \subset \mathbb{C}$ such that $\sigma(a) \subset V \cup Z$.*

*Proof.* $\Rightarrow$ is obvious.

$\Leftarrow$ Let $a \in A$, and let $Z_V := \sigma(a) \setminus V$ be countable for each polynomially convex neighborhood $V$ of $\sigma(a/I)$. Let $D \subset \mathbb{C}$ be an open disk with

$$\overline{D} \cap \mathrm{pc}(\sigma(a/I)) = \varnothing.$$

As $\mathrm{pc}(\sigma(a/I))$ is polynomially convex, its complement $C$ is connected, so it contains $E_t = \{z : |z| \geq t\}$ for sufficiently large $t > 0$, and a curve $L \subset C$ connecting $D$ with $E_t$. Let $U \subset C$ be a connected neighborhood of $E_t \cup L \cup \overline{D}$. Then the complement $V$ of $U$ is a compact neighborhood of $\sigma(a/I)$. Since the complement of $V$ is connected then $V$ is polynomially convex. By the assumption,

$$\sigma(a) \subset V \cup Z_V$$

where $Z_V$ is countable (or finite) by assumption.

The set $\mathbb{C} \backslash \mathrm{pc}(\sigma(a/I))$ is open and has therefore an inscribed covering by open disks. Taking an inscribed subcovering by disks with rational radii and coordinates of centers one can find a sequence $(D_k)_1^\infty$ of open disks such that

$$\cup_{k=1}^\infty D_k = \mathbb{C} \setminus \mathrm{pc}(\sigma(a/I)) \text{ and all } \overline{D_k} \cap \mathrm{pc}(\sigma(a/I)) = \varnothing;$$

let $V_k$ be compact neighborhoods of $\mathrm{pc}(\sigma(a/I))$ non-intersecting $D_k$. Then, as each $V_k$ lies in the complement $D_k^{\mathrm{c}}$ of $D_k$,

$$\mathrm{pc}(\sigma(a/I)) \subset \cap_{k=1}^\infty V_k \subset \cap_{k=1}^\infty D_k^{\mathrm{c}} \subset (\cup_{k=1}^\infty D_k)^{\mathrm{c}} = \mathrm{pc}(\sigma(a/I)),$$

and

$$\sigma(a) \subset \cap_{k=1}^\infty (V_k \cup Z_{V_k}) \subset (\cap_{k=1}^\infty V_k) \cup (\cup_{k=1}^\infty Z_{V_k}) = \mathrm{pc}(\sigma(a/I)) \cup Z$$

where $Z = \cup_{k=1}^\infty Z_{V_k}$ is a countable set. □

Taking into account Lemma 8.1, we see that inessential ideals of Banach algebras are thin ideals.

**Corollary 8.4.** *Let $A$ be a $Q$-algebra. Then each thin ideal $J$ is a scattered algebra.*

*Proof.* Indeed, if $a \in J$ then $\sigma(a/J) = \{0\}$, whence, by (8.1), $\sigma(a) \subset \{0\} \cup Z$ is countable. □

**Lemma 8.5.** *Let $A$ be a $Q$-algebra, and let $I \subset J$ be closed ideals of $A$. If $I$ and $J/I$ are thin ideals of $A$ and $A/I$, respectively, then $J$ is a thin ideal of $A$.*

*Proof.* Let $a \in A$ be arbitrary. Since the standard morphism $\tau : A/I \to A/J$ has the kernel $J/I$ then, by Lemma 8.3,

$$\sigma(a/I) \subset \mathrm{pc}(\sigma(a/J)) \cup Z$$

where $Z$ is a countable set. As the union of a polynomially convex set and a (bounded) countable set in $\mathbb{C}$ is polynomially convex then

$$\mathrm{pc}(\sigma(a/I)) \subset \mathrm{pc}(\sigma(a/J)) \cup Z.$$

Since $I$ is thin then

$$\sigma(a) \subset \mathrm{pc}(\sigma(a/I\ )) \cup Z'$$

where $Z'$ is countable. Thus

$$\sigma(a) \subset \mathrm{pc}(\sigma(a/J)) \cup Z \cup Z'.$$ □

The following quite general result establishes a kind of spectral continuity.

**Lemma 8.6.** *Let $A$ be a $Q$-algebra, let $(J_\alpha)_{\alpha\in\Lambda}$ be an up-directed net of closed ideals of $A$ and $J = \overline{\bigcup_{\alpha\in\Lambda} J_\alpha}$. Then*

(1) $$\sigma(a/J) = \bigcap\nolimits_{\alpha\in\Lambda} \sigma(a/J_\alpha); \tag{8.2}$$

(2) *If all ideals $J_\alpha$ are thin then $J$ is thin.*

*Proof.* (1) If $\lambda \notin \sigma(a/J)$ then there are $b \in A$ and $x \in J$ with $(a-\lambda)b = 1+x$. Hence one can find $\alpha \in \Lambda$ and $y \in J_\alpha$ with $\|x-y\| < 1$. By [P, Proposition 2.2.7], $\sum_{n>0}(y-x)^n$ converges in $A$. Therefore $1-(y-x)$ is invertible and $1+(x-y) = z^{-1}$ for some $z \in A$, whence $(a-\lambda)b = z^{-1}+y$ and

$$(a-\lambda)bz = 1+yz.$$

Therefore $\lambda \notin \sigma^l(a/J_\alpha)$. Similarly, there is $\delta \in \Lambda$ with $\lambda \notin \sigma^r(a/J_\delta)$. So $\lambda \notin \sigma(a/J_\gamma)$ for $\gamma > \alpha$ and $\gamma > \beta$. Hence

$$\bigcap_{\alpha\in\Lambda} \sigma(a/J_\alpha) \subset \sigma(a/J).$$

The converse inclusion is evident.

(2) Let $V$ be a polynomially convex neighborhood of $\sigma(a/J)$ then, by (8.2) and compactness of $\sigma(a/J)$, there is $\alpha \in \Lambda$ with $\sigma(a/J_\alpha) \subset V$. Since $J_\alpha$ is thin there is a countable set $Z \subset \mathbb{C}$ such that $\sigma(a) \subset V \cup Z$. It remains to apply Lemma 8.3. □

**Theorem 8.7.** *Let $A$ be a $Q$-algebra, and let $(J_\alpha)_{\alpha\le\gamma}$ be an increasing transfinite chain of closed ideals of $A$ with $J_0 = 0$. If each ideal $J_{\alpha+1}/J_\alpha$ is a thin ideal of $A/J_\alpha$ then all ideals $J_\alpha$ are thin ideals of $A$ and therefore scattered algebras.*

*Proof.* Using transfinite induction, assume that all $J_{\alpha'}$ with $\alpha' < \alpha$ are thin ideals. If $\alpha$ is a limit ordinal then the result follows from Lemma 8.6(2). Otherwise $\alpha = \alpha'+1$ for some $\alpha'$ and the result follows from Lemma 8.5 applied to ideals $J_{\alpha'} \subset J_{\alpha'+1}$. Therefore all ideals $J_\alpha$ are thin. By Corollary 8.4, they are also scattered. □

### 8.2. Scattered radical

Recall that the algebraic under radical $\mathrm{rad}^{\mathrm{soc}} = \mathrm{soc} * \mathrm{rad}$ sends each algebra $A$ to the ideal $\{x \in A : x/\mathrm{rad}(A) \in \mathrm{soc}(A/\mathrm{rad}(A))\}$. This ideal, the socle modulo radical, was called in [BMS, Definition F.3.1] by the *presocle* of $A$ and denoted by $\mathrm{psoc}(A)$. We preserve this notation for the map itself: $\mathrm{psoc} := \mathrm{soc} * \mathrm{rad}$.

Now we can apply the algebraic convolution procedure and obtain the hereditary algebraic radical $\mathrm{psoc}^*$ on $\mathfrak{U}_{\mathrm{a}}$.

**Theorem 8.8.** *The restriction of* $\mathrm{psoc}^*$ *to $\mathfrak{U}_q$ is a hereditary topological radical on $Q$-algebras.*

*Proof.* As $\mathrm{rad} \le \mathrm{psoc}$ then

$$\mathrm{rad} \vee \mathrm{psoc} = \mathrm{psoc}^*.$$

(Recall that $\mathrm{rad} \vee \mathrm{psoc}$ is the smallest radical that is larger than or equal to rad and psoc.) One can realize $\mathrm{rad} \vee \mathrm{psoc}$ as the action of the algebraic convolution procedure, namely

$$\mathrm{rad} \vee \mathrm{psoc} = (\mathrm{rad} * \mathrm{psoc})^*.$$

But $\text{rad} * \text{psoc} = \text{psoc}^{\text{kh}}$. So
$$\text{psoc}^* = \text{psoc}^{\text{kh}*}.$$
By Corollary 7.3, $\text{psoc}^{\text{kh}*}$ is a hereditary topological radical on $Q$-algebras. □

Define map $\mathcal{R}_s$ by
$$\mathcal{R}_s = \text{psoc}^* \quad \text{on Banach algebras;}$$
this map is a hereditary topological radical on $\mathfrak{U}_{\mathrm{b}}$; it is called the *scattered radical.*

**Lemma 8.9.** *Let $A$ be a Banach algebra. Then $\mathcal{R}_s(A)$ is a thin ideal of $A$ and a scattered algebra.*

*Proof.* Let $(R_\alpha)_\alpha$ be the convolution chain of under radicals generated by $\overline{\text{psoc}}$. As $\overline{\text{psoc}}^* = \mathcal{R}_s$ by Corollary 7.3, then there is $\gamma$ such that $R_\gamma(A) = \mathcal{R}_s(A)$ and all gap-quotients $R_{\alpha+1}(A)/R_\alpha(A) = \overline{\text{psoc}}(A/R_\alpha(A))$ are inessential ideals by [BMS, Section R], and therefore thin ideals by Lemmas 8.1 and 8.3. By Theorem 8.7, $\mathcal{R}_s(A)$ is a thin ideal of $A$ and a scattered algebra. □

Let $\mathcal{S}(A)$ be the set of all elements of $A$ with (finite or) countable spectrum.

**Theorem 8.10.** *Let $A$ be a Banach algebra. Then*

(1) *$\mathcal{R}_s(A)$ contains all one-sided, non-necessarily closed, scattered ideals of $A$;*
(2) *$\mathcal{R}_s(A)$ is the largest thin ideal and the largest scattered ideal of $A$;*
(3) *$\mathcal{R}_s(A) = \{a \in A : \sigma(ax)$ is countable for each $x \in A\}$;*
(4) *$\mathcal{R}_s(A)$ is the largest ideal of $A$ contained in $\mathcal{S}(A)$;*
(5) *$\mathcal{R}_s(A) + \mathcal{S}(A) \subset \mathcal{S}(A)$.*

*Proof.* (1) Let $J$ be a scattered left ideal in $A$. Let $B = A/\mathcal{R}_s(A)$, and let $I$ be the image of $J$ in $B$. Then $B$ is semisimple and $I$ is a scattered left ideal of $B$. If it is non-zero then it contains an element $a$ with non-zero spectrum and therefore there is an isolated non-zero point $\lambda$ in its spectrum. As $\sigma_B(a) \subset \sigma_I(a)$ and $\lambda \in \sigma_I^a(a) \subset \sigma_B^a(a)$, $\lambda$ is an isolated point of $\sigma_B(a)$. As $\lambda \neq 0$, the corresponding Riesz projection $p$ of $a$ belongs to $Ba \subset I$ (see for instance the proof of Lemma 5.7.1 of [A]). As $p \neq 0$ then $Bp$ is a closed non-zero scattered left ideal of $B$. By Barnes' Theorem [B], $\text{psoc}(Bp) \neq 0$. By Proposition 6.5, $\text{psoc}(B) \neq 0$. Thus $\mathcal{R}_s(B) \neq 0$, a contradiction. Therefore $I = 0$, whence $J \subset \mathcal{R}_s(A)$.

(2) It follows from (1) that $\mathcal{R}_s(A)$ is the *largest scattered ideal* of $A$. Let $I$ be a thin ideal of $A$. Then it is a scattered ideal of $A$ by Corollary 8.4, and so $I \subset \mathcal{R}_s(A)$.

(3) Let $a \in A$ and $J = Aa$. If $a \in \mathcal{R}_s(A)$ then $J$ is contained in $\mathcal{R}_s(A)$ and its elements have countable spectra. Conversely, if $J$ is scattered then it is contained in $\mathcal{R}_s(A)$ by Proposition 8.10.

(4) If $I$ is an ideal of $A$ contained in $\mathcal{S}(A)$ then it is a scattered ideal of $A$. By (2), $I \subset \mathcal{R}_s(A)$.

(5) Let $a \in \mathcal{S}(A)$ and $b \in \mathcal{R}_s(A)$. As $\mathcal{R}_s(A)$ is a thin ideal of $A$ then
$$\sigma(a+b) \subset \text{pc}(\sigma(a/\mathcal{R}_s(A))) \cup N$$

for some countable set $N \subset \mathbb{C}$. But $\sigma(a/\mathcal{R}_s(A))$ is countable, whence

$$\mathrm{pc}(\sigma(a/\mathcal{R}_s(A))) = \sigma(a/\mathcal{R}_s(A))$$

is countable and $a + b \in \mathcal{S}(A)$. □

**Corollary 8.11.** *$\mathcal{R}_s$ is a uniform radical on $\mathfrak{U}_{\mathrm{b}}$.*

*Proof.* Indeed, every closed subalgebra of a scattered Banach algebra is scattered. □

**Corollary 8.12.** *The radical $\mathcal{R}_s$ satisfy the condition of Banach heredity* (2.2).

*Proof.* Let $L$ be a Banach ideal of a Banach algebra $A$. This means that $L$ is an ideal of $A$ and there is an injective continuous homomorphism $f$ of a Banach algebra $B$ to $A$ with $f(B) = L$. Since $f$ is an isomorphism of algebras $B$ and $L$, and psoc* is a hereditary radical on $\mathfrak{U}_a$, we have that

$$\begin{aligned}\mathcal{R}_s((L, \|\cdot\|_B)) &= f(\mathcal{R}_s(B)) = f(\mathrm{psoc}^*(B)) = \mathrm{psoc}^*(L)\\ &= L \cap \mathrm{psoc}^*(A) = L \cap \mathcal{R}_s(A).\end{aligned}$$ □

**Corollary 8.13.** *Let $A$ be a Banach algebra, and let $I$ be a (non-necessarily closed) ideal of $A$. Then*

(1) *If $I$ is scattered then $I$ is a thin ideal of $A$;*
(2) *If $I$ and $A/I$ are scattered then $A$ is scattered;*
(3) *If $f : A \longrightarrow B$ is an algebraic morphism of Banach algebras then*

$$f(\mathcal{R}_s(A)) \subset \mathcal{R}_s(B).$$

*Proof.* (1) Indeed, $\overline{I} \subset \mathcal{R}_s(A)$ by Theorem 8.10. As $\mathcal{R}_s$ is hereditary, $\overline{I} = \mathcal{R}_s(\overline{I})$. Taking $(R_\alpha)$ as in Lemma 8.9 we see that all $R_\alpha(I)$ are thin ideals of $A$. Therefore $\mathcal{R}_s(\overline{I})$ is a thin ideal of $A$, hence $\overline{I}$ and $I$ are also thin.

(2) It follows that $\overline{I}$ and $A/\overline{I}$ are also scattered. Then $\overline{I} = \mathcal{R}_s(\overline{I})$ and $A/\overline{I} = \mathcal{R}_s\left(A/\overline{I}\right)$. By Theorem 4.6, $A = \mathcal{R}_s(A)$. Therefore $A$ is scattered.

(3) Indeed, $\mathcal{R}_s$ inherits the algebraic properties of psoc*. □

**Theorem 8.14.** *If an algebra $A$ is a subideal of a Banach algebra (see Section 2.2) or is algebraically isomorphic to the quotient of a subideal of a Banach algebra by a non-necessarily closed ideal then*

(1) $\mathrm{psoc}^*(A) = I_A$ *where* $I_A := \{a \in A : \sigma_A(ax)$ *is countable for each* $x \in A\}$;
(2) *Every scattered ideal of $A$ is a thin ideal of $A$.*

*Proof.* Note that every subideal of a Banach algebra is a normed $Q$-algebra.

Let $C$ be a $Q$-algebra for which (1) and (2) hold (under the substitution $A = C$; for instance these conditions hold for Banach algebras), and let $A$ be an ideal of $C$. Then $\mathrm{psoc}^*(A) = A \cap I_C$ by heredity of psoc*. As clearly $I_A = A \cap I_C$ then $\mathrm{psoc}^*(A) = I_A$.

Let $K$ be a scattered ideal of $A$, and let $J$ be the ideal of $C$ generated by $K$. Then

$$J^3 \subset K \subset J$$

by [AR, Lemma 1.1.5], whence $J^3$ is a scattered ideal of $C$. Then $J^3$ is a thin ideal of $C$ by (2) and therefore a thin ideal of $A$. As $K^3 \subset J^3$ then $K^3$ and $\overline{K^3}$ are thin ideals of $A$ by Lemma 8.2. As $\overline{K}/\overline{K^3}$ is nilpotent then it is a thin ideal of $A/\overline{K^3}$. By Lemma 8.5, $\overline{K}$ is a thin ideal of $A$, so $K$ is also a thin ideal of $A$.

Thus (1) and (2) hold for $A$: therefore the steps $0 \mapsto 1$ and $n \mapsto n+1$ of induction for $n$-subideals of Banach algebras are valid, for every $n$. Then the proof is completed for subideals of Banach algebras.

Let now $f : A \longrightarrow B/J$ be an algebraic isomorphism of $A$ onto $B/J$, where $B$ is a subideal of a Banach algebra and $J$ is an ideal of $B$. If $a \in A$ and $b \in B$ is any element such that $f(a) = b/J$, then clearly $\sigma(a) = \sigma(b/J)$ and it suffices to check (1) and (2) for $B/J$. As $\sigma(b/J) = \sigma\left(b/\overline{J}\right)$ for every $b \in B$ by Lemma 1.5, the proof is reduced to the case of $B/\overline{J}$. But $B/\overline{J}$ is a subideal of a Banach algebra (see for instance [ST5, Theorem 2.24]), and the result follows from the above. □

Theorem 8.14 extends the main properties of the scattered radical from Banach algebras to subideals of Banach algebras. Recall that subideals of Banach algebras form the smallest universal class $\mathfrak{U}_{\mathrm{b}}^{u}$ generated by Banach algebras. Thus we extend the denotation $\mathcal{R}_s$ for psoc* on subideals of Banach algebras.

The *regular scattered radical* $\mathcal{R}_s^r$, obtained from $\mathcal{R}_s$ by the regular procedure, extends $\mathcal{R}_s$ to normed algebras, and it is determined as

$$\mathcal{R}_s^r(A) = \left\{x \in A : \sigma_{\widehat{A}}(ax) \text{ is at most countable } \forall\, a \in \widehat{A}\right\}.$$

**Theorem 8.15.** $\mathcal{R}_{\mathrm{hc}} \le \mathcal{R}_s = \mathcal{R}_{\mathrm{hf}} \vee \mathrm{Rad} = \mathcal{R}_{\mathrm{hc}} \vee \mathrm{Rad}$ *on Banach algebras.*

*Proof.* Let $A$ be a Banach algebra. Any compact element of a Banach algebra $A$ has countable spectrum [Al] and any bicompact Banach algebra is scattered. Then any closed bicompact ideal of $A$ is scattered and contained in $\mathcal{R}_s(A)$, whence $\Sigma_{\mathrm{hc}}(A) \subset \mathcal{R}_s(A)$ and $\Sigma_{\mathrm{hc}} \le \mathcal{R}_s$ in general. As $\mathcal{R}_s$ is a radical, it follows that $\mathcal{R}_{\mathrm{hf}} \le \mathcal{R}_{\mathrm{hc}} = \Sigma_{\mathrm{hc}}^* \le \mathcal{R}_s$ by Theorems 3.1 and 3.9.

It is clear that $\mathrm{Rad} \le \mathcal{R}_s$, so $\mathcal{R}_{\mathrm{hf}} \vee \mathrm{Rad} \le \mathcal{R}_s$.

Let $A$ be $(\mathcal{R}_{\mathrm{hf}} \vee \mathrm{Rad})$-semisimple, and let $I = \mathcal{R}_s(A)$. As $A$ is semisimple and has no non-zero finite rank elements, $I$ is also semisimple and has no non-zero finite rank elements. However, if $I \neq 0$ then $I$ has a non-zero socle by Barnes' theorem [B]. Thus, by Lemma 6.6, $I$ has non-zero finite rank elements, and so $A$ has such elements, a contradiction. Therefore $I = 0$, and $A$ is $\mathcal{R}_s$-semisimple. By Theorem 2.3, $\mathcal{R}_s \le \mathcal{R}_{\mathrm{hf}} \vee \mathrm{Rad}$.

Furthermore,

$$\mathcal{R}_s = \mathcal{R}_{\mathrm{hc}} \vee \mathcal{R}_s = \mathcal{R}_{\mathrm{hc}} \vee \mathcal{R}_{\mathrm{hf}} \vee \mathrm{Rad} = \mathcal{R}_{\mathrm{hc}} \vee \mathrm{Rad}\,.$$ □

Let $A$ be an algebra, and let $\widehat{\mathrm{Irr}}(A)$ be the set of classes of equivalent strictly irreducible representations of $A$. Then $\pi \longmapsto \ker\pi$ is a map from $\widehat{\mathrm{Irr}}(A)$ onto $\mathrm{Prim}(A)$, so $\widehat{\mathrm{Irr}}(A)$ inherits the Jacobson topology from $\mathrm{Prim}(A)$: the preimages of closed sets in $\mathrm{Prim}(A)$ determine closed sets in $\widehat{\mathrm{Irr}}(A)$. The following statement says that one can identify these topological spaces for scattered Banach algebras.

**Theorem 8.16.** *Let $A$ be a Banach algebra. If $A$ is scattered then for any primitive ideal $I$, there is only one, up to the equivalence, strictly irreducible representation with kernel $I$.*

*Proof.* Any representation with kernel $I$ defines a faithful representation of $A/I$. Since $A/I$ is a primitive Banach algebra, we have only to prove that a faithful representation of a primitive scattered algebra $B$ is unique (up to the equivalence).

Since $B$ is scattered and semisimple, the socle of $B$ is non-zero. So it contains a minimal projection $p$. Taking into account that $\dim pBp = 1$ and applying Lemma 1.1, we have that there is only one strictly irreducible representation $\pi$ with $\pi(p) \neq 0$. But the last condition holds for each faithful representation. □

A topological space is called *dispersed* if it does not contain perfect subspaces, i.e., closed subsets without isolated points.

**Theorem 8.17.** *If a Banach algebra $A$ is scattered then the space* $\mathrm{Prim}(A)$ *of its primitive ideals is dispersed.*

*Proof.* Let $E$ be a closed subset in $\mathrm{Prim}(A)$, $J = \cap_{I \in E} I$ and $B = A/J$. All primitive ideals of $B$ are of the form $I/J$, $I \in E$ (because $E$ is closed) and their intersection is trivial. Thus $B$ is a semisimple scattered algebra, whence it contains a minimal projection $p$. Since $\dim pBp = 1$, there is, by Lemma 1.1, only one primitive ideal $I_0' = I_0/J$ of $B$ that does not contain $p$. It follows that $I_0$ does not contain the intersection of all $I \neq I_0$ in $E$. Hence $I_0$ is an isolated point in $E$ and $E$ is not perfect. □

**Corollary 8.18.** *If $A$ is a separable scattered Banach algebra then the spaces* $\mathrm{Prim}(A)$ *and* $\widehat{\mathrm{Irr}}(A)$ *are countable.*

*Proof.* It follows from Theorem 8.16 that if $\mathrm{Prim}(A)$ is countable then $\widehat{\mathrm{Irr}}(A)$ is countable.

In any topological space $X$ there is a decreasing transfinite chain of sets: let $X_0 = X$, $X_{\alpha+1}$ be the set of all non-isolated points in $X_\alpha$ for every ordinal $\alpha$, and let $X_\alpha = \cap_{\alpha' < \alpha} X_{\alpha'}$ if $\alpha$ is a limit ordinal. If $X$ is dispersed then $X_\delta = \varnothing$ for some $\delta$.

Let $X = \mathrm{Prim}(A)$. As $X$ is dispersed by Theorem 8.17, let $(X_\alpha)_{\alpha \leq \delta}$ be the chain of the above sets of $X$, with $X_\delta = \varnothing$. Choose $I_\alpha \in X_\alpha \backslash X_{\alpha+1}$ for every ordinal $\alpha < \delta$. By the construction, $I_\alpha \notin \overline{\{I_{\alpha'} : \alpha' > \alpha\}}$ in the Jacobson topology. Hence there are $x_\alpha \in (\cap_{\alpha' > \alpha} I_{\alpha'}) \setminus I_\alpha$. Multiplying by a constant, we can have

$$\mathrm{dist}(x_\alpha, I_\alpha) > 1.$$

Let now $\alpha' > \alpha$. Then $x_\alpha \in I_{\alpha'}$ and

$$\|x_{\alpha'} - x_\alpha\| \geq \mathrm{dist}(x_{\alpha'}, I_{\alpha'}) > 1.$$

So the last inequality holds for all ordinals $\alpha, \alpha'$ with $\alpha \neq \alpha'$. Since $A$ is separable, we obtain that $\delta$ is a countable ordinal.

Now it suffices to show that each set $X_\alpha \setminus X_{\alpha+1}$ is countable. This can be done by the same trick because each ideal in $X_\alpha \setminus X_{\alpha+1}$ is not contained in the closure of the set of the others. □

As an example let us look at the algebras $C(X)$ where $X$ is a compact set.

**Corollary 8.19.** *An algebra $C(X)$ is scattered if and only if $X$ is dispersed.*

*Proof.* It follows from Theorem 8.17 that if $C(X)$ is scattered then $X$ is dispersed. Conversely, let $X$ be dispersed. Any quotient of $C(X)$ by a closed ideal is isomorphic to $C(Y)$, where $Y$ is a compact subset of $X$. Since $Y$ is not perfect, there is an isolated point $z \in Y$. The ideal of all functions in $C(Y)$ that vanish outside $z$ is one-dimensional, hence it is thin. This allows us to construct a transfinite chain $(I_\alpha)_{\alpha\leq\gamma}$ of closed ideals of $C(X)$, such that $I_0 = 0$, $I_\gamma = C(X)$ and all gap-quotients of the chain are thin ideals. By Theorem 8.7, $C(X)$ is scattered. □

**Remark 8.20.** The above corollary gives a proof of the fact that the image of a dispersed space under a continuous map is dispersed. Indeed, if $Y = f(X)$ then $C(Y)$ is isomorphic to a closed subalgebra $B$ of $C(X)$. If $X$ is dispersed then $C(X)$ is scattered. Then $B$ is scattered, so $C(Y)$ is scattered. Hence $Y$ is dispersed.

### 8.3. Scattered radical on hereditarily semisimple Banach algebras

By Theorem 8.15, all hypocompact Banach algebras are scattered. The converse is not true in general because all Jacobson radical algebras are scattered. We will show here that if the radicals of an algebra and all its quotients are trivial that these conditions are equivalent.

Let $A$ be a [normed] algebra. Let us call $A$ *hereditarily semisimple* if all its quotients by [closed] ideals are semisimple. It is obvious that the class of all hereditarily semisimple [normed] algebras has the following properties:

(1) All quotients by [closed] ideals are in this class;
(2) If a [closed] ideal $I$ of a [normed] algebra $A$ and the quotient $A/I$ are in this class then so is $A$.

**Proposition 8.21.** *A Banach algebra $A$ is hereditarily semisimple if and only if each closed ideal of $A$ is the intersection of a family of primitive ideals.*

*Proof.* Let $J$ be a closed ideal of $A$ and let $K = \mathrm{kh}(J)$. Then $K/J = \mathrm{Rad}(A/J)$. Indeed, if $I \in \mathrm{Prim}(A/J)$ then there is $I' \in \mathrm{Prim}(A)$ with $J \subset I'$ and $I = I'/J$. Since $K \subset I'$, we have that $K/J \subset I$. Thus

$$K/J \subset \cap_{I\in \mathrm{Prim}(A/J)^1} I = \mathrm{Rad}(A/J).$$

Conversely, if $K' = q_J^{-1}(\mathrm{Rad}(A/J))$ then $K' \subset I'$ for any primitive ideal $I' \supset I$. Hence $K' \subset K$ and $\mathrm{Rad}(A/J) \subset q_J(K) = K/J$.

Now if $A$ is hereditarily semisimple then $K/J = 0$, whence $J = K$, the intersection of primitive ideals. Conversely, if $\mathrm{kh}(J) = J$ for all $J$, then $\mathrm{Rad}(A/J) = 0$, whence $A$ is hereditarily semisimple. □

In commutative case the condition that each closed ideal of a Banach algebra $A$ is the intersection of primitive ideals, is sometimes formulated as $A$ possesses the spectral synthesis. Among group algebras $L^1(G)$, only the algebras of compact groups have this property (Malliavin's Theorem, see [Rd]).

**Theorem 8.22.** *Let $A$ be a hereditarily semisimple Banach algebra. Then the following conditions are equivalent:*

(1) *$A$ is scattered;*
(2) *$A$ is hypocompact;*
(3) *$A$ is a closed-hypofinite algebra;*
(4) *Every non-zero quotient of $A$ has a minimal left ideal.*

*Proof.* (1) $\Longrightarrow$ (4) Let $A$ be scattered, and let $J$ be a closed ideal of $A$. Then $A/J$ is also scattered. So it suffices to show that a scattered semisimple Banach algebra $A$ has non-zero socle. But if $\mathrm{soc}(A) = 0$ then $\mathrm{psoc}(A) = 0$, whence $\mathrm{psoc}^*(A) = 0$, i.e., $\mathcal{R}_s(A) = 0$ while $\mathcal{R}_s(A) = A$, a contradiction.

(4) $\Longrightarrow$ (3) follows from the fact that each minimal projection is a finite rank element.

(3) $\Longrightarrow$ (2) is evident.

(2) $\Longrightarrow$ (1) follows from Theorem 8.15. □

An example of a hereditarily semisimple algebra is the algebra $C(X)$ of all continuous functions on a compact space $X$. We saw that it is scattered if and only if $X$ does not contain perfect subsets. Hence it is hypocompact under the same condition.

Since all C*-algebras are semisimple and their quotients are again C*-algebras, we have that C*-algebras are hereditarily semisimple.

Note that in the theory of C*-algebras the term "scattered" is used for a class of C*-algebras $A$ satisfying the following equivalent conditions:

($1_s$) Each positive functional on $A$ is the sum of a sequence of pure functionals;
($2_s$) $A$ is a type I C*-algebra (= GCR C*-algebra) and $\mathrm{Prim}(A)$ is dispersed in the hull-kernel topology;
($3_s$) $A$ is of type I and the maximal ideal space of its center is dispersed;
($4_s$) $A$ admits a superposition series $(I_\gamma)$ of closed ideals and each gap-quotient $I_{\gamma+1}/I_\gamma$ is isomorphic to $\mathcal{K}(H_\gamma)$ for some Hilbert space $H_\gamma$;
($5_s$) Each self-adjoint element of $A$ has countable spectrum;
($6_s$) Each C*-subalgebra of $A$ is AF (approximately finite dimensional);
($7_s$) The dual space $A^*$ of $A$ has the Radon–Nikodym property.

For the proof of the equivalence see [Ks] and the references therein. We will show that this class of algebras are exactly the scattered C*-algebras.

**Theorem 8.23.** *A C*-algebra $A$ satisfies the equivalent conditions $(1_s)$–$(7_s)$ if and only if it is scattered.*

*Proof.* Clearly each scattered C*-algebra satisfies $(6_s)$ and $(5_s)$. On the other hand, if $A$ satisfies $(4_s)$ then $A$ is hypocompact, so it is scattered by Theorem 8.15. □

Thus taking into account that C*-algebras are hereditary semisimple and using Theorem 8.22 one can add to the above list of the equivalent properties of a C*-algebra the following ones:

($8_s$) All elements of $A$ have countable spectra;
($9_s$) Each non-zero quotient of $A$ has a non-zero compact element;
($10_s$) Each non-zero quotient of $A$ has a non-zero finite element;
($11_s$) Every non-zero quotient of $A$ has a minimal projection.

It was shown in [AW] that an element a of a C*-algebra is compact if it is weakly compact, that is the map $x \longmapsto axa$ is weakly compact. So one can add also the condition:

($12_s$) Each non-zero quotient of $A$ has a non-zero weakly compact element.

## 9. Spectral continuity and radicals

Spectral continuity, that is the continuity of such functions of an operator (or an element of Banach algebra) as spectrum, special parts of spectrum, spectral radius and so on, is a very convenient property when it holds. For the information on this subject see for example [Bu] and the references therein. We consider it from the viewpoint of theory of topological radicals. An important role here is played by the scattered radical $\mathcal{R}_s$, its primitivity extension $\mathcal{R}_s^p$ and the radical $\mathcal{R}_s^{p*}$ obtained from $\mathcal{R}_s^p$ by the convolution procedure. So we begin with a study of these ideal maps.

### 9.1. $\mathcal{R}_s^a$, $\mathcal{R}_s^p$, $\mathcal{R}_s^{p*}$ and continuity of the spectrum

**Proposition 9.1.** *$\mathcal{R}_s^a$ is a hereditary topological radical.*

*Proof.* Since the radical $\mathcal{R}_s$ satisfies the condition of Banach heredity (Corollary 8.12), its centralization $\mathcal{R}_s^a$ is a hereditary topological radical on $\mathfrak{U}_b$, by Theorem 5.7 and Remark 5.8. □

The Banach heredity implies, by Theorem 7.7(3), that the primitivity extension $\mathcal{R}_s^p$ of $\mathcal{R}_s$ is a hereditary preradical on $\mathfrak{U}_b$. Recall that a Banach algebra $A$ is $\mathcal{R}_s^p$-radical if $A/I$ is $\mathcal{R}_s$-radical (= scattered), for each primitive ideal $I$ of $A$.

**Theorem 9.2.** *The classes* $\mathbf{Rad}(\mathcal{R}_s^p)\backslash\mathbf{Rad}(\mathcal{R}_s^a)$ *and* $\mathbf{Rad}(\mathcal{R}_s^a)\backslash\mathbf{Rad}(\mathcal{R}_s^p)$ *are non-empty.*

*Proof.* Let $A = C([0,1],\mathcal{K}(H))$ be the C*-algebra of all norm-continuous $\mathcal{K}(H)$-valued functions on $[0,1]$. It is well known that $\mathrm{Prim}(A)$ is isomorphic to $[0,1]$ via the map $t \mapsto I_t$ where $I_t$ is the ideal of all functions equal 0 at $t$. Since $A/I_t \cong \mathcal{K}(H)$ then $A$ is $\mathcal{R}_s^p$-radical. On the other hand, $A$ has no ideals $J$ with commutative $A/J$, so it is not $\mathcal{R}_s^a$-radical.

To show the non-voidness of $\mathbf{Rad}(\mathcal{R}_s^a)\backslash\mathbf{Rad}(\mathcal{R}_s^p)$, let us consider the Toeplitz algebra $A_t$ that is the C*-algebra generated by the unilateral shift $V$ in the space $H = l_2(\mathbb{N})$, acting by the rule $Ve_n = e_{n+1}$, where $(e_n)_{n=1}^{\infty}$ is the standard basis in $l_2(\mathbb{N})$. It is known [Dv, Theorem 5.1.5] that $A_t$ contains the ideal $\mathcal{K}(H)$ and that $A_t/\mathcal{K}(H)$ is isomorphic to the algebra $C(\mathbb{T})$ of all continuous functions on the unit circle. Since $\mathcal{K}(H) \subset \mathcal{R}_s(A_t)$, we get that $A_t \in \mathbf{Rad}(\mathcal{R}_s^a)$. On the other hand, the identity representation of $A_t$ is irreducible, while $A_t$ is not scattered, because $\sigma(V) = \mathbb{D}$, the unit disk. Therefore $A_t$ is not $\mathcal{R}_s^p$-radical. □

Now let us consider the radical $\mathcal{R}_s^{p*}$ which is obtained from $\mathcal{R}_s^p$ by the convolution procedure. We will see later that this radical plays a very important role in continuity of the spectral radius.

**Lemma 9.3.** *$\mathcal{R}_s^a < \mathcal{R}_s^{p*}$ on $\mathfrak{U}_{\mathrm{b}}$.*

*Proof.* All commutative Banach algebras belong to $\mathbf{Rad}(R_s^p)$ because their strictly irreducible representations are one-dimensional. The quotient $\mathcal{R}_s^a(A)/\mathcal{R}_s(A)$ is a commutative algebra, for every Banach algebra $A$. Thus we have that $\mathcal{R}_s^a(A)$ is an extension of an $\mathcal{R}_s^{p*}$-radical algebra by an $\mathcal{R}_s^{p*}$-radical algebra. Since $\mathcal{R}_s^{p*}$ is a radical, $\mathcal{R}_s^a(A)$ is $\mathcal{R}_s^{p*}$-radical. By Theorems 2.3 and 9.2, $\mathcal{R}_s^a < \mathcal{R}_s^{p*}$. □

Since $A_t$ is in $\mathbf{Rad}(R_s^a) \setminus \mathbf{Rad}(R_s^p)$, we conclude that $R_s^p \neq R_s^{p*}$. In other words, we obtain the following:

**Corollary 9.4.** *The under radical $\mathcal{R}_s^p$ is not a radical.*

Recall some facts about continuity of the spectrum.

Let $A$ be a Banach algebra. It is well known that $\sigma$ is upper continuous, i.e., if $a_n \to a$ then for every open neighborhood $V$ of $\sigma(a)$ there is $m > 0$ such that $\sigma(a_n) \subset V$ for every $n > m$. Moreover, Newburgh's theorem [Sn, Lemma 6.15] says that for every clopen subset $\sigma_0$ of $\sigma(a)$ and every open neighborhood $V_0$ of $\sigma_0$, there is $m_0$ such that $\sigma_n \subset V_0$ for $n > m_0$ and some clopen subsets $\sigma_n$ of $\sigma(a_n)$.

In particular, if $\sigma(a)$ is at most countable then the spectrum $\sigma$ is continuous at $a$, i.e., $\sigma(a_n) \to \sigma(a)$ by the Hausdorff metric. For continuity of $\sigma$ at $a$ it is sufficient to have that for every $\lambda \in \sigma(a)$ there is a sequence $\lambda_n \in \sigma(a_n)$ such that $\lambda_n \to \lambda$ as $n \to \infty$.

We need the following result of Zemanek [Z2, Remark 1].

**Lemma 9.5.** *Let $A$ be a Banach algebra, and let $a, a_n \in A$, $a_n \to a$ as $n \to \infty$. If $\sigma_{A/I}(a_n/I) \to \sigma_{A/I}(a/I)$, for every $I \in \mathrm{Prim}(A)$, then $\sigma_A(a_n) \to \sigma_A(a)$.*

It is convenient to use a more general result of this type.

**Lemma 9.6.** *Let $\{f_\alpha : \alpha \in \Lambda\}$ be a set of homomorphisms of a Banach algebra $A$ to some $Q$-algebras $A_\alpha$. Let $a \in A$ be such that*

$$\sigma(a) = \overline{\cup_{\alpha \in \Lambda} \sigma(f_\alpha(a))}. \tag{9.1}$$

*If $a_n \to a$ and $\sigma(f_\alpha(a_n)) \to \sigma(f_\alpha(a))$, for each $\alpha \in \Lambda$, then $\sigma(a_n) \to \sigma(a)$.*

*Proof.* Let $G = \liminf_{n\to\infty} \sigma(a_n) := \{\lambda \in \mathbb{C} : \lambda = \lim_{n\to\infty} \lambda_n$ for some $\lambda_n \in \sigma(a_n)\}$. By the upper continuity, it suffices to show that $\sigma(a) \subset G$. But $G$ is closed and contains all $\sigma(f_\alpha(a))$. So the result follows from (9.1). □

Note that apart from the case $f_\alpha = q_{I_\alpha}$, $I_\alpha \in \mathrm{Prim}(A)$, the condition (9.1) holds in many other situations, for example when $\Lambda$ is finite and $\cap_{\alpha\in\Lambda} I_\alpha \subset \mathrm{Rad}(A)$ (see Theorem 10.1).

**Theorem 9.7.** *Let $A$ be a Banach algebra. Then each element $a \in \mathcal{R}_s^p(A)$ is a point of the spectrum continuity.*

*Proof.* Let $a \in \mathcal{R}_s^p(A)$, $a_n \in A$, $a_n \to a$; by Lemma 9.5, it suffices to show that $\sigma(a_n/I) \to \sigma(a/I)$, for each $I \in \mathrm{Prim}(A)$. Since $\sigma(a/I)$ is countable of finite, the fact follows from the Newburgh theorem. □

We show now that Theorem 9.7 does not transfer to $\mathcal{R}_s^a$ and therefore to $\mathcal{R}_s^{p*}$.

**Theorem 9.8.** *The spectrum in an $\mathcal{R}_s^a$-radical Banach algebra can be discontinuous.*

*Proof.* We use the example of spectral discontinuity proposed by G. Lumer (see [H, Problem 86]). Let $H = l^2(\mathbb{Z})$, and let $\{e_n : n \in \mathbb{Z}\}$ be the natural orthonormal basis in $H$. Let us denote by $W$ the shift operator $W$ on $H$ ($We_n = e_{n+1}$). For each $k \in \mathbb{N}$, let $W_k$ be the operator acting on the basis by the formulas $W_k e_n = e_{n+1}$, for $n \neq 0$, $W_k e_0 = \frac{1}{k} e_1$. Then $W_k \to W_\infty$ where $W_\infty e_n = e_{n+1}$ for $n \neq 0$, $W_\infty e_0 = 0$. It is easy to check that $\sigma(W_k) = \mathbb{T}$ while $\sigma(W_\infty) = \mathbb{D}$, the unit disk.

Since $W_k - W$ is a rank one operator, for each $k$, all operators $W_k$, $k = 1, 2, \ldots, \infty$, belong to the C*-algebra $A$ generated by $W$ and the ideal $\mathcal{K}(H)$ of all compact operators. Then $A/\mathcal{K}(H)$ is commutative (it is generated by the image of the normal operator $W$ in the Calkin algebra $\mathcal{B}(H)/\mathcal{K}(H)$). Therefore $A \in$ **Rad** $(\mathcal{R}_s^a)$ since it is commutative modulo the scattered ideal $\mathcal{K}(H)$. □

**Corollary 9.9.** *The spectrum continuity is not a radical/semisimple property, even in Banach algebras.*

*Proof.* Indeed, assume, to the contrary, that there is a radical $P$ on Banach algebras such that **Rad** $(P)$ or **Sem** $(P)$ coincides with the class of algebras with the property of the spectrum continuity. Then this class contains commutative algebras and scattered algebras. As **Rad** $(P)$ or **Sem** $(P)$ is stable under extensions, then it contains the algebra constructed in Theorem 9.8, a contradiction. □

In the same time, **Rad** $\left(\mathrm{rad}^{\dim=\infty}|_{\mathfrak{U}_q}\right)$ is contained in the class of algebras with the property of the spectrum continuity.

**Problem 9.10.** *Is there the largest radical $P$ such that* **Rad** $(P)$ *is contained in the class of Banach algebras with the property of the spectrum continuity?*

At the moment, the largest radical $P$ about which we know that $P$-radical algebras enjoy the property of spectral continuity, is $\mathrm{rad}^{\dim=\infty}$ (which is clearly majorized by $\mathcal{R}_s^p$).

We finish by a discussion of the restrictions of the ideal maps $\mathcal{R}_s^p$, $\mathcal{R}_s^a$ and $\mathcal{R}_s^{p*}$ to the class $\mathfrak{U}_{c^*}$.

It was already mentioned that an irreducible *-representation of a C*-algebra is strictly irreducible and that for each primitive ideal $I$ of a C*-algebra $A$, the quotient $A/I$ is isometrically isomorphic to the image of $A$ in an irreducible

*-representation. These facts will be repeatedly used below. Moreover, dealing with C*-algebra we usually write representation meaning *-representation.

**Theorem 9.11.** *Let $A$ be a C*-algebra. Then $\mathcal{R}_s^{p*}(A) = \mathcal{R}_{\mathfrak{gcr}}(A)$.*

*Proof.* Since $\mathcal{R}_s^{p*}$ and $\mathcal{R}_{\mathfrak{gcr}}$ are radicals, it suffices to show (taking into account Corollary 2.4) that $\mathbf{Rad}(\mathcal{R}_{\mathfrak{gcr}}) = \mathbf{Rad}(\mathcal{R}_s^{p*}) \cap \mathfrak{U}_{c*}$.

Let $A$ be a CCR-algebra; then its image in each irreducible representation $\pi$ coincides with $\mathcal{K}(H_\pi)$, for some Hilbert space $H_\pi$; so $\pi(A)$ is scattered and therefore $\mathcal{R}_s$-radical. Hence $A$ is $\mathcal{R}_s^{p*}$-radical. Since each GCR-algebra $B$ admits an increasing transfinite chain $(I_\alpha)_{\alpha\le\gamma}$ with CCR gap-quotients and $I_0 = 0$, $I_\gamma = B$, then $\mathbf{Rad}(\mathcal{R}_s^{p*})$ contains all GCR C*-algebras.

Conversely, each $\mathcal{R}_s$-radical C*-algebra is a GCR-algebra by Theorem 8.23. If a C*-algebra $A$ is $\mathcal{R}_s^p$-radical, then, by definition, for each irreducible representation $\pi$, $\pi(A)$ is $\mathcal{R}_s$-radical, so it is a GCR-algebra. It follows that $\pi(A)$ contains a non-zero compact operator. Thus $A$ is a GCR-algebra.

Since each $\mathcal{R}_s^{p*}$-radical C*-algebra $C$ admits an increasing transfinite chain $(J)_{\alpha\le\delta}$ with $\mathcal{R}_s^p$-radical gap-quotients and $J_0 = 0$, $J_\delta = C$, then $\mathbf{Rad}(R_s^{p*}) \cap \mathfrak{U}_{c^*}$ consists of GCR C*-algebras. □

It follows from Theorem 9.11 that $\mathcal{R}_s^{p*}$ can be considered as a natural extension of the GCR-radical from the class $\mathfrak{U}_{c^*}$ to all Banach algebras.

We saw in the proof of Theorem 9.11 that the class of all $\mathcal{R}_s^p$-radical C*-algebras contains the class of all CCR-algebras. The inclusion is strict: for example it is not difficult to construct a C*-algebra $A \subset \mathcal{B}(H)$ that contains $\mathcal{K}(H)$ and such that $A/\mathcal{K}(H) \cong \mathcal{K}(H)$ – it is scattered but not CCR.

The examples in Theorem 9.2 show that CCR-algebras need not be $\mathcal{R}_s^a$-radical, and that $\mathcal{R}_s^a$-radical C*-algebras are not necessarily CCR. The full description of $\mathcal{R}_s^a$-radical C*-algebras is very non-trivial: the famous work [BDF] shows that even the classification of commutative extensions of the algebra $\mathcal{K}(H)$ is related to deep homological constructions.

Concluding this subsection let us show that spectrum is continuous *at normal elements* in a wide class of C*-algebras that are far from being GCR – for example, in C*-algebras of free groups.

A C*-algebra $A$ is called *residually finite-dimensional* (RFD, for short) if there is a family $\{\pi_\alpha : \alpha \in \Lambda\}$ of finite-dimensional representations of $A$ with $\cap_{\alpha\in\Lambda} \ker \pi_\alpha = 0$. Such algebras are $\mathrm{rad}^{\dim<\infty}$-semisimple.

**Theorem 9.12.** *Let $A$ be an RFD C*-algebra. Then every normal element $a$ of $A$ is a point of continuity of the spectrum $\sigma_A$.*

*Proof.* Let $(\pi_\alpha)_{\alpha\in\Lambda}$ be a family of finite-dimensional representations of $A$ with $\cap_{\alpha\in\Lambda} \ker \pi_\alpha = 0$. Let $(a_n)$ be a sequence of elements of $A$ tending to $a$. Since $\sigma(\pi_\alpha(a_n)) \to \sigma(\pi_\alpha(a))$ for each $\alpha$, it is sufficient to show that

$$\sigma(a) \subset \overline{\cup_{\alpha\in\Lambda}\sigma(\pi_\alpha(a))}$$

and apply Proposition 9.6.

Let $\lambda_0 \in \sigma(a)\backslash\overline{\cup_{\alpha\in\Lambda}\sigma(\pi_\alpha(a))}$, then there is a disk $D_\varepsilon(\lambda_0) = \{\lambda : |\lambda-\lambda_0| < \varepsilon\}$ which does not intersect $\cup_{\alpha\in\Lambda}\sigma(\pi_\alpha(a))$. Therefore for each $\alpha$ there is an element $b_\alpha \in \pi_\alpha(A)$ which is inverse to $\pi_\alpha(a-\lambda_0)$ and

$$\|b_\alpha\| = \left\|(\pi_\alpha(a-\lambda_0))^{-1}\right\| = \rho(\pi_\alpha(a-\lambda_0))^{-1} < \varepsilon^{-1}$$

since $a$ is normal.

Let $H = \oplus_{\alpha\in\Lambda}H_\alpha$ and $\pi = \oplus_{\alpha\in\Lambda}\pi_\alpha$, the direct sum of representations $\pi_\alpha$. Then $\ker(\pi) = 0$ and $B = \pi(A)$ is a C*-algebra isomorphic to $A$. The element $b = \oplus_{\alpha\in\Lambda}b_\alpha$ belongs to $\mathcal{B}(H)$, because $\sup_{\alpha\in\Lambda}\|b_\alpha\| < \infty$. It follows easily from the definition of $b_\alpha$ that

$$b\pi(a-\lambda_0) = \pi(a-\lambda_0)b = 1.$$

Therefore $\lambda_0 \notin \sigma_{\mathcal{B}(H)}(\pi(a))$; since $\pi(A)$ is a C*-algebra then it is an inverse-closed subalgebra of $\mathcal{B}(H)$ and

$$\lambda_0 \notin \sigma_{\pi(A)}(\pi(a)) = \sigma(a),$$

a contradiction. □

### 9.2. Continuity of the spectral radius

It turns out that the situation with the continuity of the spectral radii is different.

Let $A$ be a Banach algebra. Recall that the function $\rho : a \longmapsto \rho(a)$ is upper continuous because it is the infimum of continuous functions $a \longmapsto \|a^n\|^{1/n}$.

Let $\mathcal{V}_\rho$ be the class of all Banach algebras $A$ satisfying the following condition

($1_\rho$) For every Banach algebra $B$ containing $A$ as a closed ideal, any element $b \in B$ with $\rho(b) > \rho(b/A)$ is a point of continuity of the function $\rho$ in $B$.

**Theorem 9.13.** *Let $B$ be a Banach algebra, and let $A \in \mathcal{V}_\rho$ be its closed ideal. Then $A$ consists of points of continuity of $\rho$ in $B$.*

*Proof.* Let $a \in A$. If $\rho(a) > 0$ then $a$ is a point of continuity of $\rho$ in $B$ by ($1_\rho$). If $\rho(a) = 0$ then continuity of $\rho$ at $a$ follows from the upper continuity. □

**Lemma 9.14.** *Let $A$ be a normed algebra, and let $(J_\alpha)_{\alpha\in\Lambda}$ be an up-directed net of closed ideals of $A$, $J = \overline{\bigcup_{\alpha\in\Lambda} J_\alpha}$. Then*

$$\|M/J\| = \lim \|M/J_\alpha\| = \inf \|M/J_\alpha\|,$$
$$\rho(M/J) = \lim \rho(M/J_\alpha) = \inf \rho(M/J_\alpha)$$

*for every precompact set $M$ in $A$.*

*Proof.* This is [ST7, Lemma 4.10]. □

Let $\mathcal{F}_{\mathcal{V}_\rho}$ be the family of all under radicals whose radical classes are contained in $\mathcal{V}_\rho$.

**Theorem 9.15.** *$\vee\mathcal{F}_{\mathcal{V}_\rho}$ is a radical and $\vee\mathcal{F}_{\mathcal{V}_\rho} \in \mathcal{F}_{\mathcal{V}_\rho}$.*

*Proof.* Indeed, $\mathrm{H}_{\mathcal{F}_{\mathcal{V}_\rho}}$ is an under radical by Theorem 4.1. As $\vee\mathcal{F}_{\mathcal{V}_\rho} = \mathrm{H}^*_{\mathcal{F}_{\mathcal{V}_\rho}}$ then $\vee\mathcal{F}_{\mathcal{V}_\rho}$ is a radical by Theorem 3.1.

Let $A$ be a $(\vee\mathcal{F}_{\mathcal{V}_\rho})$-radical Banach algebra. By Theorem 4.19, there is an increasing transfinite chain $(I_\alpha)_{\alpha\leq\gamma}$ of closed ideals of $A$ such that $I_0 = 0$, $I_\gamma = A$ and for each $\alpha < \gamma$ there is $P \in \mathcal{F}_{\mathcal{V}_\rho}$ with $I_{\alpha+1}/I_\alpha = P(A/I_\alpha)$.

Let $B$ be a Banach algebra for which $A$ is an ideal. It follows that $I_1 = P_1(A)$ for some $P_1 \in \mathcal{F}_{\mathcal{V}_\rho}$. So $I_1$ is an ideal of $B$. Similarly, $I_2/I_1 = P_2(A/I_1)$ for some $P_2 \in \mathcal{F}_{\mathcal{V}_\rho}$ and therefore $I_2/I_1$ is an ideal of $B/I_1$. As $I_2 = q_{I_1}^{-1}(P_2(A/I_1))$ then $I_2$ is an ideal of $B$. Applying the transfinite induction, it is easy to check that all $I_\alpha$ are ideals of $B$.

Assume, to the contrary, that $A \notin \mathcal{V}_\rho$. Then there are a Banach algebra $B$, containing $A$ as an ideal, and an element $b \in B$ with $\rho(b) > \rho(b/A)$ which is a point of discontinuity of $\rho$. So there is a sequence $(b_n) \subset B$ such that $b_n \to b$ as $n \to \infty$, but $\limsup \rho(b_n) < \rho(b)$.

Take the first ordinal $\alpha'$ for which $\rho(b) \neq \rho(b/I_{\alpha'})$. By Lemma 9.14, $\alpha'$ is not a limit ordinal. So there is an ordinal $\alpha < \gamma$ such that

$$\rho(b) = \rho(b/I_\alpha) > \rho(b/I_{\alpha+1}). \tag{9.2}$$

Let $C = B/I_\alpha$, $J = I_{\alpha+1}/I_\alpha$ and $x = b/I_\alpha$. Then one can rewrite (9.2) as

$$\rho(x) > \rho(x/J).$$

As $J \in \mathcal{V}_\rho$ is a closed ideal of $C$ then $x$ is a point of continuity of $\rho$. Then

$$\liminf \rho(b_n) \geq \liminf \rho(b_n/I_\alpha) = \rho(b/I_\alpha) = \rho(b) > \limsup \rho(b_n),$$

a contradiction. □

The radical $\vee\mathcal{F}_{\mathcal{V}_\rho}$ is denoted by $\mathcal{R}_{\vec{\rho}}$ and is called the *$\rho$-continuity radical.* Now we extend our knowledge about $\mathbf{Rad}(\mathcal{R}_{\vec{\rho}})$.

**Theorem 9.16.** $\mathcal{R}_s^{p*} \leq \mathcal{R}_{\vec{\rho}}$.

*Proof.* Let $B$ be a Banach algebra, let $A$ be a closed $\mathcal{R}_s^p$-radical ideal of $B$, and let $b \in B$ be such that $\rho(b) > \rho(b/A)$. Let a sequence $b_n$ tend to $b$ as $n \to \infty$.

Assume, to the contrary, that $\limsup \rho(b_n) < \rho(b)$. Choose $\varepsilon > 0$ such that

$$\rho(b) - \varepsilon > \max\{\rho(b/A), \limsup \rho(b_n)\}. \tag{9.3}$$

As $\rho(b) = \sup_{I\in\mathrm{Prim}(B^1)} \rho(b/I)$ then there is a primitive ideal $I$ of $B$ such that

$$\rho(b/I) > \rho(b) - \varepsilon.$$

Thus

$$\rho(b/I) > \rho(b/A) \geq \rho(q_I(b)/\overline{q_I(A)}).$$

Since $q_I(A)$ is an $\mathcal{R}_s$-radical ideal of $B/I$ then so is $\overline{q_I(A)})$. It follows from Theorem 9.13 that $b/I$ is a point of continuity of $\rho$. In particular,

$$\liminf(\rho(b_n)) \geq \liminf(\rho(b_n/I)) = \rho(b/I),$$

a contradiction with (9.3).

We proved that any $\mathcal{R}_s^p$-radical algebra lies in $\mathcal{V}_\rho$. So $\mathcal{R}_s^p \in \mathcal{F}_{\mathcal{V}_\rho}$ and $\mathcal{R}_s^p \leq \mathcal{R}_{\vec{\beta}}$. As $\mathcal{R}_s^{p*}$ is the smallest radical that is larger than or equal to $\mathcal{R}_s^p$ then $\mathcal{R}_s^{p*} \leq \mathcal{R}_{\vec{\beta}}$. □

**Corollary 9.17.** *$\mathcal{R}_s^a \leq \mathcal{R}_{\vec{\beta}}$; in particular, all $\mathcal{R}_s^a$-radical ideals of a Banach algebra consist of the points of continuity for $\rho$ in the algebra.*

*Proof.* It follows from Lemma 9.3 and Theorem 9.16. □

**Corollary 9.18.** *Let $A$ be a $C^*$-algebra, $J = \mathcal{R}_{\mathfrak{gcr}}(A)$, the largest GCR-ideal of $A$. Then the spectral radius is continuous at every point $a \in J$.*

*Proof.* It follows from Theorem 9.11 and Theorem 9.16. □

**Problem 9.19.** *Is $\mathcal{R}_{\vec{\beta}}$ the largest radical among all radicals $P$ for which $P$-radical ideals consist of the points of continuity for the spectral radius?*

**Problem 9.20.** *If the spectral radius is continuous on a $C^*$-algebra $A$, is $A$ a GCR-algebra?*

### 9.3. Continuity of the joint spectral radius

Let $A$ be a normed algebra. Here it will be convenient to denote by $\rho_j$ (instead of $\rho$) the function $M \longmapsto \rho(M)$ defined on bounded sets of $A$. It is upper continuous with respect to Hausdorff's distance [ST1, Proposition 3.1], that is

$$\limsup_{k\to\infty} \rho(M_k) \leq \rho(M)$$

if $M_k \to M$ in the sense that $\operatorname{dist}(M_k, M) \to 0$ as $k \to \infty$. If $\lim_{k\to\infty} \rho(M_k) = \rho(M)$ for each sequence $M_k \to M$ then we say that $M$ is a *point of continuity for the joint spectral radius.*

Let $\mathcal{V}_{\rho_j}$ be the class of all normed algebras $A$ satisfying the following condition

($1_j$) For every normed algebra $B$ containing $A$ as a closed ideal, any precompact set $M \subset B$ with $\rho(M) > \rho(M/A)$ is a point of continuity of the function $\rho_j$.

Let $\mathcal{F}_{\mathcal{V}_{\rho_j}}$ be the family of all topological under radicals whose radical classes are contained in $\mathcal{V}_{\rho_j}$.

**Theorem 9.21.** *Let $\mathcal{R}_{\vec{\rho_j}} = \vee \mathcal{F}_{\mathcal{V}_{\rho_j}}$. Then*

1. *$\mathcal{R}_{\vec{\rho_j}}$ is a radical and $\mathcal{R}_{\vec{\rho_j}} \in \mathcal{F}_{\mathcal{V}_{\rho_j}}$;*
2. *For every normed algebra $A$, every precompact subset $M$ of $\mathcal{R}_{\vec{\rho_j}}(A)$ is a point of continuity of $\rho_j$;*
3. *For every normed algebra $A$, $\rho(M) = \sup\{\rho(K) : K \subset M \text{ is finite}\}$ for every precompact set $M$ in $\mathcal{R}_{\vec{\rho_j}}(A)$;*
4. *$\mathcal{R}_{\mathrm{hc}}^r \vee \mathcal{R}_{\mathrm{cq}}^a \leq \mathcal{R}_{\vec{\rho_j}}$.*

*Proof.* (1) and (4) Using Lemma 9.14 and repeating the argument in Theorem 9.15, we obtain that $\mathcal{R}_{\vec{\rho_j}}$ is a radical and all $\mathcal{R}_{\vec{\rho_j}}$-radical algebras lie in $\mathcal{V}_{\rho_j}$. It was proved in [ST7, Theorem 6.3] that for $\mathcal{R}_{\mathrm{hc}}^r \vee \mathcal{R}_{\mathrm{cq}}^a$-radical ideals the condition ($1_j$) holds. Therefore (4) is valid.

(2) is similar to the proof of Theorem 9.15.
(3) follows from (2). □

We call $\mathcal{R}_{\overrightarrow{\rho_j}}$ the $\rho_j$*-continuity radical.*

**Corollary 9.22.** *The joint spectral radius is continuous on precompact subsets of any scattered $C^*$-algebra.*

In Section 10 this result will be extended to all GCR-algebras.

### 9.4. Continuity of the tensor radius

The function $\rho_t : N \longmapsto \rho_t(N)$ defined on summable families $N = (a_m)_1^\infty$ of elements of $A$ is upper continuous with respect to the metric $\mathrm{d}(N', N) = \sum_1^\infty \|a'_m - a_m\|$ where $N' = (a'_m)_1^\infty$ (see [ST6, Proposition 3.12]). We say that a family $N$ is a *point of continuity of the tensor spectral radius* if $\rho_t(N_n) \to \rho_t(N)$, for any sequence $N_n$ of summable families in $A$ that tends to $N$ with respect to the metric d.

Let $\mathcal{V}_{\rho_t}$ be the class of all normed algebras $A$ satisfying the following condition

$(1_t)$ For every normed algebra $B$ containing $A$ as a closed ideal, any summable family $N$ in $B$ with $\rho_t(N) > \rho_t(N/A)$ is a point of continuity of the function $\rho_t$.

**Lemma 9.23.** *Let $A$ be a normed algebra, let $(J_\alpha)_{\alpha\in\Lambda}$ be an up-directed net of closed ideals of $A$ and $J = \overline{\bigcup_{\alpha\in\Lambda} J_\alpha}$. Then*

$$\|N/J\|_+ = \lim \|N/J_\alpha\|_+ = \inf \|N/J_\alpha\|_+ , \tag{9.4}$$

$$\rho_t(N/J) = \lim \rho_t(N/J_\alpha) = \inf \rho_t(N/J_\alpha) \tag{9.5}$$

*for every summable family $N = (a_m)_1^\infty$ in $A$.*

*Proof.* As $\|N/J\|_+ \le \|N/J_\alpha\|_+$ then

$$\rho_t(N/J) = \inf_n \|N^n/J\|_+^{1/n} \le \inf_\alpha \inf_n \|N^n/J_\alpha\|_+^{1/n} = \inf_\alpha \rho(N/J_\alpha)$$

and

$$\|N/J\|_+ \le \inf_\alpha \|N/J_\alpha\|_+ \le \liminf_\alpha \|N/J_\alpha\|_+ .$$

By our assumption, for every $a_m$ and $\varepsilon_m > 0$ there exists $\alpha = \alpha(a_m, \varepsilon_m)$ such that

$$\|a_m/J_\alpha\| \le \|a_m/J\| + \varepsilon_m. \tag{9.6}$$

For $\varepsilon > 0$, take $k > 0$ such that $\mathrm{d}\left(N|_1^k, N\right) < \varepsilon$. Let $N_k = N|_1^k$. Take $\varepsilon_m > 0$ such that $\sum_1^k \varepsilon_m < \varepsilon$. Then $\mathrm{d}(N_k/J_\gamma, N/J_\gamma) \le \mathrm{d}(N_k, N) < \varepsilon$, and

$$\|N/J_\gamma\|_+ \le \mathrm{d}(N_k/J_\gamma, N/J_\gamma) + \|N_k/J_\gamma\|_+ \le \|N_k/J\|_+ + \varepsilon \le \|N/J\|_+ + \varepsilon$$

for $\gamma \ge \max\{\alpha(a_m, \varepsilon_m) : m = 1, \dots, k\}$ by (9.6). Therefore

$$\inf_\alpha \|N/J_\alpha\|_+ \le \limsup_\alpha \|N/J_\alpha\|_+ \le \|N/J\|_+ \tag{9.7}$$

that implies (9.4). Take $n > 0$ such that $\|N^n/J\|_+^{1/n} \le \rho(N/J)+\varepsilon$. It follows from (9.7) applied to $N^n$ that

$$\inf_\alpha \rho_t(N/J_\alpha) \le \limsup_\alpha \rho_t(N/J_\alpha) \le \limsup_\alpha \|N^n/J_\alpha\|_+^{1/n} \le \|N^n/J\|_+^{1/n} \le \rho_t(N/J)+\varepsilon.$$

This implies (9.5). □

**Lemma 9.24.** *Let $A$ be a commutative Banach algebra. Then $\rho_t$ is uniformly continuous on $B$ with respect to the metric* d.

*Proof.* One can assume that $A$ is unital. Let $M = (a_n)_1^\infty$ and $N = (b_n)_1^\infty$ be summable families in $A$, and let $\mathcal{F}$ be the set of all multiplicative functionals $f$ on $A$ with $\|f\| = f(1) = 1$. By Theorem 5.29,

$$\rho_t(M) = \sup\left\{\sum_1^\infty |f(a_n)| : f \in \mathcal{F}\right\}$$

and, by [ST6, Propositions 3.3 and 3.4], $\rho_t$ is subadditive on $A$, whence

$$|\rho_t(M) - \rho_t(N)| \le \rho_t(M-N) = \sup\left\{\sum_1^\infty |f(a_n - b_n)| : f \in \mathcal{F}\right\}$$
$$\le \sum_1^\infty \|a_n - b_n\| = \mathrm{d}(M,N). \qquad \square$$

Let $\mathcal{F}_{\mathcal{V}_{\rho_t}}$ be the family of all topological under radicals whose radical classes are contained in $\mathcal{V}_{\rho_t}$.

**Theorem 9.25.** *Let $\mathcal{R}_{\overrightarrow{\rho_t}} = \vee\mathcal{F}_{\mathcal{V}_{\rho_t}}$. Then*

(1) *$\mathcal{R}_{\overrightarrow{\rho_t}}$ is a radical and $\mathcal{R}_{\overrightarrow{\rho_t}} \in \mathcal{F}_{\mathcal{V}_{\rho_t}}$;*
(2) *For every normed algebra $A$, every summable family $N$ of $\mathcal{R}_{\overrightarrow{\rho_t}}(A)$ is a point of continuity of $\rho_t$;*
(3) *For every normed algebra $A$, $\rho_t(N) = \sup_k \rho_t\left(N|_1^k\right)$ for every summable family $N$ in $\mathcal{R}_{\overrightarrow{\rho_t}}(A)$;*
(4) *$\mathcal{R}_t^a \le \mathcal{R}_{\overrightarrow{\rho_t}}$.*

*Proof.* (1) Using Lemma 9.23 and repeating the argument in Theorem 9.15, we obtain that $\mathcal{R}_{\overrightarrow{\rho_t}}$ is a radical and all $\mathcal{R}_{\overrightarrow{\rho_t}}$-radical algebras lie in $\mathcal{V}_{\rho_t}$.

(2) is similar to Theorem 9.15.

(3) follows from (2).

(4) It is sufficient to check the condition $(1_t)$ for any $\mathcal{R}_t$-semisimple Banach algebra $B$. Let $A$ be a closed $\mathcal{R}_t^a$-radical ideal of $B$. As $A$ is $\mathcal{R}_t$-semisimple, it is a central ideal. By Theorem 5.19,

$$\rho_t(N) = \rho_t(\mathrm{L}_N) \tag{9.8}$$

for every summable family $N = (a_m)_1^\infty$ in $B$. As $A$ is a closed invariant subspace for $\mathrm{L}_B$ then

$$\rho_t(\mathrm{L}_N) = \max\left\{\rho_t\left(\mathrm{L}_N|_{B/A}\right), \rho_t\left(\mathrm{L}_N|_A\right)\right\} \tag{9.9}$$

by Theorem 5.18. It is clear that

$$\rho_t\left(\mathrm{L}_N|_{B/A}\right) = \rho_t\left(\mathrm{L}_{N/A}\right) = \rho_t(N/A). \tag{9.10}$$

Let now $\rho_t(N) > \rho_t(N/A)$. It follows from (9.8), (9.9) and (9.10) that

$$\rho_t(N) = \rho_t\left(\mathrm{L}_N|_A\right)$$

Let $(N_n)$ be a sequence of summable families of $B$ such that $N_n \to N$ in the metric d as $n \to \infty$. Then $\mathrm{L}_{N_n}|_A \to \mathrm{L}_N|_A$. It is easy to check that the algebra $\mathrm{L}_B|_A$ is commutative. By Lemma 9.24, $\rho_t$ is continuous on $\mathrm{L}_B|_A$, whence

$$\liminf \rho_t(N_n) \geq \liminf \rho_t\left(\mathrm{L}_{N_n}|_A\right) = \rho_t\left(\mathrm{L}_N|_A\right) = \rho_t(N)$$

and $\lim \rho_t(N_n) = \rho_t(N)$.

This implies that $\mathbf{Rad}(\mathcal{R}_t^a) \subset \mathcal{V}_{\rho_t}$. As $\mathcal{R}_{\overrightarrow{\rho_t}}$ is the largest radical with such property, $\mathcal{R}_t^a \leq \mathcal{R}_{\overrightarrow{\rho_t}}$. □

We call $\mathcal{R}_{\overrightarrow{\rho_t}}$ the *$\rho_t$-continuity radical.*

## 10. Estimations of the joint spectral radius

In this section we study how various spectral characteristics of an element or a subset of a Banach algebra can be expressed via its images in quotients of the algebra, and how the characteristics of these images depend on the corresponding ideals. The main attention is devoted to the joint spectral radius.

For example, it is well known that for each element $a$ of any Banach algebra $A$

$$\rho(a) = \sup\{\rho(a/I) : I \in \mathrm{Prim}(A)\}. \tag{10.1}$$

Can $\mathrm{Prim}(A)$ be changed here by arbitrary family of (primitive or not) ideals whose intersection is contained in $\mathrm{Rad}(A)$? Is it possible to extend (10.1) to $\rho(M)$, where $M$ is a precompact subset of $A$? How in general depend $\rho(a/I)$ and $\rho(M/I)$ on $I$? Questions of this type often arise in the spectral theory.

In what follows by an *operator* we mean a bounded linear operator.

### 10.1. Finite families of ideals with trivial intersection

We begin with the spectrum.

**Theorem 10.1.** *Let $A$ be a Banach algebra, $I_1, \ldots, I_n$ be closed ideals of $A$, and let $\cap_{k=1}^n I_k \subset \mathrm{Rad}(A)$. Then*

$$\sigma(a) = \cup_{k=1}^n \sigma(a/I_k),$$

*for any element $a \in A$.*

*Proof.* Let $I, J$ be ideals of $A$. Assuming that $0 \notin \sigma(a/I) \cup \sigma(a/J)$, we prove that $0 \notin \sigma(a/K)$ where $K = I \cap J$. Indeed, since $a$ is invertible in $A/I$ and $A/J$ then there are $b, c \in A$ with $ba = 1 + i$ and $ca = 1 + j$ where $i \in I$, $j \in J$. Therefore

$$(ba - 1)(ca - 1) = ij \in K.$$

It follows that $(bac - c - b)a \in 1 + K$ that is $a$ is left invertible in $A/K$. Similarly, $a$ is right invertible in $A/K$. Thus $a$ is invertible in $A/K$.

Now using induction, we obtain that

$$\sigma(a/\cap_{k=1}^{n} I_k) \subset \cup_{k=1}^{n} \sigma(a/I_k).$$

In our assumptions this implies that

$$\sigma(a) = \sigma(a/\operatorname{Rad}(A)) \subset \cup_{k=1}^{n} \sigma(a/I_k);$$

the converse inclusion is evident. □

Theorem 10.1 implies that if $\mathcal{F}$ is a finite set of closed ideals with intersection in $\operatorname{Rad}(A)$ then

$$\rho(a) = \sup_{J \in \mathcal{F}} \rho(a/J) \text{ for each } a \in A. \tag{10.2}$$

Our next aim is to establish a similar result for the joint spectral radius.

We use the following statement proved in [ST1, Corollary 4.3].

**Lemma 10.2.** *Let $M$ be a bounded set of operators on a Banach space $X$, and let $X \supset Z_1 \supset \cdots \supset Z_n$ be a chain of closed subspaces invariant for all operators in $M$. Then*

$$\rho(M) = \max\left\{\rho\left(M|_{X/Z_1}\right), \rho\left(M|_{Z_1/Z_2}\right), \dots, \rho\left(M|_{Z_{n-1}/Z_n}\right), \rho\left(M|_{Z_n}\right)\right\},$$

*where by $M|_{Z_k/Z_{k+1}}$ we mean the family of operators induced by $M$ in the quotient Banach space $Z_k/Z_{k+1}$.*

Let $A$ be a Banach algebra. Recall (cf. Section 1.2.7) that by a *Banach ideal* of $A$ we call any ideal $I$ of $A$ which is complete with respect to a norm $\|\cdot\|_I$ such that

$$\|x\|_I \geq \|x\|$$

for all $x \in I$; recall also that $\mathrm{W}_a := \mathrm{L}_a \mathrm{R}_a$, for $a \in A$, and $\mathrm{W}_M := \{\mathrm{W}_a : a \in M\}$, for $M \subset A$. It is clear that all ideals of $A$ are invariant subspaces for the operators $\mathrm{W}_a$.

Let $\rho_{(A;\|\cdot\|)}(M)$ denote the joint spectral radius of a bounded set $M$ in the algebra $(A; \|\cdot\|)$.

**Lemma 10.3.** *Let $A$ be a Banach algebra, and let $I$ be a Banach ideal of $A$. Then*

1. *All operators $\mathrm{W}_a|_I$, $a \in A$, are bounded in the norm $\|\cdot\|_I$;*
2. *For each bounded subset $M$ of $A$,*

$$\rho_{(I;\|\cdot\|_I)}(\mathrm{W}_M|_I) \leq \rho_{(A;\|\cdot\|)}(M)^2. \tag{10.3}$$

*Proof.* (1) It was proved in [B] that there is a constant $C > 0$ such that

$$||axb||_I \leq C||a||_A||x||_I||b||_A$$

for all $a, b \in A^1$, $x \in I$. Therefore $\|\mathrm{W}_a|_I\|_I \leq C\|a\|^2$ which proves (1).

(2) As $||\mathrm{W}_{M^n}||_I \leq C\left(||M^n||_A\right)^2$ then $||\mathrm{W}_{M^n}||_I^{1/n} \leq C^{1/n}\left(||M^n||_A\right)^{2/n}$, and taking $n \to \infty$ we obtain (2). □

**Proposition 10.4.** *Let $A$ be a Banach algebra, let $I_1, \ldots, I_n$ be closed ideals of $A$, and let $J = \cap_{i=1}^n I_i$. Then*

$$\rho(M)^2 = \max\{\max_{1\leq i\leq n} \rho(M/I_i)^2, \rho(\mathrm{W}_M|_J)\}, \tag{10.4}$$

*for each bounded subset $M$ of $A$.*

*Proof.* We will prove $\leq$ because the converse inequality is trivial.

For $k \leq n$, let $J_k = I_1 \cap I_2 \cap \cdots \cap I_k$. Applying Lemma 10.2 to the family $\mathrm{W} = \mathrm{W}_M$ of operators on $A$ and the chain $A \supset J_1 \supset J_2 \supset \cdots \supset J_n$, we obtain that

$$\begin{aligned}\rho(M)^2 &= \rho(\mathrm{W}_M)\\ &= \max\left\{\rho(\mathrm{W}_M|_{A/J_1}), \rho(\mathrm{W}_M|_{J_1/J_2}), \ldots, \rho(\mathrm{W}_M|_{J_{n-1}/J_n}), \rho(\mathrm{W}_M|_{J_n})\right\}.\end{aligned}$$

So it suffices to show that

$$\rho(\mathrm{W}_M|_{J_{k-1}/J_k}) \leq \rho(M/I_k)^2 \tag{10.5}$$

for each $k$ (assuming $J_0 = A$).

For $k = 1$, the inequality (10.5) is in fact an equality:

$$\rho(\mathrm{W}_M|_{A/I_1}) = \rho(M/I_1)^2.$$

For a fixed $k$ with $1 < k \leq n$, let $B = A/I_k$ and $q = q_{I_k} : A \to B$. Then the algebraic isomorphism $\phi$ of $I := q(J_{k-1}) = (J_{k-1} + I_k)/I_k$ onto the Banach algebra $C := J_{k-1}/(I_k \cap J_{k-1}) = J_{k-1}/J_k$ allows us to supply $I$ with a new norm

$$\|x\|_I = \|\phi(x)\|_C,$$

and it is easy to check that $I$ is a Banach ideal of $B$ in this norm. It follows from the definition of the norm $\|\cdot\|_I$ that $\|\mathrm{W}_a|_{J_{k-1}/J_k}\| = \|\mathrm{W}_{q(a)}|_I\|_I$, for $a \in A$. Therefore

$$\rho(\mathrm{W}_M|_{J_{k-1}/J_k}) = \rho_{(I;\|\cdot\|_I)}(\mathrm{W}_{q(M)}|_I). \tag{10.6}$$

Applying Lemma 10.3 to the subset $q(M)$ of $B$ we get:

$$\rho_{(I;\|\cdot\|_I)}(\mathrm{W}_{q(M)}|_I) \leq \rho_B(q(M))^2 = \rho(M/I_k)^2.$$

By (10.6), this is a reformulation of (10.5). □

**Corollary 10.5.** *Let $A$ be a Banach algebra, let $\mathcal{F} = \{I_1, \ldots, I_n\}$ be a finite family of closed ideals of $A$, such that $\cap_{i=1}^n I_i \subset \mathcal{R}_{\mathrm{cq}}(A)$. Then*

$$\rho(M) = \max\left\{\rho(M/I_1), \ldots, \rho(M/I_n)\right\}, \tag{10.7}$$

*for each precompact subset $M$ of $A$.*

*Proof.* The equality (10.7) is a consequence of Theorem 10.4, if $\cap_{i=1}^n I_i = 0$. It follows that in general

$$\rho(M/(\cap_{i=1}^n I_i)) = \max\{\rho(M/I_1),\dots,\rho(M/I_n)\}. \tag{10.8}$$

Indeed, setting $J = \cap_{i=1}^n I_i$, $J_i = I_i/J$ and $\widetilde{M} = M/J$, we have that

$$\cap_{i=1}^n J_i = 0$$

whence $\rho(\widetilde{M}) = \max_i \rho(\widetilde{M}/J_i)$, and it suffices to note that $\widetilde{M}/J_i$ corresponds to $M/I_i$ with respect to the standard isomorphism of $A/I_i$ onto $(A/J)/J_i$.

Now, since $\cap_{i=1}^n I_i \subset \mathcal{R}_{\mathrm{cq}}(A)$ and (2.7) holds for precompact sets, we obtain that

$$\rho(M) = \rho(M/\mathcal{R}_{\mathrm{cq}}(A)) \le \rho(M/(\cap_{i=1}^n I_i)) = \max\{\rho(M/I_1),\dots,\rho(M/I_n)\}.$$

The converse inequality is evident. □

### 10.2. Arbitrary families of ideals with trivial intersection

The result of Theorem 10.1 does not extend to arbitrary families of ideals. It suffices to show that the equality (10.2) fails in general.

**Example 10.6.** Let $\{e_n : 1 \le n < \infty\}$ be an orthonormal basis in a Hilbert space $H$, and let $A$ be the algebra of all operators on $H$ preserving the subspaces $H_n = \mathrm{span}(e_1,\dots,e_n)$. Let $K_n = \{T \in A : T|_{H_n} = 0\}$. Then all $K_n$ are closed ideals of $A$, and

$$\cap_{n=1}^\infty K_n = 0.$$

Let $S$ be the backward shift: $Se_n = e_{n-1}$, $Se_1 = 0$. Then $S \in A$ and all elements $S/K_n$ are nilpotent,

$$\rho(S/K_n) = 0$$

while $\sigma(S) = \{\lambda \in \mathbb{C} : |\lambda| \le 1\}$ and

$$\rho(S) = 1.$$

Let us first of all give a slight extension of Theorem 10.1.

**Proposition 10.7.** *Let $A$ be an algebra, and let $\mathcal{F} = (I_\alpha)_{\alpha\in\Lambda}$ be a family of ideals of $A$. For each $\alpha \in \Lambda$, set $J_\alpha = \cap_{\beta\ne\alpha} I_\beta$. If $\cap_{\alpha\in\Lambda} I_\alpha \subset \mathrm{rad}\,(A)$ then*

$$\sigma(a) = (\cup_{\alpha\in\Lambda}\sigma(a/I_\alpha)) \cup (\cap_{\alpha\in\Lambda}\sigma(a/J_\alpha)),$$

*for each $a \in A$*

*Proof.* For each $\alpha \in \Lambda$, $I_\alpha \cap J_\alpha \subset \mathrm{rad}\,(A)$, whence

$$\sigma(a) = \sigma(a/I_\alpha) \cup \sigma(a/J_\alpha)$$

by Theorem 10.1. Now if $\lambda \in \sigma(a)$ and $\lambda \notin \cup_{\alpha\in\Lambda}\sigma(a/I_\alpha)$ then $\lambda \in \sigma(a/J_\alpha)$ for every $\alpha$. Therefore $\lambda \in \cap_{\alpha\in\Lambda}\sigma(a/J_\alpha)$. □

The following result shows that for scattered Banach algebras the situation is sufficiently satisfactory.

**Theorem 10.8.** *Let $A$ be a Banach algebra. If the spectrum of an element $a$ of $A$ is the closure of its isolated points then $\sigma(a) = \overline{\cup_{\alpha\in\Lambda}\sigma(a/I_\alpha)}$ for each family $\mathcal{F} = (I_\alpha)_{\alpha\in\Lambda}$ of closed ideals with $\cap_{\alpha\in\Lambda} I_\alpha \subset \mathrm{Rad}(A)$.*

*Proof.* It suffices to show that $\lambda \in \cup_{\alpha\in\Lambda}\sigma(a/I_\alpha)$ for each isolated point $\lambda \in \sigma(a)$. Assume, to the contrary, that

$$\lambda \notin \sigma(a/I_\alpha)$$

for each $\alpha$. Let $p$ be the Riesz projection of $a$ corresponding to $\lambda$. Then $(a-\lambda)p$ is quasinilpotent whence $(a-\lambda)p/I_\alpha$ is quasinilpotent for each $\alpha$. But $(a-\lambda)/I_\alpha$ is an invertible element of the algebra $A/I_\alpha$ that commutes with $p/I_\alpha$. Therefore $p/I_\alpha$ is quasinilpotent. As it is idempotent, $p/I_\alpha = 0$ whence $p \in I_\alpha$. Then

$$p \in \cap_{\alpha\in\Lambda} I_\alpha \subset \mathrm{Rad}(A),$$

a contradiction. □

Let us now come to consideration of the joint spectral characteristics of subsets in $A$. Recall that the *Berger–Wang spectral radius* $r(M)$ of a bounded family $M \subset A$ is defined by

$$r(M) = \limsup_{n\to\infty} \sup_{a\in M^n} \rho(a)^{1/n}.$$

**Proposition 10.9.** *Let $A$ be a Banach algebra, and let $\mathcal{F}$ be a family of closed ideals such that*

$$\rho(a) = \sup_{J\in\mathcal{F}} \rho(a/J)$$

*for each $a \in A$. Then*

$$r(M) = \sup_{I\in\mathcal{F}} r(M/I) \tag{10.9}$$

*for each bounded set $M \subset A$.*

*Proof.* By [ST7, (2.1)],

$$r(M) = \sup_n \sup_{a\in M^n} \rho(a)^{1/n}.$$

Hence by (10.2),

$$r(M) = \sup_n \sup_{a\in M^n} \sup_{I\in\mathcal{F}} \rho(a/I)^{1/n} = \sup_{I\in\mathcal{F}} \sup_n \sup_{a\in M^n} \rho(a/I)^{1/n} = \sup_{I\in\mathcal{F}} r(M/I). \quad \square$$

**Corollary 10.10.** *Let $A$ be a Banach algebra, and let $\mathcal{F}$ be a finite family of closed ideals. If $\cap\{I : I \in \mathcal{F}\} \subset \mathrm{Rad}(A)$ then*

$$r(M) = \sup_{I\in\mathcal{F}} r(M/I)$$

*for each bounded set $M \subset A$.*

Now we consider the joint spectral radius $\rho(M)$.

**Proposition 10.11.** *Let $A$ be a Banach algebra, let $\mathcal{F}$ be a family of closed ideals of $A$, and let $\mathcal{E}$ be the set of all finite intersections of ideals from $\mathcal{F}$. Then*

$$\rho(M)^2 = \max\left\{\max_{I\in\mathcal{F}} \rho(M/I)^2, \inf_{J\in\mathcal{E}} \rho(\mathrm{W}_M|_J)\right\}$$

*for every bounded set $M$ in $A$.*

*Proof.* Only the inequality $\leq$ should be proved. If $\inf_{J\in\mathcal{E}} \rho(\mathrm{W}_M|_J) < \rho(M)^2$ then there are ideals $I_1, \dots, I_n \in \mathcal{F}$ with

$$\rho(\mathrm{W}_M|_J) < \rho(M)^2$$

where $J = I_1 \cap I_2 \cap \cdots \cap I_n$. By Proposition 10.4,

$$\rho(M)^2 = \max\left\{\max_{1\leq i\leq n} \rho(M/I_i)^2, \rho(\mathrm{W}_M|J)\right\} = \max_{1\leq i\leq n} \rho(M/I_i)^2 \leq \max_{I\in\mathcal{F}} \rho(M/I)^2$$

and we are done. □

Now we shall estimate the number $\inf_{J\in\mathcal{E}} \rho(W_M|J)\}$. Let $\mathcal{E}$ be a family of closed subspaces of a Banach space $X$, and let

$$\begin{aligned}
\mathrm{Alg}(\mathcal{E}) &= \{T \in \mathcal{B}(X) : TY \subset Y \text{ for all } Y \in \mathcal{E}\},\\
\|T\|_{\mathcal{E}} &= \inf\{\|T|_Y\| : Y \in \mathcal{E}\} \text{ for any } T \in \mathrm{Alg}(\mathcal{E}),\\
\ker\mathcal{E} &= \{T \in \mathrm{Alg}(\mathcal{E}) : \|T\|_{\mathcal{E}} = 0\}.
\end{aligned}$$

For a bounded set $N \subset \mathrm{Alg}(\mathcal{E})$, we write

$$\|N\|_{\mathcal{E}} = \sup_{T\in N} \|T\|_{\mathcal{E}}\,.$$

**Lemma 10.12.** *Let $X$ be a Banach space, and let $\mathcal{E}$ be a family of closed subspaces of $X$, closed under finite intersections and satisfying the condition $\cap_{Y\in\mathcal{E}} Y = 0$. Then*

($1_e$) *$\|\cdot\|_{\mathcal{E}}$ is a seminorm on $\mathrm{Alg}(\mathcal{E})$;*
($2_e$) *$\|TS\|_{\mathcal{E}} \leq \|T\|_{\mathcal{E}} \|S\|_{\mathcal{E}}$ for any $T, S \in \mathrm{Alg}(\mathcal{E})$;*
($3_e$) *$\|T\|_{\mathcal{E}} \leq \|T\|$ for all $T \in \mathrm{Alg}(\mathcal{E})$;*
($4_e$) *$\|T\|_{\mathcal{E}} = 0$ for each operator $T \in \mathrm{Alg}(\mathcal{E})$ which is compact on some $Y \in \mathcal{E}$.*

*Proof.* ($1_e$) and ($2_e$) follow from the fact that $\|T|_{Y_1\cap Y_2}\| \leq \min\{\|T|_{Y_1}\|, \|T|_{Y_2}\|\}$,

($3_e$) is evident.

($4_e$) Let $T \in \mathrm{Alg}(\mathcal{E})$ be compact on $Y \in \mathcal{E}$. Let $Z_\odot$ be the closed unit ball of $Z \in \mathcal{E}$, and $K_Z = \overline{T(Z_\odot)}$. If $\|T\|_{\mathcal{E}} > 0$, choose a number $t$ such that $0 < t < \|T\|_{\mathcal{E}}$. For each $Z \in \mathcal{E}$, let $D_Z = \{\zeta \in K_Z : \|\zeta\| \geq t\}$. Then $(D_{Z\cap Y})_{Z\in\mathcal{E}}$ is a centered family of compact sets, so there is an element $z$ in their intersection. It is non-zero and belongs to $\cap\{Y \cap Z : Z \in \mathcal{E}\} = 0$, a contradiction. □

It follows from ($3_e$) that $\|\cdot\|_{\mathcal{E}}$ is continuous with respect to $\|\cdot\|$ on $\mathrm{Alg}(\mathcal{E})$. This together with ($2_e$) yield that $\ker\mathcal{E}$ is a closed ideal of $\mathrm{Alg}(\mathcal{E})$.

We return to the conditions of Proposition 10.11. Let $A$ be a Banach algebra, and let $J$ be a closed ideal of $A$. Let $\mathbb{K}_J(A)$ be the set of all closed ideals $I$ of $A$

such that $I$ is generated by a compact element of $J$. (Recall that $a$ is a *compact element* of $A$ if $\mathrm{W}_a$ is a compact operator on $A$.) Set

$$\mathbb{K}_{\mathcal{E}}(A) = \cup_{J\in\mathcal{E}}\mathbb{K}_J(A).$$

**Lemma 10.13.** *Let $A$ be a Banach algebra, let $\mathcal{F}$ be a family of closed ideals of $A$ with zero intersection, and let $\mathcal{E}$ be the set of all finite intersections of ideals from $\mathcal{F}$. If $K \in \mathbb{K}_{\mathcal{E}}(A)$ then $\mathrm{L}_K\mathrm{R}_K \subset \ker\mathcal{E}$ and*

$$\inf_{J\in\mathcal{E}} \rho(\mathrm{W}_M|_J) \leq \rho(M) \inf_{K\in\mathbb{K}_{\mathcal{E}}(A)} \rho(M/K)$$

*for every bounded set $M$ in $A$.*

*Proof.* Let $L \in \mathcal{E}$, let $a$ be a compact element of $L$, $I$ the ideal of $A$ generated by $a$, and $K = \overline{I}$. Then $\mathrm{W}_a \in \mathrm{Alg}(\mathcal{E})$ and it is compact on $L$. By Lemma 10.12, $\|\mathrm{W}_a\|_{\mathcal{E}} = 0$. Note that the operator $\mathrm{L}_x\mathrm{R}_y$ for any $x, y \in I$ is represented as a finite sum $\sum \mathrm{L}_{u_i}\mathrm{R}_{v_i}\mathrm{W}_a\mathrm{L}_{s_i}\mathrm{R}_{t_i}$ for some $u_i, v_i, s_i, t_i \in A^1$ whence

$$\|\mathrm{L}_x\mathrm{R}_y\|_{\mathcal{E}} = \left\|\sum \mathrm{L}_{u_i}\mathrm{R}_{v_i}\mathrm{W}_a\mathrm{L}_{s_i}\mathrm{R}_{t_i}\right\|_{\mathcal{E}} \leq \sum \|\mathrm{L}_{u_i}\mathrm{R}_{v_i}\|_{\mathcal{E}}\,\|\mathrm{W}_a\|_{\mathcal{E}}\,\|\mathrm{L}_{s_i}\mathrm{R}_{t_i}\|_{\mathcal{E}} = 0.$$

So $\mathrm{L}_K\mathrm{R}_K \subset \ker\mathcal{E}$.

Let $M$ be a bounded set in $A$. Denote $\|M/K\|$ by $d$. For $a, b \in M$ and $\varepsilon > 0$ choose $u, v \in K$ with

$$\|a - u\| \leq d + \varepsilon \text{ and } \|b - v\| \leq d + \varepsilon.$$

Then $\|\mathrm{L}_u\mathrm{R}_v\|_{\mathcal{E}} = 0$ whence

$$\begin{aligned}
\|\mathrm{L}_a\mathrm{R}_b\|_{\mathcal{E}} &= \|\mathrm{L}_u\mathrm{R}_v + \mathrm{L}_u\mathrm{R}_{b-v} + \mathrm{L}_{a-u}\mathrm{R}_v + \mathrm{L}_{a-u}\mathrm{R}_{b-v}\|_{\mathcal{E}} \\
&\leq \|\mathrm{L}_u\mathrm{R}_{b-v}\|_{\mathcal{E}} + \|\mathrm{L}_{a-u}\mathrm{R}_v\|_{\mathcal{E}} + \|\mathrm{L}_{a-u}\mathrm{R}_{b-v}\|_{\mathcal{E}} \\
&\leq \|u\|(d+\varepsilon) + \|v\|(d+\varepsilon) + (d+\varepsilon)^2 \\
&\leq (\|a\| + d + \varepsilon)(d+\varepsilon) + (\|b\| + d + \varepsilon)(d+\varepsilon) + (d+\varepsilon)^2.
\end{aligned}$$

Taking $\varepsilon \to 0$ we get that $\|\mathrm{L}_a\mathrm{R}_b\|_{\mathcal{E}} \leq d(\|a\| + \|b\| + 3d) \leq 5\|M/K\|\|M\|$. Therefore

$$\|\mathrm{W}_M\|_{\mathcal{E}} \leq 5\|M/K\|\|M\|. \tag{10.10}$$

Changing $M$ by $M^n$ in (10.10) we have $\|\mathrm{W}_{M^n}\|_{\mathcal{E}} \leq 5\|M^n/K\|\|M^n\|$. Therefore, there is $J \in \mathcal{E}$ with

$$\|\mathrm{W}_{M^n}|_J\| \leq 6\|M^n/K\|\|M^n\|$$

whence $\rho(\mathrm{W}_M|_J) \leq \|\mathrm{W}_{M^n}|_J\|^{1/n} \leq 6^{1/n}\|M^n/K\|^{1/n}\|M^n\|^{1/n}$. Thus

$$\inf_{J\in\mathcal{E}} \rho(\mathrm{W}_M|_J) \leq \inf_n \left(6^{1/n}\|M^n/K\|^{1/n}\|M^n\|^{1/n}\right) = \rho(M/K)\rho(M)$$

for any $K \in \mathbb{K}(A)$. This is what we need. $\square$

**Theorem 10.14.** *Let $A$ be a Banach algebra, let $\mathcal{F}$ be a family of closed ideals of $A$ such that $\cap_{I\in\mathcal{F}} I = 0$, and let $\mathcal{E}$ be the set of all finite intersections of ideals from $\mathcal{F}$. Then, for any bounded subset $M$ of $A$,*

$$\rho(M) = \max\left\{\inf_{K\in\mathbb{K}_{\mathcal{E}}(A)} \rho(M/K), \max_{I\in\mathcal{F}} \rho(M/I)\right\}. \tag{10.11}$$

*Proof.* Using the result of Lemma 10.13, we get from Proposition 10.11 that

$$\begin{aligned}\rho(M)^2 &\le \max\left\{\rho(M) \inf_{K\in\mathbb{K}_{\mathcal{E}}(A)} \rho(M/K), \max_{I\in\mathcal{F}} \rho(M/I)^2\right\}\\ &\le \rho(M)\max\left\{\inf_{K\in\mathbb{K}_{\mathcal{E}}(A)} \rho(M/K), \max_{I\in\mathcal{F}} \rho(M/I)\right\}\end{aligned}$$

whence we obtain the inequality $\le$. The opposite inequality is trivial. □

### 10.3. The joint spectral radius and primitive ideals

We already mentioned and used the fact that the equality (10.2) holds for $\mathcal{F} = \mathrm{Prim}(A)$. Since $\cap\{I : I \in \mathrm{Prim}(A)\} = \mathrm{Rad}(A)$, it follows from Proposition 10.9 that

$$r(M) = \sup_{I\in\mathrm{Prim}(A)} r(M/I) \text{ for each bounded } M \subset A. \tag{10.12}$$

Since $\rho(\pi(a)) \le \rho(a/\ker\pi)$ the following well-known equality extends (10.2), for the case $\mathcal{F} = \mathrm{Prim}(A)$:

$$\rho(a) = \sup_{\pi\in\mathrm{Irr}(A)} \rho(\pi(a)), \text{ for each } a \in A. \tag{10.13}$$

Arguing as in the proof of Proposition 10.9 we obtain a more strong version of (10.12):

**Proposition 10.15.** *Let $A$ be a Banach algebra. Then*

$$r(M) = \sup_{\pi\in\mathrm{Irr}(A)} r(\pi(M))$$

*for each bounded subset $M$ of $A$.*

We are looking for the conditions that provide the validity of similar statements for the joint spectral radius:

$$\rho(M) = \sup_{I\in\mathrm{Prim}(A)} \rho(M/I) \text{ for each precompact } M \subset A, \tag{10.14}$$

$$\rho(M) = \sup_{\pi\in\mathrm{Irr}(A)} \rho(\pi(M)) \text{ for each precompact } M \subset A. \tag{10.15}$$

Recall that a bounded subset $M$ of a normed algebra $A$ is a *point of continuity* of the joint spectral radius if $\rho(M_n) \to \rho(M)$, for each sequence of bounded subsets $M_n \subset A$ that tends to $M$ in the Hausdorff metric.

**Lemma 10.16.** *Let $A$ be a normed algebra, and let $M$ be a bounded set of $A$ such that $\rho(M) = r(M)$. If $M$ consists of the points of continuity of the (usual) spectral radius then $M$ is a point of continuity of the joint spectral radius.*

*Proof.* Let $M_n \to M$. Since the joint spectral radius is upper semicontinuous [ST1, Proposition 3.1], we have only to show that $\liminf_{n\to\infty} \rho(M_n) \ge \rho(M)$.

Assume, to the contrary, that

$$\lim_{n\to\infty} \rho(M_n) < 1 < \rho(M).$$

Since $\rho(M) = r(M)$, there are $k \in \mathbb{N}$ and $T \in M^k$ such that $\rho(T) > 1$. Clearly there are $T_n \in M_n^k$ with $T_n \to T$; since $\rho$ is continuous at $T$ then $\rho(T_n) \to \rho(T)$. But

$$\rho(T_n) \le \rho(M_n^k) = \rho(M_n)^k < 1$$

for sufficiently big $n$, a contradiction. □

Recall that a normed algebra $A$ is a *Berger–Wang algebra* if $\rho(M) = r(M)$, for each precompact subset $M$ of $A$. It follows immediately from Proposition 10.15 that (10.14) and (10.15) hold for every Berger–Wang algebra.

**Theorem 10.17.** *Let $A$ be a Berger–Wang Banach algebra. Then every precompact subset of $\mathcal{R}_s^{p*}(A)$ is a point of continuity of the joint spectral radius.*

*Proof.* Follows from Theorem 9.16 and Lemma 10.16. □

Recall that a non-necessarily Hausdorff topological space $T$ is called *quasicompact* if each its open covering contains a finite subcovering. A function $\phi : T \longrightarrow \mathbb{R}$ is *lower (upper) semicontinuous* if for each $\lambda \in \mathbb{R}$, the set $\{t \in T : \phi(t) \le \lambda\}$ (respectively $\{t \in T : \phi(t) \ge \lambda\}$) is closed.

The following result is a variation of the classical Dini Theorem:

**Theorem 10.18.** *Let $f_n$ be a decreasing sequence of functions on a quasicompact space $T$ pointwise converging to a function $f$. If all $f_n$ are upper semicontinuous then*

$$\sup_{t\in T} f_n(t) \to \sup_{t\in T} f(t) \text{ as } n \to \infty.$$

*Proof.* Choose a number $d > \sup_{t\in T} f(t)$. If $\sup_{t\in T} f_n(t) > d$, for all $n$, then the sets $E_n = \{t \in T : f_n(t) \ge d\}$ are non-empty and closed. Since $E_n \subset E_{n-1}$ and $T$ is quasicompact, there is a point $t_0 \in \cap_n E_n$. Thus $f_n(t_0) \ge d$ for all $n$, and therefore

$$f(t_0) = \lim_{n\to\infty} f_n(t_0) \ge d > \sup_{t\in T} f(t),$$

a contradiction.

It follows that

$$\lim_{n\to\infty} \sup_{t\in T} f_n(t) = \inf_{n\to\infty} \sup_{t\in T} f_n(t) \le \sup_{t\in T} f(t).$$

On the other hand,

$$\lim_{n\to\infty} \sup_{t\in T} f_n(t) \ge \sup_{t\in T} f(t)$$

because $f_n(t) \ge f(t)$ for all $t$. □

We consider the following properties which a Banach algebra $A$ can have:

($1_c$) For each $a \in A$, $\|a\| = \sup\{\|a/I\| : I \in \mathrm{Prim}(A)\}$;
($2_c$) For each $a \in A$, the map $I \longmapsto \|a/I\|$ is upper semicontinuous on $\mathrm{Prim}(A)$.

**Lemma 10.19.** *Let $A$ be a Banach algebra, and let $M$ be a precompact subset of $A$. Then*

(1) *If $A$ has the property $(1_c)$ then $\|M\| = \sup_{I\in\mathrm{Prim}(A)} \|M/I\|$;*
(2) *If $A$ has the property $(2_c)$ then the map $I \longmapsto \|M/I\|$ is upper semicontinuous on* $\mathrm{Prim}(A)$.

*Proof.* (1) For each $\varepsilon > 0$, choose $a \in M$ with $\|a\| > \|M\| - \varepsilon$. Using $(1_c)$, choose $I_0 \in \mathrm{Prim}(A)$ with $\|a/I_0\| > \|a\| - \varepsilon$. Then

$$\|M\| < \|a/I_0\| + 2\varepsilon \le \|M/I_0\| + 2\varepsilon \le \sup_I \|M/I\| + 2\varepsilon.$$

Taking $\varepsilon \to 0$ we get that $\|M\| \le \sup_I \|M/I\|$, the converse is evident.

(2) If $N \subset M$ is finite then $I \longmapsto \|N/I\|$ is upper semicontinuous because the set

$$\{I \in \mathrm{Prim}(A) : \|N/I\| \ge \lambda\} = \cup_{a\in N}\{I \in \mathrm{Prim}(A) : \|a/I\| \ge \lambda\}$$

is closed being a finite union of closed sets. Now, for each $\varepsilon > 0$, let $N_\varepsilon$ be a finite $\varepsilon$-net in $M$. Then $E_\varepsilon := \{I \in \mathrm{Prim}(A) : \|N_\varepsilon/I\| \ge \lambda - \varepsilon\}$ is closed and therefore

$$\{I \in \mathrm{Prim}(A) : \|M/I\| \ge \lambda\} = \cap_{\varepsilon>0} E_\varepsilon$$

is closed. □

**Theorem 10.20.** *Let $A$ be a Banach algebra satisfying $(1_c)$ and $(2_c)$, and let $M$ be a precompact subset of $A$. If $\rho(M/I) = r(M/I)$ for all $I \in \mathrm{Prim}(A)$, then*

$$\rho(M) = \sup_{I\in\mathrm{Prim}(A)} \rho(M/I) = r(M). \tag{10.16}$$

*Proof.* The functions $I \longmapsto f_n(I) := \|M^{2^n}/I\|^{1/2^n}$ are upper semicontinuous by Lemma 10.19(2). Moreover, they decrease ($f_{n+1}(I) \le f_n(I)$) and

$$\lim_n f_n(I) = \rho(M/I) = r(M/I).$$

Thus by Theorem 10.18, $\sup_I f_n(I) \to \sup_I \rho(M/I)$. Using Lemma 10.19(1), one gets

$$\sup_I f_n(I) = \sup_I \|M^{2^n}/I\|^{1/2^n} = \|M^{2^n}\|^{1/2^n} \to \rho(M).$$

We proved that

$$\rho(M) = \sup_{I\in\mathrm{Prim}(A)} \rho(M/I);$$

the second equality follows from Theorem 10.15. □

It is convenient to formulate an analogue of Theorem 10.20 for algebras of vector-valued functions on arbitrary compacts.

**Theorem 10.21.** *Let $B$ be a Berger–Wang Banach algebra, and $A = C(T, B)$, the algebra of all continuous $B$-valued functions from a quasicompact space $T$ to $B$, supplied with the sup-norm. Then $A$ is a Berger–Wang algebra.*

*If the spectral radius function $x \longmapsto \rho(x)$ is continuous on $B$ then the same is true for $A$, and moreover, the joint spectral radius is continuous on $B$.*

*Proof.* We argue as in the proof of Theorem 10.20 with the change of $\mathrm{Prim}(A)$ by $T$. The analogue of $(1_c)$ follows from the definition of the norm in $A$. The analogue of $(2_c)$ holds by definition (functions are continuous, so the norms are continuous).

Let $M$ be a precompact subset of $A$; for each $t \in T$, let $M(t) = \{a(t) : a \in M\}$. Arguing as above, we prove that

$$\rho(M) = \sup_{t \in T} \rho(M(t)) = r(M). \tag{10.17}$$

Let $x \longmapsto \rho(x)$ be continuous on $B$. To check the continuity of $\rho$ on $A$, note that for $a \in A$

$$\rho(a) = \sup_{t \in T} \rho(a(t))$$

(a special case of (10.17)). Now if $a_\lambda \to a$ in $A$ then $a_\lambda(t) \to a(t)$ in $B$, for each $t \in T$, whence

$$\rho(a(t)) = \lim_\lambda \rho(a_\lambda(t)) \le \liminf_\lambda \rho(a_\lambda).$$

Therefore $\rho(a) \le \liminf_\lambda \rho(a_\lambda)$. Since upper semicontinuity holds in general, we get the continuity of $a \longmapsto \rho(a)$ on $A$.

The continuity of the joint spectral radius follows from Lemma 10.16. □

One can take for $B$ an arbitrary algebra of compact operators on a Banach space $X$ (for this case, the equality $\rho(M) = r(M)$ was established in [ST1]).

### 10.4. A C*-algebra version of the joint spectral radius formula

In [ST7] it was shown that for each Banach algebra $A$ and each precompact set $M \subset A$,

$$\rho(M) = \max\{r(M), \rho(M/\mathcal{R}_{\mathrm{hc}}(A))\}. \tag{10.18}$$

It is an interesting and important question if dealing with C*-algebras one can change in (10.18) the ideal $\mathcal{R}_{\mathrm{hc}}(A)$ by the much larger ideal $\mathcal{R}_{\mathfrak{gcr}}(A)$. In other words, we study the validity of the equality

$$\rho(M) = \max\{r(M), \rho(M/\mathcal{R}_{\mathfrak{gcr}}(A))\}. \tag{10.19}$$

Our approach is based on the consideration of primitive ideals; in particular we study C*-algebras for which the equality (10.16) holds.

The results of previous sections give some valuable information on this question. Note that the condition $(1_c)$ holds for each C*-algebra. The condition $(2_c)$ holds for C*-algebras with Hausdorff spectra. Indeed, it is proved in [Dm, Proposition 3.3.7] that for each element $a \in A$, the set $\{I \in \mathrm{Prim}(A) : \|a/I\| \ge \lambda\}$ is quasicompact; thus if $\mathrm{Prim}(A)$ is Hausdorff then this set is closed, whence the function $I \longmapsto \|a/I\|$ is upper semicontinuous (in fact, continuous because it is always lower semicontinuous [Dm, Proposition 3.3.2]). Thus we obtain from Theorem 10.20 that the equality (10.16) holds for each CCR C*-algebra $A$ with Hausdorff $\mathrm{Prim}(A)$.

We will obtain much more general results here. Let us say that a topological space has the *property* $(QC)$ if the intersection of any down-directed net of non-empty quasicompact subsets is non-empty. Of course, each Hausdorff space has this property.

**Lemma 10.22.** *Let $T$ be a topological space. Then*

(1) *If $T = \cup_{\lambda\in\Lambda} U_\lambda$, where the net $\{U_\lambda : \lambda \in \Lambda\}$ is up-directed, all $U_\lambda$ are open and have property $(QC)$, then $T$ has property $(QC)$;*
(2) *If $T = F \cup U$, where $F$ is closed and Hausdorff, $U$ is open and has property $(QC)$, then $T$ has property $(QC)$;*
(3) *If $T$ has the property $(QC)$ then each subset of $T$ has this property.*

*Proof.* Let $\{E_\omega : \omega \in \Omega\}$ be a down-directed set of quasicompact subsets.

(1) For some $\omega_1 \in \Omega$, let $\lambda_1, \dots, \lambda_n \in \Lambda$ be such that $E_{\omega_1} \subset \cup_{k=1}^n U_{\lambda_k}$. Since the net $\{U_\lambda : \lambda \in \Lambda\}$ is up-directed,

$$E_{\omega_1} \subset U_{\lambda_0}$$

for some $\lambda_0 \in \Lambda$. Then all $E_\omega$ with $\omega > \omega_1$ are contained in the space $U_{\lambda_0}$ which has the property $(QC)$. By definition, their intersection is non-empty whence $\cap_{\omega\in\Omega} E_\omega$ is non-empty.

(2) Let us check firstly that the intersection of a quasicompact set $K$ with a closed set $W$ is quasicompact. Indeed, if $\{U_\lambda : \lambda \in \Lambda\}$ is a family of open subsets in $T$ with $K \cap W \subset \cup_{\lambda\in\Lambda} U_\lambda$ then

$$K \subset (\cup_{\lambda\in\Lambda} U_\lambda) \cup (T \setminus W);$$

choosing a finite subcovering and removing $T \setminus W$ we obtain a finite subcovering of $K \cap W$.

Now in assumptions of (2) all sets $E_\omega \cap F$ are quasicompact subsets of a Hausdorff space, so they are compact. Hence if all $E_\omega \cap F$ are non-empty then they have non-empty intersection.

On the other hand, if $E_{\omega_0} \cap F = \varnothing$, for some $\omega_0 \in \Omega$, then $E_{\omega_0} \subset U$, and the same is true for all $\omega > \omega_0$, so the intersection is non-empty because $U$ has property $(QC)$.

(3) Let $W$ be an arbitrary subset of $T$. It follows easily from the definition that if $E \subset W$ is quasicompact in $W$ then it is quasicompact in $T$. Thus a down-directed net of quasicompact subsets of $W$ is a down-directed net of quasicompact subsets of $T$; by the assumptions, the intersection is non-void. □

Let $\mathfrak{C}_{\mathfrak{qc}}$ denote the class of all C*-algebras $A$ for which $\mathrm{Prim}(A)$ has the property $(QC)$.

**Corollary 10.23.** *Let $A$ be a C*-algebra. Then*

(1) *If $(J_\lambda)_{\lambda\in\Lambda}$ is an up-directed net of closed ideals of $A$ such that $\cup_{\lambda\in\Lambda} J_\lambda$ is dense in $A$ and $J_\lambda \in \mathfrak{C}_{\mathfrak{qc}}$ for all $\lambda$, then $A \in \mathfrak{C}_{\mathfrak{qc}}$;*
(2) *If $A$ has an ideal $J \in \mathfrak{C}_{\mathfrak{qc}}$ such that $\mathrm{Prim}(A/J)$ is Hausdorff, then $A \in \mathfrak{C}_{\mathfrak{qc}}$;*
(3) *If $A \in \mathfrak{C}_{\mathfrak{qc}}$ then all closed ideals and all quotients of $A$ belong to $\mathfrak{C}_{\mathfrak{qc}}$.*

*Proof.* (1) One has that

$$\mathrm{Prim}(A) \cong \overline{\cup_{\lambda\in\Lambda} \mathrm{Prim}(J_\lambda)},$$

if $\mathrm{Prim}(J_\lambda)$ is identified with the set of all $I \in \mathrm{Prim}(A)$ that do not contain $J_\lambda$. All $\mathrm{Prim}(J_\lambda)$ are open and have the property $(QC)$. Using Lemma 10.22(1) we get that $\mathrm{Prim}(A)$ has property $(QC)$. So $A \in \mathfrak{C}_{\mathfrak{qc}}$.

(2) In this case $\mathrm{Prim}(A) \cong \mathrm{Prim}(J) \cup \mathrm{Prim}(A/J)$ and it remains to apply Lemma 10.22(2).

(3) Follows from Lemma 10.22(3) because, for each closed ideal $J$ of $A$, the space $\mathrm{Prim}(J)$ (respectively, $\mathrm{Prim}(A/J)$) is homeomorphic to an open (respectively, closed) subset of $\mathrm{Prim}(A)$. □

**Corollary 10.24.** *All GCR $C^*$-algebras belong to $\mathfrak{C}_{\mathfrak{qc}}$.*

*Proof.* It is known [Dm, Proposition 4.5.3 and Theorem 4.5.5] that if $A$ is a GCR algebra then there is an increasing transfinite chain $(J_\alpha)_{\alpha\leq\delta}$ of closed ideals such that $J_0 = 0$, $J_\delta = A$ and all gap-quotients $J_{\alpha+1}/J_\alpha$ have Hausdorff space of primitive ideals.

Assume, to the contrary, that there is a smallest ordinal $\gamma$ for which $J_\gamma \notin \mathfrak{C}_{\mathfrak{qc}}$. If $\gamma$ is limit then $J_\gamma \in \mathfrak{C}_{\mathfrak{qc}}$ by Lemma 10.23(1); if not, then $\gamma = \alpha + 1$ for some ordinal $\alpha$, whence $J_\gamma \in \mathfrak{C}_{\mathfrak{qc}}$ by Lemma 10.23(2), a contradiction. □

At the moment we do not know an example of a C*-algebra which does not belong to $\mathfrak{C}_{\mathfrak{qc}}$.

**Lemma 10.25.** *Let $A$ be a $C^*$-algebra, and let $M$ be a precompact subset of $A$. Then, for any $t > 0$, the set $\{I \in \mathrm{Prim}(A) : \|M/I\| \geq t\}$ is quasicompact.*

*Proof.* We may argue as in [Dm, Proposition 3.3.7], where the statement is proved for one-element sets, if we establish that for each closed ideal $J$ of $A$, there is a primitive ideal $I \supset J$ with $\|M/I\| = \|M/J\|$. To show the existence of $I$, let $x_n \in M$ be such that

$$\|M/J\| = \lim_{n\to\infty} \|x_n/J\|.$$

By precompactness of $M$ we may assume that $x_n \to x \in A$, and it remains to take $I$ with $\|x/I\| = \|x/J\|$ ([Dm, Lemma 3.3.6]). □

**Theorem 10.26.** *Let $A$ be a $C^*$-algebra. If $A \in \mathfrak{C}_{\mathfrak{qc}}$ then*

$$\rho(M) = \sup_{I\in\mathrm{Prim}(A)} \rho(M/I)$$

*for each precompact subset $M$ of $A$.*

*Proof.* For each $n$, let $K_n = \left\{I \in \mathrm{Prim}(A) : \|M^{2^n}/I\| \geq \rho(M)^{2^n}\right\}$. Since

$$\|M^{2^n}\| \geq \rho(M)^{2^n},$$

all $K_n$ are non-zero, and they are quasicompact by Lemma 10.25. Since the sequence $\|M^{2^n}/I\|^{1/2^n}$ decreases, the inclusion

$$K_n \supset K_{n+1}$$

holds for each $n$. It follows that there is a primitive ideal $I \in \cap_{n=1}^{\infty} K_n$. Thus

$$\|M^{2^n}/I\|^{1/2^n} \geq \rho(M)$$

for all $n$. Taking the limit we get that $\rho(M/I) \geq \rho(M)$. □

**Corollary 10.27.** *Let $A$ be a $C^*$-algebra, and let $M$ be a precompact subset of $A$. If $A \in \mathfrak{C}_{\mathfrak{qc}}$ and $\rho(M/I) = r(M/I)$ for all $I \in \mathrm{Prim}(A)$ then*

$$\rho(M) = r(M).$$

*Proof.* By Theorem 10.26, we obtain that

$$\rho(M) = \sup_{I \in \mathrm{Prim}(A)} \rho(M/I) = \sup_{I \in \mathrm{Prim}(A)} r(M/I) \leq r(M).$$ □

The following result establishes a C*-version of the joint spectral radius formula for algebras satisfying the condition (10.14) with all their quotients.

**Proposition 10.28.** *Let $A$ be a $C^*$-algebra. If, for any closed ideal $K$ of $A$,*

$$\rho(N) = \sup_{I \in \mathrm{Prim}(A/K)} \rho(N/I)$$

*for every precompact subset $N$ of $A/K$ then*

$$\rho(M) = \max\{r(M), \rho(M/\mathcal{R}_{\mathfrak{gcr}}(A))\}$$

*for every precompact subset $M$ of $A$.*

*Proof.* We show that

$$\rho(M) = \max\{r(M), \rho(M/J)\} \tag{10.20}$$

for each closed GCR ideal $J$ of $A$.

Let us firstly prove (10.20) for the case that $J$ is a CCR-ideal (assuming only that (10.14) is true for $A$). By (10.14), it suffices to show that

$$\rho(M/I) \leq \max\{\rho(M/J), r(M)\}$$

for each $I \in \mathrm{Prim}(A)$. Let $\pi$ be an irreducible representation of $A$ on a Hilbert space $H$, with $\ker \pi = I$. Then

$$\rho(M/I) = \rho(\pi(M)).$$

If $\pi(J) = 0$ then $\pi$ is a representation of $A/J$ and the inequality

$$\rho(\pi(M)) \leq \rho(M/J)$$

is evident. Otherwise $\pi(J) = \mathcal{K}(H)$ and

$$\rho_e(\pi(M)) = \rho(\pi(M)/\pi(J)) \leq \rho(M/J).$$

Applying the operator version of the joint spectral radius formula (2.10) to $\pi(M)$ we obtain the needed inequality.

In general, we have an increasing transfinite chain $(J_\alpha)_{\alpha \leq \delta}$ such that $J_0 = 0$, $J_\delta = J$ and all gap-quotients $J_{\alpha+1}/J_\alpha$ are CCR-algebras.

Assume, to the contrary, that there is an ordinal $\gamma$ which is the smallest one among ordinals $\alpha$ for which (10.20) is not true with $J = J_\alpha$. If $\gamma$ is not limit then $\gamma = \alpha + 1$ for some $\alpha$. As $M/J_\gamma \cong (M/J_\alpha)/(J_\gamma/J_\alpha)$ then

$$\rho(M/J_\alpha) = \max\{r(M/J_\alpha), \rho(M/J_\gamma)\}.$$

Since also $\rho(M) = \max\{r(M), \rho(M/J_\alpha)\}$ then

$$\rho(M) = \max\{r(M), \rho(M/J_\gamma)\},$$

a contradiction.

Let now $\gamma$ be a limit ordinal. By Lemma 9.14,

$$\begin{aligned}\max\{r(M), \rho(M/J_\gamma)\} &= \max\{r(M), \inf_{\alpha'<\gamma} \rho(M/J_{\alpha'})\} \\ &= \inf_{\alpha'<\gamma} \max\{r(M), \rho(M/J_{\alpha'})\} = \rho(M),\end{aligned}$$

a contradiction, so the equality holds for all ordinals. □

**Corollary 10.29.** *Let $A$ be a $C^*$-algebra. If $A \in \mathfrak{C}_{\mathfrak{qc}}$ then*

$$\rho(M) = \max\{r(M), \rho(M/\mathcal{R}_{\mathfrak{gcr}}(A))\}$$

*for every precompact subset $M$ of $A$.*

*Proof.* Indeed, any quotient of a C*-algebra from $\mathfrak{C}_{\mathfrak{qc}}$ belongs to $\mathfrak{C}_{\mathfrak{qc}}$ (see Corollary 10.23(3)). So the result follows from Theorems 10.28 and 10.26. □

**Corollary 10.30.** *Each GCR $C^*$-algebra is a Berger–Wang algebra.*

*Proof.* Follows from Corollary 10.29. □

**Corollary 10.31.** *Let $A$ be a $C^*$-algebra. Then any precompact subset of $\mathcal{R}_{\mathfrak{gcr}}(A)$ is a point of continuity of the joint spectral radius.*

*Proof.* Note that $\mathcal{R}_{\mathfrak{gcr}}(A)$ is the largest GCR-ideal of $A$ by Theorem 2.23. So the result follows from Corollary 10.30 and Theorems 9.11 and 10.17. □

**Acknowledgment**

The authors are grateful to Matej Brešar and Peng Cao for useful discussions and information, and to Edward Kissin for his encouraging interest in this work. They express special thanks to the referee for numerous useful remarks and suggestions.

## References

[AW] C. Akemann, S. Wright, *Compact actions on $C^*$-algebras*, Glasgow Math. J., **21** (1980) 143–149.

[Al] J.C. Alexander, *Compact Banach algebras*, Proc. London Math. Soc. (3) **18** (1968) 1–18.

[AR] V.A. Andrunakievich, Yu.M. Ryabuhin, Radicals of algebras and structure theory, Moscow, "Nauka", 1979. [In Russian]

[A] B. Aupetit, Primer to Spectral Theory, Springer-Verlag, N.Y., Berlin, 1991.

[AM] B. Aupetit, M. Mathieu, *The continuity of Lie homomorphisms*, Studia Math. 138 (2000) 193–199.

[B] B.A. Barnes, *On the existence of minimal ideals in a Banach algebra*, Trans. AMS, **133** (1968) 511–517.

[BMS] B.A. Barnes, G.J. Murphy, M.R.F. Smyth, T.T. West, Riesz and Fredholm theory in Banach algebras, Pitman Publ. Inc., Boston, 1982.

[BS] D. Beltiţa, M. Şabak, Lie algebras of bounded operators, Basel, Boston, Berlin, Birkhäuser, 2001.

[BW] A.M. Berger and Y. Wang, *Bounded semigroups of matrices*, Linear Algebra and Appl., **166** (1992) 21–27.

[BD] F.F. Bonsall, J. Duncan, Complete normed algebras, Springer-Verlag, Berlin, Heidelberg, N.Y., 1973.

[Bo] N. Bourbaki, Theories Spectrales, Hermann, 1967.

[BE] M. Brešar, D. Eremita, *Lower socle and finite rank elementary operators*, Communications in Algebra, **31**, 3 (2003) 1485–1497.

[BT] M. Brešar, Yu. V. Turovskii, *Compactness conditions for elementary operators*, Studia Math. **178** (1) (2007) 1–18.

[BDF] L.G. Brown, R.G. Douglas and P.A.Fillmore, *Unitary equivalence modulo the compact operators and extensions of $C^*$-algebras*, Proceedings of a conference on operator theory, Halifax, Nova Scotia 1973, Lect. Notes in Math. 345 (Berlin Heidelberg N.Y.: Springer-Verlag), 58–128.

[Br] L.G. Brown, *Extension of $C^*$-algebras,* Operator Algebras and Applications, Proc. Symp. Pure Math. **38** (1981) 175–176, Amer. Math. Soc., Providence.

[Bu] L. Burlando, *Spectral continuity in some Banach algebras*, Rocky Mountain J. Math. **23** (1993) 17–39.

[C] P. Cao, *Quasinilpotent operators in operator Lie algebras III*, J. Math. Anal. Appl.**386** (2012) 709–717.

[Dv] K.R. Davidson, C*-algebras by Examples, Fields Institute Monographs, 6, AMS, Providence, RI, 1996.

[Dm] J. Dixmier, Les C*-algèbres et leur représentations, Gauthier-Villars, Paris, 1964.

[D1] P.G. Dixon, *Semiprime Banach algebras*, J. London Math. Soc. (2) **6** (1973) 676–678.

[D2] P.G. Dixon, *A Jacobson-semisimple Banach algebra with a dense nil subalgebra*, Colloq. Math. **37** (1977) 81–82.

[D3] P.G. Dixon, *Topologically irreducible representations and radicals in Banach algebras*, Proc. London Math. Soc. (3) **74** (1997) 174–200.

[D4] P.G. Dixon, *Non-closed sums of closed ideals in Banach algebras*, Proc. Amer. Math. Soc. **128**, 2 (2000) 3647–3654.

[Ds] A.A. Dosiev, *Cartan–Slodkowski spectra, splitting elements and noncommutative spectral mapping theorems*, J. Funct. Anal. **230** (2006) 446–493.

[G] T.W. Gamelin, Uniform algebras, Prentice-Hall, 1969.

[Gr] L.T. Gardner, *On isomorphisms of $C^*$-algebras*, Amer. J. Math. **87** (1985) 384–396.

[GR] S. Giotopoulos, M. Roumeliotis, *Algebraic ideals of semiprime Banach algebras*, Glasgow Math. J. **33** (1991) 359–363.

[H] R.E. Harte. Invertibility and singularity, Dekker, 1988.

[F] A.S. Fainshtein, *Taylor joint spectrum for families of operators generating nilpotent Lie algebras*, J. Oper. Theory, **29** (1993) 3–27.

[FK] I. Feldman, N. Krupnik, *On the continuity of the spectrum in certain Banach algebras*, Int. Equ. Oper. Theory **38** (2000) 284–301.

[Fr] E. Formanek, *The Nagata-Higman Theorem*, Acta Applic. Math., **21**(1990) 185–192.

[H] P. Halmos, Hilbert space problem book, Van Nostrand, 1967.

[J] B.E. Johnson, *The uniqueness of the (complete) norm topology*, Bull. Amer. Math. Soc. **42** (1967) 537–539.

[KS] E. Kissin, V.S. Shulman, Representations on Krein spaces and derivations of C*-algebras, Pitman Monogr. and Surveys in Pure and Applied Math. 89, Addison Wesley Longman, London, 1997.

[KST] E. Kissin, V.S. Shulman, Yu.V. Turovskii, *Topological radicals and the Frattini theory of Banach Lie algebras*, Integr. Equ. Oper. Theory **74** (2012), 51–121.

[Ks] M. Kusuda, *A characterization of scattered $C^*$-algebras and its applications to $C^*$-crossed products*, J. Operator Theory, **63**:2 (2010) 417–424.

[Ln] S. Lang, Algebra, Addison-Wesley, 1965.

[M] V. Müller, Spectral theory of linear operators and spectral systems in Banach algebras, 2ed., Birkhäuser, 2007.

[P] T.W. Palmer, Banach algebras and the general theory of *-algebras, v. 1, Cambridhe Univ. Press, 1994.

[PZ] V. Ptak, J. Zemanek, *On uniform continuity of the spectral radius in Banach algrebras*, Manuscripta Math. **20** (1977) 177–189.

[RS] G.-C. Rota, W.G. Strang, *A note on the joint spectral radius*, Indag. Math. **22** (1960) 379–381.

[Rd] W. Rudin, Harmonic analysis on groups, Int. Sci. Tracts in Pure and Appl. Math., No 12, Wiley, N.Y., 1962.

[Sn] A.M. Sinclair, Automatic continuity of linear operators, Cambridge Univ. Press, 1976.

[S] V.S. Shulman, *On invariant subspaces of Volterra operators*, Funct. Anal. Appl. **18** (1984) 160–161.

[ST1] V.S. Shulman, Yu.V. Turovskii, *Joint spectral radius, operator semigroups, and a problem of W. Wojtyński*, J. Funct. Anal. **177** (2000) 383–441.

[ST2] V.S. Shulman, Yu.V. Turovskii, *Radicals in Banach algebras and some problems in the theory of radical Banach algebras*, Funct. Anal. and its Appl. **35** (2001) 312–314.

[ST3] V.S. Shulman, Yu.V. Turovskii, *Formulae for joint spectral radii of sets of operators*, Studia Math. **149** (2002) 23–37.

[ST4] V.S. Shulman, Yu.V. Turovskii, *Invariant subspaces of operator Lie algebras and Lie algebras with compact adjoint action*, J. Funct. Anal. **223** (2005) 425–508.

[ST5] V.S. Shulman, Yu.V. Turovskii, *Topological radicals, I. Basic properties, tensor products and joint quasinilpotence*, Banach Center Publications, **67** (2005) 293–333.

[ST6] V.S. Shulman, Yu.V. Turovskii, *Topological radicals, II. Applications to the spectral theory of multiplication operators,* Operator Theory: Advances and Applications, **212** (2010) 45–114.

[ST7] V.S. Shulman, Yu.V. Turovskii, *Topological radicals and the joint spectral radius*, Funct. Anal. and its Appl., **46** (2012) 287–304.

[Tk1] M. Takesaki, On the cross norms of the direct product of C*-algebras, Tôhoku Mathematical J. **16** (1964) 111–112.

[Tk2] M. Takesaki, Theory of Operator algebras, III, Encycl. Math. Sci., 127, Springer-Verlag, 2002.

[T] A.W. Tullo, *Conditions on Banach algebras which imply finite dimensionality*, Proc. Edinburgh Math. Soc. (2) **20** (1976) 1–5.

[T1] Yu.V. Turovskii, *Spectral properties of certain Lie subalgebras and the spectral radius of subsets of a Banach algebra*, Spectral theory of operators and its applications, **6** (1985), "Elm", Baku, 144–181. [in Russian]

[T2] Yu.V. Turovskii, *Volterra semigroups have invariant subspaces*, J. Funct. Anal. **182** (1999) 313–323.

[Ws] S. Wasserman, Exact C*-algebras and related topics, Seoul National Univ. Lecture Notes, Research Institute of Math., 1994, 1–92.

[W] W. Wojtyński, *Quasinilpotent Banach–Lie algebras are Baker–Campbell–Hausdorff*, J. of Functional Anal **153** (1998) 405–413.

[Z1] J. Zemanek, *Spectral radius characterizations of commutativity in Banach algebras*, Studia Math. **61** (1977) 257–268.

[Z2] J. Zemanek, *Spectral characterization of two-sided ideals in Banach algebras*, Studia Math. **67** (1980) 1–12.

Victor S. Shulman
Vologda State Technical University
Lenin St. 15
Vologda 160000, Russia
e-mail: shulman.victor80@gmail.com

Yurii V. Turovskii
Institute of Mathematics and Mechanics
National Academy of Sciences of Azerbaijan
F. Agaev St. 9
Baku AZ1141, Azerbaijan
e-mail: yuri.turovskii@gmail.com

Operator Theory:
Advances and Applications, Vol. 233, 281–294


# A Brief Survey on
# 1. Infinite-dimensional Schur–Horn Theorems and Infinite Majorization Theory with Applications to Operator Ideals
# 2. $B(H)$-subideals of Operators

Gary Weiss

**Abstract.** This paper surveys the study of diagonals of operators, particularly strictly positive compact operators, in various kinds of orbits like unitary orbits, and its interplay with the development of infinite majorization theory, stochastic matrices and arithmetic mean ideals of operators. It includes a description of the author's main results joint with V. Kaftal in [13] and [14] and a brief description of the history including contributions from Hardy–Littlewood–Pólya, Schur and Horn, Gohberg and Markus, A. Neumann, Antezana–Massey–Ruiz–Stojanoff, and Arveson and Kadison. For a more extensive list of historical and related references, see [13] and [14].

This survey on $B(H)$-subideals [22], which paper is joint with S. Patnaik, is a brief description of the 1983 Fong–Radjavi characterization of those principal ideals in $K(H)$ that are also $B(H)$ ideals and our recent generalization of this to arbitrary $B(H)$-ideals along with characterization of their structure. Central to this is a notion of soft ideals.

**Mathematics Subject Classification (2010).** Primary: 15A51, 47A75, 47L20, 47B10, 47B07; Secondary: 47B47, 47B37, 13C05, 13C12.

**Keywords.** Schur–Horn theorem, majorization, stochastic matrices, operator ideals, principal ideals, subideals.

## 1. Infinite-dimensional Schur–Horn theorems, majorization and applications to ideals

A main result of [13] is the extension of the Schur–Horn Theorem [24], [10] to infinite sequences:

For two nonincreasing nonsummable strictly positive sequences $\xi$ and $\eta$ that converge to 0, there exists a positive compact operator $A$ with eigenvalues $\eta$ and diagonal sequence $\xi$ (i.e., $\xi$ appears as the diagonal of an operator in the unitary orbit of $A$) if and only if $\sum_{j=1}^{n} \xi_j \leq \sum_{j=1}^{n} \eta_j$ for every $n$ if and only if $\xi = Q\eta$ for some orthostochastic matrix $Q$. When $\xi, \eta$ are summable sequences, requiring also equality of their sums yields the same conclusion, extending a theorem of Arveson and Kadison [2] (see also [11], [12]).

Our proof depends on the construction and analysis of an infinite product of T-transform matrices. Further results on majorization for infinite sequences [14] providing "intermediate" sequences generalize known results from the finite case. Majorization properties and invariance under various classes of stochastic matrices are then used to characterize arithmetic mean closed operator ideals.

The Schur–Horn Theorem, Theorem 1.1 below, characterizes the diagonals of (finite) selfadjoint matrices via sequence majorization. Finite sequence majorization (order relation $\xi \preccurlyeq \eta$) for $\xi, \eta \in \mathbb{R}^N$ is:

$$\xi \preccurlyeq \eta : \quad \sum_{j=1}^{n} \xi_j^* \leq \sum_{j=1}^{n} \eta_j^* \text{ for } 1 \leq n \leq N \text{ with equality at the end.}$$

$\xi^*, \eta^*$ denotes their decreasing rearrangement.

For nonsummable (so infinite) monotone decreasing sequences, the generalized $\xi \preccurlyeq \eta$ condition we call "strong majorization," $\underline{\lim} \sum_{j=1}^{n} (\eta_j - \xi_j) = 0$, retains many of the key properties of "equality at the end" for finite and for summable sequences and is essential for our extension of the Schur–Horn Theorem. So also is $\xi \prec \eta$, majorization applied in the same way to infinite sequences but without requiring any kind of end condition like "equality at the end."

Majorization theory evolved in the early 1900s from seemingly unrelated topics including wealth distribution (Lorenz [18]), convex inequalities (Hardy, Littlewood and Pólya [9]) (e.g., characterizing when $\phi(\xi) \leq \phi(\eta)$ for all convex $\phi$), and Birkoff's Theorem that doubly stochastic matrices are convex combinations of permutation matrices [3], to the more recent work (and central to [13]) on doubly stochastic matrices with relations between eigenvalue lists and diagonals of selfadjoint matrices, to [14] on relationships between convex combinations, stochastic matrices, diagonal invariance of $B(H)$-ideals ($\operatorname{diag} I \subset I$) and arithmetic mean closed ideals (${}_a(I_a) = I$ from general commutator theory), and to infinite-dimensional extensions to $c_o^*$ of "density" facts from Hardy Littlewood and Pólya [9]: $\xi \prec \eta \Rightarrow \exists\, \zeta, \rho$ for which $\xi \prec \zeta \preccurlyeq \eta$, $\xi \preccurlyeq \rho \prec \eta$. Recall that $c_o^*$ denotes the cone of nonnegative sequences tending to zero which is $\Sigma(K(H))$, Calkin's characteristic set of the ideal of compact operators; the arithmetic mean ideal $I_a$ is characterized by its characteristic set $\Sigma(I_a) := \{x \in c_o^* \mid x = O(y_a) \text{ for some } y \in \Sigma(I)\}$ where $y_a$ denotes the arithmetic mean sequence $\langle \frac{y_1 + \cdots + y_n}{n} \rangle$ and ${}_aI$ is characterized by its characteristic set $\Sigma({}_aI) = \{x \in c_o^* \mid x_a \in \Sigma(I)\}$; and for Theorem 1.1 below $E(A)$ denotes the traditional conditional expectation for a fixed basis, i.e., it denotes the diagonal operator taking the diagonal entries from the diagonal of the matrix $A$.

**History**

**Theorem 1.1.** *Let* $\xi, \eta \in \mathbb{R}^N$.

(i) **Hardy, Littlewood and Pólya Theorem** [9].

$$\xi \preccurlyeq \eta \quad \textit{if and only} \quad \xi = P\eta \quad \textit{for some doubly stochastic matrix } P.$$

(ii) **Horn Theorem** [10, Theorem 4].

$$\xi \preccurlyeq \eta \quad \textit{if and only if} \quad \xi = Q\eta \quad \textit{for some orthostochastic matrix } Q,$$

*i.e., the Schur-square of an orthogonal matrix* ($Q_{ij} = (U_{ij})^2 \ \forall\ i,j$ *for some real entried unitary matrix* $U$).

(iii) **Schur–Horn Theorem** [24], [10]. *For a selfadjoint* $N \times N$ *matrix* $A$ *having eigenvalue list* $\eta$, (*when* $A$ *is positive this is merely the s-number sequence for* $A$, $s(A) = \eta$),

*there is a basis in which* $A$ *has diagonal entries* $\xi$ *if and only if* $\xi \preccurlyeq \eta$.

Linking orthostochasticity with how basis change impacts the diagonal is the following lemma.

**Lemma 1.2.** *Let* $\xi, \eta \in c_o^*$ *and let* $Q_{ij} = |L_{ij}|^2$ *for all* $i, j$ *for some contraction* $L \in B(H)$. *Then*

$$\operatorname{diag} \xi = E(L \operatorname{diag} \eta\, L^*) \qquad \textit{if and only if} \qquad \xi = Q\eta.$$

*Proof.* $(L \operatorname{diag}\, \eta\, L^*)_i = (L \operatorname{diag}\, \eta\, L^* e_i, e_i) = \sum_k L_{ik}\eta_k \overline{L}_{ik} = \sum_k |L_{ik}|^2 \eta_k = (Q\eta)_i$. □

One important class of stochastic matrices come from "Schur-squares" ($Q_{ij} = |L_{ij}|^2$) of contractions $L$. Another is the orthostochastic matrices $Q$, Schur-squares of unitaries, in which case $\xi = Q\eta$ if and only if $\operatorname{diag} \xi \in E(U \operatorname{diag} \eta\, U^*)$ for some unitary $U$ (i.e., the diagonal from a unitary orbit). Other larger more well-known related classes of stochastic matrices (by definition positive entried) are substochastic (row and column sums $\leq 1$; column (row)-stochastic (substochastic with column (row) sums $= 1$; doubly stochastic (column & row stochastic; and block stochastic (direct sums of finite doubly stochastic matrices).

For infinite majorization it is nontrivial to extend the fact that $Q\eta \prec \eta$ for substochastic $Q$. It was proved by A. Markus [19] using Ostrowski identity:

$$\begin{aligned}\sum_{i=1}^{n}(\eta_i - (Q\eta)_i) &= \sum_{j=1}^{n}\left(1 - \sum_{i=1}^{n} Q_{ij}\right)(\eta_j - \eta_n) + \sum_{j=n+1}^{\infty}\sum_{i=1}^{n} Q_{ij}(\eta_n - \eta_j) \\ &\quad + \left(n - \sum_{i=1}^{n}\sum_{j=1}^{\infty} Q_{ij}\right)\eta_n \\ &\geq \left(1 - \sum_{i=1}^{n} Q_{ij}\right)(\eta_j - \eta_n) \geq 0 \quad \text{for all } n \geq j.\end{aligned}$$

In recent years infinite majorization arose in the study of commutator ideals of compact operators $[I, B(H)]$ [6] and in outgrowths of this. From [17]: $[I, B(H)]^+ = ({}_aI)^+$ when diag $\langle \frac{1}{n} \rangle \in I$ (when not contained, then a slightly more complicated equality), and from [6]: $[I, B(H)] = I$ which is equivalent to $I$ possessing no traces (unitary invariant linear functionals) and to the arithmetic mean ideal condition $I_a = I$. Hence the birth of arithmetic mean ideals: $I_a = I$ (am-stability, that is, regularity of sequence generators in the principal ideal case), the arithmetic mean closed property ${}_a(I_a) = I$ (am-closed) and a host of others with relations. In comparing two arithmetic mean ideals, that is, determining when $I_a \subset J_a$, one needs to know if every diag $\xi \in I$ has its arithmetic mean $\xi_a = O(\eta_a)$ for some diag $\eta \in J$, that is, whether or not

$$\frac{\xi_1 + \cdots + \xi_n}{n} \leq M \frac{\eta_1 + \cdots + \eta_n}{n} \quad \text{for every } n, \text{ i.e., } \xi \prec M\eta.$$

(Voila, majorization.)

Also invariance under various classes of stochastic matrices of the diagonals of diagonal operators in an ideal provides equivalences to the ideal being am-closed [14, Section 4, esp., Theorem 4.5–Theorem 4.9].

**Historical stochasticity**

In 1964, two less-known papers with two almost disjoint approaches, Markus [19] and Gohberg–Markus [8] found infinite-dimensional versions of the Hardy, Littlewood and Pólya Theorem 1.1(i) and an extension to the summable case of Horn's Theorem [10, Theorem 4], (Theorem 1.1(ii)). More recently, Arveson and Kadison obtained other characterizations in [2] by different methods.Summarizing,
If $\xi, \eta \in c_o^*$, then

$$\begin{aligned} &\xi \prec \eta \Longleftrightarrow \xi = Q\eta \text{ for some substochastic } Q \\ &\qquad (Q \text{ row-stochastic if } \xi_n > 0\ \forall n) \quad \text{[Markus]} \\ &\xi \prec \eta \Longleftrightarrow \xi = Q\eta \text{ with } Q_{ij} = |W_{ij}|^2 \text{ for some co-isometry } W \\ &\qquad \text{[Gohberg–Markus]} \\ &\xi \prec \eta \Longleftrightarrow \xi = Q\eta \text{ with } Q_{ij} = |L_{ij}|^2 \text{ for some contraction } L \\ &\qquad \text{[Arveson–Kadison]} \end{aligned} \tag{1}$$

If $\xi, \eta \in (\ell^1)^+$, then

$$\begin{aligned} &\xi \preccurlyeq \eta \Leftrightarrow \xi = Q\eta \text{ with } Q_{ij} = |U_{ij}|^2 \text{ for some unitary } U \\ &\qquad \text{[Gohberg–Markus]} \\ &\xi \preccurlyeq \eta \Leftrightarrow \xi = Q\eta \text{ with } Q_{ij} = |W_{ij}|^2 \text{ for some isometry } W \\ &\qquad \text{[Arveson–Kadison]} \end{aligned} \tag{2}$$

In [13] we modify slightly-then matrix model-the Markus construction [19, Lemma 3.1] he used to prove $\xi \prec \eta \Leftrightarrow \xi = P\eta$ for some substochastic $P$. Our analysis, after modification, identifies a distinguished sequence of orthogonal matrices

(T-transforms) underlying the construction. An analysis of their properties and infinite products permits us to obtain:

(i) If $\xi, \eta \in c_o^*$, $\xi_n > 0\ \forall n$, and $\xi \prec \eta$, then there is a canonical co-isometry with real entries $W(\xi,\eta)$ for which $\xi = Q(\xi,\eta)\eta$ with $Q(\xi,\eta)_{ij} = (W(\xi,\eta)_{ij})^2$ [13, Theorem 3.7].

(ii) If $\xi, \eta \in c_o^*$, then $\xi \preccurlyeq \eta \Leftrightarrow \xi = Q\eta$ with **orthostochastic** $Q_{ij} = |U_{ij}|^2$ for some orthogonal matrix $U$ [ibid, Theorem 3.9].

This construction applied to finite sequences provides another proof of the (finite) Horn Theorem.

**Historical study of diagonals of operators in unitary orbits**

[13] extends some of the results of Gohberg and Marcus in [8] and Arveson and Kadison in [2]. It has limited overlaps with A. Neumann [21] and Antezana–Massey–Ruiz–Stojanoff [1] as these authors characterize the closures of the expectation of the unitary orbit of a selfadjoint not necessarily compact operator – we do not take closures. For sequences $\xi, \eta \in c_o^*$, the following is a comparison of these different results. Here the unitary orbit of an operator is denoted $\mathcal{U}(A) := \{U^*AU \mid U \text{ is a unitary operator}\}$ and for below its partial isometry orbit is denoted $\mathcal{V}(A) := \{V^*AV \mid V \text{ is a partial isometry}\}$.

If $\xi \notin (\ell^1)^*$, then

$$\xi \prec \eta \Leftrightarrow \operatorname{diag}\xi \in \begin{cases} \overline{E(\mathcal{U}(\operatorname{diag}\eta))}^{\|\cdot\|} & \text{[A. Neumann]} \\ E\{L \operatorname{diag}\eta L^* \mid L \in B(H)_1\} & \text{[Arveson–Kadison]} \\ E(\mathcal{U}(\operatorname{diag}\eta)) & \text{[13, Proposition 6.6(ii)]} \end{cases}$$

If $\xi \in (\ell^1)^*$, then

$$\xi \preccurlyeq \eta \Leftrightarrow \operatorname{diag}\xi \in \begin{cases} \overline{E(\mathcal{U}(\operatorname{diag}\eta))}^{\|\cdot\|_1} & \text{[Antezana–Massey–Ruiz–Stojanoff]} \\ E(\overline{\mathcal{U}(\operatorname{diag}\eta)}^{\|\cdot\|_1}) & \text{[Arveson–Kadison]} \\ E(\mathcal{U}(\operatorname{diag}\eta)) & \text{[Gohberg–Markus]} \end{cases}$$

**Diagonals of operators in unitary orbits/partial isometry orbits, proof descriptions and open question**

For the rest of Part I, we list some of the main results of [13] with a brief indication of their proofs and an open question.

**Theorem ([13, Theorem 3.9]).** *If $\xi, \eta \in c_o^*$ and $\xi \preccurlyeq \eta$, then $\xi = Q\eta$ for some orthostochastic matrix $Q$.*

**Theorem ([13, Theorem 5.3]).** *If $\xi, \eta \in c_o^*$ and $\xi \notin (\ell^1)^*$, then $\xi \prec \eta$ if and only if $\xi = Q\eta$ for some orthostochastic matrix $Q$.*

Applying these to Lemma 1.2 obtains directly the infinite dimension extension of the Schur–Horn Theorem for positive compact operators $A$, [13, Theorem 6.4] and [13, Proposition 6.6]:

**Theorem ([13, Theorem 6.4]).** *Partial isometry orbits*

$$E(\mathcal{V}(A)) = \begin{cases} \{B \in \mathcal{D} \cap K(H)^+ \mid s(B) \prec s(A)\} \setminus \mathcal{L}_1 & \textit{if } \operatorname{Tr}(A) = \infty \\ \{B \in \mathcal{D} \cap K(H)^+ \mid s(B) \preccurlyeq s(A)\} & \textit{if } \operatorname{Tr}(A) < \infty \end{cases}$$

**Theorem ([13, Proposition 6.6]).** *If $A \in K(H)^+$ then*

(i) $E(\mathcal{U}(A)) \subset E(\mathcal{V}(A)) \cap \{B \in \mathcal{D} \mid \operatorname{Tr}(R_B^\perp) \le \operatorname{Tr}(R_A^\perp)\}$.

(ii) *If additionally $R_A = I$ then*

$$\begin{aligned} E(\mathcal{U}(A)) &= E(\mathcal{V}(A)) \cap \{B \in \mathcal{D} \mid R_B = I\} \\ &= \begin{cases} \{B \in \mathcal{D} \cap K(H)^+ \mid s(B) \prec s(A), R_B = I\} \setminus \mathcal{L}_1 & \textit{if } \operatorname{Tr}(A) = \infty \\ \{B \in \mathcal{D} \cap K(H)^+ \mid s(B) \preccurlyeq s(A), R_B = I\} & \textit{if } \operatorname{Tr}(A) < \infty. \end{cases} \end{aligned}$$

(iii) *The inclusion in (i) is proper unless $R_A = I$ or $A$ has finite rank.*

**Open problem for positive compact operators**

What is $E(\mathcal{U}(A))$ when range projection $R_A \neq I$ nor finite, that is, when $A = \operatorname{diag} \eta \oplus 0_n$ $(1 \le n \le \infty)$?

We know $E(\mathcal{U}(\operatorname{diag} \eta)) \neq E(\mathcal{U}(0 \oplus \operatorname{diag} \eta))$ [13, Proposition 6.6(iii)] so majorization as we know it, i.e., the partial isometry orbit conditions, do not suffice to characterize it.

**Idea of proofs of [13, Theorems 3.9, 5.3]**

[13, Theorem 5.3]: For $\xi \prec \eta$, $\xi \not\preccurlyeq \eta$ and nonsummable: with care one can split $\mathbb{N}$ disjointly so that restricted to each piece, one has the summable case $\langle \xi_{n_k} \rangle \preccurlyeq \langle \eta_{n_k} \rangle$, then direct sum the unitaries provided by [13, Theorem 3.9].

For the case $\xi \preccurlyeq \eta$, use [13, Theorem 3.9]: Infinite products of T-transforms described next. Fortunately these T-transforms are real unitaries so their Schur-squares are orthostochastic.

As mentioned above, we modify slightly then model the Markus construction used to prove $\xi \prec \eta \Leftrightarrow \xi = P\eta$ for some substochastic $P$. See [13, Section 3, §2] for a detailed comparative discussion to Markus's construction. How does this work?

The first step suffices to explain the process. Because

$$\xi \prec \eta \Rightarrow \xi_1 \le \eta_1,$$

so $\xi_1 \in (\eta_k, \eta_{k+1}]$ hence $\xi_1 = t\eta_k + (1-t)\eta_{k+1}$ for $0 < t \le 1$. For this to make sense one needs to choose $n_o = 0$. Then this choice of $t, k$ is unique and canonical, unlike Markus, and is the slight modification we indicated.

It yields:

$$\begin{pmatrix} \eta_1 \\ \eta_2 \\ \vdots \\ t\eta_k + (1-t)\eta_{k+1} \\ (1-t)\eta_k + t\eta_{k+1} \\ \vdots \\ \eta_n \\ \vdots \end{pmatrix} = \begin{pmatrix} \eta_1 \\ \eta_2 \\ \vdots \\ \xi_1 \\ (1-t)\eta_k + t\eta_{k+1} \\ \vdots \\ \eta_n \\ \vdots \end{pmatrix} \prec \eta$$

and better

$$\xi = \begin{pmatrix} \xi_1 \\ \xi_2 \\ \vdots \end{pmatrix} \prec \begin{pmatrix} \xi_1 \\ \eta_1 \\ \eta_2 \\ \vdots \\ (1-t)\eta_k + t\eta_{k+1} \\ \vdots \\ \eta_n \\ \vdots \end{pmatrix} \prec \eta.$$

Notice density at play, although the middle vector lost monotonicity, its truncation remains monotone:

$$\begin{pmatrix} \xi_2 \\ \vdots \end{pmatrix} \prec \begin{pmatrix} \eta_1 \\ \eta_2 \\ \vdots \\ (1-t)\eta_k + t\eta_{k+1} \\ \vdots \\ \eta_n \\ \vdots \end{pmatrix}.$$

Now iterate.

This process for every $\xi_k$ uniquely determines its associated $(m_k, t_k)$ and the $\eta$ related vectors just described.

For every integer $m \geq 1$ and $0 < t \leq 1$, define the $(m+1)\times(m+1)$ orthogonal matrix $\oplus I$ and hence its Schur-square

$$V(m,t) := \begin{pmatrix} 0 & 0 & \dots & 0 & \sqrt{t} & -\sqrt{1-t} \\ 1 & 0 & \dots & 0 & 0 & 0 \\ 0 & 1 & \dots & 0 & 0 & 0 \\ \vdots & \vdots & \vdots & \vdots & \vdots & \vdots \\ 0 & 0 & \dots & 1 & 0 & 0 \\ 0 & 0 & \dots & 0 & \sqrt{1-t} & \sqrt{t} \end{pmatrix} \oplus I,$$

$$Q(m,t) := \begin{pmatrix} 0 & 0 & \dots & 0 & t & 1-t \\ 1 & 0 & \dots & 0 & 0 & 0 \\ 0 & 1 & \dots & 0 & 0 & 0 \\ \vdots & \vdots & \vdots & \vdots & \vdots & \vdots \\ 0 & 0 & \dots & 1 & 0 & 0 \\ 0 & 0 & \dots & 0 & 1-t & t \end{pmatrix} \oplus I.$$

**Infinite products of T-transforms**

$$W(\{m_k, t_k\}) := \prod_{k=1}^{\infty} (I_k \oplus V(m_k, t_k) \oplus I)$$

for use in

$$\text{diag}\ \xi = E(W(\{m_k, t_k\})^* (\text{diag}\ \eta) W(\{m_k, t_k\}))$$

or, equivalently for its Schur-square,

$$\xi = Q(\{m_k, t_k\})\, \eta.$$

**Example.** The case $m_k \equiv 1$ is simplest for determining matrices $W(\{m_k, t_k\})$ and Schur-square $Q(\{m_k, t_k\})$. Infinite product

$$W(\{1, t_k\}) = \prod_{k=1}^{\infty} (I_k \oplus V(1, t_k) \oplus I)$$

and Q (modeling the modified Markus construction):

$W(\{1, t_k\})$

$$= \begin{pmatrix} \sqrt{t_1} & -\sqrt{1-t_1} & 0 & 0 & \dots \\ \sqrt{t_2(1-t_1)} & \sqrt{t_2 t_1} & -\sqrt{1-t_2} & 0 & \dots \\ \sqrt{t_3(1-t_2)(1-t_1)} & \sqrt{t_3(1-t_2)t_1} & \sqrt{t_3 t_2} & -\sqrt{1-t_3} & \dots \\ \vdots & \vdots & \vdots & \vdots & \vdots \\ \sqrt{t_k \prod_{i=1}^{k-1}(1-t_i)} & \sqrt{t_k t_1 \prod_{i=2}^{k-1}(1-t_i)} & \sqrt{t_k t_2 \prod_{i=3}^{k-1}(1-t_i)} & \sqrt{t_k t_3 \prod_{i=4}^{k-1}(1-t_i)} & \dots \\ \vdots & \vdots & \vdots & \vdots & \vdots \end{pmatrix}$$

$$Q(\{1,t_k\}) = \begin{pmatrix} t_1 & 1-t_1 & 0 & 0 & \cdots \\ t_2(1-t_1) & t_2t_1 & 1-t_2 & 0 & \cdots \\ t_3(1-t_2)(1-t_1) & t_3(1-t_2)t_1 & t_3t_2 & 1-t_3 & \cdots \\ \vdots & \vdots & \vdots & \vdots & \vdots \\ t_k\prod_{i=1}^{k-1}(1-t_i) & t_kt_1\prod_{i=2}^{k-1}(1-t_i) & t_kt_2\prod_{i=3}^{k-1}(1-t_i) & t_kt_3\prod_{i=4}^{k-1}(1-t_i) & \cdots \\ \vdots & \vdots & \vdots & \vdots & \vdots \end{pmatrix}.$$

**Analysis of the infinite products for the W's and Q's.**
**When is W unitary, i.e., when is its Schur-square Q orthostochastic?**

**Theorem ([13, Theorem 4.7]).** *Let $\xi \prec \eta$ for some $\xi, \eta \in c_o^*$ with $\xi_n > 0$ for every $n$. Then TFAE.*

(i) $\xi \preccurlyeq \eta$.
(ii) $\sum\{t_k \mid m_k = 1\} = \infty$.
(iii) *$Q(\xi, \eta)$ is orthostochastic.*

Which is what happens when $\eta = s(A)$ is strictly positive, i.e., the range projection of $A$ is identity (recall the condition $R_A = I$).

## 2. $B(H)$-subideals of operators

A $B(H)$-subideal is an ideal of an ideal of $B(H)$ and a principal subideal is a principal ideal of an ideal of $B(H)$. In [22] we determine necessary and sufficient conditions for a principal subideal to be an ideal of $B(H)$. This generalizes to arbitrary ideals the 1983 work of Fong and Radjavi [7] characterizing principal subideals of the ideal of compact operators that are also ideals of $B(H)$. We then provide a characterization of all principal subideals of $B(H)$. We also investigate the lattice structure of subideals as part of the general study of ideal lattices such as the often studied lattice structure of ideals of $B(H)$. The study of subideals is closely related to the study of elementary operators with constraints.

For a short survey on subideals we reproduce here [22, Introduction] without proofs. And for a discussion of lattice properties, see [22, Remark 2.2, esp. (iv) and Sections 3–5].

[22] investigates the subideal structure of $B(H)$ following the spirit of Calkin's well-known singular number characterization of ideals of $B(H)$ [4]. A subideal is an ideal of an ideal $J$ (henceforth called a $J$-ideal) for the $B(H)$-ideal $J$. Recall for general rings, an ideal is an additive commutative subgroup which is closed under left and right multiplication by elements of the ring. Ideals in the ring $B(H)$ are ubiquitous throughout operator theory. Some well-known $B(H)$-ideals are the compact operators, $K(H)$, the finite rank operators, $F(H)$, principal ideals, Banach ideals, the Hilbert–Schmidt class, the trace class, Orlicz ideals, Marcinkiewicz

ideals and Lorentz ideals, to name a few. Definitions of these ideals may be found in [6]. Here and throughout this paper $H$ denotes a separable infinite-dimensional complex Hilbert space, $B(H)$ the algebra of all bounded linear operators on $H$, and $\mathbb{C}, \mathbb{R}, \mathbb{N}, \mathbb{Z}$, respectively, the classes of complex numbers, real numbers, positive integers and integers.

There are three natural kinds of *principal J-ideals*: classical principal $J$-ideals $(S)_J$ which we call principal linear $J$-ideals, principal $J$-ideals $\langle S\rangle_J$, and principal real linear $J$-ideals $(S)_J^{\mathbb{R}}$ ([22, Definition 2.1]). Standard notation then dictates that we denote $(S) = (S)_{B(H)}$. *It is immediate that $B(H)$-ideals are always $J$-ideals but, as we shall see later, often not conversely* ([7], *see also Example* 2.3 *below*).

The main results of this paper, generalizing the 1983 work of Fong and Radjavi [7] and characterizing the principal subideals of $B(H)$, are summarized in the following two theorems.

For $S, T \in J$, $s(S)$ denotes the sequence of singular numbers (s-numbers) for $S$, and the product $s(S)s(T)$ denote their pointwise product.

**Theorem 2.1.** *For $S \in J$, the following are equivalent.*

(i) *Any of the three types of principal $J$-ideals generated by $S$, $(S)_J$, $\langle S\rangle_J$ or $(S)_J^{\mathbb{R}}$ is a $B(H)$-ideal.*

(ii) *The principal $B(H)$-ideal $(S)$ is $J$-soft, i.e., $(S) = J(S)$ (equivalently, $(S) = (S)J$).*

(iii) $S = AS + SB + \sum_{i=1}^{m} A_i S B_i$ *for some $A_i$, $B_i$, $A$, $B \in J$, $m \in \mathbb{N}$.*

(iv) $s(S) = O(D_k(s(S))s(T))$ *for some $T \in J$ and $k \in \mathbb{N}$.*

Denoting $JS + SJ + J(S)J := \{AS + SB + CS'D \mid A, B, C, D \in J, S' \in (S)\}$,

**Theorem 2.2.** *The principal $J$-ideal, the principal linear $J$-ideal and the principal real linear $J$-ideal generated by $S \in J$ are respectively given by*

$$\langle S\rangle_J = \mathbb{Z}S + JS + SJ + J(S)J$$

$$(S)_J = \mathbb{C}S + JS + SJ + J(S)J$$

$$(S)_J^{\mathbb{R}} = \mathbb{R}S + JS + SJ + J(S)J.$$

*Also*

$$J(S)J \subseteq JS + SJ + J(S)J \subseteq \langle S\rangle_J \cap (S)_J \cap (S)_J^{\mathbb{R}}$$

*which first two are respectively a $B(H)$-ideal and a $J$-ideal, so visually each of these kinds of principal ideals have the common $B(H)$-ideal "nucleus" $J(S)J$, with the common $J$-ideal $JS + SJ + J(S)J$ containing it.*

*All these principal ideals, $\langle S\rangle_J \subseteq (S)_J^{\mathbb{R}} \subseteq (S)_J$, are distinct except under the following equivalent conditions. They all collapse to merely*

$$J(S) = (S) = \langle S\rangle_J = (S)_J = (S)_J^{\mathbb{R}}$$

- *if and only if the principal $B(H)$-ideal $(S)$ is $J$-soft, i.e., $(S) = J(S)$*
- *if and only if any, and hence all of them, is a $B(H)$-ideal.*

This extension of the ideal notion called soft-edged is essential for our generalization of Fong-Radjavi's work [7]. Soft-edged ideals (soft ideals for short), that is, ideals $I$ for which $IK(H) = I$, were first introduced in [15] and [17, Section 3, esp., Definition 3.1] and studied further in [16] as part of a study on traces motivated in part by Dixmier's implicit use of softness to construct the so-called Dixmier trace [5]. However, these softness notions involving $K(H)$ appeared some years earlier with different terminology in works of Mityagin focusing on Hilbert space and Pietsch in Banach spaces [20], [23].

**Background**

In 1941, Calkin [4] characterized $B(H)$-ideals via his lattice preserving isomorphism between ideals of $B(H)$ and characteristic sets $\Sigma \subseteq c_0^*$: $I \to \Sigma(I)$ induced by $I \ni X \to s(X) \in \Sigma(I)$. Here $c_0^*$ denotes the cone of non-negative sequences decreasing to zero; characteristic sets $\Sigma$ are those subsets of $c_0^*$ that are additive, hereditary (solid) and ampliation invariant (invariant under each $m$-fold ampliation $D_m\xi := \langle \xi_1, \ldots, \xi_1, \xi_2, \ldots, \xi_2, \ldots \rangle$ with each entry $\xi_i$ repeated $m$ times); the characteristic set $\Sigma(I) := \{\eta \in c_0^* \mid \operatorname{diag} \eta \in I\}$; and $s(X)$ denotes the $c_0^*$-sequence of s-numbers of compact operator X.

Motivated by this characterization, a natural question to ask and the subject of this paper is:

*What can be said about subideals, i.e., is it possible to classify them in some way?*

In 1983, Fong and Radjavi [7] investigated those subideals that are principal $K(H)$-ideals, perhaps in part because of the distinguished role $K(H)$ plays as the unique norm closed proper ideal of $B(H)$. They considered only principal linear $K(H)$-ideals. Though unstated in [7], this was recently clarified to us (private communications). They found principal linear $K(H)$-ideals that are not $B(H)$-ideals (Example 2.3) by determining necessary and sufficient conditions for a principal linear $K(H)$-ideal to be a $B(H)$-ideal [7, Theorem 2]. And in doing so, at least for the authors of this paper, they initiated the study of subideals.

**Theorem ([7, Theorem 2]).** *Let $T$ be a compact operator of infinite rank and let $P = (T^*T)^{\frac{1}{2}}$. Let $\mathcal{T}$ and $\mathcal{P}$ be the ideals in $K(H)$ generated by $T$ and $P$, respectively. Then the following are mutually equivalent.*

(i) *$\mathcal{T}$ is an ideal in $B(H)$.*

(ii) *$\mathcal{P}$ is an ideal in $B(H)$.*

(iii) *$T = A_1TB_1 + \cdots + A_kTB_k$ for some $k$ and some $A_i \in K(H)$, $B_i \in B(H)$.*

(iv) *$T = A_1TB_1 + \cdots + A_kTB_k$ for some $k$ and some $A_i \in K(H)$, $B_i \in K(H)$.*

Fong and Radjavi proved this via the positive case employing the Lie ideal condition (ii) below.

**Theorem ([7, Theorem 1]).** *Let $P$ be a positive compact operator of infinite rank, and let $\mathcal{I}$ be the ideal in $K(H)$ generated by $P$. Then the following are equivalent.*

(i) *$\mathcal{I}$ is an ideal in $B(H)$.*

(ii) *$\mathcal{I}$ is a Lie ideal in $B(H)$.*

(iii) *$P = A_1PA_1^* + \cdots + A_kPA_k^*$ for some $k$ and some compact operators $A_i$.*

(iv) *$P = A_1PB_1 + \cdots + A_kPB_k$ for some $k$ and some compact operators $A_i$ and $B_i$.*

(v) *$P = A_1PB_1 + \cdots + A_kPB_k$ for some $k$, where $A_i$, $B_i \in B(H)$ and either the $A_i$ or the $B_i$ are compact.*

(vi) *For some integer $k > 1$, $s_{nk}(P) = o(s_n(P))$ as $n \to \infty$.*

**Example 2.3.** *Condition* (vi) *of* [7, Theorem 1] *shows by direct computation that if the singular number sequence of the operator $P$ is given by $s(P) = \langle \frac{1}{2^n} \rangle$, then the principal linear $K(H)$-ideal generated by $P$ is a $B(H)$-ideal. But if $s(P) = \langle \frac{1}{n} \rangle$, then the principal linear $K(H)$-ideal generated by $P$ is not a $B(H)$-ideal.*

**In summary**

[22] fully generalizes [7, Theorem 2] to arbitrary ideals and all but the Lie ideal condition in [7, Theorem 1]. We investigate all three types of principal $J$-ideals, whereas Fong–Radjavi considered only the principal linear $J$-ideals for the case $J = K(H)$ [7]. We determine necessary and sufficient conditions for when a principal $J$-ideal is a $B(H)$-ideal and we employ these conditions to characterize them (Theorems 2.1–2.2). We also investigate the lattice structure of subideals and principal subideals (building blocks of subideals just as principal ideals are building blocks for ideals in all rings (see [22, Remark 2.2(iv)]). Our methods are largely purely algebraic.

Motivated by the advances on ideals of the last decade (for example the semiring structure of the lattice of $B(H)$-ideals, esp. their additive and multiplicative structure, see [ibid]), we found that bypassing the Lie ideal considerations of Fong–Radjavi [7, Theorem 1], the positive operator case, we can prove the main theorem [7, Theorem 2] more generally, and we believe more simply and directly. These advances, to be sure, evolved from works such as [7]. Indeed, the proof of [6, Lemma 6.3] shares some of the attributes of the proof of [7, Theorem 1], in particular, the use of a unitary from $H \to H^{\oplus m}$, and their use of [22, Proposition 2.3(i)] for principal $K(H)$-ideals is implicit in their proofs.

## References

[1] Antezana, J., Massey, P., Ruiz, M., and Stojanoff, D., *The Schur–Horn Theorem for operators and frames with prescribed norms and frame operator*, Illinois J of Math., to appear.

[2] Arveson, W. and Kadison, R.V., *Diagonals of self-adjoint operators.* Operator theory, operator algebras, and applications, Contemp. Math. **414** Amer. Math. Soc., Providence, RI (2006), 247–263.

[3] Birkhoff, G., *Tres observaciones sobre el algebra lineal*, Univ. Nac. Tucuman Rev. Ser. A **5** (1946), 147–151.

[4] Calkin, J.W., *Two-sided ideals and congruences in the ring of bounded operators in Hilbert space*, Ann. of Math. **42** (2)(1941), 839–873.

[5] Dixmier, J., *Existence de traces non normales*, C. R. Acad. Sci. Paris Sér. A-B 262 (1966), A1107–A1108.

[6] Dykema, K., Figiel, T., Weiss, G., and Wodzicki, M., *The commutator structure of operator ideals*, Adv. Math., **185** (1) (2004), 1–79.

[7] Fong, C.K. and Radjavi, H., *On ideals and Lie Ideals of Compact Operators*, Math. Ann. **262**, 23–28 (1983).

[8] Gohberg, I.C., and Markus, A.S., *Some relations between eigenvalues and matrix elements of linear operators*, Mat. Sb. **64** (106) (1964), 481–496 (Russian); Amer. Math. Soc. Transl. (2) **52** (1966) 201–216 (English).

[9] Hardy, G.H., Littlewood, J.E., and Pólya, G., *Inequalities.* 2d ed. Cambridge University Press, 1952.

[10] Horn, R.A. *Doubly stochastic matrices and the diagonal of a rotation matrix.* Amer. J. Math. **76** (1954), 620–630.

[11] Kadison, R., *The Pythagorean Theorem I: the finite case*, Proc. Natl. Acad. Sci. USA **99** (7) (2002), 4178–4184.

[12] Kadison, R., *The Pythagorean Theorem II: the infinite discrete case*, Proc. Natl. Acad. Sci. USA **99** 8 (2002), 5217–5222.

[13] Kaftal, V. and Weiss, G., *An infinite dimensional Schur-Horn theorem and majorization theory*, J. Functional Analysis, 259 (2010) 3115–3162.

[14] Kaftal, V. and Weiss, G., *Majorization and arithmetic mean ideals*, Indiana University Math J., to appear. Preprint posted at http://www.iumj.indiana.edu/IUMJ/Preprints/4603.pdf

[15] Kaftal, V. and Weiss, G., *Traces, ideals, and arithmetic means*, Proc. Nat. Acad. Sci. U.S.A. 99(2002), 7356–7360.

[16] Kaftal, V. and Weiss, G., *Soft ideals and arithmetic mean ideals*, Integral equations and Operator Theory 58 (2007), 363–405.

[17] Kaftal, V. and Weiss, G., *Traces on operator ideals and arithmetic means*, J. Operator Theory, 63 Issue 1, Winter 2010, 3–46.

[18] Lorenz, M.O., *Methods of measuring concentration of wealth*, J. Amer. Statist. Assoc. **9** (1905), 209–219.

[19] Markus, A.S., *The eigen- and singular values of the sum and product of linear operators*, Uspekhi Mat. Nauk 4 (118) (1964), 93–123.

[20] Mityagin, B.S., *Normed ideals of intermediate type*, Amer. Math. Soc. Transl. (2) 63 (1967), 180–194.

[21] Neumann, A., *An infinite-dimensional generalization of the Schur–Horn convexity theorem.* J. Funct. Anal., **161** (2) 1999, 418–451.

[22] Patnaik, S. and Weiss, G., *Subideals of operators*, J. Operator Theory, to appear.

For later advances, see also its sequel:

S. Patnaik and G. Weiss, *Subideals of operators* II, J. Integral Equations and Operator Theory, December 2012, Volume 74, Issue 4, 587–600.

Also:

S. Patnaik and G. Weiss, *A survey on subideals of operators and an introduction to subideal-traces*, submitted 2013.

[23] Pietsch, A., Operator Ideals, Deutsch. Verlag Wiss., Berlin, 1978; North-Holland, Amsterdam–London–New York–Tokyo, 1980.

[24] Schur, I., *Über eine Klasse von Mittelbildungen mit Anwendungen auf der Determinantentheorie*, Sitzungsber. Berliner Mat. Ges., ( **22**) (1923), 9–29.

Gary Weiss
University of Cincinnati
Department of Mathematics
Cincinnati, OH, 45221-0025, USA
e-mail: gary.weiss@math.uc.edu

Birkhäuser | www.birkhauser-science.com

# Operator Theory: Advances and Applications (OT)

This series is devoted to the publication of current research in operator theory, with particular emphasis on applications to classical analysis and the theory of integral equations, as well as to numerical analysis, mathematical physics and mathematical methods in electrical engineering.

**Edited by**
Joseph A. Ball (Blacksburg, VA, USA), Harry Dym (Rehovot, Israel), Marinus A. Kaashoek (Amsterdam, The Netherlands), Heinz Langer (Vienna, Austria), Christiane Tretter (Bern, Switzerland)

■ **OT 238: Edmunds, D.E. / Evans, W.D.**, Representations of Linear Operators Between Banach Spaces (2013).
ISBN 978-3-0348-0641-1

■ **OT 237: Kaashoek, M.A. / Rodman, L. / Woerdeman, H.J.** (Eds.), Advances in Structured Operator Theory and Related Areas. The Leonid Lerer Anniversary Volume (2013).
ISBN 978-3-0348-0638-1

■ **OT 236: Cepedello Boiso, M. / Hedenmalm, H. / Kaashoek, M.A. / Montes Rodríguez, A. / Treil, S.** (Eds.), Concrete Operators, Spectral Theory, Operators in Harmonic Analysis and Approximation. 22nd International Workshop in Operator Theory and its Applications, Sevilla, July 2011 (IWOTA11) (2013).
ISBN 978-3-0348-0647-3

■ **OT 234/OT235: Eidelman, Y. / Gohberg, I. / Haimovici, I.** (Eds.), Separable Type Representations of Matrices and Fast Algorithms.
Vol. 1. Basics. Completion problems. Multiplication and inversion algorithms (2013).
ISBN 978-3-0348-0605-3
Vol 2. Eigenvalue method (2013).
ISBN 978-3-0348-0611-4

■ **OT 233: Todorov, I.G. / Turowska, L.** (Eds.), Algebraic Methods in Functional Analysis. The Victor Shulman Anniversary Volume (2013).
ISBN 978-3-0348-0501-8

■ **OT 232: Demuth, M. / Kirsch, W.** (Eds.), Mathematical Physics, Spectral Theory and Stochastic Analysis (2013).
ISBN 978-3-0348-0590-2

■ **OT 231: Molahajloo, S. / Pilipović, S. / Toft, J. / Wong, M.W.** (Eds.), Pseudo-Differential Operators, Generalized Functions and Asymptotics (2013).
ISBN 978-3-0348-0584-1

■ **OT 230: Brown, B.M. / Eastham, M.S.P. / Schmidt, K.M.**, Periodic Differential Operators (2013).
ISBN 978-3-0348-0527-8

■ **OT 229: Almeida, A. / Castro, L. / Speck, F.-O.** (Eds.), Advances in Harmonic Analysis and Operator Theory. The Stefan Samko Anniversary Volume (2013).
ISBN 978-3-0348-0515-5

■ **OT 228: Karlovich, Y.I. / Rodino, L. / Silbermann, B. / Spitkovsky, I.M.** (Eds.), Operator Theory, Pseudo-Differential Equations, and Mathematical Physics (2013).
ISBN 978-3-0348-0536-0

■ **OT 227: Janas, J. / Kurasov, P. / Laptev, A. / Naboko, S.** (Eds.), Operator Methods in Mathematical Physics. Conference on Operator Theory, Analysis and Mathematical Physics (OTAMP) 2010, Bedlewo, Poland (2013).
ISBN 978-3-0348-0530-8

■ **OT 226: Alpay, D. / Kirstein, B.**, Interpolation, Schur Functions and Moment Problems II (2012).
ISBN 978-3-0348-0427-1

■ **OT 225: Sakhnovich, L.A.**, Levy Processes, Integral Equations, Statistical Physics: Connections and Interactions (2012).
ISBN 978-3-0348-0355-7

Printed by Publishers' Graphics LLC
KSO131115.15.18.18